DO NOT REMOVE
THIS BOOK FROM
THE LIBRARY

D1065311

Reagents for Organic Synthesis

PROOFREADER
Armin Burghart

INDEXER
Honor Ho

Fiesers'

Reagents for Organic Synthesis

VOLUME NINETEEN

Tse-Lok Ho
National Chaio Tung University
Republic of China

A WILEY-INTERSCIENCE PUBLICATION
JOHN WILEY & SONS, INC.
NEW YORK / CHICHESTER / WEINHEIM / BRISBANE / SINGAPORE / TORONTO

This book is printed on acid-free paper. ♾

Published simultaneously in Canada.

For ordering and customer service, call 1-800-CALL-WILEY.

Library of Congress Cataloging in Publication Data:

ISBN 0-471-32709-3
ISSN 0271-616X

Printed in the United States of America.

10 9 8 7 6 5 4 3 2 1

PREFACE

This volume examines literature of mostly the 1995–1996 period. Due to the proliferation of publications, I have decided not to cite full papers extending from those preliminary communications already included in previous volumes of the series and containing sufficient details for synthetic applications of the reagents. Deciding what to cover is always a dilemma, but my policy is to report the newest and significant reagents and reactions when they first appear. For less significant work or an old reagent with a single new use, I might delay the discussion until a later volume. As I have mentioned previously, I shall strive to amend my mistakes in missing important work. In volumes prepared by the Fiesers, the nomenclature of compounds does not always conform to the IUPAC or the CA system, probably for the sake of convenience to the reader. My arrangement is in the same spirit, and actually, I have tried to place cognate reagents near each other so that related information becomes more immediately available to a browser. Thus, azidotrimethylsilane, chlorotrimethylsilane, and other Me_3SiX are grouped under T, near trimethylsilyl triflate. Similarly, titanocenes with modified ligands are assembled together.

TSE-LOK HO

CONTENTS

GENERAL ABBREVIATIONS

Ac	acetyl
acac	acetylacetonate
ADDP	1,1′-(azodicarbonyl)dipiperidine
AIBN	2,2′-azobisisobutyronitrile
aq	aqueous
Ar	aryl
Bn	benzyl
Boc	t-butoxycarbonyl
Bu	*n*-butyl
Bz	benzoyl
18-c-6	18-crown-6
c-	cyclo
cat	catalytic
Cp	cyclopentadienyl
Cy	cyclohexyl
DABCO	1,4-diazabicyclo[2.2.2]octane
DAST	(diethylamino)sulfur trifluoride
DBN	1,5-diazabicyclo[4.3.0]non-5-ene
DCC	N,N′-dicyclohexylcarbodiimide
DDQ	2,3-dichloro-5,6-dicyano-1,4-benzoquinone
de	diastereomer excess
DIBAH	diisobutylaluminum hydride
DMAP	4-(dimethylamino)pyridine
DMD	dimethyldioxirane
DME	1,2-dimethoxyethane
DMF	N,N-dimethylformamide
DMPU	N,N′-dimethylpropyleneurea
DMSO	dimethyl sulfoxide
dppb	1,4-bis(diphenylphosphino)butane
dppe	1,2-bis(diphenylphosphino)ethane
dppf	1,2-bis(diphenylphosphino)ferrocene
dppp	1,2-bis(diphenylphosphino)propane
E	COOMe
ee	enantiomer excess
Et	ethyl
EVE	ethyl vinyl ether
HMPA	hexamethylphosphoric triamide

hv light
Ipc isopinocampheyl
iPr isopropyl
kbar kilobar
L ligand
LAH lithium aluminum hydride
LDA lithium diisopropylamide
LTMP lithium 2,2,6,6-tetramethylpiperidide
lut 2,6-lutidine
M metal (alkali)
MCPBA m-chloroperoxybenzoic acid
Me methyl
Ms mesyl (methanesulfonyl)
MTO methylrhodium trioxide
MVK methyl vinyl ketone
NBS N-bromosuccinimide
NCS N-chlorosuccinimide
NIS N-iodosuccinimide
NMO N-methylmorpholine oxide
Nu nucleophile
Ctc octyl
PCC pyridinium chlorochromate
PDC pyridinium dichromate
PEG polyethylene glycol
Ph phenyl
Pht phthaloyl
Piv pivaloyl
Pr *n*-propyl
py pyridine
Q^+ quaternary onium ion
RAMP (R)-1-amino-2-methoxymethylpyrrolidine
RaNi Raney nickel
R^f perfluoroalkyl
(s) solid
SAMP (*S*)-1-amino-2-methoxymethylpyrrolidine
sens. photosensitizer
TBAF tetrabutylammonium fluoride
TBS =TBDMS, *t*-butyldimethylsilyl
TEMPO 2,2,6,6,-tetramethylpiperidinoxy
TES triethylsilyl
THF tetrahydrofuran

TIPS	triisopropylsilyl
TMEDA	*N,N,N,N'*-tetramethylethylenediamine
TMS	trimethylsilyl
Ts	tosyl (*p*-toluenesulfonyl)
Δ	heat
))))	microwave

REFERENCE ABBREVIATIONS

ACR	Acc. Chem. Res.
ACS	Acta Chem. Scand.
ACIEE	Angew. Chem. Int. Ed. Engl.
AJC	Aust. J. Chem.
AOMC	Appl. Organomet. Chem.
BBB	Biosc. Biotech. Biochem.
BCSJ	Bull. Chem. Soc. Jpn.
BSCB	Bull. Soc. Chim. Belg.
BSCF	Bull. Soc. Chim. Fr.
BRAS	Bull. Russ. Acad. Sci.
CB	Chem. Ber.
CC	Chem. Commun.
CCCC	Collect. Czech. Chem. Commun.
CEJ	Chem. Eur. J.
CJC	Can. J. Chem.
CL	Chem. Lett.
CPB	Chem. Pharm. Bull.
CR	Carbohydr. Res.
DC	Dokl. Chem. (Engl. Trans.)
G	Gazz. Chim. Ital.
H	Heterocycles
HC	Heteroatom Chem.
HCA	Helv. Chim. Acta
HX	Huaxue Xuebao
IJC(B)	Indian J. Chem., Sect. B
IJS(B)	Int. J. Sulfur Chem., Part B
JACS	J. Am. Chem. Soc.
JCC	J. Carbohydr. Chem.
JCCS(T)	J. Chin. Chem. Soc. (Taipei)
JCR(S)	J. Chem. Res. (Synopsis)
JCS(P1)	J. Chem. Soc. Perkin Trans. 1
JFC	J. Fluorine Chem.
JHC	J. Heterocycl. Chem.
JMC	J. Med. Chem.
JNP	J. Nat. Prod.
JOC	J. Org. Chem.
JOMC	J. Organomet. Chem.

JOCU	J. Org. Chem. USSR (Engl. Trans.)
LA	Liebigs Ann. Chem.
MC	Mendeleev Commun.
NKK	Nippon Kagaku Kaishi
OM	Organometallics
PAC	Pure Appl. Chem.
PSS	Phosphorus Sulfur Silicon
RJOC	Russian J. Org. Chem.
RTC	Recl. Trav. Chim. Pays-Bas
S	Synthesis
SC	Synth. Commun.
SL	Synlett
SOC	Synth. Org. Chem. (Jpn.)
T	Tetrahedron
TA	Tetrahedron:Asymmetry
TL	Tetrahedron Lett.
YH	Youji Huaxue

Reagents for Organic Synthesis

A

Acetic acid.

Cleavage of p-methoxybenzyl ethers.[1] Alcohols are liberated from the ethers on warming with HOAc (7 examples, 88–96%).

[1]Hodgetts, K.J., Wallace, T.W. *SC* **24**, 1151 (1994).

Acetone cyanohydrin.

Nitriles from alcohols.[1] Using this reagent as a donor in the Mitsunobu reaction successfully completes the preparation of alkyl nitriles.

[1]Aesa, M.C., Baan, G., Novak, L., Szantay, C. *SC* **26**, 909 (1996).

Acetonitrile. 15, 1; **18,** 2

β-Hydroxy nitriles.[1] Acetonitrile protonates acyllithium species, which are formed from RLi and CO. Subsequent reaction of the aldehydes with the cyanomethyl anion affords the β-hydroxy nitriles.

[1]Li, N.-S., Yu, S., Kabalka, G.W. *JOC* **60**, 5973 (1995).

***N*-Acetyl-*N*-acyl-3-aminoquinazolinones.**

Acetylation of primary amines.[1] Secondary amines are not affected by these reagents.

[1]Atkinson, R.S., Barker, E., Sutcliffe, M.J. *CC* 1051 (1996).

Acetyl chloride. 18, 2

β-Chlorosulfides.[1] Sulfenyl chlorides (RSCl) are formed when sulfenate esters (RSOR′) are treated with acetyl chloride (or other acid chlorides). The reactive species functionalize olefins in situ.

Deprotection of α-halo aldehyde dimethyl acetals.[2] Acetyl chloride in combination with acetic anhydride and sodium acetate regenerates the aldehydes (9 examples, 83–98%).

[1]Brown, C., Evans, G.R. *TL* **37**, 679 (1996).
[2]Benincasa, M., Boni, M., Ghelfi, F., Pagnoni, U.M. *SC* **25**, 1843 (1995).

N^1-Acylbenzotriazoles.

Sulfoxides.[1] Reaction of the activated amides with arenesulfinate anions results in sulfoxides. α-Sulfinyl carboxylic acids are likely the intermediates.

DMF, 3 h, Δ; PhS(O)ONa

75%

β-Lactones.[2] The amides undergo aldolization with ketones and aldehydes, furnishing β-lactones in one step. Both transformations imply ketene formation in the initial step.

LDA / THF, -95°; R'COR"

R = H
R"+R" = $(CH_2)_5$ 80%

[1]Katritzky, A.R., Yang, B., Qian, Y. *SL* 701 (1996).
[2]Wedler, C., Kleiner, K., Kunath, A., Schick, H. *LA* 881 (1996).

Acyl cyanides.

Cyanohydrin esters.[1] Reaction of acyl cyanides with aldehydes in the presence of K_2CO_3 in aqueous acetonitrile leads to α-cyanohydrin esters.

K_2CO_3, CH_3CN / H_2O, 0° -> rt, 1 h

79 - 96%

[1]Okimoto, M., Chiba, T. *S* 1188 (1996).

N-Acyl-2-methylimidazoles.

Acylsilanes.[1] Acylimidazoles are electrochemically reduced on a Pt cathode, and the ensuing acyl anion equivalents can be trapped with Me_3SiCl.

[1]Kise, N., Kaneko, H., Unemoto, N., Yoshida, J. *TL* **36**, 8839 (1995).

2-Acyloxyacroleins.

Dienophiles.[1] These compounds are available from 2,2-dimethyl-1,3-dioxan-5-one on enolacetylation and thermolysis of the resulting enol esters at 100°. They serve as dienophiles in Diels–Alder reactions.

100°, hydroquinone, CH_2Cl_2; OCOPh; H; OCOPh; i-Pr_3SiO; CH_2Cl_2, 105°, 20 h; CHO; OCOPh; i-Pr_3SiO; 69%

[1]Funk, R.L., Yost, K.J. *JOC* **61**, 2598 (1996).

Alkenylboronic acids.

α-Amino acids.[1] A three-component condensation involving an alkenylboronic acid, an amine, and an α-oxo acid proceeds in uniformly good yields. Products having a natural substitution pattern are formed by using benzylic amines and glyoxylic acid.

Ph, $B(OH)_2$; Ph, H_2N, OH; OHC-COOH, CH_2Cl_2; Ph, HN, OH, Ph, COOH; 78% (> 99%de); H_2/ Pd-C, MeOH, HCl; $\overset{+}{N}H_3$ Cl^-, Ph, COOH; 76% (> 99%de)

[1]Petasis, N.A., Zavialov, I.A. *JACS* **119**, 445 (1997).

1-[*N*-(Alkoxyoxalyl)-*N*-methylamino]-3-methylimidazolium salts.

α-Keto esters.[1] A general method for the synthesis of α-keto esters is by Grignard reaction of the salts.

[1]De las Heras, M.A., Vaquero, J.J., Garcia-Navio, J.L., Alvarez-Builla, J. *JOC* **61**, 9009 (1996).

Alkynyl triflones.

Alkynylation.[1] The introduction of an alkynyl group to an unactivated position in a carbon skeleton is a remarkable achievement. With alkynyl triflones in the presence of a radical initiator, good yields of α-alkynyl ethers or 1-alkynyladamantanes are formed from the corresponding ethers and adamantane, respectively. Alkenes mainly undergo addition to give β-trifluoromethylalkyl alkynes.

(peroxide) 25°, 3 h — 90%

Alkenyl triflones.[2] The (*Z*)-alkenyl triflones can be made from alkynyl triflones by the addition of HI followed by Stille coupling. Some other alkenyl triflones are available from organocopper reactions.

Alkenyl and dienyl triflones also insert into unactivated C–H bonds.

[1]Gong, J., Fuchs, P.L. *JACS* **118**, 4486 (1996).
[2]Xiang, J., Fuchs, P.L. *JACS* **118**, 11986 (1996).

Allenyl *n*-butyl telluride.

Homopropargylic alcohols.[1] On successive treatment with BuLi and an aldehyde, the telluride transfers its allenyl group as a propargyl residue to the latter compound.

[1]Dabdoub, M.J., Rotta, J.C.G. *SL* 526 (1996).

Allenyldiphenylphosphine oxide.

α,β-Unsaturated oximes.[1] The addition of hydroxylamine to the allenylphosphine oxides affords oximes of α-phosphinoyl ketones, which may be used to olefinate ketones.

NH_2OH, $CHCl_3$ — 80%; MeLi ; cyclohexanone — 74%

[1]Palacios, F., Aparicio, D., de los Santos, J.M., Rodriguez, E. *TL* **37**, 1289 (1996).

Allyl *N*-arenesulfonyloxy carbamates.

Allyl carbamates.[1] Allyl carbamates are formed by displacement of the *N*-arenesulfonyloxy group of the reagents with organocopper compounds.

[1]Greck, C., Bischoff, L., Ferreira, F., Genet, J.P. *JOC* **60**, 7010 (1995).

Allylbarium reagents.

Homoallylic amines.[1] The regioselectivity for the addition of γ-substituted allylic reagents to imines is dependent on reaction temperatures. γ-Adducts are formed at –78°, whereas α-adducts are obtained at 0°.

1,5-Dienes.[2] The coupling of allylbarium reagents with allylic bis-(2,2,2-trifluoroethyl)phosphates proceeds at α and α′ positions and is thus different from that of Grignard reagents (α,γ′ cross coupling). Hence, little transposition occurs with the use of the allylic phosphate esters in these reactions.

[1]Yanagisawa, A., Ogasawara, K., Yasue, K., Yamamoto, H. *CC* 367 (1996).
[2]Yanagisawa, A., Yasue, K., Yamamoto, H. *SL* 842 (1996).

Allyl benzotriazol-1-yl carbonate.

Allyl carbonates.[1] Mixed carbonates derived from carbohydrates are readily prepared from this reagent in the presence of Et_3N. Primary hydroxyl groups react preferentially.

[1]Harada, T., Yamada, H., Tsukamoto, H., Takahashi, T. *JCC* **14**, 165 (1995).

Allylboranes and allylboronic acid derivatives.

Preparation.[1] One method of preparation of allylboronates involves the Pd(0)-catalyzed replacement of allylic acetate with bis(pinacolato)diboron (**1**).

(dba)$_2$Pd / DMSO, 50°, 60 h

(**1**) 89%

Allylation. γ-Selective allylation of aldehydes using chiral reagents formed in situ from tartrate esters and allyldiisopropoxyboranes[2] shows 37–85% ee. On the other hand, the reaction with allylic silanes follows a pathway leading to hydroxyallylation of the double bond, and primary alcohols are obtained from 5-silyl-1,3-dienes.[3]

BCl_2 + Me_3Si — -45° -> 25° ; H_2O_2 - NaOH, THF — Me_3Si, OH

70%

[1]Ishiyama, T., Ahiko, T., Miyaura, N. *TL* **37**, 6889 (1996).
[2]Yamamoto, Y., Hara, S., Suzuki, A. *SL* 883 (1996).
[3]Singleton, D.A., Waller, S.C., Zhang, Z., Frantz, D.E., Leung, S.-W. *JACS* **118**, 9986 (1996).

(η^1-Allyl)dimethylgold complexes.

Allylation.[1] These allylating agents react with aromatic aldehydes in a γ- and *anti*-selective manner.

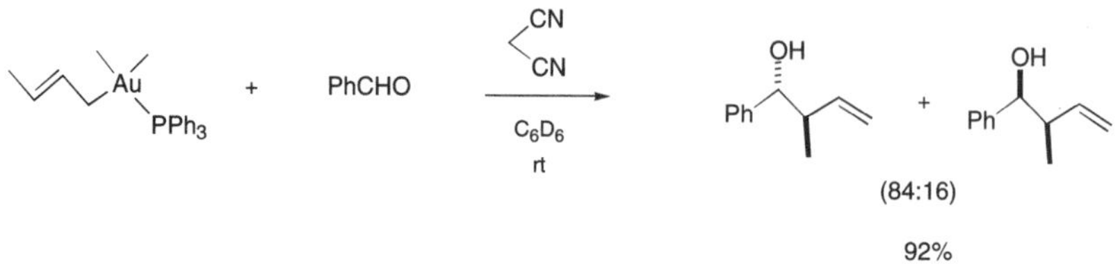

(84:16)

92%

[1]Sone, T., Ozaki, S., Kasuga, N.C., Fukuoka, A., Komiya, S. *BCSJ* **68**, 1523 (1995).

Allylindium reagents.

Carbonindation of alkynes.[1] The reaction of allylindium reagents with propargyl or homopropargyl alcohols shows regio- and stereoselectivities, leading to *(E)*-allylic or -homoallylic alcohols.

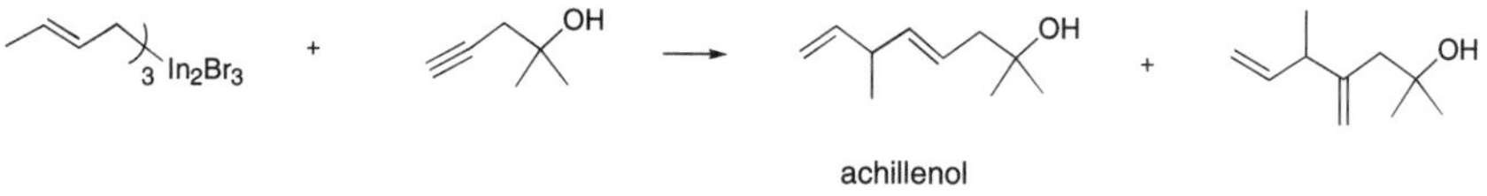

achillenol

[1]Araki, S., Imai, A., Shimizu, K., Yamada, M., Mori, A., Butsugan, Y. *JOC* **60**, 1841 (1995).

Allyl isothiocyanate.

N-Allyl carboxamides.[1] These amides are formed in good yields by heating fatty acids with $CH_2{=}CHCH_2N{=}C{=}S$ in the presence of Amberlyst A26–OH resin at 100°. The formation of adducts that liberate COS is implicated.

[1]Delaveau, V., Mouloungui, Z., Gaset, A. *SC* **26**, 2341 (1996).

Allylmanganese reagents.

Homoallylic alcohols.[1] The reagents are formed in situ by treating allylic phenyl sulfides with lithium 4,4′-di-*t*-butylbiphenylide and then $MnCl_2$. Reaction with aldehydes gives the alcohols.

[1]Ahn, Y., Doubleday, W.W., Cohen, T. *SC* **25**, 33 (1995).

(π-Allyl)palladium chloride dimer.

Alkenes.[1] With a bidentate ligand and in the presence of *t*-BuOK, the Pd catalyst promotes the addition of carbon acids to a terminal carbon of an allene, except for intramolecular reaction which forms a cyclohexane.

OR OEt OEt + $PhSO_2$ $PhSO_2$ —$PdCl/2$ MeO OMe P 2 P 2 THF, 80° $PhSO_2$ $PhSO_2$ RO OEt OEt 71% (R=H)

OR O O O O —$PdCl/2$ dppf / THF, 80° RO O O O O R = TBDPS 85%

Hydrosilylative dimerization of 1-alkynes.[2] A synthesis of 2,4-dialkyl-1-silyl-1,3-butadienes with good stereocontrol is realized. With the use of $HSiCl_3$ as reagent, the products are isolated after Grignard reactions ($Cl_3SiR \rightarrow (EtO)_3SiR$).

C_5H_{11} + $HSiCl_3$ —$PdCl/2$ Ph_3P / CH_2Cl_2 , 50° 14 h C_5H_{11} Cl_3Si C_5H_{11} 76% + C_5H_{11} Cl_3Si C_5H_{11} 15%

[1]Trost, B.M., Gerusz, V.J. *JACS* **117**, 5156 (1995).
[2]Kawanami, Y., Yamamoto, K. *SL* 1232 (1995).

Allylphosphine oxides and allylphosphonates.

1,3-Dienes. The Horner–Wittig reactions of allyl(diphenyl)phosphine oxides are *(E)*-selective.[1] However, different bases may change the outcome of the reactions with allylphosphonates, and therefore, a judicious choice according to the nature of the allyl group is critical.[2] Thus, NaH is employed for phosphonates bearing a *P*-allyl group, *t*-BuOK for those bearing 2-methallyl and prenyl groups, and BuLi-pentamethyldiethylenepentamine for crotylphosphonates.

[1]Liu, R.-Q., Schlosser, M. *SL* 1195 (1996).
[2]Liu, R.-Q., Schlosser, M. *SL* 1197 (1996).

Allylsamarium bromide.

Homoallylic amines.[1] Displacement of benzotriazole from the readily available *N*-aminoalkylbenzotriazoles proceeds rapidly to afford the products (8 examples, 72–93%).

[1]Wang, J., Zhou, J., Zhang, Y. *SC* **26**, 3395 (1996).

Allylsilanes. 13, 11–13; **14**, 18–19; **15**, 8; **18**, 14, 15–16

N-Tosyl-2-aryl-4-pentenylamines.[1] Lewis acid-catalyzed opening of *N*-tosyl-2-arylaziridines in the presence of allyltrimethylsilane gives a mixture of the pentenylamines and 4-aryl-2-trimethylsilylmethylpyrrolidines. A homogeneous product is obtained on treating such a mixture with TBAF.

Cl, NTs, +, $SiMe_3$, $BF_3 \cdot OEt_2$, CH_2Cl_2, 0°, Cl, NTs, $SiMe_3$, TBAF, +, Cl, NHTs

Allylation of α-diketones.[2] Allyltrifluorosilanes react with α-diketones in the presence of Et_3N at room temperature. *syn*-Diols are obtained as the major products. γ,γ-Disubstituted allylsilanes give α-ketols under the same conditions.

Allyltriarylbismuthonium salts.[3] These compounds are prepared by Lewis acid-catalyzed reaction of allylsilanes with Ar_3BiF_2 at −78°. They are thermally unstable, decomposing into allylarenes at room temperature.

Allylsilylation. The distributive addition of the allyl and silyl moieties to a double[4] or triple bond[5] is catalyzed by $AlCl_3$. Results from the latter reaction show that these are both regio- and stereoselective and that *syn*-addition is indicated.

[1]Schneider, M.-R., Mann, A., Taddei, M. *TL* **37**, 8493 (1996).
[2]Gewald, R., Kira, M., Sakurai, H. *S* 111 (1996).
[3]Matano, Y., Yoshimune, M., Suzuki, H. *TL* **36**, 7475 (1995).
[4]Yeon, S.H., Lee, B.W., Yoo, B.R., Suk, M.-Y., Jung, I.N. *OM* **14**, 2361 (1995).
[5]Yeon, S.H., Han, J.S., Hong, E., Do, Y., Jung, I.N. *JOMC* **499**, 159 (1995).

Allyl tolyl sulfone.

Extrusion of sulfur dioxide.[1] Allyl tolyl sulfone acts as a scavenger for the free radicals derived from alkyl allyl sulfones ($R{-}SO_2{-}CH_2CH{=}CH_2$) such that extrusion of SO_2 to give $R{-}CH_2CH{=}CH_2$ becomes a synthetically useful process. In other words, processes other than SO_2 extrusion and radical recombination are suppressed.

The addition of R· and allyl radical generated from $R{-}SO_2{-}CH_2CH{=}CH_2$ to alkenes is feasible. The relay allyl tolyl sulfone is omitted in these cases, but the allyl sulfone is added in about five fold excess.

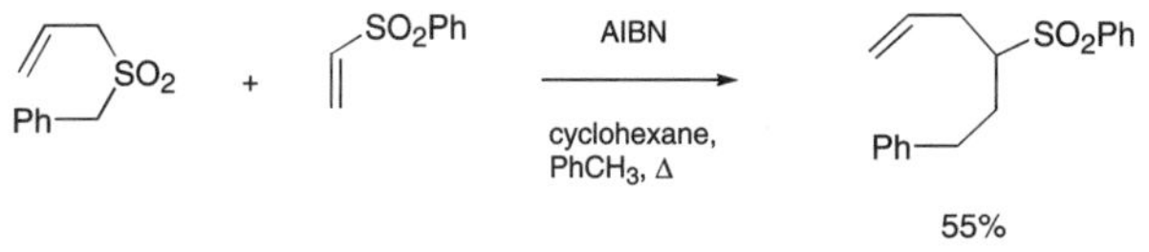

AIBN; cyclohexane, $PhCH_3$, Δ — 55%

[1]Quiclet-Sire, B., Zard, S.Z. *JACS* **118**, 1209 (1996).

Allyltributylstannane. 13, 10; **14**, 14–17; **16**, 7–9; **17**, 12–13

Preparation.[1] From allyl alcohol, the preparation involves mesylation (BuLi; MsCl) and reaction with Bu_3SnLi. Various substituted analogs are obtained in yields ranging from 71% to quantitative.

Displacement of activated halides. α-Halo esters,[2] amides,[3] and acetals[4] undergo radical allylation, which is initiated by AIBN or light. A related transformation, such as displacement of α-seleno esters (e.g., that of β-hydroxy-α-phenylseleno esters)[5] requires a Lewis acid (e.g., Me_3Al) to give a high *anti*-selectivity.

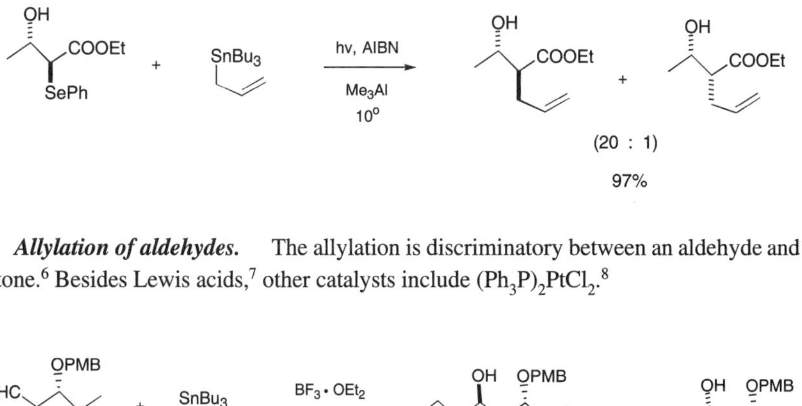

hν, AIBN; Me_3Al, 10° — (20 : 1) 97%

Allylation of aldehydes. The allylation is discriminatory between an aldehyde and a ketone.[6] Besides Lewis acids,[7] other catalysts include $(Ph_3P)_2PtCl_2$.[8]

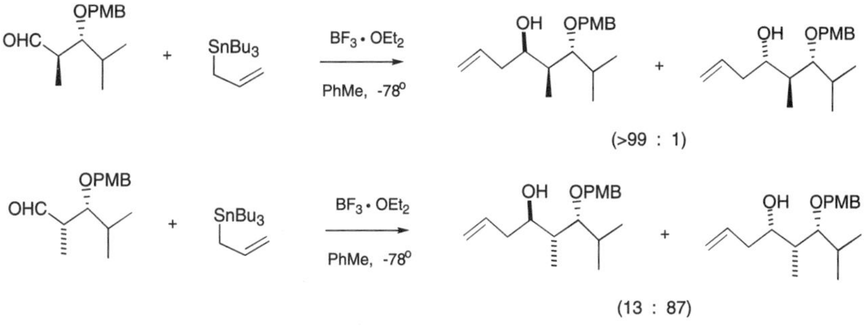

$BF_3 \cdot OEt_2$; PhMe, -78° — (>99 : 1)

$BF_3 \cdot OEt_2$; PhMe, -78° — (13 : 87)

Homoallylic amines. The reaction of aldimines with allyltributylstannane is catalyzed by Me_3SiCl[9] and by the π-allylpalladium chloride dimer.[10]

[Note that sulfonyl azides and azidoformate esters form *N*-allyl sulfonamides and carbamates, respectively, in the free-radical reactions (using $Ph_3SnCH_2CH{=}CH_2$ as the allyl donor).[11]]

[1]Weigand, S., Brückner, R. *S* 475 (1996).
[2]Landais, Y., Planchenault, D. *T* **51**, 12097 (1995).
[3]Hanessian, S., Yang, H., Schaum, R. *JACS* **118**, 2507 (1996).
[4]Nagano, H., Azuma, Y. *CL* 845 (1996).
[5]Gerster, M., Audergon, L., Moufid, N., Renaud, P. *TL* **37**, 6335 (1996).
[6]Kim, S., Kim, S.H. *TL* **36**, 3723 (1995).
[7]Evans, D.A., Dart, M.J., Duffy, J.L., Yang, M.G., Livingston, A.B. *JACS* **117**, 6619 (1995).
[8]Nakamura, H., Asao, N., Yamamoto, Y. *CC* 1273 (1995).
[9]Wang, D.-K., Dai, L.-X., Hou, X.-L. *TL* **36**, 8649 (1995).
[10]Nakamura, H., Iwama, H., Yamamoto, Y. *CC* 1459 (1996).
[11]Dang, H.-S., Roberts, B.P. *JCS(P1)* 1493 (1996).

Alumina. 14, 20–21; **16**, 9–10; **18**, 16–17

Aroyldiazomethanes.[1] Diaroyldiazomethanes decompose at room temperature to generate aroyldiazomethanes in the presence of alumina.

Wittig reactions.[2] The synthesis of unsaturated esters is conveniently performed.

Piperazines.[3] Twofold *N*-alkylation of *N*-arylamines with carbamates of bis(2-bromoethyl)amine on basic alumina at 150° forms *N*-arylpiperazines. The *N*-protecting group is also removed during the reaction.

Friedel–Crafts acylation.[4] An improved procedure for the acylation of aromatic ethers is to carry it out with acids in the presence of $(CF_3CO)_2O$ on an alumina surface.

[1]Korneev, S., Richter, C. *S* 1248 (1995).
[2]Dhavale, D.D., Sindkhedkar, M.D., Mali, R.S. *JCR(S)* 414 (1995).
[3]Mishani, E., Dence, C.S., McCarthy, T.J., Welch, M.J. *TL* **37**, 319 (1996).
[4]Ranu, B.C., Ghosh, K., Jana, U. *JOC* **61**, 9546 (1996).

Aluminum. 18, 17–18

1,2-Diols and 1,2-diamines. Aromatic carbonyl compounds and imines are rapidly converted to 1,2-diols[1] and 1,2-diamines,[2] respectively, with Al-KOH in methanol (ArCHO, 15 examples, 52–93%; ArCOR, 9 examples, 65–93%; ArCH=NAr′, 12 examples, 65–90%).

Trisubstituted allenes.[3] Propargylic sulfones undergo reductive elimination with Al/Hg in aqueous THF (11 examples, 74–92%).

Al-Hg / THF - H_2O → [ ] → 91%

Reductive alkylation of styrenes.[4] Using a sacrificing aluminum anode, the electrochemical reductive alkylation of styrenes with alkyl halides in DMF (60°) takes place at the β-carbon. If proper α,ω-dihaloalkanes are used as the alkylating agents, arylcycloalkanes are formed.

α,N,N-Tris(trimethylsilyl) enamines.[5] Electrolysis of α-siloxy nitriles with a sacrificing aluminum anode in the presence of Me_3SiCl results in reductive silylation (8 examples, 59–82%).

OSiMe3 / CN —[e⁻ (Al anode); Me3SiCl; THF - HMPA]→ SiMe3 / N(SiMe3)2 82%

[1]Khurana, J.M., Sehgal, A., Gogia, A., Manian, A., Maikap, G.C. *JCS(P1)* 2213 (1996).
[2]Baruah, B., Prajapati, D., Sandhu, J.S. *TL* **36**, 6747 (1995).
[3]Baldwin, J.E., Adlington, R.M., Crouch, N.P., Hill, R.L., Laffey, T.G. *TL* **36**, 7925 (1995).
[4]Leonel, E., Paugam, J.P., Nedelec, J.-Y., Perichon, J. *JCR(S)* 278 (1995).
[5]Constantieux, T., Picard, J.-P. *OM* **15**, 1604 (1996).

Aluminum bis(triflamide).

Acetylation. The triflamide is a highly efficient catalyst for the acetylation of phenols and aliphatic alcohols (10 examples, 90–99%). But some other metal triflamides and triflates are also effective.

[1]Mikami, K., Kotera, O., Motoyama, Y., Sakaguchi, H., Maruta, M. *SL* 171 (1996).

Aluminum chloride. 13, 15–17; **14**, 21–22; **15**, 10; **16**, 10–11; **17**, 15; **18**, 18–20

Acylation. Some noteworthy examples of acylation are the formation of δ-chloro-α-allenyl ketones,[1] phenyl esters of β,γ-unsaturated acids,[2] and substituted cyclopentenones.[3]

—[AcCl / AlCl3; CH2Cl2, -60°]→ C=C=, Cl, O (60%) + Cl, O (19%)

PhO, O, Cl —[Me3Si (allyl); AlCl3, PhMe; 20°, 2 h]→ PhO, O 82%

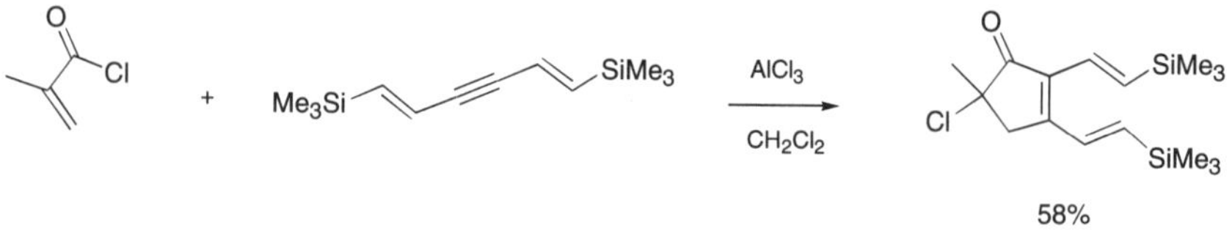

Friedel–Crafts alkylations. α-(1-Acetoxyalkyl)acrylic acid derivatives that are readily obtained via the Baylis–Hillman reaction are good alkylating agents. The substitution follows predominantly an S_N2' pattern.[4] Alkylation of aromatic hydrocarbons with benzalazines provides unsymmetrical diarylmethanes (10 examples, 63–73%).[5]

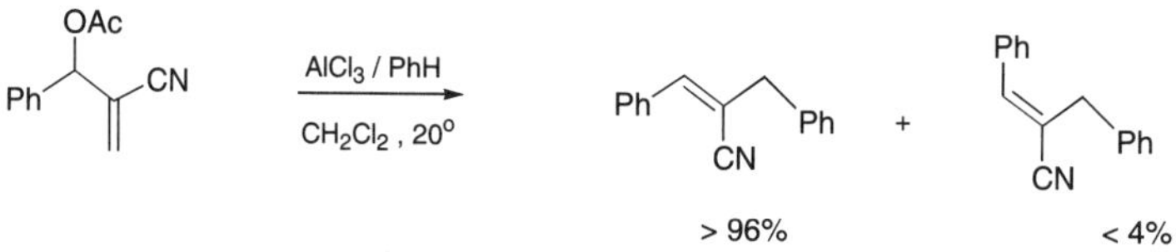

Formyldestannylation. A trimethylstannyl group attached to an aromatic ring[6] or an alkenyl position[7] is liable to replacement on reaction with dichloromethyl methyl ether. Aldehydes are formed.

Hydrosilylation.[8] Alkynes react with hydrosilanes to give *(Z)*-silylalkenes as a result of *anti*-addition.

Thia–Fries rearrangement.[9] Benzenesulfinylphenols are obtained on treatment of the aryl benzenesulfinates with $AlCl_3$ in CH_2Cl_2 at room temperature.

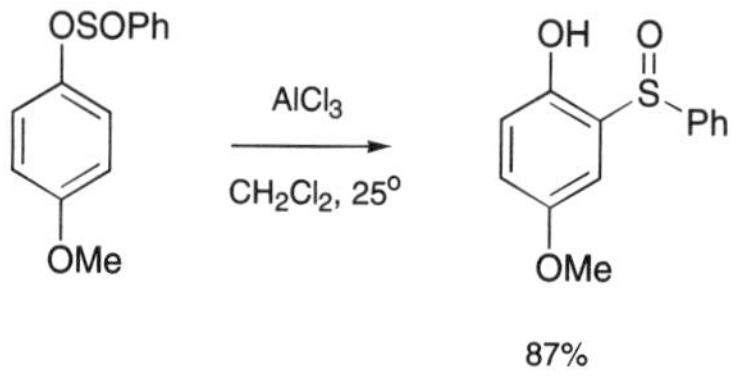

[1]Santelli-Rouvier, C., Lefrere, S., Santelli, M. *JOC* **61**, 6678 (1996).
[2]Mayr, H., Gabriel, A.O., Schumacher, R. *LA* 1583 (1995).
[3]Fiandanese, V., Marchese, G., Punzi, A., Ruggieri, G. *TL* **37**, 8455 (1996).
[4]Basavaiah, D., Pandiaraju, S., Padmaja, K. *SL* 393 (1996).
[5]Mathew, F., Bhattacharjee, S., Myrboh, B. *SC* **25**, 1795 (1995).
[6]Niestroj, M., Neumann, W.P. *CB* **129**, 45 (1996).
[7]Niestroj, M., Neumann, W.P., Mitchell, T.N. *JOMC* **519**, 45 (1996).
[8]Asao, N., Sudo, T., Yamamoto, Y. *JOC* **61**, 7654 (1996).
[9]Jung, M.E., Lazarova, T.I. *TL* **37**, 7 (1996).

Aluminum isopropoxide—trifluoroacetic acid.

Meerwein–Ponndorf–Verley reduction. The presence of CF_3COOH accelerates the reduction.[1] Both reagents are used in catalytic amounts. (*i*-PrOH supplies the hydride.)[2]

[1]Akamanchi, K.G., Varalakshmy, N.R. *TL* **36**, 3571 (1995).
[2]Akamanchi, K.G., Noorani, V.R. *TL* **36**, 5085 (1995).

Aluminum tris(2,6-diphenylphenoxide). [ATPH] 18, 21

Conjugate addition to carbonyl compounds. Complexation to the bulky Lewis acid prevents attack at the carbonyl group, and the addition occurs at the *o*- or *p*-positions of aromatic aldehydes and ketones.[1] With relatively bulky nucleophiles, such as *s*-BuLi and *t*-BuLi, products from addition in the *p*-position prevail.

Nonaromatic conjugate carbonyl compounds also undergo chemoselective 1,4-reduction[2] with *i*-BuAlH–BuLi at –78° after complexation with ATPH.

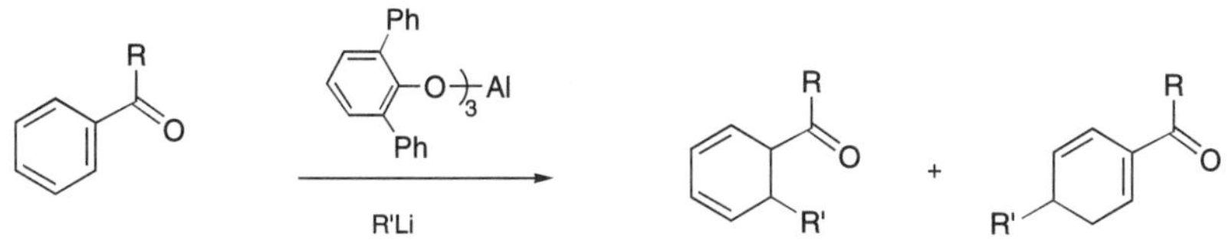

Diels–Alder reactions.[3] The stereoselectivity of the Diels–Alder reaction between cyclopentadiene and conjugate ketones and acrylonitrile is completely changed from endo to exo in the presence of ATPH. The bulky complexes are less sterically demanding in the exo orientation. The difference in an intramolecular reaction is also remarkable.

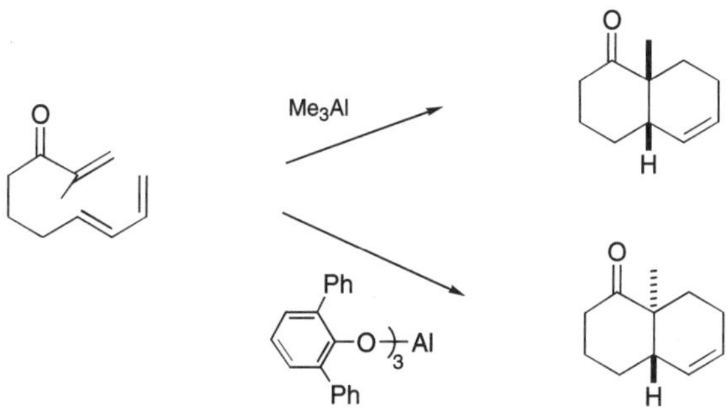

Claisen rearrangement. Complexation of the ethereal oxygen with ATPH prior to rearrangement ensures a high *(E):(Z)* ratio of 4-alkenals. ATPH is more effective than methylaluminum bis(2,6-diphenylphenoxide), but less effective than aluminum tris[(2-α-naphthyl-6-phenyl)phenoxide]. Two chiral analogs of the latter have been synthesized from 2-hydroxy-3-arylbinaphthalenes; (**1**) induces high degrees of asymmetry in the rearrangement.[4]

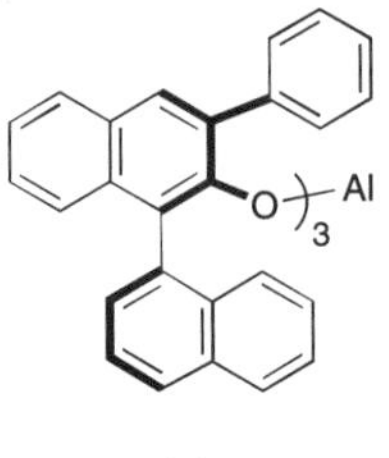

(**1**)

Aluminum tris(4-bromo-2,6-diphenylphenoxide) is also an excellent catalyst.[5]

[1]Maruoka, K., Ito, M., Yamamoto, H. *JACS* **117**, 9091 (1995).
[2]Saito, S., Yamamoto, H. *JOC* **61**, 2928 (1996).
[3]Maruoka, K., Imoto, H., Yamamoto, H. *JACS* **116**, 12115 (1994).
[4]Maruoka, K., Saito, S., Yamamoto, H. *JACS* **117**, 1165 (1995).
[5]Saito, S., Shimada, K., Yamamoto, H. *SL* 720 (1996).

N-Amino-2-methoxymethylpyrrolidine formaldehyde hydrazone.

Masked formyl anion.[1] With a silyl triflate as catalyst, the hydrazone acts as Michael donor to various enones. Thus, the method is valuable for the synthesis of 4-ketoaldehydes and 4-ketonitriles.

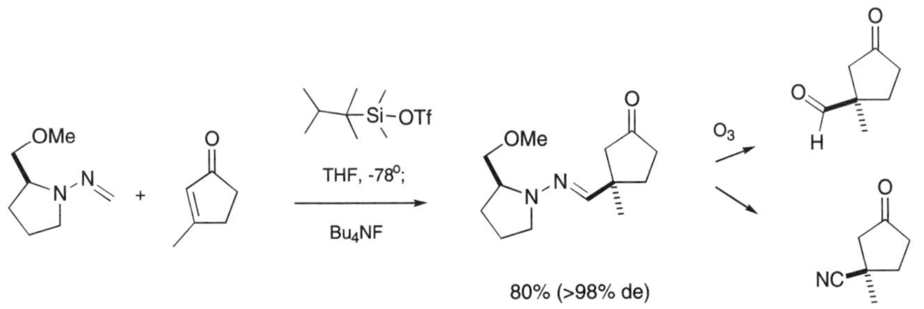

THF, -78°; Bu$_4$NF

80% (>98% de)

[1]Lassaletta, J.-M., Fernandez, R., Martin-Zamora, E., Diez, E. *JACS* **118**, 7002 (1996).

Antimony(III) chloride. 18, 25

Conjugate additions.[1] A catalytic activity of $SbCl_3$ is shown in the Pd-catalyzed transfer of an aryl group from $NaBPh_4$ or $ArB(OH)_2$ to enals and enones. The former reaction, involving $NaBPh_4$, proceeds via Ph_2B–PdPh and is suppressed by a phosphine.

However, the reaction employing $ArB(OH)_2$ can be induced by a wide range of catalysts; thus, besides $SbCl_3$, other metal chlorides, such as $TiCl_4$, $AsCl_3$, and, to a lesser extent, $BiCl_3$, $AlCl_3$, $SnCl_4$, $MoCl_5$, and $CeCl_3$, are active.

[1]Cho, C.S., Motofusa, S.-I., Ohe, K., Uemura, S., Shim, S.C. *JOC* **60**, 883 (1995).

Antimony(III) ethoxide.

Macrolactamization.[1] Carboxylic esters containing polyamine units undergo cyclization facilitated by preorganization with $Sb(OEt)_3$. Based on the ready preparation of a phenyl-substituted 17-membered lactam by this method in 90% yield, synthesis of several spermine alkaloids (verbacine, verbaskine, and verbascenine) is achieved in a straightforward manner.

Sb(OEt)3

PhH Δ

90%

verbascenine

verbaskine

verbacine

[1]Ishihara, K., Kuroki, Y., Hanaki, N., Ohara, S., Yamamoto, H. *JACS* **118**, 1569 (1996).

Antimony(V) fluoride.

Enones.[1] The reaction of 1-alkynes with aldehydes furnishes enones directly by the catalysis of SbF_5.

SbF5 / MeCN

0°, 5 min

73%

[1]Hayashi, A., Yamaguchi, M., Hirama, M. *SL* 195 (1995).

Arene(triphenylphosphine)ruthenium dichlorides—ammonium hexafluorophosphate.

Cycloaromatization.[1] These catalysts effect isomerization/cyclization of conjugated dienynes to form a benzene ring. One of the double bonds may be part of a five-membered heterocycle (furan, thiophene, or *N*-alkylpyrrole). The reactions probably proceed by electrocyclization of Ru^+-1,3,5-trienylcarbenoid intermediates.

(cymene)Ru(PPh_3)Cl_2, NH_4PF_6, CH_2Cl_2 Δ, 10 h → [$[Ru]^+$] → 89%

[1]Merlic, C.A., Pauly, M.E. *JACS* **118**, 11319 (1996).

4-Azidobenzyl 4′-nitrophenyl carbonate.

Amino group protection.[1] The reagent reacts with amines to form carbamates. Release of the amines is accomplished by reduction of the azido group with dithiothreitol. The 4-aminobenzyl carbamates thus generated undergo spontaneous fragmentation.

O_2N … N_3 —(RNH_2, py / THF, 25°, 12 h)→ RNH … N_3

84% (R = Bn)

[1]Griffin, R.J., Evers, E., Davison, R., Gibson, A.E., Layton, D., Irwin, W.J. *JCS(P1)* 1205 (1996).

ω-Azidoalkanols.

Schmidt reaction. The Lewis acid-catalyzed reaction of 2-azidoethanol or 3-azidopropanol with ketones leads to amides or lactams. These reagents obviate the problem experienced by using ordinary alkyl azides, which have severe limitations, such as requiring a strong Lewis acid ($TiCl_4$), and even so, the Schmidt reaction under such conditions is not general at all. The bifunctional azides form *N*-diazonio-1,3-oxazolidine intermediates, from which rearrangement occurs. Using *(R)*-2-azido-2-phenylethanol or (2*R*,4*S*)-4-azido-2-pentanol, the reaction with 4-*t*-butylcyclohexanone gives rise to chiral lactams.[1]

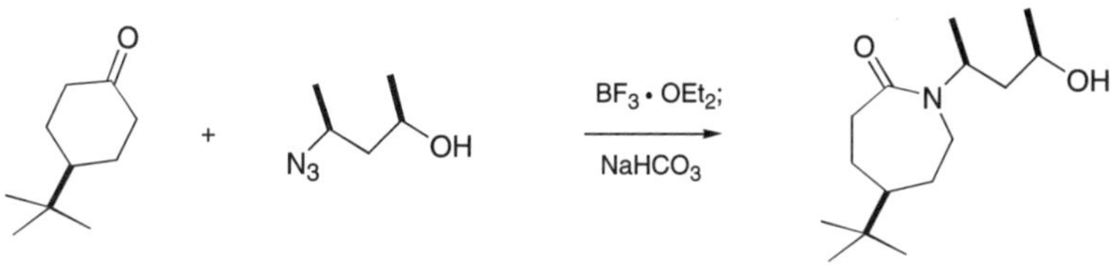

BF$_3$ • OEt$_2$;
NaHCO$_3$

The synthetic utility of this method is increased by intercepting the oxazolinium salts with various nucleophiles to prepare lactams that bear an ω-functionalized carbon chain. It is also possible to form azacycles containing a 2-dicyanomethylene group. Direct reduction of the oxazolinium salts with $NaBH_4$ gives amines.[2]

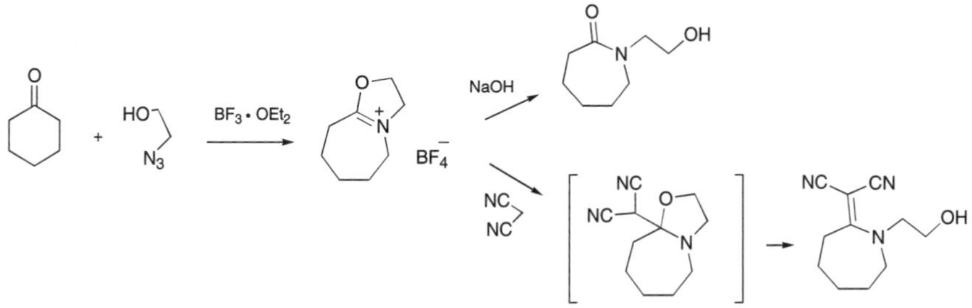

BF$_3$ • OEt$_2$

NaOH

[1]Gracias, V., Milligan, G.L., Aube, J. *JACS* **117**, 8047 (1995).
[2]Gracias, V., Milligan, G.L., Aube, J. *JOC* **61**, 10 (1996).

B

Baker's yeast. 18, 28

Carbonyl reduction. Reduction of α-diketones,[1] α-ketoesters,[2] β-ketoesters,[3] and 1-acetoxy-2,4-alkanediones[4] in high ee have been reported. γ-Nitroketones afford (S)-4-nitro alcohols.[5]

Reduction of nitrogenous compounds. Azidoarenes are converted to anilines in good yields, except for 4-azidophenol.[6] In organic solvent systems, the saturation of the double bond of nitrostyrenes is accomplished (10 examples, 8–82%).[7]

[1]Nakamura, K., Kondo, S.I., Kawai, Y., Hida, K., Kitano, K., Ohno, A. *TA* **7**, 409 (1996).
[2]Rotthaus, O., Kruger, D., Demuth, M., Schaffner, K. *T* **53**, 935 (1997).
[3]North, M. *TL* **37**, 1699 (1996).
[4]Utaka, M., Ito, H., Mizumoto, T., Tsuboi, S. *TA* **6**, 685 (1995).
[5]Guarna, A., Occhiato, E.G., Spinetti, L.M., Vallecchi, M.E., Scarpi, D. *T* **51**, 1775 (1995).
[6]Baruah, M., Boruah, A., Prajapati, D., Sandhu, J.S. *SL* 1193 (1996).
[7]Bak, R.R., McAnda, A.F., Smallridge, A.J., Trewhella, M.A. *AJC* **49**, 1257 (1996).

Barium. 17, 25–26

1,5-Dienes.[1] Active barium prepared by treating BaI_2 with couples allylic halides in a reaction promoted by biphenyllithium under Ar at –78°.

Carbocyclization.[2] *(E,E)*-1,4-Diphenyl-1,3-butadiene reacts with dichloroalkanes in the presence of active barium or strontium, forming 3-, 5-, 6-membered rings.

Ph Ph — Ba ; Cl Cl → Ph Ph 99%

[1]Yanagisawa, A., Hibino, H., Habaue, S., Hisada, Y., Yasue, K., Yamamoto, H. *BCSJ* **68**, 1263 (1995).
[2]Sell, M.S., Rieke, R.D. *SC* **25**, 4107 (1995).

Barium hydroxide. 18, 28–29

4-Arylpyridines.[1] Suzuki coupling of arylboronic acids with 4-iodopyridines takes place in the presence of $Ba(OH)_2$. This strong base helps to overcome problems associated with sterically hindered substrates.

[1]Godard, A., Rocca, P., Pomel, V., Thomas-dit-Dumont, L., Rovera, J.C., Thaburet, J.F., Marsais, F., Queguiner, G. *JOMC* **517**, 25 (1996).

Benzenethiol. 16, 327–329

Radical addition.[1] In the context of a synthesis of *ent*-lycoricidine, the addition of PhSH to a dehydrocinnamyl alcohol leading to a cyclohexene unit is the key step. The use of $Bu_3SnH–Et_3B$ is not viable, because of reverse regioselectivity, and there is no cyclization.

PhSH, hν; 91%; *ent*-lycoricidine

[1]Keck, G.E., Wager, T.T. *JOC* **61**, 8366 (1996).

Benzisoselenazol-3(2*H*)-ones.

Oxidation catalyst.[1] These compounds (**1**), as well as bis(2-carbamoyl)phenyl diselenides, are very effective catalysts for the oxidation of sulfides, dimethylhydrazones (to nitriles), and azines (to carbonyl compounds).

(**1**) + PhSPh —H_2O_2 / MeOH, rt, 3 h→ PhS(O)Ph 92%

[1]Mlochowski, J., Giurg, M., Kubicz, E., Said, S.B. *SC* **26**, 291 (1996).

Benzothiazol-2-ylsulfonyl chloride.

Amine protection.[1] The reagent (**1**) and related heteroarenesulfonyl chlorides sulfonylate amino acids on nitrogen in the presence of aqueous NaOH (pH 10–10.5). The products are useful for peptide coupling, even in the case of the sensitive phenylglycine system activated with $SOCl_2$. Deprotection is achieved using Zn/HOAc–EtOH or 50% H_3PO_2. Some other heteroarene-2-sulfonyl chlorides have the same capability.

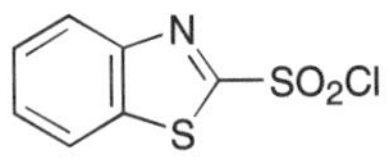

(1)

[1]Vedejs, E., Lin, S., Klapars, A., Wang, J. *JACS* **118**, 9796 (1996).

Benzotriazole.

N-Styrylamides.[1] The condensation of benzamides with aldehydes is mediated by benzotriazole. Thus, after the mixture is heated in toluene and then treated with NaH, the enamides are obtained.

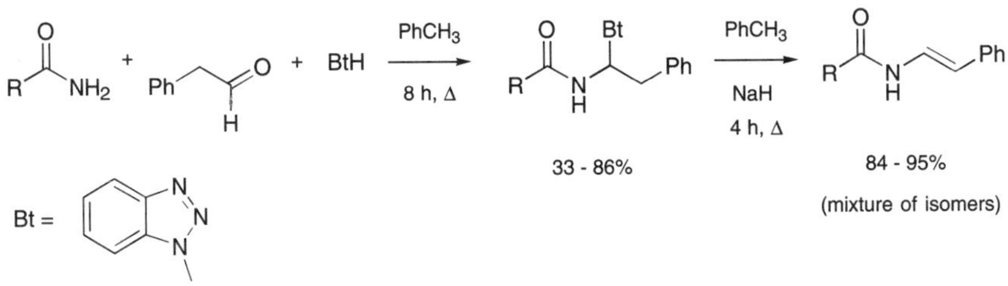

Aryl ketones. Aromatic aldehydes are converted to ketones[2] by condensation with benzotriazole in the presence of triethyl orthoformate, alkylation of the resulting ethers [ArCH(OEt)Bt], and treatment with dilute HCl to expel benzotriazole and ethanol. Silyl ketones are similarly prepared, using chlorosilanes as alkylating agents.[3]

[1]Katritzky, A.R., Ignatchenko, A.V., Lang, H. *SC* **25**, 1197 (1995).
[2]Katritzky, A.R., Lang, H., Wang, Z., Zhang, Z., Song, H. *JOC* **60**, 7619 (1995).
[3]Katritzky, A.R., Wang, Z., Lang, H. *OM* **15**, 486 (1996).

Benzotriazole-1-carboxamidinium tosylate.

Guanidines.[1] Amines are converted to guanidines (8 examples, 55–86%) by this reagent at room temperature in DMF (with Hünig's base).

[1]Katritzky, A.R., Parris, R.L., Allin, S.M., Steel, P.J. *SC* **25**, 1173 (1995).

Benzotriazol-1-yl derivatives. 18, 32–36

Insertion into aldehydes and ketones.[1] The α-lithiated 1-substituted benzotriazoles react smoothly with carbonyl compounds. The adducts undergo rearrangement with expulsion of benzotriazole on heating with $ZnBr_2$. Certain functional groups on the substituent of the benzotriazole nucleus do not interfere with the transformation, but actually render the method more versatile.

BuLi ; ZnBr2 [- Bt] OH Ph 60%

[1]Katritzky, A.R., Xie, L., Toader, D., Serdyuk, L. *JACS* **117**, 12015 (1995).

3-(α-Benzotriazol-1-yl)-3-ethoxypropene.

Ketones. The reagent (**1**) is a propenoyl anion synthon. After alkylation, enones are released.[1] On the other hand, the adducts undergo allylic displacement with Grignard reagents, and on hydrolytic workup, ketones are formed.[2] The possibility of creating α-bromoalkyl ketones, 1,4-diketones, 2-ethoxy-2-cyclopentenones, and α-keto enamines bespeaks for the versatility of this synthesis.

BuLi EtI BrMg ; HCl EtO (1) 81% (overall yield)

Cyclopropanes.[3] The anion behaves as a 1,1-dipole toward conjugated carbonyl compounds, forming cyclopropane derivatives in moderate to good yields.

[1]Katritzky, A.R., Zhang, G., Jiang, J. *JOC* **60**, 7589 (1995).
[2]Katritzky, A.R., Zhang, G., Jiang, J. *JOC* **60**, 7605 (1995).
[3]Katritzky, A.R., Jiang, J. *JOC* **60**, 7597 (1995).

α-(Benzotriazol-1-yl)methyl phenyl sulfides.

Homoallyl sulfides.[1] The benzotriazole unit is displaced by an allyl group on reaction with allyltrimethylsilane in the presence of $ZnBr_2$.

[1]Katritzky, A.R., Chien, J., Belyokov, S.A. *TL* **37**, 6631 (1996).

N-(α-Benzotriazol-1-yl)methyl-*N*-phenylamines.

1,2,3,4-Tetrahydroquinolines.[1] These amines condense with acetaldehyde to form 4-benzotriazolyl-1,2,3,4-tetrahydroquinolines. The benzotriazole group can be replaced with nucleophiles.

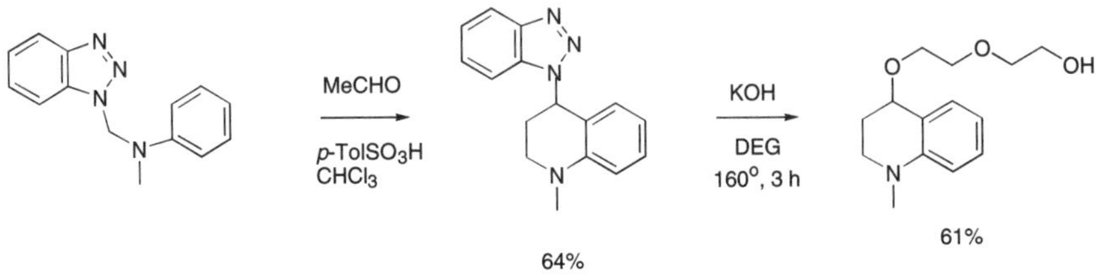

MeCHO, p-TolSO$_3$H, CHCl$_3$ → 64%; KOH, DEG, 160°, 3 h → 61%

[1]Katritzky, A.R., Rachwal, B., Rachwal, S. *JOC* **60**, 7631 (1995).

1-(Benzoylamino)-3-methylimidazolium chlorochromate.

Oxidation.[1] Allylic and benzylic alcohols are selectively oxidized to carbonyl compounds by this mild reagent, a yellow-orange, air-stable powder. Its preparation (92% yield) involves treatment of the imidazolium ylide with CrO_3 and HCl.

[1]Martinez, Y., de las Heras, M.A., Vaquero, J.J., Garcia-Navio, J.L., Alvarez-Builla, J. *TL* **36**, 8513 (1995).

1-Benzyl-4-aza-1-azoniabicyclo[2.2.2]octane borohydride.

Reductions.[1] Carbonyl compounds, including conjugated members, acid chlorides, azides, epoxides, and disulfides, are reduced. This salt (**1**) is a more selective reducing agent than tetrabutylammonium borohydride.

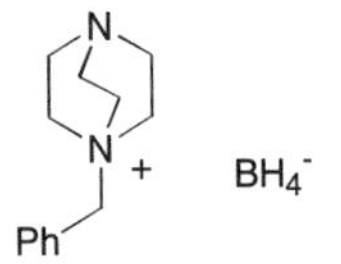

BH_4^-

(**1**)

Cleavage of trimethylsilyl ethers.[2] Desilylation with (**1**) is accomplished in refluxing *t*-butanol (11 examples, 80–100%).

[1]Firouzabadi, H., Afsharifar, G.R. *BCSJ* **68**, 2595 (1995).
[2]Firouzabadi, H., Afsharifar, G.R. *SC* **26**, 1065 (1996).

Benzyltriethylammonium permanganate.

Oxidation of enamines.[1] A method for the synthesis of chiral 1,4-dihydropyridines from bicyclic oxazolidines involves oxidation to the oxazolidinones and saponification.

MeOOC COOEt Ar N O $(BnNEt_3)^+ MnO_4^-$ CH_2Cl_2 0° -> reflux MeOOC COOEt Ar N O O 26% (Ar = 3-Cl-C_6H_4) KOH MeOH rt, 6 h MeOOC COOEt Ar N H 43%

[1]Caballero, E., Puebla, P., Sanchez, M., Medarde, M., Moran del Prado, L., San Feliciano, A. *TA* **7**, 1985 (1996).

Benzyltrimethylammonium isopropoxide.

Alkylation of α-substituted aldehydes.[1] The direct alkylation proceeds in good yields. The products were isolated as 2,4-DNPs (11 examples, 60–100%).

[1]Valenta, Z., MaGee, D.I., Setiadji, S. *JOC* **61**, 9076 (1996).

Beryllium chloride.

Dealkylation.[1] Highly selective demethylation of ArOMe in refluxing toluene is achieved (17 examples, 90–95%). For example, 2,4-dimethoxybenzophenone gives 2-hydroxy-4-methoxybenzophenone in 90% yield.

[1]Sharghi, H., Tamaddon, F. *T* **52**, 13623 (1996).

1,1′-Bi-2,2′-naphthol (BINOL) aluminum complexes.

Michael additions.[1] The complex, prepared from *i*-Bu_2AlH and BINOL in THF at 0° followed by treatment with BuLi, is an excellent catalyst for asymmetric Michael and tandem Michael–aldol reactions.

O + COOEt COOEt + Li O Al O O O THF O H COOEt COOEt

84% (91% ee)

α-Hydroxy phosphonates.[2] The addition of dimethyl phosphonate to aldehydes is subject to asymmetric induction by the lithium BINOL–aluminate (10 examples, 39–95%, > 55% ee).

Hetero Diels–Alder reactions.[3] Methylaluminum BINOLoxide catalyzes cycloadditions such as those involving glyoxylic esters. Products with excellent ee values of up to 97% are routinely obtained.

Polymeric BINOL aluminum chloride.[4] Prepared by Ni(0)-catalyzed cross- coupling of chiral 6,6′-dibromo-BINOL diacetate, hydrolysis, and treatment with Et_2AlCl, the chiral catalyst is effective for the Mukaiyama aldol reaction.

[1]Arai, T., Sasai, H., Aoe, K.I., Okamura, K., Date, T., Shibasaki, M. *ACIEE* **35**, 104 (1996).
[2]Arai, T., Bougauchi, M., Sasai, H., Shibasaki, M. *JOC* **61**, 2926 (1996).
[3]Graven, A., Johannsen, M., Jorgensen, K.A. *CC* 2373 (1996).
[4]Hu, Q.-S., Zheng, X.-F., Pu, L. *JOC* **61**, 5200 (1996).

1,1′-Bi-2,2′-naphthol–lanthanide complexes. 17, 28–30; **18**, 41–42

Reduction.[1] Secondary alcohols with moderate ee values are obtained from the reduction of ketones with borane in the presence of such a complex.

Nef reactions. The nitroaldol reactions from prochiral reagents show very high *syn*-selectivity and enantioselectivity. Catalysts derived from 6,6-disubstituted BINOLs appear to be of special value.[2] α,α-Difluoro aldehydes also undergo this condensation.[3]

Tetrahydroquinolines.[4] A chiral ytterbium–BINOL complex formed from $Yb(OTf)_3$, BINOL, and DBU has been employed in the aza Diels–Alder reaction involving alkylideneanilines to produce chiral tetrahydroquinolines.

α-Amino phosphonates.[5] At room temperature, imines form chiral adducts with dialkyl phosphonates (8 examples, yield 47–87%) in modest to high ee.

Baylis–Hillman reaction.[6] Acceleration of the reaction conducted in the presence of DABCO by $Ln(OTf)_3$–BINOL is observed. Best results are obtained with Ln = La, Sm.

[1]Zhang, F.-Y., Yip, C.-W., Chan, A.S.C. *TA* **7**, 2463 (1996).
[2]Sasai, H., Tokunaga, T., Watanabe, S., Suzuki, T., Itoh, N., Shibasaki, M. *JOC* **60**, 7388 (1995).
[3]Iseki, K., Oishi, S., Sasai, H., Shibasaki, M. *TL* **37**, 9081 (1996).
[4]Ishitani, H., Kobayashi, S. *TL* **37**, 7357 (1996).
[5]Sasai, H., Arai, S., Tahara, Y., Shibasaki, M. *JOC* **60**, 6656 (1995).
[6]Aggarwal, V.K., Tarver, G.J., McCague, R. *CC* 2713 (1996).

1,1′-Bi-2,2′-naphthol–tin(IV) chloride. 18, 42–43

Enantioselective protonation.[1] A chiral BINOL derivative is capable of assisting protolytic cleavage of enol silyl ethers enantioselectively.

[1]Ishihara, K., Nakamura, S., Kaneeda, M., Yamamoto, H. *JACS* **118**, 12854 (1996).

1,1′-Bi-2,2′-naphthol–titanium complexes. 15, 26–27; **16**, 24–25; **17**, 28–30; **18**, 43–44

Allylation. The highly enantioselective addition of β-substituted allylstannanes to aldehydes[1] has been observed, although the process appears to be slow. The reaction is accelerated by *i*-PrSSiMe.[2] A catalyst prepared from TiF_4 and chiral BINOL shows excellent activity (yield up to 93%, ee up to 94%) in promoting allylation with allylsilanes.[3]

Reduction. Chiral deuterated alcohols (6 examples, 75–90%, 90–97% ee) are available by reduction[4] of aldehydes with Bu_3SnD in the presence of a catalyst derived from $(i\text{-PrO})_4Ti$, (–)-BINOL , CF_3COOH, and 4 Å molecular sieves in refluxing ether.

For asymmetric reduction of ketones[5] using $(EtO)_3SiH$ as hydride source, the ee values are lower (10–55%).

Aldol condensation.[6] δ-Hydroxy-β-keto ester precursors are readily acquired by the reaction of 4-trimethylsiloxy-6-methylene-1,3-dioxins with aldehydes (6 examples, 32–93%).

PhCHO + [OSiMe3 dioxin] + [BINOL–Ti(OPr^i)$_2$] —THF, -78° -> rt, 10 h→ [product]

93% (92% ee)

A preparation of β-hydroxy thioesters is best carried out in ether within a narrow range of catalyst concentration (ca. 20 mol %) for maximum conversion and enantiomer excess.[7]

Dihydropyrones.[8] The Danishefsky condensation gives chiral products in the presence of a BINOL–Ti complex and CF_3COOH.

[1]Weigand, S., Brückner, R. *CEJ* **2**, 1077 (1996).
[2]Yu, C.-M., Choi, H.-C., Jung, W.-H., Lee, S.-S. *TL* **37**, 7095 (1996).
[3]Gauthier, D.R., Carreira, E.M. *ACIEE* **35**, 2363 (1996).
[4]Keck, G.E., Krishnamurthy, D. *JOC* **61**, 7638 (1996).
[5]Imma, H., Mori, M., Nakai, T. *SL* 1229 (1996).
[6]Sato, M., Sunami, S., Sugita, Y., Kaneko, C. *H* **41**, 1435 (1995).
[7]Keck, G.E., Krishnamurthy, D. *JACS* **117**, 2363 (1995).
[8]Keck, G.E., Li, X.-Y., Krishnamurthy, D. *JOC* **60**, 5998 (1995).

1,1′-Bi-2,2′-naphthol–zirconium complexes.

Allylation.[1] A chiral zirconate complex is prepared from BINOL and commercial $(i\text{-PrO})_4Zr{\cdot}i\text{-PrOH}$. The catalyst promotes the enantioselective addition of allylstannanes to aldehydes.

Kinetic resolution.[2] By an enantioselective Grignard reaction of cyclic allylic ethers in the presence of BINOL and a chiral ethylenebis(4,5,6,7-tetrahydro-1-indenyl)-zirconium(IV) dichloride, resolution of the important synthetic intermediates is achieved.

[1]Bedeschi, P., Casolari, S., Costa, A.L., Tagliavini, E., Umani-Ronchi, A. *TL* **36**, 7897 (1995).
[2]Visser, M.S., Harrity, J.P.A., Hoveyda, A.H. *JACS* **118**, 3779 (1996).

1,1′-Bi-2,2′-naphthyl methyl phosphate.

Asymmetric methylation.[1] The reaction of an allylic Grignard reagent with the chiral phosphate introduces a new stereogenic center at the γ-carbon.

Ph⁄⁄MgCl + (binaphthyl phosphate, P=O, OMe) —THF, -78°, 30 min→ Ph-branched product + Ph linear product (97 : 3) 83%

[1]Yanagisawa, A., Nomura, N., Yamada, Y., Hibino, H., Yamamoto, H. *SL* 841 (1995).

Bis(acetonitrile)dichloropalladium(II). 13, 33, 211, 236; **14**, 35–36; **15**, 28–29; **16**, 25–26; **17**, 30–31

Cleavage of TBS ethers. *t*-Butyldimethylsilyl ethers of both phenols[1] and aliphatic alcohols[2] are cleaved using catalytic amounts of $PdCl_2\cdot(MeCN)_2$ in aqueous acetone. One-pot desilylation–oxidation is also possible, except that the reaction conditions do not tolerate THP or TES ethers; aryl bromides also suffer reduction.

Carbonylation of aryl iodides.[3] A synthesis of aroic acids under atmospheric CO in aqueous butanol is facilitated by adding sodium dodecyl sulfate.

Dehydrative cyclization.[4] Intramolecular displacement of a cyclic allylic alcohol can set up spirocyclization.

MeOOC, HO, O, OTHP —$(MeCN)_2PdCl_2$, Me_2CO, 25°, 24 h→ COOMe, O, O 75%

[1]Wilson, N.S., Keay, B.A. *TL* **37**, 153 (1996).
[2]Wilson, N.S., Keay, B.A. *JOC* **61**, 2918 (1996).
[3]Cheprakov, A.V., Ponomareva, N.V., Beletskaya, I.P. *JOMC* **486**, 297 (1995).
[4]Tenaglia, A., Kammerer, F. *SL* 576 (1996).

N,N'-Bis(alkoxycarbonyl)-1,3-propylenethiourea.

Carbamates.[1] The reagents (**1**) are donors of alkoxycarbonyl groups for amines. Thus, N-Boc and N-Cbz derivatives are readily prepared in refluxing dioxane (7 examples, 65–97%).

NH + BocN(C=S)NBoc (**1**) → (dioxane, Δ 7 h) NBoc 97%

[1]Matsumura, N., Noguchi, A., Kitayoshi, A., Inoue, H. JCS(P1) 2953 (1995).

Bis(benzonitrile)dichloropalladium(II). 13, 34; **15**, 29; **18**, 46–47

Enediynes.[1] A route to functionalized enediynes has been developed; it depends on the cross-coupling of chloroenynes with 1-alkynes by $PdCl_2 \cdot (PhCN)_2$, CuI, and piperidine.

Insertion of C=O into silacyclobutanes.[2] Acid chlorides react with silacyclobutanes to form cyclic silyl enol ethers or 3-(chlorosilyl)propyl ketones.

α-Vinyl-β-lactams.[3] Imines undergo [2+2]-cycloaddition with ketenes, which are generated in situ from triallyl phosphate and CO by a carbonylation-elimination pathway.

PhCOCl + SiMe$_2$ (silacyclobutane) → ($(PhCN)_2PdCl_2$, Et_3N / PhMe, 80°, 4 h) Ph, O, SiMe$_2$ (cyclic silyl enol ether) 97%

2-Methylene-1,4-alkanediols.[4] The $PdCl_2 \cdot (PhCN)_2$ – $SnCl_2$ couple transforms an allylic alcohol into an allyl anion equivalent. 2-Methylene-1,3-propanediol can be used to prepare 2-methylene-1,4-alkanediols on reaction with carbonyl compounds.

HO–CH$_2$C(=CH$_2$)CH$_2$–OH + PhCHO → ($(PhCN)_2PdCl_2$ - $SnCl_2$, DMF, rt) HO–CH$_2$C(=CH$_2$)CH$_2$CH(OH)Ph 73%

Hetero Diels–Alder reactions.[5] A related cationic complex $Pd(dppp)(PhCN)_2(BF_4)_2$ is able to promote the condensation of dienes with aldehydes to form dihydropyrans.

Claisen rearrangement.[6] The *anti*-selectivity is in contrast to the purely thermal reaction.

PhMe

100°	11	:	89
$(PhCN)_2PdCl_2$	90	:	10

[1]Mladenova, M., Alami, M., Linstrumelle, G. *SC* **26**, 2831 (1996).
[2]Tanaka, Y., Yamashita, H., Tanaka, M. *OM* **15**, 1524 (1996).
[3]Zhou, Z., Alper, H. *JOC* **61**, 1256 (1996).
[4]Masuyama, Y., Kagawa, M., Kurusu, Y. *CC* 1585 (1996).
[5]Oi, S., Kashiwagi, K., Terada, E., Ohuchi, K., Inoue, Y. *TL* **37**, 6351 (1996).
[6]Sugiura, M., Nakai, T. *CL* 697 (1995).

N,N'-Bis(benzyloxycarbonyl)-*S*-methylisothiourea.

Guanidines.[1] The methylthio group of the reagent is replaced on reaction with an amine at room temperature in a system that contains $HgCl_2$ and Et_3N in DMF.

[1]Chandrakumar, N.S. *SC* **26**, 2613 (1996).

Bis(benzyltriethylammonium) tetrathiomolybdate.

Reductions.[1] Nitrones and *N*-oxides are deoxygenated by $(BnNEt_3)_2MoS_4$. Acyl azides give amides. Alkyl azides undergo a homocoupling reaction to form imines, whereas stabilized azides, such as acyl, sulfonyl, and aryl azides, undergo reductive elimination of N_2.[2]

Reductive coupling of thiocyanates furnishes disulfides, including those containing a macrocycle by an intramolecular reaction.[3]

$$PhCH_2SCN \xrightarrow[CH_3CN]{(BnNEt_3)_2MoS_4} PhCH_2S\text{-}SCH_2Ph \quad 88\%$$

Thioamides.[4] Conversion of amides to thioamides is accomplished by exposing the corresponding chloroiminium salts to the tetrathiomolybdate reagent.

Cleavage of propargyl esters.[5] With the use of this salt, the selective removal of the propargyl group is possible without affecting other esters derived from allyl, benzyl, and alkyl alcohols.

[1]Ilankumaran, P., Chandrasekaran, S. *TL* **36**, 4881 (1995).
[2]Ramesha, A.R., Bhat, S., Chandrasekaran, S. *JOC* **60**, 7682 (1995).
[3]Prabhu, K.R., Ramesha, A.R., Chandrasekaran, S. *JOC* **60**, 7142 (1995).
[4]Ilankumaran, P., Ramesha, A.R., Chandrasekaran, S. *TL* **36**, 8311 (1995).
[5]Ilankumaran, P., Manooj, N., Chandrasekaran, S. *CC* 1957 (1996).

***Bis(sym-*collidine)iodine(I) salts.** **15**, 30; **17**, 155; **18**, 49

1,n-Diols.[1] 2-Alkenyl-1,3-dioxolanes undergo ring expansion to give 1,4-dioxacycloalkane derivatives. Asymmetric induction occurs in the cases of chiral acetals. Thus, optically active 1,4- and 1,5-diols can be prepared using this reaction as a linchpin. A formal synthesis of solenopsin has been accomplished.

$(coll)_2ICIO_4$; $MeOCH_2CH_2OH$; CH_2Cl_2, -78°-> rt

90%

Cyclic silyl ethers.[2] Alkenyl silanols cyclize in the same manner as the all-carbon analogues.

Difunctionalization of cyclopentadienes. The stereoselective elaboration of an aminocyclopentenol completes the first stage of a synthesis of (+)-trehazolin.[3] Introduction of the *cis*-1,3 functionalities is initiated by a reaction of a trichloroacetimidate with $(sym\text{-collidine})_2I^+ClO_4^-$.

$(coll)_2ICIO_4$; $NaHCO_3$; MeCN - H_2O

(+)-trehazolin

[1]Fujioka, H., Kitagawa, H., Nagotimi, Y., Kita, Y. *JOC* **61**, 7309 (1996).
[2]Takaku, K., Shinokubo, H., Oshima, K. *TL* **37**, 6781 (1996).
[3]Ledford, B.E., Carreira, E.M. *JACS* **117**, 11811 (1995).

Bis(1,5-cyclooctadiene)nickel(0). **13**, 35; **14**, 36–37; **15**, 30–32, 131–132; **16**, 29; **17**, 32; **18**, 49–50

Cycloadditions. The Ni(0)-catalyzed intramolecular [4+2]-cycloaddition[1] generates heterocycles in a novel and efficient manner. Norbornadiene undergoes [2+2+2]-cycloaddition with electron-deficient alkenes, while some substituted norbornadienes (e.g., 2-TMS) can form [2+2]-cycloadducts.[2]

$Ni(cod)_2$; $(ArO)_3P$ / THF; rt → 91%; DDQ; dioxane → 88%

Ni(cod)$_2$ - 2 Ph$_3$P

ClCH$_2$CH$_2$Cl, 80°

99%

Bicyclic cyclopentenones.[3] Cyclization of the Pauson–Khand type from enynes is achievable with Ni(cod)$_2$ (12 examples, 38–85%). A bulky bis-ketimine ligand and a "CO" equivalent are present in the reaction medium. For the latter $(i\text{-Pr})_3$SiCN is adequate. The silyl cyanide is in equilibrium with the isocyanide. The primary products are the cyclopentenone imines, which undergo hydrolysis on workup.

Additive cyclization of enones. Enones containing an unsaturation linkage that is reactive to organozinc/nickel species readily undergo cyclization.[4,5] Two new C–C bonds are formed.

Ph$_2$Zn-PhZnCl

Ni(cod)$_2$

65%

Bu$_2$Zn-BuZnCl

Ni(cod)$_2$

60-90%

A more significant result is the twofold cyclization that completes the pentacyclic skeleton of certain Strychnos alkaloids.[6] The product is converted to akuammicine in two more steps.

Ni(cod)$_2$

Et$_3$N - LiCl

MeCN - DMF

40%

akuammicine

Hydroalumination.[7] Certain bridged oxabicycles are cleaved by i-Bu$_2$AlH. The regioselectivity can be different when Ni(cod)$_4$ and Ph$_3$P are present.

i-Bu$_2$AlH
Ni(cod)$_2$ - PPh$_3$
PhMe
OTBS
HO
OH
(9.4 : 1)
71%

[1]Wender, P.A., Smith, T.E. *JOC* **61**, 824 (1996).
[2]Lautens, M., Edwards, L.G., Tam, W., Lough, A.J. *JACS* **117**, 10276 (1995).
[3]Zhang, M., Buchwald, S.L. *JOC* **61**, 4498 (1996).
[4]Montgomery, J., Savchenko, A.V. *JACS* **118**, 2099 (1996).
[5]Savchenko, A.V., Montgomery, J. *JOC* **61**, 1562 (1996).
[6]Sole, D., Bonjoch, J., Bosch, J. *JOC* **61**, 4194 (1996).
[7]Lautens, M., Chiu, P., Ma, S., Rovis, T. *JACS* **117**, 532 (1995).

Bis(1,5-cyclooctadiene)rhodium(I) chloride.

Cycloadditions.[1] An intramolecular [4+2]-cycloaddition of a diene and an allene unit allows the assembly of cyclic systems with desirable structural features. The different regioselectivities exerted by Rh(I) and Ni(0) catalysts are remarkable.

OTBS
[(cod)RhCl]$_2$
(ArO)$_3$P / THF
45°
90%
Ni(cod)$_2$
(ArO)$_3$P / THF
25°
97%

Hydrosilylation and silylformylation of carbonyls. Condensation of RR′SiH$_2$ with carbonyl compounds provides chiral silyl ethers[2] in the presence of [RhCl(cod)]$_2$ and a chiral ligand. In the presence of CO, α-siloxy carbonyl compounds are obtained.[3]

Aroyl chlorides + alkynes.[4] Both decarbonylative addition and dechlorinative cycloaddition have been observed in the Rh-catalyzed reactions.

PhCOCl + Ph—≡ → [(cod)RhCl]$_2$ - PPh$_3$, octane, 140°; 20 h
Ph, Cl, Ph
91%

PhCOCl + Pr—≡—Pr → [(cod)RhCl]$_2$ - PPh$_3$ / Na$_2$CO$_3$, *o*-xylene / 145°; 24 h

73%

[1]Wender, P.A., Jenkins, T.E., Suzuki, S. *JACS* **117**, 1843 (1995).
[2]Ohta, T., Ito, M., Tsuneto, A., Takaya, H. *CC* 2525 (1994).
[3]Wright, M.E., Cochran, B.B. *OM* **15**, 317 (1996).
[4]Kokubo, K., Matsumasa, K., Miura, M., Nomura, M. *JOC* **61**, 6941 (1996).

Bis(cyclopentadienyl)dihydridomolybdenum.

Reduction of imines.[1] Together with a protonic acid (e.g., HOAc), Cp_2MoH_2 effects the reduction of imines to amines at room temperature (4 examples, 48–89%).

[1]Minato, M., Fujiwara, Y., Ito, T. *CL* 647 (1995).

Bis(cyclopentadienyl)dimethyluranium.

Imines.[1] The uranocene derivative $Cp_2^*UMe_2$ catalyzes the addition of amines to 1-alkynes (7 examples, 50–95%).

[1]Haskel, A., Straub, T., Eisen, M.S. *OM* **15**, 3773 (1996).

Bis(dibenzylideneacetone)palladium. 13, 103; **14**, 38; 299; **15**, 302; **17**, 144

Deallylation of amines.[1] A practical method for the synthesis of secondary amines is by deallylation with $(dba)_2Pd$ in the presence of dppb and *o*-mercaptobenzoic acid. Diallylamines undergo selective monodeallylation.

Isomerization of alkynes.[2] Alkynes substituted with a perfluoroalkyl group undergo isomerization to give 1-perfluoroalkyl-1,3-dienes in refluxing toluene. The system also includes Ph_3P and HOAc.

Addition to allenes. Malonate anions and amines add to the chain terminus of 1,2-alkadienes.[3] When the proton source is excluded and a vinyl triflate is added to trap the initially formed vinylpalladium species, certain specially constituted dienes are accessible.[4]

OTf + C + → NaH / THF ; (dba)$_2$Pd - PPh$_3$ / DMSO, 65°, 72 h

77%

1,4-Carbosilylation.[5] 1,3-Dienes are converted into allylsilanes (29 examples, 47–95%) on reaction with acyl chlorides and hexaalkyldisilanes in the presence of $(dba)_2Pd$. Decarbonylation of the acyl chloride occurs prior to its incorporation into the product.

Vinylcyclopropanes.[6] A highly stereoselective tandem alkylation and an S_N2' process occur when a mixture of an active methylene compound, 1,4-dichloro-2-butene, and a base is exposed to $(dba)_2Pd–Ph_3P$.

Cl⁄⁄Cl + NC–CH₂–N=CPh₂ → [NaH / THF ; $(dba)_2Pd$ - PPh_3, rt, 15 min] → vinylcyclopropane (CN, N=CPh_2)

74% (de 100%)

Displacement reactions.[7] Haloarenes in which the *p*-position carries an electron-withdrawing group (e.g., CHO, COPh, and CN) undergo displacement by alkoxide ions in the presence of $(dba)_2Pd$–dppf. Perhaps palladium alkoxides are involved.

A variant of conventional Suzuki coupling is represented by the reaction of 9-organothio-9-borabicyclo[3.3.1]nonanes with organic electrophiles to form unsymmetrical sulfides.[8]

[1]Lemaire-Audoire, S., Savignac, M., Dupuis, D., Genet, J.P. *BSCF* **132**, 1157 (1995).
[2]Wang, Z., Lu, X. *T* **51**, 11765 (1995).
[3]Besson, L., Gore, J., Cazes, B. *TL* **36**, 3853, 3857 (1995).
[4]Gauthier, V., Cazes, B., Gore, J. *BSCF* **133**, 563 (1996).
[5]Obora, Y., Tsuji, Y., Kawamura, T. *JACS* **117**, 9814 (1995).
[6]Franzone, G., Carle, S., Dorizon, P., Olliver, J., Salaün, J. *SL* 1067 (1996).
[7]Mann, G., Hartwig, J.F. *JACS* **118**, 13109 (1996).
[8]Ishiyama, T., Mori, M., Suzuki, A., Miyaura, N. *JOMC* **525**, 225 (1996).

o-Bis(diethylhydrosilyl)benzene.

1,1,3,3-Tetraethyl-1,3-disilaisoindolines.[1] These derivatives of primary amines can be subjected to flash chromatography on silica gel. They are prepared from disilane and amines by any of the following three methods: CsF in DMPU (at 120°), $PdCl_2$ in refluxing xylene, or successive treatment with Br_2 and Et_3N. Deprotection of the amines is accomplished on prolonged treatment with HOAc in ether.

[1]Davis, A.P., Gallagher, P.J. *TL* **36**, 3269 (1995).

2,2′-Bis(diphenylphosphino)-1,1′-binaphthyl (BINAP). 13, 36–37; **14**, 38–44; **15**, 34; **16**, 32–36; **17**, 34–38; **18**, 39–41

Palladium complexes.

Asymmetric Heck reaction. The chiral product from an intramolecular process can be trapped. Thus, the method has been applied to a synthesis of (–)-$\Delta^{9,12}$-capnellene[1] and halenaquinone.[2] Hypervalent alkenyliodonium salts can be used[3] instead of triflates.

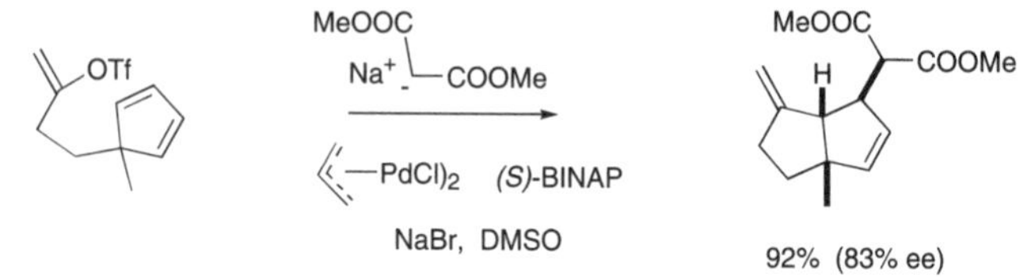

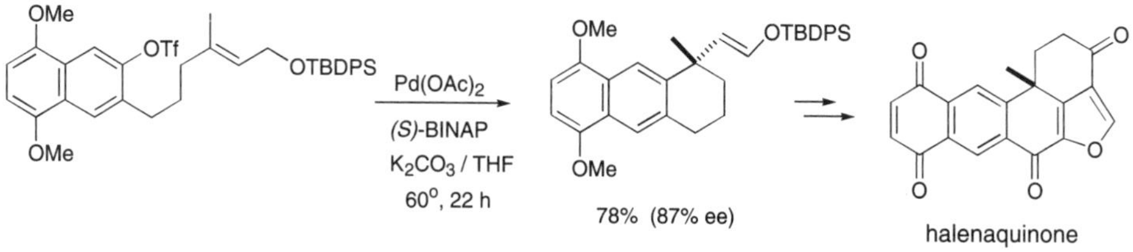

Asymmetric Mukaiyama aldol reaction. The cationic Pd complex generated from [*(R)*-BINAP]$PdCl_2$ and AgOTf in wet DMF in the presence of 4 Å molecular sieves is highly effective for the aldol reaction at room temperature.[4] A reaction proceeding via a Pd(II) enolate was identified for the first time.

[1]Ohshima, T., Kagechika, K., Adachi, M., Sodeoka, M., Shibasaki, M. *JACS* **118**, 7108 (1996).
[2]Kojima, A., Takemoto, T., Sodeoka, M., Shibasaki, M. *JOC* **61**, 4876 (1996).
[3]Kurihara, Y., Sodeoka, M., Shibasaki, M. *CPB* **42**, 2357 (1994).
[4]Sodeoka, M., Ohrai, K., Shibasaki, M. *JOC* **60**, 2648 (1995).

Rhodium(I) complexes.

Hydroboration.[1] Reaction of alkenylboronic esters with catecholborane in the presence of $Rh(COD)_2BF_4$ and *(R)*-BINAP, followed by oxidative workup, affords chiral glycols (8 examples, 9–87%, ee > 79%).

Hydroformylation.[2] The asymmetric version of hydroformylation has been studied using the $Rh(acac)(CO)_2$–*(R,S)*-BINAPHOS system. Substituent effects on the olefins on the regioselectivity of the reaction were evaluated.

[1]Wiesauer, C., Weissensteiner, W. *TA* **7**, 5 (1996).
[2]Nozaki, K., Nanno, T., Takaya, H. *JOMC* **527**, 103 (1997).

Ruthenium(II) complexes.

Asymmetric hydrogenation.[1] *N*-Tosylimines give the tosylamines. The yields are variable (< 5–82%). Alkenes without heteroatom functionalities are reduced satisfactorily.[2]

Reduction of β-keto derivatives.[3] The asymmetric hydrogenation of β-keto phosphonates using $RuCl_2(binap)(dmf)_n$ as catalyst is highly effective (7 examples, 96–99%), giving alcohols of high ee. The method has been used in a synthesis of the antibiotic fosfomycin. β-Ketoesters are reduced at atmospheric pressure (8 examples, 80–100%).[4]

[1]Charette, A.B., Giroux, A. *TL* **37**, 6669 (1996).
[2]Ohta, T., Ikegami, H., Miyake, T., Takaya, H. *JOMC* **502**, 169 (1995).
[3]Kitamura, M., Tokunaga, M., Noyori, R. *JACS* **117**, 2931 (1995).
[4]Genet, J.P., Ratovelomanana-Vidal, V., Cano de Andrade, M.C., Pfister, X., Guerreiro, P., Lenoir, J. Y. *TL* **36**, 4801 (1995).

Silver(I) complexes.

Asymmetric allylation.[1] Allylation of aldehydes with allytributyltin in the presence of AgOTf–BINAP furnishes homoallylic alcohols with good ee. Other combinations, such as BINAP and $AgNO_3$, $AgClO_4$, $AgOCOCF_3$, or other chiral phosphines with AgOTf, are inferior.

[1]Yanagisawa, A., Nakashima, H., Ishiba, A., Yamamoto, H. *JACS* **118**, 4723 (1996).

2,2′-Bis(6-halomethyl)-3,4-dihydropyrans.

Protection and resolution of glycols.[1] Upon reaction of the bisenol ether with glycols, bisaetals are formed. By virtue of the anomeric effect, the adducts assume a particular conformation. 1,3-Diaxial interactions between the bulky substituent of the glycol disfavor the formation of adducts from one of the enantiomers.

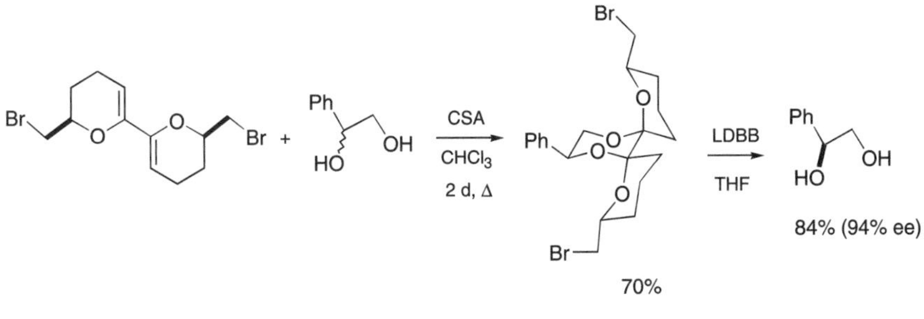

Br O O Br + Ph HO OH —CSA, $CHCl_3$, 2 d, Δ→ Ph (bisacetal, Br, Br) 70% —LDBB, THF→ Ph HO OH 84% (94% ee)

[1]Ley, S.V., Mio, S., Meseguer, B. *SL* 789, 791 (1996).

Bis(2-ketooxazolidin-3-yl)phosphonyl chloride.

Macrolactamization. Closure of the 23-membered ring is the most critical step for a synthesis of (–)-madumycin II.[1] Using the Bop-Cl[2] in the presence of $i\text{-}Pr_2NEt$ provides a solution. Despite the modest yield (32%), the method is serviceable.

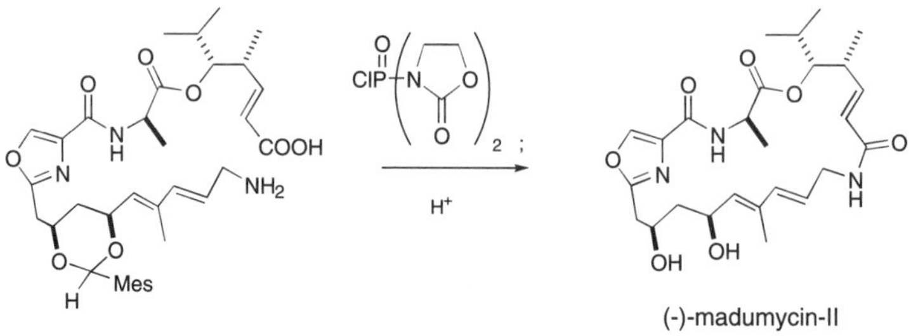

(-)-madumycin-II

[1]Tavares, F., Lawson, J.P., Meyers, A.I. *JACS* **118**, 3303 (1996).

Bis[(1*R*, 3*S*, 4*S*)-menth-3-yl)methyl]boron halide.

Aldol reactions.[1] This reagent (**1**) prepared from (–)-menthone and the (1*S*, 3*R*, 4*R*)-isomer (**2**) from (+)-menthone separately induce the formation of 3,4-*anti* and 3,4-*syn* 3-hydroxy-4-amino thioesters, respectively, from the condensation of ketene *S*-*t*-butylthioacetal and α-amino aldehydes. The method is well suited for the synthesis of (3*S*, 4*S*)-statine.

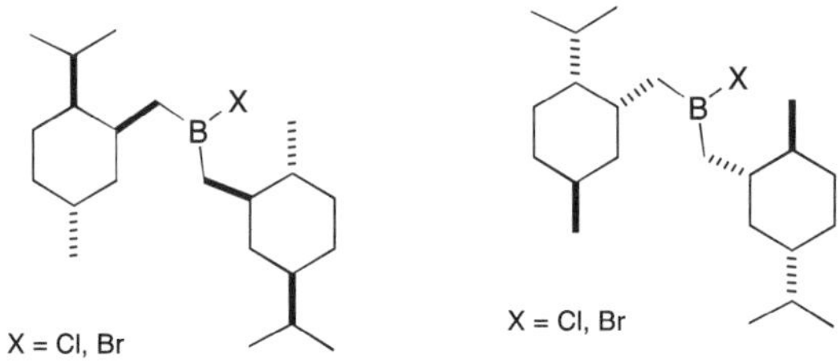

X = Cl, Br

(**1**)

X = Cl, Br

(**2**)

[1]Gennari, C., Pain, G., Moresca, D. *JOC* **60**, 6248 (1995).

Bismuth. **13**, 39; **18**, 52

Reduction of nitroarenes. Nitroarenes containing fluorine atoms undergo reduction to amines[1] by Bi in liquid HF-CH_2Cl_2. Other metals, such as Sn and Pb, are effective as well. Bismuth also catalyzes the reduction of nitroarenes to azoxybenzenes by $NaBH_4$.[2]

[1]Tordeux, M., Wakselman, C. *JFC* **74**, 251 (1995).
[2]Ren, P., Pan, S., Dong, T., Wu, S. *SC* **26**, 3903 (1996).

Bismuth(III) chloride. 15, 37; **18**, 52

Dithioacetalization. $BiCl_3$, as well as $Bi_2(SO_4)_3$, is an efficient catalyst for converting carbonyl compounds into dithioacetals (20 examples, 85–100%)[1] at room temperature. Interestingly, the cleavage of such groups can be effected by a catalytic amount of $Bi(NO_3)_3 \cdot 5H_2O$ in the presence of air and water (15 examples, 72–98%).[2]

[1]Komatsu, N., Uda, M., Suzuki, H. *SL* 984 (1995).
[2]Komatsu, N., Taniguchi, A., Uda, M., Suzuki, H. *CC* 1847 (1996).

Bismuth(III) chloride—zinc.

α,β-Dihydroxy ketones.[1] Aldehydes and α-diketones undergo reductive cross-coupling. The reaction occurs even in aqueous media.

Ph‿CHO + MeCOCOMe —($BiCl_3$ - Zn, THF Δ)→ Ph...CH(OH)C(OH)(Me)COMe (two diastereomers) 85%

[1]Miyoshi, N., Fukuma, T., Wada, M. *CL* 999 (1995).

Bismuth(III) chloride–zinc iodide. 18, 52

Allyl ketones. The metal halide combination is the first efficient system for catalyzing the acylation of allylsilanes at room temperature.

[1]Le Roux, C., Dubac, J. *OM* **15**, 4646 (1996).

Bis(*p*-nitrobenzyloxy) disulfide.

S_2-Transfer.[1] The reagent is stable at room temperature. However, when it is heated to 100° in the presence of a diene, dienophilic S_2 is transferred, forming a 1,2-dithia-4-cyclohexene.

2,3-dimethylbutadiene + (p-$NO_2C_6H_4CH_2O$-S-S-$OCH_2C_6H_4NO_2$-p) —(PhMe, ~ 100°, 24 h)→ 4,5-dimethyl-1,2-dithia-4-cyclohexene 79%

[1]Tardif, S.L., Williams, C.R., Harpp, D.N. *JACS* **117**, 9067 (1995).

N,N′-Bis(o-nitrobenzenesulfonyl) selenodiimide.

Allylamines.[1] An ene-type reaction between the reagent NsN=Se=NNs (prepared from NsNHNa, $NsNCl_2$, and Se) and an alkene at room temperature gives the allylic sulfonamide. Under similar conditions, 1,3-dienes afford 1,2-diamino-3-alkenes.

[1]Bruncko, M., Khuong, T.-A.V., Sharpless, K.B. *ACIEE* **35**, 454 (1996).

Bis(pentafluorophenyl)tin dibromide.

Cyanohydrins.[1] The organotin compound catalyzes the Mukaiyama aldol reaction, the derivatization of carbonyl compounds with Me_3SiCN, and other Lewis acid-catalyzed processes at −78°. It is prepared from diallyltin dibromide by reaction with pentafluorophenylmagnesium bromide followed by bromolysis.

[1]Chen, J., Sakamoto, K., Orita, A., Otera, J. *SL* 877 (1996).

Bis(pentamethylcyclopentadienyl)[bis(trimethylsilyl)methyl]samarium.

Pyrrolines and pyrrolizines. Cyclization of 4-alkynylamines is readily promoted by $(Cp^*)_2SmCH(TMS)_2$, giving cyclic imines by a hydroamination/cyclization process.[1] When secondary amines containing two unsaturated bonds are subjected to the reaction conditions, a tandem ring formation process is realized.[2]

$(Me_5C_5)_2SmCH(SiMe_3)_2$ → 92%

$(Me_5C_5)_2SmCH(SiMe_3)_2$ → 91%

Hydrogenation and hydrosilylation.[3] Substituted methylenecycloalkanes are saturated or hydrosilylated with hydrogen or $PhSiH_3$, respectively. The silyl group is placed at the terminal position of the double bond.

[1]Li, Y., Marks, T. *JACS* **118**, 9295 (1996).
[2]Li, Y., Marks, T. *JACS* **118**, 707 (1996).
[3]Molander, G.A., Winterfeld, J. *JOMC* **524**, 275 (1996).

Bis(pentamethylcyclopentadienyl)methylytterbium.

Hydrosilylation. The addition of $PhSiH_3$ to alkynes furnishes vinylsilanes.[1] Dienes form 5- and 6-membered ring products.[2]

[1]Molander, G.A., Retsch, W.H. *OM* **14**, 4570 (1995).
[2]Molander, G.A., Nichols, P. J. *JACS* **117**, 4415 (1995).

Bis(pentamethylcyclopentadienyl)samarium bis(tetrahydrofuran).

Acetylation.[1] Alcohols and amines are acetylated by vinyl acetate in the presence of $(Cp^*)_2Sm(thf)_2$.

Redox coupling and addition.[2] The Tishchenko reaction products of aldehydes, which are formed by the action of the organosamarium species, are trapped by vinyl esters in situ. For the Tishchenko reaction, $(Cp^*)_2LaCH(TMS)_2$ is also effective.[3]

CHO + ; $Cp^*_2Sm(thf)_2$, $PhCH_3$, 0° -> rt, 3 h; 77%

[1]Ishii, Y., Takeno, M., Kawasaki, Y., Muromachi, A., Nishiyama, Y., Sakaguchi, S. *JOC* **61**, 3088 (1996).
[2]Takeno, M., Kikuchi, S., Morita, K.I., Nishiyama, Y., Ishii, Y. *JOC* **60**, 4974 (1995).
[3]Onozawa, S.-Y., Sakakura, T., Tanaka, M., Shiro, M. *T* **52**, 4291 (1996).

Bis(titanocene chloride).

Glycals.[1] Substituted glycosyl bromides undergo the elimination of bromine and a group at the adjacent carbon on brief treatment with $(Cp_2TiCl)_2$ in THF at room temperature. Glycals are obtained in good yields.

Pinacol formation.[2] Stereoselective coupling of carbonyl compounds (with the *dl* form predominant) in aqueous media is observed. It is important that Cl^- be present.

Coupling of tungsten carbene complexes with epoxides. α-(Titanoxy) radicals are formed when epoxides react with $(Cp_2TiCl)_2$. These radicals add to unsaturated tungsten carbene complexes to give tetrahydropyran derivatives.[3]

[Cp2TiCl]2
OTi(Cp)2Cl
OTi(Cp)2Cl
W(CO)5
Ph
W(CO)5
OMe
Ph OMe
H O W(CO)5 + H O W(CO)5
H Ph H Ph
56% (3.5 : 1)

Homoallylic alcohols. C–C Bond formation occurs when an ω-alkynyl epoxide is treated with $(Cp_2TiCl)_2$. The β-alkoxy radical generated by epoxide cleavage adds to the triple bond. A synthesis of ceratopicanol[4] featured this process as an annulation step.

O
[Cp2TiCl]2
THF
OH
OH
ceratopicanol

[1]Cavallaro, C.L., Schwartz, J. *JOC* **60**, 7055 (1995); Spencer, R.P., Schwartz, J. *TL* **37**, 4357 (1996).
[2]Barden, M.C., Schwartz, J. *JACS* **118**, 5484 (1996).
[3]Merlic, C.A., Xu, D., Nguyen, M.C., Truong, V. *TL* **34**, 227 (1993).
[4]Clive, D.L.J., Magnuson, S.R. *TL* **36**, 15 (1995).

Bis(tributyltin) oxide. 13, 41–42; **15**, 39; **18**, 54

Organostannanes.[1] The sonochemical Barbier reaction of an organohalogen compound, Mg, 1,2-dibromoethane and $(Bu_3Sn)_2O$ is a good method for the preparation of organostannanes.

Cleavage of steroid esters.[2] On heating steroid esters with $(Bu_3Sn)_2O$ in THF at 110°, sterols are recovered.

[1]Lee, A.S.-Y., Dai, W.-C. *T* **53**, 859 (1997).
[2]Perez, M.G., Maier, M.S. *TL* **36**, 3311 (1995).

[Bis(trifluoroacetoxy)iodo]benzene.

Oxidative heterocyclization.[1] Sulfur compounds undergo oxidation, and when the sulfur atom is separated from a polymethoxylated aromatic nucleus by three, four, or five bonds, cyclization ensues (8 examples, 53–98%).

BTIB, $BF_3 \cdot OEt_2$
CH_2Cl_2 ;
aq. $MeNH_2$

98%

[1]Kita, Y., Egi, M., Ohtsubo, M., Saiki, T., Takada, T., Tohma, H. *CC* 2225 (1996).

Bis(2,2,2-trifluoroethyl) alkanesulfinylmethylphosphonates.

(Z)-Alkenyl sulfoxides.[1] The reaction of $RS(O)CH_2P(O)(OCH_2CF_3)_2$ with aldehydes is highly stereoselective. However, the corresponding sulfenyl reagents show a trend toward the formation of *(E)*-isomers. Thus, for the synthesis of *(Z)*-alkenyl sulfides, this method is useful in view of the fact that deoxygenation of the sulfoxides with $Bu_3P\text{-}CCl_4$ proceeds with retention of configuration.

[1]Kokin, K., Tsuboi, S., Motoyoshiya, J., Hayashi, S. *S* 637 (1996).

Bis(2,2,2-trifluoroethyl) 1,3-dithian-2-ylphosphonate.

Homologation.[1] The lower basicity of the anion derived from (**1**), compared with that of the dimethyl phosphonate, is useful for homologation of ketones containing acidic α-hydrogens, including cyclopentanone and β-tetralones. The products are ketene dithioacetals.

(**1**)

[1]Mink, D., Deslongchamps, G. *SL* 875 (1996).

Bis(trimethylphosphine)titanocene.

Reductive cyclization.[1,2] Unsaturated carbonyl compounds having the double bond at a remote position give cycloalkanols when a hydrosilane is present.

Fused lactones.[3,4] The foregoing reaction is diverted toward the formation of lactones when CO is added instead of a hydrosilane.

$Cp_2Ti(PMe_3)_2$ / pentane / 25°, 2 h → $TiCp_2$ 91% → CO → $TiCp_2$ 83% → air → 60%

Pauson–Khand-type reaction.[5] Titanacycles formed from enynes can be converted to *N*-silylimines of bicyclic enones on reaction with R_3SiCN.

[1]Kablaoui, N.M., Buchwald, S.L. *JACS* **118**, 3182 (1996).
[2]Crowe, W.E., Rachita, M.J. *JACS* **117**, 6787 (1995).
[3]Crowe, W.E., Vu, A.T. *JACS* **118**, 1557 (1996).
[4]Kablaoui, N.M., Hicks, F.A., Buchwald, S.L. *JACS* **118**, 5818 (1996).
[5]Hicks, F.A., Berk, S.C., Buchwald, S.L. *JOC* **61**, 2713 (1996).

Bis(trimethylsilyl) chromate—silica gel.

Oxidation.[1] The reagent is generated from CrO_3 and $(Me_3Si)_2O$ in CH_2Cl_2 and supported on silica gel. It has similar oxidizing power toward alcohols as do other chromates (22 examples, 71–99%).

[1]Lee, J.G., Lee, J.A., Sohn, S.Y. *SC* **26**, 543 (1996).

Bis[trinitratocerium(IV)] chromate.

Oxidation of trimethylsilyl ethers.[1] Refluxing the silyl ethers in benzene with this reagent affords carbonyl products (8 examples, 80–92%). The reaction is sluggish when trinitratocerium(IV) chromate dihydrate is used.

[1]Firouzabadi, H., Shiriny, F. *SC* **26**, 649 (1996).

Bis(triphenylphosphonio)oxide bistriflate. 18

Dienes from epoxides.[1] In refluxing dichloroethane in the presence of triethylamine, epoxides undergo a double-elimination reaction, giving conjugated dienes in good yields. Crowded epoxides remain unchanged; thus, the mixture of *cis-* and *trans*-cyclododecene oxides affords a clean mixture of *(1E, 3Z)*-cyclododecadiene and *trans*-cyclododecene oxide (50% yield).

[1]Hendrickson, J.B., Walker, M.A., Varvak, A., Hussoin, M.S. *SL* 661 (1996).

Borane. 18, 58

1,2-Diarylhydrazines.[1] Azoarenes are reduced at 0° by $BH_3 \cdot THF$ (12 examples, 48–88%).

Protection of oxazoles.[2] 2-Lithiooxazoles are liable to undergo ring opening; thus, alkylation of the heterocycle via the lithiated species is problematic. On the other hand, oxazole–borane complexes behave normally during deprotonation. The borane moiety is easily removed after the alkylation by treatment with HOAc.

Activation of benzylamines.[3] Metallation at the benzylic position of benzylamines is facilitated after forming the borane complexes.

Phosphine complexes.[4] Maintenance of a tetrahedral state of the phosphorus atom in such complexes and enhancement of acidity of the hydrogen atoms attached to an adjacent carbon enable the enantioselective deprotonation by the *s*-BuLi/(–)–sparteine complex. Oxidative coupling of the lithio derivatives and removal of the borane units accomplish the synthesis of C_2-symmetric *P*-chiral diphosphines.

s-BuLi - (-)-sparteine; Cu(OPiv)$_2$ / THF → + meso isomer 72%; Et_2NH, 55° → 100%

[1]Akula, M.R., Kabalka, G.W. *SC* **26**, 3821 (1996).
[2]Vedejs, E., Monahan, S.D. *JOC* **61**, 5192 (1996).
[3]Ebden, M.R., Simpkins, N.S., Fox, D.N.A. *TL* **36**, 8697 (1995).
[4]Muci, A.R., Campos, K.R., Evans, D.A. *JACS* **117**, 9075 (1995).

Borane–amines. 13, 42; **18**, 58

Reductive amination.[1] Treatment of aldehydes or ketones with amines and the borane–pyridine complex in methanol in the presence of 4 Å molecular sieves leads to secondary amines.

Deoxygenation of aromatic ketones.[2] Borane-dimethylamine, together with $TiCl_4$, is a mild and efficient reagent for achieving the reduction.

Acetal cleavage.[3] Interestingly, the regioselectivity in the reductive ring opening of 4,6-*O*-benzylidene glucose and glucosamine derivatives by borane-dimethylamine-boron trifluoride etherate, is dependent on the solvent used.

+ $BH_3 \cdot Me_2NH$ / $BF_3 \cdot OEt_2$ → MeCN, 0°, 1 h: 79%; CH_2Cl_2, 0°, 1 h: 72%

[1]Bomann, M.D., Guch, I.C., DiMare, M. *JOC* **60**, 5995 (1995).
[2]Dehmlow, E.V., Niemann, T., Kraft, A. *SC* **26**, 1467 (1996).
[3]Oikawa, M., Liu, W.-C., Nakai, Y., Koshida, S., Fukase, K., Kusumoto, S. *SL* 1179 (1996).

Borane–dimethyl sulfide. 14, 53; **15**, 44; **17**, 50–51; **18**, 59

Hydrogenation.[1] Hindered olefins are saturated by the borane complex in the presence of an electron-transfer agent (15 examples, 48–97%).

$BH_3 \cdot SMe_2$ / CH_2Cl_2

OMe

OMe

0°, 1 h

88%

Ar = *p*-MeC_6H_2-2,5-$(OMe)_2$

Reductive acetal cleavage.[2] The reagent is a combination of $BH_3{\cdot}SMe_2$–$BF_3{\cdot}OEt_2$. Acetals derived from 1,2- or 1,3-diols bearing a neighboring hydroxy group can be selectively cleaved at the acetal C–O bond proximate to the hydroxy group.

Protection and activation of amines.[3] Chiral borane–amine complexes, such as those derived from alanine, are formed by an exchange reaction. The complexes are useful for asymmetric alkylation.

[1]Rathore, R., Weigand, U., Kochi, J.K. *JOC* **61**, 5246 (1996).
[2]Saito, S., Kuroda, A., Tanaka, K., Kimura, R. *SL* 231 (1996).
[3]Ferey, V., Toupet, L., Le Gall, T., Mioskowski, C. *ACIEE* **35**, 430 (1996).

Borohydride exchange resin—nickel boride.

Reductions. The reagent system can be used to selectively hydrogenate 1-alkenes at 0° in MeOH in the presence of disubstituted alkenes and conjugated acid derivatives. At 65°, disubstituted alkenes are saturated, leaving trisubstituted alkenes untouched.[1] Semihydrogenation of alkynes is also achieved.[2]

Primary amines are obtained (8 examples, 68–94%)[3] on exposing oximes to the borohydride exchange resin–$Ni(OAc)_2$ (nickel boride should be formed upon admixture) in methanol. Mixed anhydrides of the type RC(O)OC(O)OEt are reduced selectively to alcohols.[4]

Disulfides.[5] Thioacetates are cleaved (through methanolysis), giving thiols, and subsequently dimerized.

Radical reactions. The nickel boride adsorbed on the resin induces the formation of radicals from alkyl iodides, which can be intercepted in situ by alkenes such as acrylic esters[6] and vinyl ethers.[7]

[1]Choi, J., Yoon, N.M. *S* 597 (1996).
[2]Choi, J., Yoon, N.M. *TL* **37**, 1057 (1996).
[3]Bandgar, B.P., Nikat, S.M., Wadgaonkar, P.P. *SC* **25**, 863 (1995).
[4]Bandgar, B.P., Modhave, R.K., Wadgaonkar, P.P., Sande, A.R *JCS(P1)* 1993 (1996).
[5]Choi, J., Yoon, N.M. *SL* 1073 (1995).
[6]Sim, T.B., Choi, J., Yoon, N.M. *TL* **37**, 3137 (1996).
[7]Ahn, J.H., Lee, D.W., Joung, M.J., Lee, K.H., Yoon, N.M. *SL* 1224 (1996).

Boron tribromide. **13**, 43; **14**, 53–54; **18**, 59

Chiral bromoboranes.[1] The chiral reagents, such as (**1**), can be prepared from BBr_3, Me_3SiH, and chiral terpenes (e.g., α-pinene).

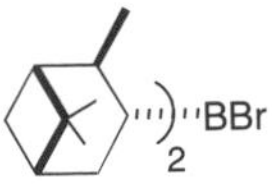

(**1**)

[1]Dhokte, U.P., Soundararajan, R., Ramachandran, P.V., Brown, H.C. *TL* **37**, 8345 (1996).

Boron trichloride. **13**, 43; **14**, 54; **15**, 44; **18**, 59–60

Mediation of reduction.[1] The reduction of β-hydroxy ketones with borane-pyridine is rendered *syn*-selective by BCl_3 (5 examples, 77–96%, *syn/anti* > 90:10). $TiCl_4$ has the same effect.

(E)-Alkenyl boronates.[2] (*E*)-Alkenylsilanes undergo hetero-group exchange on treatment with BCl_3 in CH_2Cl_2 at 0°. Subsequent reaction with catechol in benzene at room temperature furnishes the boronic esters (4 examples, 74–89%).

Prenylation.[3] Prenyltributylstannane is activated by conversion to the tertiary (dimethylallyl)dichloroborane on reaction with BCl_3. Prenylation of ketones and indole (at C-2) is readily achieved. The method is crucial for the synthesis of tryprostatin-B. It is important that the dichloroborane be trapped immediately. This prevents it from attaining equilibrium, and, therefore, the generation of regioisomers (with respect to the allyl residue) is avoided.

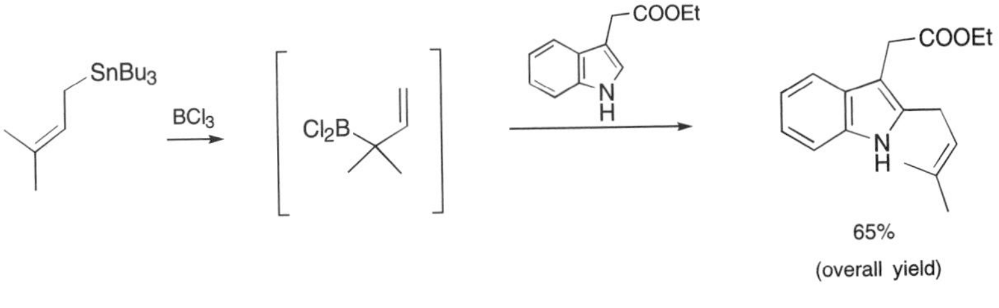

65%
(overall yield)

Homologation of propargylic alcohols.[4] Lithiation of the mixed acetals of propargylic alcohols, treatment with BH_3 and BCl_3 in succession, and then treatment with alkaline H_2O_2 give homopropargylic alcohols. The functions of BCl_3 are to induce fragmentation of the acetal unit and to catalyze the alkylation step.

Ph OBu BuLi; BH3 Ph OBu BCl3 Ph BH2 HCHO rt, NaOOH Ph OH 89% (overall yield)

[1]Sarko, C.R., Collibee, S.E., Knorr, A.L., DiMare, M. *JOC* **61**, 868 (1996).
[2]Farinola, G.M., Fiandanese, V., Mazzone, L., Naso, F. *CC* 2523 (1995).
[3]Depew, K.M., Danishefsky, S.J., Rosen, N., Sepp-Lorenzino, L. *JACS* **118**, 12463 (1996).
[4]Carrie, D., Carboni, B., Vaultier, M. *TL* **36**, 8209 (1995).

Boron trifluoride etherate. 13, 43–47; **14**, 54–56; **15**, 45–47; **16**, 44–47; **17**, 52–53; **18**, 60–63

Mannich reactions.[1] The reaction of methoxymethyl(dibenzyl)amine with silyl enol ethers proceeds at low temperatures in the presence of $BF_3{\cdot}OEt_2$. A bulky silyl group at the α′-position of the enol ether can serve as a stereocontrol element.

OSiMe3 Si Bn2NCH2OMe BF3 • OEt2 CH2Cl2, -95° O NBn2 Si 95% (92% de) NH4F Bu4NF O NBn2 96% (91% ee)

Hetero-Diels–Alder reaction.[2] Dihydropyran synthesis, in the manner reported by Danishefsky, from monoactivated dienes and aldehydes is best catalyzed by $BF_3{\cdot}OEt_2$. The Mukaiyama aldol condensation pathway does not operate in such cases.

[3+2]Cycloadditions. Methylenecyclopentanes are formed in the catalyzed reaction of 2-trimethylsilylmethyl-3-benzoyloxypropene derivatives with electron-rich alkenes.[3] The condensation of benzoquinone bisimines and styrenes furnishes 2-aryl(dihydro)indoles.[4]

SiMe3 BzO SC5H11 + SPh BF3 • OEt2 -23° -> 0° SC5H11 SPh 75%

[3,3]Sigmatropic rearrangements. The amino-Claisen rearrangement is catalyzed by $BF_3{\cdot}OEt_2$,[5] although it still requires high temperatures, and the yields are moderate at best. The conversion of *N*-vinylsulfinylanilines to indoles can be effected under milder conditions when $BF_3{\cdot}OEt_2$ or $Et_3O^+BF_4^-$ is present.[6]

1,3-Oxazacycles.[7] Aldehydes combine with 1,2- and 1,3-azido alcohols in the presence of $BF_3{\cdot}OEt_2$ to give heterocyclic products (21 examples, 18–96%). Mechanistically, the intramolecular reaction of oxonium intermediates with the azido group is akin to the Schmidt reaction.

Aziridines.[8] The Lewis acid-catalyzed reaction of imines with diazoacetic esters generates aziridines in which the *cis*-isomers predominate.

Silylated enynes.[9] Condensation of 3,3-bis(trimethylsilyl)propyne with aldehydes provides predominantly *(E)*-1-trimethylsilylalk-3-en-1-ynes (8 examples, 48–60%).

Me_3Si Me_3Si + Et—CHO $BF_3 \cdot OEt_2$ CH_2Cl_2 -60° -> -20° 45 min Et $SiMe_3$ 57%

1,2-Alkyl shift. Exposure of methyl *(E)*-3-hydroxy-3-methyl-2-oxo-5-dodecenoate to $BF_3{\cdot}OEt_2$ induces a 1,2-migration of the nonenyl group.[10] It is important that in the transition state the original hydroxyl and the ketone be *syn*-aligned as if they form a chelate to give the product with the desired absolute configuration. The resulting α-hydroxy β-ketoester is suitable for synthesis of the bisindole metabolite (+)-K252a.

COOMe N_2 OH C_6H_{13} $Rh_2(OAc)_4$ C_6H_{13} OH COOMe $BF_3 \cdot OEt_2$ COOMe HO C_5H_{11} O_3; MeOH, TsOH MeO O OMe MeOOC OH H N O N N O MeOOC OH

(+)-K252a

Mukaiyama aldol reaction. Double stereodifferentiation using chiral reactants in the presence of $BF_3 \cdot OEt_2$ is observed, even though such a reaction adopts open transition states.[11] *syn*-Aldol diastereomers are obtained from both *(Z)*- and *(E)*-siloxyalkenes in excellent yields; the absolute stereochemistry is controlled by existing α- as well as β-substituents.[12]

TBSO OR OSiMe3 + H (aldehyde, oxazolidinone, Ph) → $BF_3 \cdot OEt_2$, CH_2Cl_2, -78° → TBSO OR O OH (product, oxazolidinone, Ph) 83%

[1]Enders, D., Ward, D., Adam, J., Raabe, G. *ACIEE* **35**, 981 (1996).
[2]Mujica, M.T., Afonso, M.M., Galindo, A., Palenzuela, J.A. *T* **52**, 2167 (1996).
[3]Takahashi, Y., Tanino, K., Kuwajima, I. *TL* **37**, 5943 (1996).
[4]Engler, T.A., Meduna, S.P., La Tessa, K.O., Chai, W. *JOC* **61**, 8598 (1996).
[5]Anderson, W.K., Lai, G. *S* 1287 (1995).
[6]Baudin, J.-B., Commenil, M.-G., Julia, S.A., Lorne, R., Mauclaire, L. *BSCF* **133**, 329 (1996).
[7]Badiang, J.G., Aube, J. *JOC* **61**, 2484 (1996).
[8]Casarrubios, L., Perez, J.A., Brookhart, M., Templeton, J.L. *JOC* **61**, 8358 (1996).
[9]Pornet, J., Princet, B., Mevaa, L.M., Miginiac, L. *SC* **26**, 2099 (1996).
[10]Wood, J.L., Stoltz, B.M., Dietrich, H.-J. *JACS* **117**, 10413 (1995).
[11]Evans, D.A., Yang, M.G., Dart, M.J., Duffy, J.L., Kim, A.S. *JACS* **117**, 9598 (1995).
[12]Evans, D.A., Dart, M.J., Duffy, J.L., Yang, M.G., Livingston, A.B. *JACS* **117**, 6619 (1995).

Bromine. 13, 47; **14**, 56–57; **15**, 47; **18**, 64

Disulfides.[1] Thiols undergo oxidative coupling when treated with bromine at room temperature in the neat.

Thio- and selenosulfonates.[2] Bromine mediates the reaction of disulfides and diselenides with sodium alkanesulfinates. Thus, new thio- and selenoesters of triflic acids ($F_3C–SO_2–XR$) are readily accessible.

[1]Wu, X., Rieke, R.D., Zhu, L. *SC* **26**, 191 (1996).
[2]Billard, T., Langlois, B.R., Large, S., Anker, D., Roidot, N., Roure, P. *JOC* **61**, 7545 (1996).

Bromine trifluoride. 18, 64

Acyl fluorides.[1] Primary alcohols are directly oxidized to RCOF at slightly below room temperature.

[1]Rozen, S., Ben-David, I. *JFC* **76**, 145 (1996).

B-Bromo-9-borabicyclo[3.3.1]nonane.

(Z)-Dialkylboron enolates.[1] Enol silyl ethers give the boron enolates that undergo *syn*-selective aldol reactions.

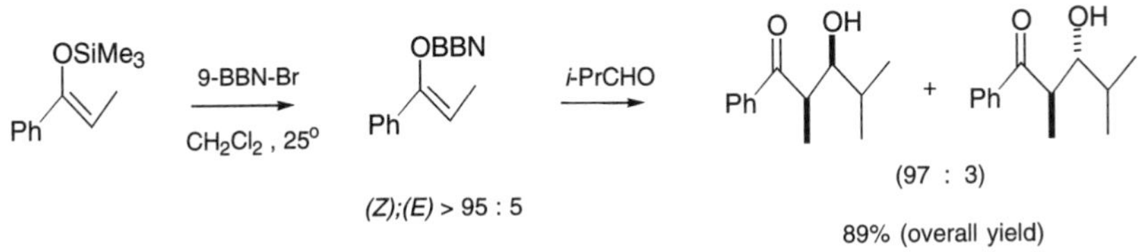

(Z);(E) > 95 : 5

(97 : 3)

89% (overall yield)

[1]Duffy, J.L., Yoon, T.P., Evans, D.A. *TL* **36**, 9245 (1995).

Bromochloromethane.

Methoxymethylation of thiols.[1] Under phase transfer conditions and the presence of methanol, thiols are protected.

[1]Toste, F.D., Still, I.W.J. *SL* 159 (1995).

Bromomethyl(dimethyl)silyl chloride.

Reductive hydroxymethylation.[1] α,β-Unsaturated acids undergo the transformation shown below. The substitution pattern is critical for the regioselectivity.

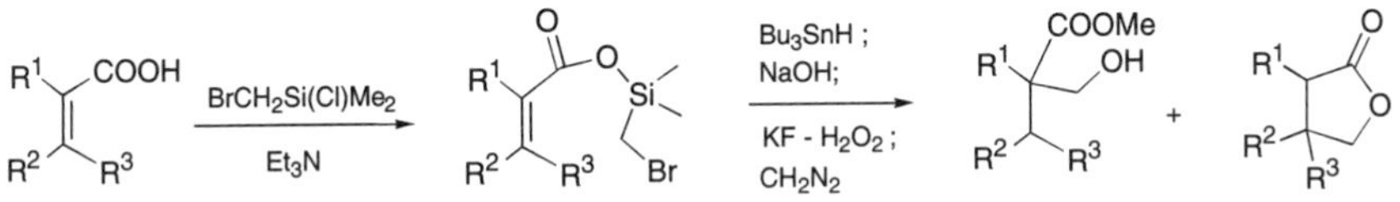

R^1 = H, R^2= R^3 = Me	74%	-
R^1 =R^2 = Me, R^3 = H	-	67%

[1]Linker, T., Maurer, M., Rebien, F. *TL* **37**, 8363 (1996).

Bromopentacarbonylrhenium(I).

Friedel–Crafts acylation.[1] $ReBr(CO)_5$ is a novel catalyst for acylation of arenes using acyl halides. Indanones and tetralones are generated in the intramolecular reactions.

[1]Kusama, H., Narasaka, K. *BCSJ* **68**, 2379 (1995).

***N*-Bromosuccinimide.** **13**, 49; **14**, 57–58; **15**, 50–51; **16**, 49; **18**, 65–67

Oxidation. The oxidation of benzyl alcohols, 1,2-diols, and α-hydroxy carboxylic acids using the 1:1 complex of NBS–Bu_4NI is selective.[1]

Ring opening to afford γ-bromo ketones attends the reaction of cyclopropylcarbinols with NBS in *t*-butanol (7 examples, 40–60%).[2]

Methoxymethyl sulfides undergo oxidative cleavage on treatment with NBS in methanol, generating methyl sulfinates (8 examples, 53–98%).[3]

Alkylation of amines.[4] Secondary and tertiary amines are synthesized from amines and alcohols, the latter being transformed into alkoxyphosphonium salts by NBS+Ph_3P.

Alkenyl bromides.[5] Alkenyl boronic acids are easily converted by NBS to the corresponding bromides at room temperature (13 examples, 62–86%).

Methyl carbamates.[6] A modified Hofmann rearrangement involves heating primary amides with NBS and NaOMe in methanol (8 examples, 85–100%).

Methyl α-oxo thiolcarboxylates.[7] Rapid oxidative removal of two methylthio units from trimethyl 2-oxo-orthothioesters occurs when the latter is treated with NBS in aqueous THF at room temperature.

Hunsdiecker reaction.[8] Cinnamic acids give β-bromostyrenes or α-(dibromomethyl)benzenemethanols on reaction with NBS (one or two equivalents, respectively) in the presence of $Mn(OAc)_2$.

COOH

NBS - $Mn(OAc)_2$

MeCN - H_2O

rt, 16 h

Br + Br Br OH

NBS : 1 equiv 83%

NBS : 2 equiv 82%

Transannular cyclization.[9] *m*-Cyclophane-lactams undergo oxidative cyclization. Certain members give indoles.

COOMe MeO O NH O

NBS - AIBN

CCl_4 Δ

MeO COOMe N O O + Br MeO COOMe N O O

[1]Beebe, T.R., Boyd, L., Fonkeng, S.B., Horn, J., Mooney, T.M., Saderholm, M.J., Skidmore, M.V. *JOC* **60**, 6602 (1995).
[2]Cossy, J., Furet, N. *TL* **36**, 3691 (1995).
[3]Ko, Y.K., Koo, D.W., Kim, J.S., Kim, D.-W. *SC* **25**, 2871 (1995).
[4]Froyen, P., Juvvik, P. *TL* **36**, 9555 (1995).
[5]Petasis, N.A., Zavialov, I.A. *TL* **37**, 567 (1996).
[6]Huang, X., Keillor, J.W. *TL* **38**, 313 (1997).
[7]Degani, I., Dughera, S., Fochi, R., Gatti, A. *S* 467 (1996).
[8]Chowdhury, S., Roy, S. *TL* **37**, 2623 (1996).
[9]Rezaie, R., Bremner, J.B. *SL* 1061 (1996).

n-Butylammonium chlorochromate–18-crown-6.

Selective oxidation.[1] A secondary alcohol is selectively oxidized in the presence of a primary alcohol. The reagent is as efficient as PCC, while being more selective.

[1]Kasmai, H.S., Mischke, S.G., Blake, T.J. *JOC* **60**, 2267 (1995).

t-Butyldimethylsilyl triflate. **13**, 50–51; **15**, 54–55; **18**, 70

Ether cleavage and exchange.[1] *t*-Alkyl ethers are cleaved by catalytic amounts of TBS-OTf, whereas they are converted into TBS ethers on successive treatment with slightly more than one equivalent of the silyl triflate and 2,6-lutidine.

Conjugate addition.[2] TBS-OTf promotes the conjugate addition of organo-aluminates to enones.

α-Silyl ketones.[3] SAMP/RAMP hydrazones undergo enantioselective *C*-silylation, providing an entry to optically active α-silyl ketones after ozonolysis of the products (10 examples, 89–96%).

[1]Franck, X., Figadere, B., Cave, A. *TL* **36**, 711 (1995).
[2]Kim, S., Park, J.H. *SL* 163 (1995).
[3]Enders, D., Lohray, B.B., Burkamp, F., Bhusan, V., Hett, R. *LA* 189 (1996).

t-Butyl hydroperoxide. **16**, 53–54; **17**, 56–57; **18**, 71–73

Epoxidation.[1] Concomitant epoxidation and alkyl peroxide formation occurs when ω-vinyl-ω-hydroxylactams are exposed to *t*-BuOOH–$SnCl_2$ in CH_2Cl_2.

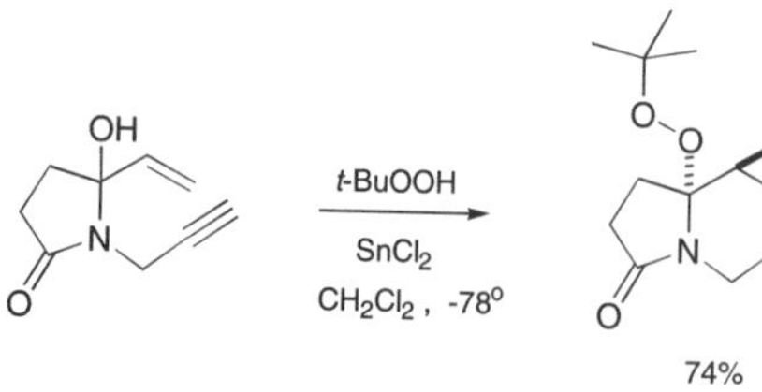

The difficulties in the epoxidation of perfluoroalkyl-substituted α,β-unsaturated esters are overcome by using a reagent derived from *t*-BuOOH and BuLi at −78°.[2]

The enantioselective epoxidation of a series of ω-bromo-2-methylalkenes catalyzed by chloroperoxidase (oxidant: *t*-BuOOH) is correlated somewhat with the molecular dissymmetry.[3]

Oxidation of alcohols. Molecular sieves[4] and other additives, including $(Ph_3P)_2CoCl_2$,[5] have been found useful in catalyzing the oxidation of allylic and benzylic alcohols. On the other hand, the system containing $(t\text{-BuO})_4Zr$ [6] is capable of oxidizing other primary and secondary alcohols.

Phenols → quinones. Useful catalysts include polymer-supported vanadium salts[7] and $(Ph_3P)_3RuCl_2$.[8] The Rh-catalyzed reaction can oxidize *p*-substituted phenols, giving 4-*t*-butylperoxy-2,5-cyclohexadienones, which undergo rearrangement (migration of the substituent) to afford the quinones on treatment with $TiCl_4$.

The oxidation of 2-allyl-1-naphthols leads to 4-allyl-1,2-naphthoquinones (4 examples, 65–73%) as a result of hydroxylation followed by oxy-Cope rearrangement.[9]

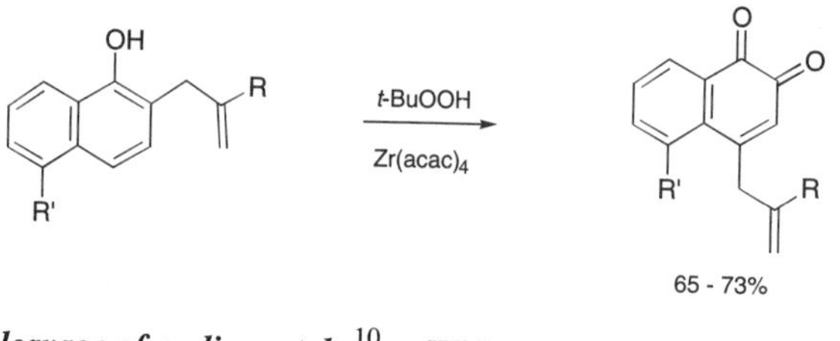

Oxidative cleavage of cyclic acetals.[10] With $VO(OAc)_2$ as catalyst, *t*-BuOOH effects the transformation of cyclic acetals to hydroxyalkyl esters.

Reaction of alkenes. *t*-BuOOH acts as a coaddend for alkenes to give mixed peroxides under certain conditions (e.g., in the presence of β-cyclodextrin and water[11] or with $Cu(OAc)_2$ as catalyst to generate the tetrahydrofuranyl α-radical).[12]

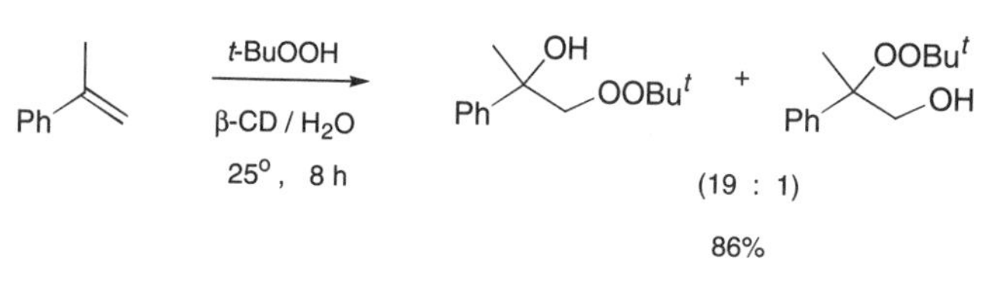

Allylic oxidation of unsaturated steroids to enones[13] has been achieved with *t*-BuOOH–$RuCl_3$, and the cleavage of silyl enol ethers is catalyzed by Ti-containing silicalite.[14] In the latter reaction, oxidative cleavage of the double-bond results in the formation of carboxylic acids.

Allylamines.[15] Amines are allylated by alkenes (with double-bond transposition) in an Mo-catalyzed reaction with *t*-BuOOH.

Silanes to alcohols.[16] Hindered silanes are cleaved by treatment with *t*-BuOOH-KH in NMP, followed by TBAF.

[1]Hursthouse, M.B., Khan, A., Marson, C.M., Porter, R.A. *TL* **36**, 5979 (1995).
[2]Lanier, M., Haddach, M., Pastor, R., Riess, J.G. *TL* **34**, 2469 (1993).
[3]Lakner, F.J., Cain, K.P., Hager, L.P. *JACS* **119**, 443 (1997).
[4]Palombi, L., Arista, L., Lattanzi, A., Bonadies, F., Scettri, A. *TL* **37**, 7849 (1996).
[5]Iyer, S., Varghese, J.P. *SC* **25**, 2261 (1995).
[6]Krohn, K., Vinke, I., Adam, H. *JOC* **61**, 1467 (1996).
[7]Suresh, S., Skaria, S., Ponrathnam, S. *SC* **26**, 2113 (1996).
[8]Murahashi, S.-I., Naota, T., Miyaguchi, N., Noda, S. *JACS* **118**, 2509 (1996).
[9]Krohn, K., Bernhard, S. *S* 699 (1996).
[10]Choudary, B.M., Reddy, P.N. *SL* 959 (1995).
[11]Bhat, S., Chandrasekaran, S. *TL* **37**, 3581 (1996).
[12]Araneo, S., Fontana, F., Minisci, F., Recupero, F., Serri, A. *CC* 1399 (1995).
[13]Miller, R.A., Li, W., Humphrey, G.R. *TL* **37**, 3429 (1996).
[14]Raju, S.V.N., Upadhya, T.T., Ponrathnam, S., Daniel, T., Sudalai, A. *CC* 1969 (1996).
[15]Srivastava, R.S., Nicholas, K.M. *CC* 2335 (1996).
[16]Smitrovich, J.H., Woerpel, K.A. *JOC* **61**, 6044 (1996).

t-Butyl hydroperoxide–dialkyl tartrate-titanium(IV) isopropoxide. 13, 51–53; **14**, 61–62; **15**, 55–56; **16**, 54–55; **17**, 57–58; **18**, 73–74

Cyclobutanones.[1] Epoxidation of an allylic methylene cyclopropane is followed by Lewis acid-catalyzed rearrangement of the strained oxaspiranes in situ. Such cyclobutanones are valuable for the synthesis of certain natural products (e.g., aplysin).

OH
OTBS
t-BuOOH
(i-PrO)$_4$Ti - (+)-DIPT
- 50°
OH
O
OR
98% (95% ee)

[1]Fukumoto, K., Nemoto, H., Nagamochi, M., Ishibashi, H. *JOC* **59**, 74 (1994).

t-Butyl hypochlorite. 13, 55; **18**, 74

***N*-Chloroaldimines.**[1] After admixture of aromatic aldehydes with aqueous ammonia, bisbenzalaminals are formed. Further treatment with t-BuOCl completes the overall chloroimidation.

$$\text{Ph–CH(N=CHPh)}_2 \xrightarrow[\text{-65° -> rt, 15 min}]{t\text{-BuOCl, MeOH}} \text{Ph–CH=NCl} \quad 53\%$$

[1]Kupfer, R., Brinker, U.H. *JOC* **61**, 4185 (1996).

Butyllithium. 13, 56; **14**, 63–68; **15**, 59–61; **17**, 59–60; **18**, 74–77

Eliminations. γ-Functionalized α,β-unsaturated dimethyl acetals undergo elimination of methanol to provide bifunctional 1,3-dienes.[1,2]

Primary alcohols are converted to terminal alkenes by way of dehydrobenzyloxylation with BuLi.[3] Other groups (e.g., trityl) are not affected.

Exposure of 3-sulfolenes to BuLi and then *N*-chlorosuccinimide accomplishes the synthesis of substituted 1,3-butadiene-1-sulfonyl chlorides.[4]

Alkylations and silylations. Moderately stabilized allylic and propargylic anions are readily generated by treating substituted alkenes and alkynes with BuLi. 1-(Benzotriazol-1-yl)propargyl ethyl ethers are a convenient source of alkynyl ketones[5] by virtue of their facile alkylation.

An asymmetric synthesis of functionalized cyclopropanes[6] by condensation of conjugated ketones and esters with chiral γ-chloroallylphosphonic bis(amide)s involves the generation of the *P(O)*-stabilized allyl anions.

BuLi; THF, -78° — 73%

Arylmethyl methyl ethers can be alkylated (BuLi, hexane–THF, –40°; RX),[7] whereas *N,N*-diethylbenzyl and cinnamyl carbamates undergo disilylation[8] using BuLi as the base.

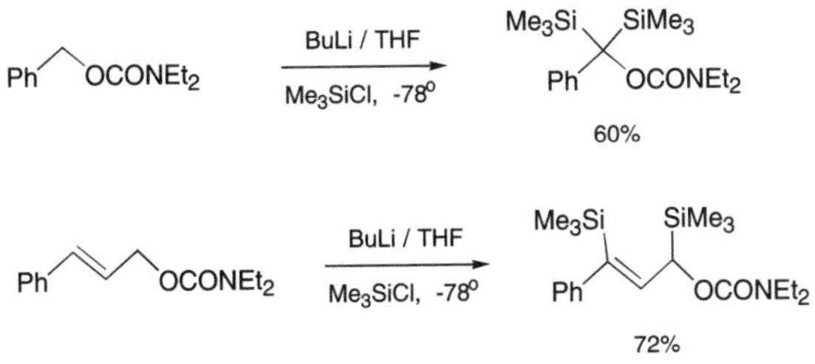

The reaction of deprotonated trimethylsilyldiazomethane with CO generates the silylethynolate anion,[9] which can be silylated (to give disilylketenes) and alkylated. The *t*-butyldimethylsilylethynolate anion obtained from deprotonation of the ketene[10] undergoes silylation either at oxygen or carbon, according to the bulkiness of the silylating agent.

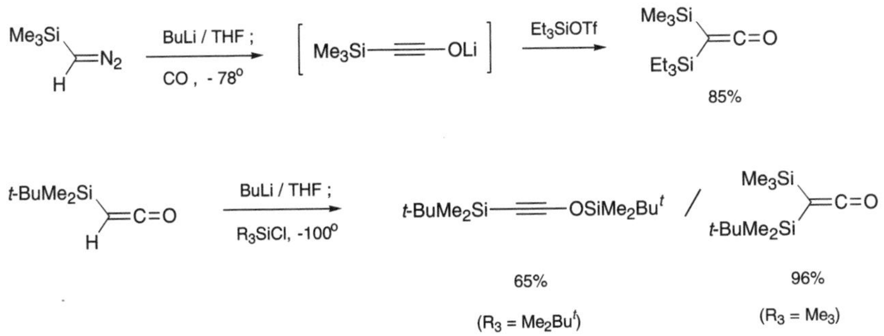

Rearrangements. α-Hydroxy epoxides undergo 1,2-rearrangement[11] to give enones via a carbenoid pathway.

The Still–Wittig rearrangement is highly *(Z)*-selective if assisted by a proximal oxy functionality.[12] A sila-version[13] of the rearrangement may have advantages in avoiding the toxic tin compounds.

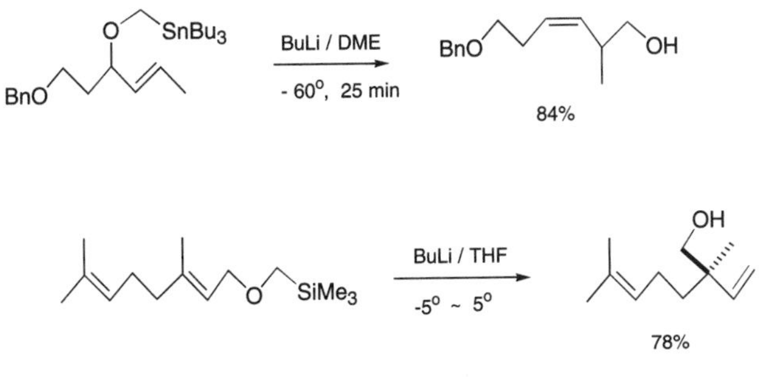

The [1,2]–Wittig rearrangement has been exploited in the conversion of *O*-glycosides to *C*-glycosides.[14] The configuration at the anomeric center is retained.

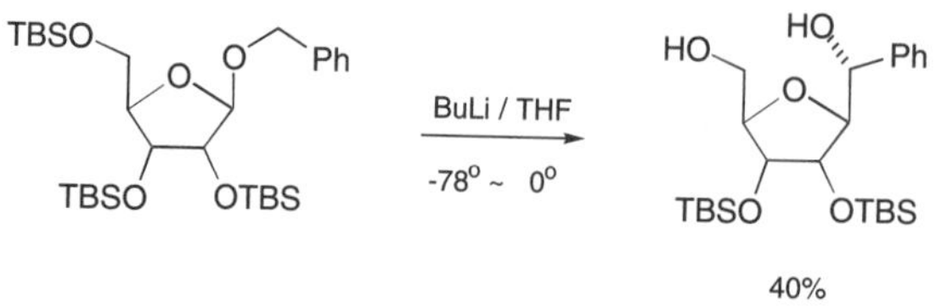

An intriguing diastereoselective synthesis of α-hydroxyalkylphosphonate esters is shown below.[15]

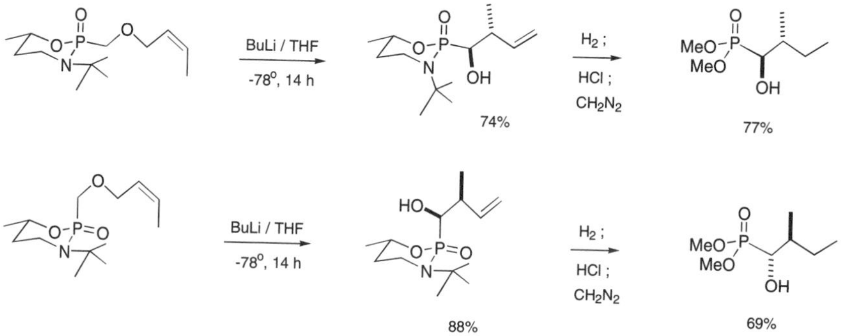

Halogen–lithium exchange. With hexane as solvent, alkenyl- and allenyllithiums are prepared[16] by the exchange reaction at room temperature. In ether, the formation of enolates from α-haloketones is possible through treatment with BuLi.[17] α-Bromoenolates, also accessible from α,α-dibromoketones by this method, can be used in a synthesis of epoxy ketones by reaction with carbonyl compounds.[18]

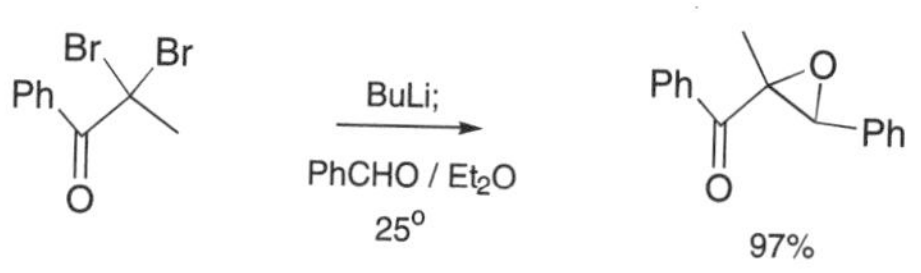

BuLi mediates the silylation and alkylation of diethyl trichloromethylphosphonate, allowing the preparation of 1-formylalkylphosphonates[19] and 1-alkynylphosphonates.[20]

(EtO)2P(O)CCl3 —BuLi, Me3SiCl / THF→ [(EtO)2P(O)C(Cl)(SiMe3)Li] —RX; BuLi; HCOOEt, Me3SiCl; HCl→ (EtO)2P(O)CH(R)CHO

↓ PhCHO

[(EtO)2P(O)C(Cl)=CHPh] —NaN(SiMe3)2→ (EtO)2P(O)C≡CPh 89%

Chalcogen–lithium exchange. The preparation[21] of bis(lithiomethyl) sulfide is conveniently carried out by the reaction of the bistelluride with BuLi. This reagent is a colorless solid that ignites on contact with traces of air, but it can be stored at –20° for six months.

S(CH2Br)2 —BuTeLi / THF, -78° ~ rt→ S(CH2TeBu)2 83% —BuLi / Et2O, - 30°→ S(CH2Li)2

A dramatic solvent effect has been revealed in the carbocyclization of ω-alkenylbenzyllithiums, which are formed from the corresponding selenides.[22] Selenobenzophenones undergo coupling to give symmetrical tetraarylethenes[23] on treatment with BuLi.

Ph(Me)C(SeMe)CH2CH2CH2CH=CH2 —BuLi→ [Ph(Me)C(Li)CH2CH2CH2CH=CH2 → cyclopentyl-CH2Li (cis) + cyclopentyl-CH2Li (trans)] —MeOH→

	cis	trans
THF , -78°	95	5
pentane, 0°	5	95

α-Amino alkyllithiums. These species are prepared by Sn–Li exchange. Some of them have remarkable configurational stability, which is highly significant in synthesis. For example, an approach to derivatives of (+)-pseudoheliotridane is cleared by treatment of *(S)-N*-3-butenyl-2-tributylstannylpyrrolidine with BuLi at –78°.[24] Anionic cyclization of the lithium compound ensues on warming to room temperature.

$SnBu_3$ — BuLi / hexane-Et_2O ; Me_3SiCl → $SiMe_3$ product, 63%

α-Heterosubstituted alkenyllithiums. *(Z)*-1-Chloro-1-haloalk-1-en-3-ynes are versatile precursors of chloroenynes and enetriynes. Their availability from *(E)*-1-chloroenynes via deprotonation[25] at –100° has facilitated synthetic work in the area. Another useful reagent is 2-lithio-3-benzenesulfonyl-4,5-dihydrofuran.[26] Its employment in the elaboration of 1,6-dioxaspiro[4.5]decanes in one step has been developed.

C_5H_{11}, Cl — BuLi / THF-Et_2O ; I_2 , -100° → C_5H_{11}, Cl, I, 84%

Wittig reactions.[27] A synthesis of 3-alkylidenecyclobutanols from methyltriphenylphosphonium bromide, epichlorohydrin, and aldehydes is very useful.

$Ph_3\overset{+}{P}CH_3$ Br^- — BuLi ; epichlorohydrin (Cl) → [OLi, $\overset{+}{P}Ph_3$ I^-] — BuLi ; RCHO, -40° → OH, R; 62% (R = Ph)

[1]Maddaluno, J., Gaonac'h, O., Le Gallic, Y., Duhamel, L. *TL* **36**, 8591 (1995).
[2]Guillam, A., Maddaluno, J., Duhamel, L. *CC* 1295 (1996).
[3]Matsushita, M., Nagaoka, Y., Hioki, H., Fukuyama, Y., Kodama, M. *CL* 1039 (1996).
[4]Lee, Y.S., Ryu, E.K., Yun, K.-Y., Kim, Y.H. *SL* 247 (1996).

[5]Katritzky, A.R., Lang, H. *JOC* **60**, 7612 (1995).
[6]Hanessian, S., Andreotti, D., Gomtsyan, A. *JACS* **117**, 10393 (1995).
[7]Azzena, U., Demartis, S., Fiori, M.G., Melloni, G., Pisano, L. *TL* **36**, 5641 (1995).
[8]Mason, P.H., Yoell, D.K., van Staden, L.F., Emslie, N.D. *SC* **25**, 3347 (1995).
[9]Kai, H., Iwamoto, K., Chatani, N., Murai, S. *JACS* **118**, 7634 (1996).
[10]Akai, S., Kitagaki, S., Naka, T., Yamamoto, K., Tsuzuki, Y., Matsumoto, K., Kita, Y. *JCS(P1)* 1705 (1996).
[11]Doris, E., Dechoux, L., Mioskowski, C. *JACS* **117**, 12700 (1995).
[12]Fujii, K., Hara, O., Fujita, Y., Sakagami, Y. *TL* **37**, 389 (1996).
[13]Mulzer, J., List, B. *TL* **37**, 2403 (1996).
[14]Tomooka, K., Yamamoto, H., Nakai, T. *JACS* **118**, 3317 (1996).
[15]Denmark, S.E., Miller, P.C. *TL* **36**, 6631 (1995).
[16]Shinokubo, H., Miki, H., Yokoo, T., Oshima, K., Utimoto, K. *T* **51**, 11681 (1995).
[17]Aoki, Y., Oshima, K., Utimoto, K. *CL* 463 (1995).
[18]Aoki, Y., Oshima, K., Utimoto, K. *SL* 1071 (1995).
[19]Zanella, Y., Berte-Verrando, S., Diziere, R., Savignac, P. *JCS(P1)* 2835 (1995).
[20]Diziere, R., Savignac, P. *TL* **37**, 1783 (1996).
[21]Strohmann, C. *ACIEE* **35**, 528 (1996).
[22]Krief, A., Bousbaa, J. *SL* 1007 (1996).
[23]Okuma, K., Koda, G., Okumura, S., Ohno, A. *CL* 609 (1996).
[24]Coldham, I., Hufton, R., Snowden, D.J. *JACS* **118**, 5322 (1996).
[25]Alami, M., Crousse, B., Linstrumelle, G. *TL* **36**, 3687 (1995).
[26]Carretero, J.C., Rojo, J., Diaz, N., Hamdouchi, C., Poveda, A. *T* **51**, 8507 (1995).
[27]Okuma, K., Tsubakihara, K., Tanaka, Y., Koda, G., Ohta, H. *TL* **36**, 5591 (1995).

Butyllithium—hexamethylphosphoric triamide.

Furans. 1,1-Diphenoxy-2-(benzenesulfonyl)cyclopropane has been used for a furan synthesis by way of acylation.[1]

BuLi / THF-HMPA ; PhCOCl, -78°; 62%; TsOH, CH_2Cl_2; 82%

Aza–Brook rearrangement.[2] (α-Silylallyl)amines undergo the rearrangement to generate the functionalized allyl anions, which, on alkylation and acid hydrolysis, lead to substituted aldehydes.

NHPh; $SiPh_2Bu^t$; BuLi / THF-HMPA ; $BnO(CH_2)_3I$, -78° ~ 0°; aq. HCl; CHO; O Bn; 67%

[1]Pohmakotr, M., Takampon, A., Ratchataphusit, J. *T* **52**, 17149 (1996).
[2]Honda, T., Mori, M. *JOC* **61**, 1196 (1996).

Butyllithium—*N,N,N′,N′*-tetramethylethylenediamine.

o-Lithiation of benzamides.[1] Secondary amides can be masked as the *N*-(1-methoxy-2,2-dimethylpropyl) derivatives before lithiation. The protecting group is removed under mild conditions.

Shapiro–Suzuki reaction combination.[2] The two reactions can be done in sequence such that arylalkenes are obtained.

BuLi/ TMEDA ; $Pd(OAc)_2$ - Na_2CO_3, Ph_3P , PhBr Δ

55% 20%

Vinylsilanes. On γ-alkylation of allyl(*t*-butyldiphenyl)silane, which starts from deprotonation with BuLi-TMEDA, bulky vinylsilanes are synthesized.[3]

[1]Phillion, D.P., Walker, D.M. *JOC* **60**, 8417 (1995).
[2]Passaparo, M.S., Keay, B.A. *TL* **37**, 429 (1996).
[3]Blanco, F.J., Cuadrado, P., Gonzalez, A.M., Pulido, F.J. *S* 42 (1996).

***s*-Butyllithium. 14**, 69; **16**, 56; **18**, 77–79

Deprotonation. After *N,N*-diethyl-1-lithiovinylcarbamate, which is obtained by a deprotonation process (with *s*-BuLi as base), is converted into the organozinc reagent, it can be used to couple with aryl halides (e.g., bromides) and triflates in the presence of (dppf)$PdCl_2$ to afford the enol carbamates of the acetophenones. The overall transformation can be considered an anionic equivalent of Friedel–Crafts acylation.[1]

$OCONEt_2$ — *s*-BuLi, -78° ; $ZnBr_2$; (dppf)$PdCl_2$, ArBr — $OCONEt_2$

Ar = 2-Np 84%

N,N′-Bisacylimidazolidines [e.g., (**1**)] are acyl anion equivalents; they undergo reaction with various electrophiles at C-2 after deprotonation with *s*-BuLi.[2]

(1)

s-BuLi / THF, -78°; El+

Amide α-anions. The $BocN(Me)CH_2Li$ species (prepared by lithiation with *s*-BuLi/TMEDA) can be converted into cuprate reagent and coupled with enol triflates. This method is useful for the synthesis of allylic amines.[3]

[1]Superchi, S., Sotomayor, N., Miao, G., Joseph, B., Snieckus, V. *TL* **37**, 6057 (1996).
[2]Coldham, I., Houdayer, P.M.A., Judkins, R.A., Witty, D.R. *SL* 1109 (1996).
[3]Dieter, R.K., Dieter, J.W., Alexander, C.W., Bhinderwala, N.S. *JOC* **61**, 2930 (1996).

***t*-Butyllithium. 13**, 58; **15**, 64–65; **16**, 56–57; **18**, 79–81

Deprotonation. Lithiation at C-2 of *N*-methylindole serves to functionalize at that position, e.g., via the triethylborate.[1] Allyl 3-furylmethyl ether undergoes rearrangement according to the site of deprotonation.[2]

t-BuLi, THF, -78°

(>95 : <5)

Triflamides behave differently toward *n*-BuLi and *t*-BuLi, being attacked at the sulfonyl group and the α-hydrogen, respectively, to form *n*-butanesulfonamides and α-(*t*-butyl)alkylamines.[3] In the latter process, a base-induced elimination of CF_3SO_2H to form imines is followed by addition of the *t*-Bu group.

Alkylation of 4-Boc-2-oxopiperazines at C-3 is initiated by treatment with *t*-BuLi.[4] Bis(dimethylamino)methane undergoes double lithiation.[5]

t-BuLi, C_5H_{12}, -78°; $SiCl_4$, THF

37%

Halogen–lithium exchange. Various types of organohalides can undergo exchange. The resulting lithio derivatives may participate in cyclization resulting from intramolecular displacement,[6] addition to an unsaturated linkage,[7–10] or elimination.[11]

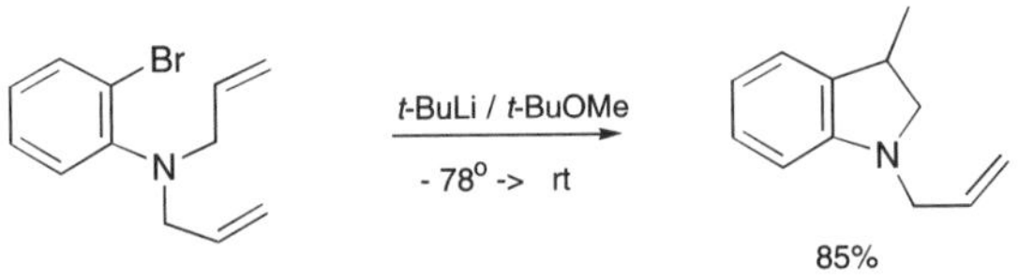

t-BuLi / *t*-BuOMe, - 78° -> rt; 85%

The alkenyllithium species from 2-bromo-1-trimethylsilylprop-2-ene is prone to isomerization.[12] The derived organozinc reagent prepared by treatment with *t*-BuLi and $ZnCl_2$ has the structure $CH_2{=}C(SiMe_3)CH_2ZnCl$. A propene 2,3-dianion synthon, 2-trimethylgermyl-3-trimethylstannylpropene, $CH_2{=}C(GeMe_3)CH_2SnMe_3$, is obtained from 2-bromo-2-propen-1-ol by stepwise treatment with *t*-BuLi, Me_3GeBr, TsCl, and Me_3SnLi.[13]

Sila–Wittig rearrangements.[14] The *Si*-lithio derivatives of *N*-silyl allylamines undergo rearrangement on addition of 12-crown-4 at −45°. The lithio species are made from the corresponding trimethylstannyl compounds by treatment with *t*-BuLi. A similar reaction is possible in the oxygen- (not N-) based systems, but the *Si*-lithiosilyl allyl ethers can be formed only by transmetallation from tertiary allyl ethers.

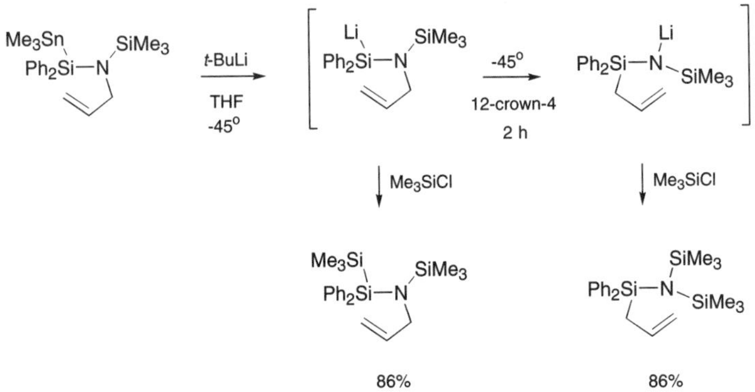

t-BuLi, THF, -45°; -45°, 12-crown-4, 2 h; Me_3SiCl; 86%; 86%

[1]Ishikura, M., Terashima, M. *JOC* **59**, 2634 (1994).
[2]Tsubuki, M., Okita, H., Honda, T. *CC* 2135 (1995).
[3]Bozec-Ogor, S., Salou-Guiziou, V., Yaouanc, J.J., Handel, H. *TL* **36**, 6063 (1995).
[4]Schanen, V., Cherrier, M.-P., Jose de Melo, S., Auirion, J.-C., Husson, H.-P. *S* 833 (1996).
[5]Karsch, H.H. *CB* **129**, 483 (1996).
[6]Bailey, W.F., Aspris, P.H. *JOC* **60**, 754 (1995).

[7]Zhang, D., Liebeskind, L.S. *JOC* **61**, 2594 (1996).
[8]Bailey, W.F., Jiang, X.-L. *JOC* **61**, 2596 (1996).
[9]Bailey, W.F., Wachter-Jurcsak, N.M., Pineau, M.R., Ovaska, T.V., Warren, R.R., Lewis, C.E. *JOC* **61**, 8216 (1996).
[10]Comins, D.L., Zhang, Y.-m. *JACS* **118**, 12248 (1996).
[11]Satoh, T., Takano, K. *T* **52**, 2349 (1996).
[12]Eshelby, J.J., Parsons, P.J., Crowley, P.J. *JCS(P1)* 191 (1996).
[13]Piers, E., Kaller, A.M. *SL* 549 (1996).
[14]Kawachi, A., Doi, N., Tamao, K. *JACS* **119**, 233 (1997).

t-Butyl perbenzoate. 13, 58

Dehydrogenation.[1] Oxazolines and thiazolines are converted to the fully aromatic heterocycles on treatment with PhC(O)OOBu^t–CuBr in refluxing benzene (12 examples, 55–83%).

Allylic oxidation.[2] The formation of allyl benzoates from alkenes by reaction with PhC(O)OOBu^t is efficiently promoted by the Cu(OTf)$_2$–DBN/DBU complex.

[1]Meyers, A.I., Tavares, F.X. *JOC* **61**, 8207 (1996).
[2]Sekar, G., Dattagupta, A., Singh, V.K. *TL* **37**, 8435 (1996).

1-(*t*-Butylperoxy)-1,2-benziodoxol-3(1*H*)-one.

Oxidation of benzyl and allyl ethers.[1] At room temperature and in the presence of alkaline metal carbonates, the hypervalent iodine compound (**1**) initiates an attack by free radicals on the ethers to form esters. Benzene and cyclohexane are the solvents of choice.

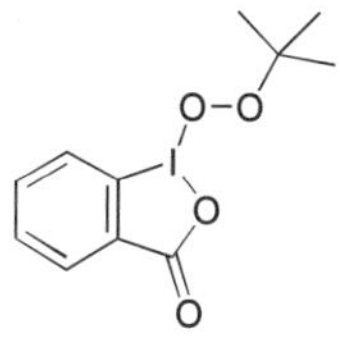

(**1**)

[1]Ochiai, M., Ito, T., Takahashi, H., Nakanishi, A., Toyonari, M., Sueda, T., Goto, S., Shiro, M. *JACS* **118**, 7716 (1996).

N-t-Butyl-γ-trimethylsilylcrotonaldimine.

Homologation.[1] Conjugated dienals are readily prepared in one pot by reaction of the silylated crotonaldimine with aldehydes in the presence of CsF and subsequent treatment with aqueous $ZnCl_2$.

CHO + Me_3Si N

CsF / DMSO ;

$ZnCl_2$ / H_2O -Et_2O

CHO

94%

[1]Bellassoued, M., Salemkour, M. *T* **52**, 4607 (1996).

C

Cadmium. 13, 60; **18**, 83

β,γ-Unsaturated ketones.[1] Allylation of acid chlorides with allyl bromides is mediated by cadmium.

Cleavage of Troc groups.[2] The 2,2,2-trichloroethoxycarbonyl protecting group of amines is removed in neutral conditions using 10% Cd–Pb in THF containing NH_4OAc (5 examples, 89–94%).

[1]Baruah, B., Boruah, A., Prajapati, D., Sandhu, J.S. *TL* **37**, 9087 (1996).
[2]Dong, Q., Anderson, C.E., Ciufolini, M.A. *TL* **36**, 5681 (1995).

Carbon dioxide.

Carbonates.[1] Phosgene can be avoided in the preparation of carbonates by using a guanidine as catalyst to condense alcohols with CO_2, after which *O*-alkylation ensues. The yields are quite respectable (7 examples, 53–97%).

i-PrOH + BnCl → ($[Me_2N]_2C{=}NCy$; CO_2, NMP; 55°, 18 h) → i-PrO–C(=O)–OBn 85%

[1]McGhee, W., Riley, D. *JOC* **60**, 6205 (1995).

Carbonyldihydridotris(triphenylphosphine)ruthenium(II).

(Triethoxysilyl)ethylation. Introduction of a silylethyl group to aromatic ketones and esters at an ortho position is highly efficient, employing a reaction with triethoxy(vinyl)silane in the presence of $RuH_2(CO)(PPh_3)_3$ in toluene.[1] Interesting directing effects of substituents on this insertion reaction have been observed.[2]

Ar-ketone (R) + vinyl-Si(OEt)3 → ($(Ph_3P)_3Rh(CO)H_2$; PhMe, 135°, 3 h) → ortho-(2-(EtO)3Si-ethyl) products

R = OMe	83%	10%
R = Me	3%	93%

An analogous (β-silyl)ethylation reaction of conjugated esters and ketones by insertion of the C–H bond at the β-carbon is also possible.[3,4]

[1]Kakiuchi, F., Sekine, S., Tanaka, Y., Kamatani, A., Sonoda, M., Chatani, N., Murai, S. *BCSJ* **68**, 62 (1995).
[2]Sonoda, M., Kakiuchi, F., Chatani, N., Murai, S. *JOMC* **504**, 151 (1995).
[3]Kakiuchi, F., Tanaka, Y., Sato, T., Chatani, N., Murai, S. *CL* 679, 681 (1995).
[4]Trost, B.M., Imi, K., Davies, I.W. *JACS* **117**, 5371 (1995).

Carbonylhydridotris(triphenylphosphine)rhodium(I).

Enynes.[1] 1-Alkynes add to unactivated allenes to give *(E)*-enynes under the influence of $RhH(CO)(PPh_3)_3$ (21 examples, 43–91%).

$(Ph_3P)_3Rh(CO)H$, Et_3P / THF, Δ 9 h

68%

Addition to imines.[2] With the use of malononitrile derivatives as nucleophiles, synthetically valuable C–C-bond-forming reactions are performed.

4-Aryl-2,3-dihydropyrroles.[3] A facile synthesis of these heterocyclic compounds involves Rh catalyzed cyclization and reductive incorporation of CO into *N*-tosylcinnamylamines.

$(Ph_3P)_3Rh(CO)H$, Ph_3P - CO - H_2; H^+

83%

β-Arylthio α,β-unsaturated aldehydes.[4] The thioformylation of 1-alkynes is performed with the Rh(I) catalyst in MeCN at 120°. Usually, the reaction yields the *(Z)*-isomers selectively, unless a coordinating group (e.g., OH) is present at a proximal position in the carbon chain. The carbonylation is not effected by $Pd(OAc)_2$, which catalyzes the addition of thiols to alkynes in the Markovnikov sense. Phosphine-ligated Pd species promote the formation of many types of products, while Ni, Ir, and Ru complexes exhibit no catalytic activity.

$$C_6H_{13}C{\equiv}CH + PhSH + CO \xrightarrow[\text{MeCN, }120^\circ]{(Ph_3P)_3Rh(CO)H} C_6H_{13}C(SPh){=}CHCHO \quad 82\%$$

[1]Yamaguchi, M., Omata, K., Hirama, M. *TL* **35**, 5689 (1994).
[2]Yamamoto, Y., Kubota, Y., Honda, Y., Fukui, H., Asao, N., Nemoto, H. *JACS* **116**, 3161 (1994).
[3]Busacca, C.A., Dong, Y. *TL* **37**, 3947 (1996).
[4]Ogawa, A., Takeba, M., Kawakami, J., Tyu, I., Kambe, N., Sonoda, N. *JACS* **117**, 7564 (1995).

3-Carboxypyridinium chlorochromate.

Hydrolytic cleavage of C=N bonds.[1] This reagent derived from nicotinic acid converts phenylhydrazones, *p*-nitrophenylhydrazones, semicarbazones, and azines in refluxing dichloromethane to the parent carbonyl compounds in good yields.

[1]Baltork, I.M., Pouranshirvani, S. *SC* **26**, 1 (1996).

Catecholborane. 16, 65–66; **17**, 67–68; **18**, 85

Aldehydes.[1] Partial reduction of nitriles with catecholborane in THF at room temperature (15 examples, 85–100%) offers an alternative to the more commonly used procedure using diisobutylaluminum hydride.

Hydroboration. Elevated temperatures (70–100°) are usually required for the hydroboration of alkenes using catecholborane. An improvement is to add 10–20 mol% *N,N*-dimethylacetamide, which can promote the reaction at room temperature.[2]

[1]Cha, J.S., Chang, S.W., Kwon, O.O., Kim, J.M. *SL* 165 (1996).
[2]Garrett, C.E., Fu, G.C. *JOC* **61**, 3224 (1996).

Cerium(IV) ammonium nitrate. 13, 67–68; **14**, 74–75; **15**, 70–72; **16**, 66; **17**, 68; **18**, 85–87

Deprotection. The use of CAN in the cleavage of TBS ethers,[1] trityl ethers,[2] and *t*-Boc-amines[3] has been reported.

Enones. Oxidative desilylation of cyclic triisopropylsilyl enol ethers with CAN in DMF leads to enones.[4,5]

Functionalization of alkenes. An unusual reaction involving dimethyl malonate and styrene gives a keto diester and a lactone ester as the major products.[6]

$$PhCH{=}CH_2 + CH_2(COOMe)_2 \xrightarrow[\text{MeOH, }20^\circ]{\text{CAN}} PhCOCH_2CH(COOMe)_2\ (42\%) + \text{5-phenyl-3-(methoxycarbonyl)-}\gamma\text{-butyrolactone}\ (29\%)$$

A synthesis of α-azido ketones[7] consists of the oxidation of triisopropylsilyl enol ethers in the presence of NaN_3. Nitroalkanes act as aldol acceptors when their anions are exposed to CAN in the presence of silyl enol ethers.[8] Conjugated ketones are obtained on subsequent treatment with triethylamine.

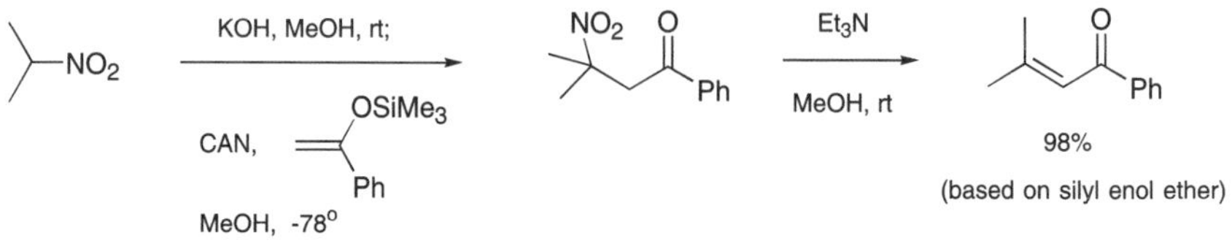
NO2
KOH, MeOH, rt;
OSiMe3
CAN,
Ph
MeOH, -78°
NO2
O
Ph
Et3N
MeOH, rt
O
Ph
98%
(based on silyl enol ether)

The addition of sulfinate anion concomitant with ring expansion of 1-vinylcyclobutanols[9] is an interesting process.

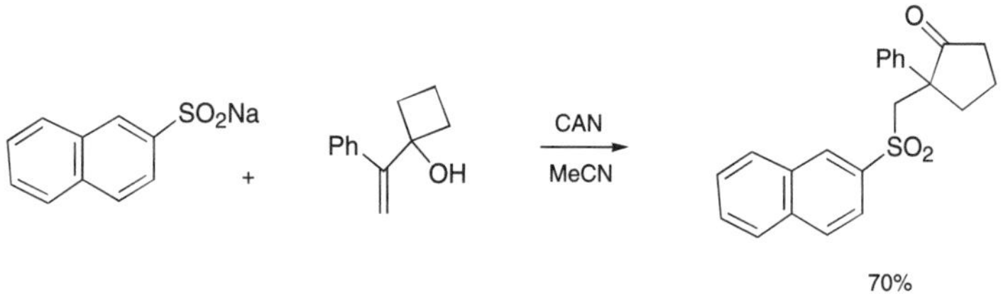
SO2Na
+
Ph
OH
CAN
MeCN
O
Ph
SO2
70%

Nitration of alkenes, including allylsilanes, is accomplished with CAN-$NaNO_2$-HOAc as reagent.[10]

A three-component assembly of a 2,3-disubstituted cycloalkanone[11] from an electron-rich alkene (e.g., an enol ether), a 2-cycloalkenone, and another nucleophile is illustrated by the following process in which a 1,1-dialkoxycyclopropane acts as a latent homoenolate anion.

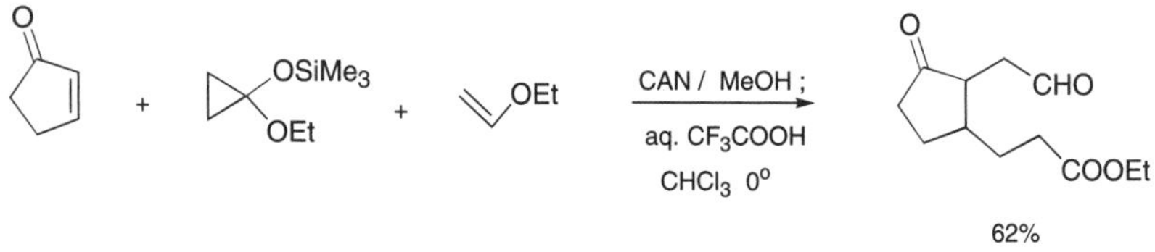
O
+
OSiMe3
OEt
+
OEt
CAN / MeOH ;
aq. CF3COOH
CHCl3 0°
O
CHO
COOEt
62%

In another condensation,[12] 1,3-butadiene is used to assemble two nucleophiles, the oxidation in one of which initiates the coupling.

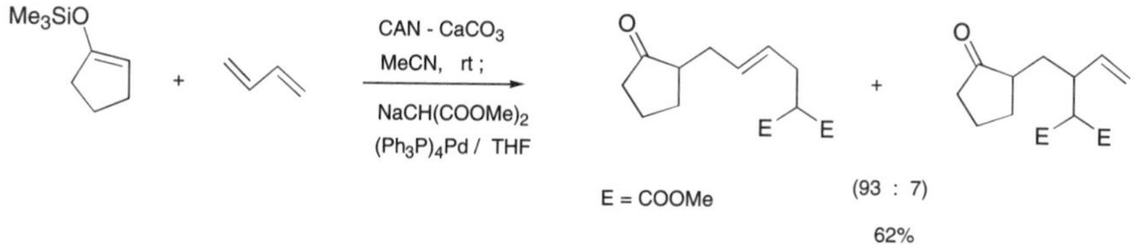

Addition to 1,3-dicarbonyl compounds. The CAN oxidation of these compounds in the presence of allyltrimethylsilane results in *C*-allylation.[13] The oxidative addition of 1,3-dicarbonyl compounds to alkenes to form dihydrofurans is a rather general reaction.[14,15]

2-Oxazolines.[16] α-Nitro ketones form nitrile oxides, which are trapped by dipolarophiles.

Preparation of sulfur compounds. CAN acts as a catalyst for dithioacetalization[17] and the transformation of epoxides to episulfides (using NH_4SCN).[18]

Carbonylation of triarylstilbines.[19] Ester formation in the Pd-catalyzed reaction also requires CAN.

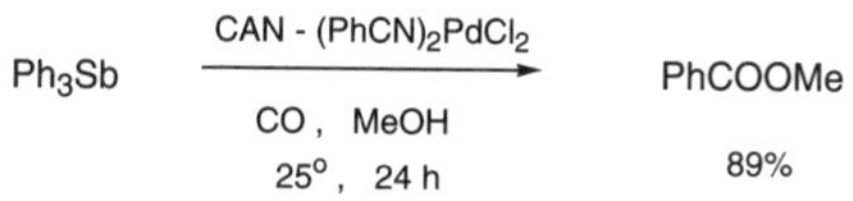

[1]Gupta, A.D., Singh, R., Singh, V.K. *SL* 69 (1996).
[2]Hwu, J.R., Jain, M.L., Tsay, S.-C., Hakimelahi, G.H. *CC* 545 (1996).
[3]Hwu, J.R., Jain, M.L., Tsay, S.-C., Hakimelahi, G.H. *TL* **37**, 2035 (1996).
[4]Evans, P.A., Longmire, J.M., Modi, D.P. *TL* **36**, 3985 (1995).
[5]Evans, P.A., Nelson, J.D. *JOC* **61**, 7600 (1996).
[6]Nair, V., Mathew, J. *JCS(P1)* 1881 (1995).
[7]Magnus, P., Barth, L., Lacour, J., Coldham, I., Mugrage, B., Bauta, W.B. *T* **51**, 11075 (1995).
[8]Arai, N., Narasaka, K. *CL* 987 (1995).
[9]Michizuki, T., Hayakawa, S., Narasaka, K. *BCSJ* **69**, 2317 (1996).
[10]Hwu, J.R., Chen, K.-L., Ananthan, S., Patel, H.V. *OM* **15**, 499 (1996).
[11]Paolobelli, A.B., Ruzziconi, R. *JOC* **61**, 6434 (1996).
[12]Paolobelli, A.B., Ceccherelli, P., Pizzo, F., Ruzziconi, R. *JOC* **60**, 4954 (1995).
[13]Hwu, J.R., Chen, C.N., Shiao, S.-S. *JOC* **60**, 856 (1995).
[14]Nair, V., Mathew, J., Radhakrishnan, K.V. *JCS(P1)* 1487 (1996).
[15]Kobayashi, K., Mori, M., Uneda, T., Morikawa, O., Konishi, H. *CL* 451 (1996).
[16]Sugiyama, T. *AOMC* **9**, 399 (1995).
[17]Mandal, P.K., Roy, S.C. *T* **51**, 7823 (1995).
[18]Iranpoor, N., Kazemi, F. *S* 821 (1996).
[19]Cho, C.S., Tanabe, K., Itoh, O., Uemura, S. *JOC* **60**, 274 (1995).

Cesium acetate.

Inversion of alcohols.[1] By chloromesylation (with $ClCH_2SO_2Cl$/py), treatment with CsOAc/18-crown-6, and hydrolysis, a secondary alcohol is converted into its enantiomer.

[1]Shimizu, T., Hiranuma, S., Nakata, T. *TL* **37**, 6145 (1996).

Cesium carbonate. 13, 70; **14**, 77–78; **15**, 73–75; **18**, 87–88

Phenyl ethers.[1] $Cs_2(CO)_3$ has been used in the alkylation of phenols.

Intramolecular Michael addition.[2] Addition of a β-keto ester moiety to an enone to form a macrocycle (e.g., a 14-membered ring) using $Cs_2(CO)_3$ as the base is efficient.

Diazo transfer.[3] The introduction of a diazo group from tosyl azide to active methylene compounds requires that the reaction take place at room temperature (8 examples, 91–99%) in the presence of $Cs_2(CO)_3$.

[1]Lee, J.C., Yuk, J.Y., Cho, S.H. *SC* **25**, 1367 (1995).
[2]Berthiaume, G., Deslongchamps, P. *BSCF* **132**, 371 (1995).
[3]Lee, J.C., Yuk, J.Y. *SC* **25**, 1511 (1995).

Cesium fluoride. 13, 68; **14**, 79; **15**, 75–76; **16**, 69–70; **17**, 68; **18**, 88–89

As base. The *O*-alkylation of tetronic acids and the alkylation of 2-pyridone (at O with primary halides and at N with secondary halides) are mediated by CsF.[1] In each case the displacement of secondary mesylates with carboxylic acids[2] and malonic esters[3] proceeds stereoselectively using CsF as the base, thereby permitting the preparation of chiral substances.

The *trans*-2,3-dimethylchroman-4-one system can be constructed from *o*-hydroxyaryl 1-methyl-1-propenyl ketones by an intramolecular Michael reaction. While the ring closure is catalyzed by both acids and bases, the best diastereoselectivity is obtained with CsF.[4]

OMe HO O O O CsF / THF 60°, 1 d OMe O O O O

quant. (*trans:cis* 95:5)

Desilylative reactions. A new version of the Peterson olefination employs a combination of ethyl trimethylsilylacetate, an aldehyde, and CsF.[5] Similarly, epoxide synthesis from carbonyl compounds avoids strongly basic conditions by using $[Ph_2SCH_2SiMe_3]^+ TfO^-$ as a source of the sulfonium methylide.[6]

α-Carbanions of phosphine oxides are formed from the α-trimethylsilyl derivatives on exposure to CsF/18-crown-6. Their reaction with benzaldehyde is *anti*–selective.[7] In a 1,3-dipolar cycloaddition to form tetrahydrofuran derivatives, the carbonyl ylides are readily generated from α-chloro-α′-trimethylsilyl ethers.[8]

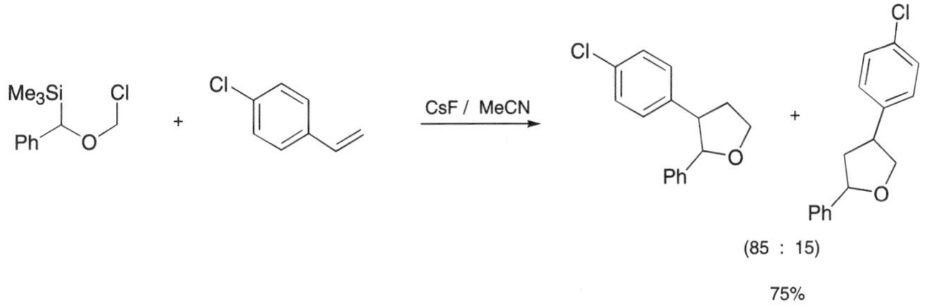

Me_3Si Cl Ph O + Cl — CsF / MeCN → Cl Ph O + Cl Ph O

(85 : 15)

75%

A synthetically significant application is the CsF-induced closure of the enediyne bridge characteristic of dynemicin through the intramolecular addition of silylated alkynes to aldehydes.[9]

Suzuki coupling.[10] A preparation of β-aryl enones from 1,3-diborylbutadienes is initiated by a coupling at the terminal carbon. CsF is present to facilitate the reaction.

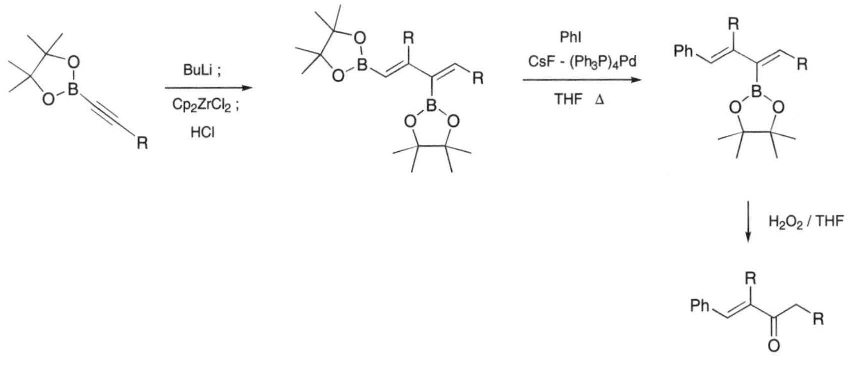

O B O R — BuLi ; Cp_2ZrCl_2 ; HCl → O B O R R O B O — PhI, CsF - $(Ph_3P)_4Pd$, THF Δ → Ph R R O B O — H_2O_2 / THF → Ph R R O

[1]Sato, T., Yoshimatsu, K., Otera, J. *SL* 843 (1996).
[2]Sato, T., Otera, J. *SL* 336 (1995).
[3]Sato, T., Otera, J. *JOC* **60**, 2627 (1995).
[4]Ishikawa, T., Oku, Y., Kotake, K.-I., Ishii, H. *JOC* **61**, 6484 (1996).
[5]Bellassoued, M., Ozanne, N. *JOC* **60**, 6582 (1995).
[6]Hioki, K., Tani, S., Sato, Y. *S* 649 (1995).
[7]O'Brien, P., Warren, S. *SL* 579 (1996).
[8]Hojo, M., Ishibashi, N., Hosomi, A. *SL* 234 (1996).

[9]Wender, P.A., Beckham, S., Mohler, D.L. *TL* **36**, 209 (1995).
[10]Desurmont, G., Dalton, S., Giolando, D.M., Srebnik, M. *JOC* **61**, 7943 (1996).

Chiral auxiliaries and catalysts. 18, 89–97

Due to the extensive literature, only a portion of these reagents is mentioned. The chief criterion is their efficiency (e.g., mostly > 90% ee), unless the utility or reaction is novel.

Proton transfer. Protonation of prostereogenic enolates with the γ-hydroxyselenoxides, such as **1**, sometimes gives excellent ee.[1] The $SnCl_4$ complex of a methyl ether of chiral BINOL can be used in catalytic amounts to protonate silyl enol ethers, affording ketones in high optical yields.[2] A catalytic enantioselective deprotonation to form a bromoalkene[3] is achieved by KH in the presence of *N*-methylephedrine.

Acylation. The C_2-symmetric phosphine **2** is useful for enantioselective acylation of alcohols.[4] Chiral mixed carbonates are prepared from **3** and secondary alcohols, allowing a kinetic resolution of the alcohols.[5]

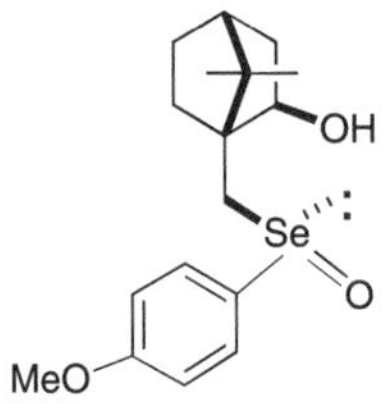

(1) 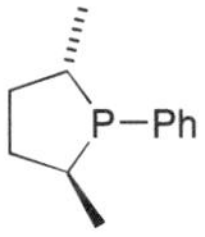 **(2)** 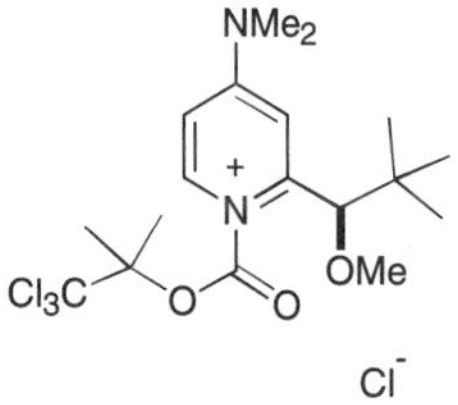 **(3)**

The conversion of *meso*-anhydrides to chiral monoisopropyl esters can be performed by reacting with diisopropoxytitanium TADDOLates.[5a]

Addition to double bonds. Hydrotrichlorosilylation of 1-alkenes, which provides intermediates for chiral alcohols, is Pd catalyzed with a binaphthylphosphine (**4**) as ligand.[6,7]

Hydroboration with catecholborane catalyzed by a rhodium complex of 1-(2-diphenylphosphino-1-naphthyl)isoquinoline (**5**)[8] is accompanied by high enantioselectivity. Bis(1-neomenthylindenyl)zirconium dichloride (**6**) promotes alkylalumination[9] of 1-alkenes through a noncyclic mechanism.

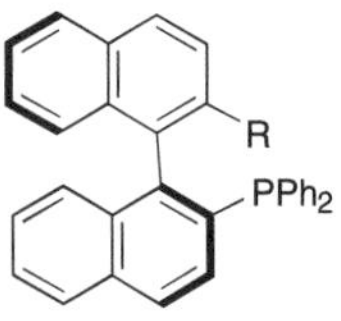

R = OMe (**4A**)
R = H (**4B**)

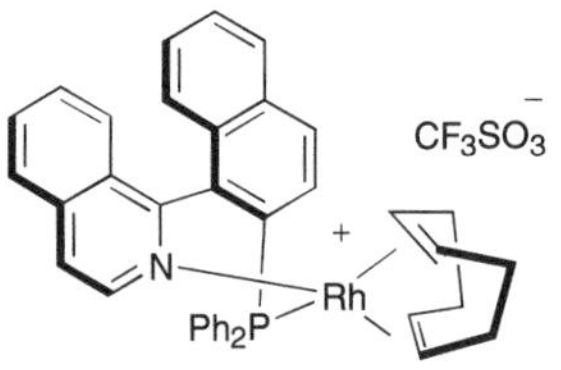

(5) 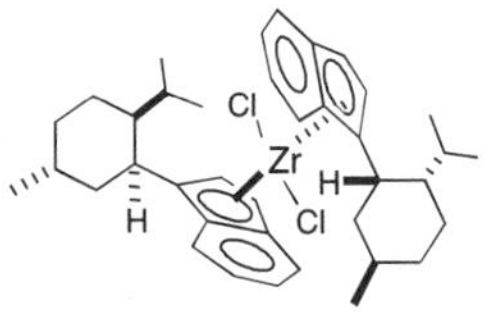 **(6)**

A member of the new ligand class for the asymmetric dihydroxylation is the bis(dihydroquinidine) ether of 1,4-dihydroxy-9,10-anthraquinone.[10] Cinchona alkaloid ligands bound to soluble polymer supports[11] are effective catalysts for asymmetric dihydroxylation.

Selenoalkoxylation and -lactonization using chiral areneselenenyl derivatives such as **7**[12] and **8**[13], the ferrocenyl diselenide **9**[14], and the camphor-based species (**10**)[15] are efficient processes. The diselenides are activated by Br_2 to form reactive selenenyl bromides.

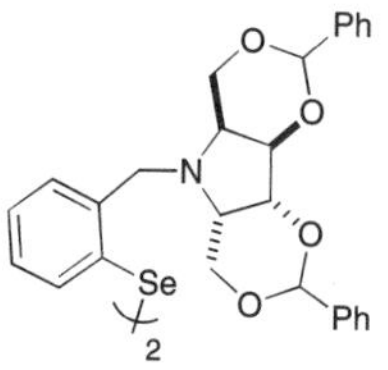

(7) 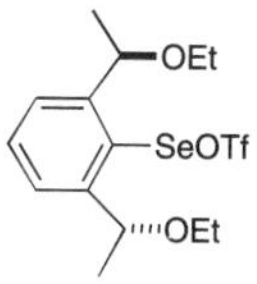 **(8)** 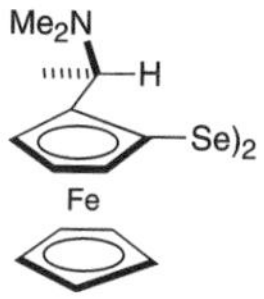 **(9)** 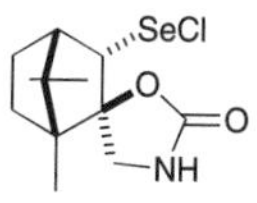 **(10)**

Heck reactions. The *o*-oxazolinylphenyl(diphenyl)phosphine ligand **11** is an excellent ligand to direct enantioselective Heck reactions.[16] On the other hand, the dioxazolinylmethane derivative **12** has been used for the annulation of allenes by means of functionalized aryl iodides.[17]

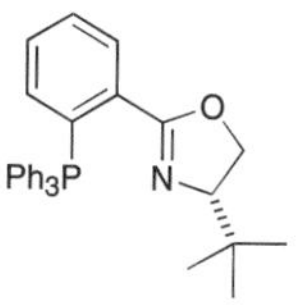

(11) 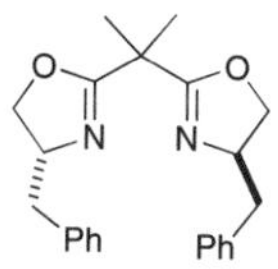 **(12)**

Aldol reactions. The copper complex **13** is a Lewis acid that catalyzes the reaction of silylketene acetals with (benzyloxy)acetaldehyde.[18]

An approach to 5-hydroxy-3-ketoesters by the condensation of diketene with aldehydes involves a titanium alkoxide derived from the complex Schiff base **14**.[19]

An imino–BINOL complex **15** is effective for an analogous reaction.[20]

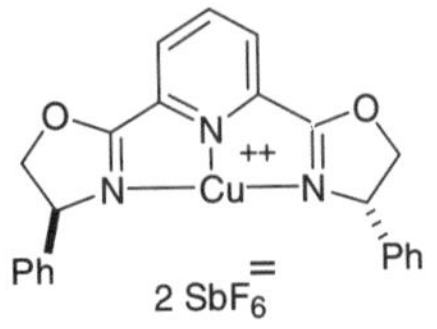
Ph
2 SbF6
Ph

(13)

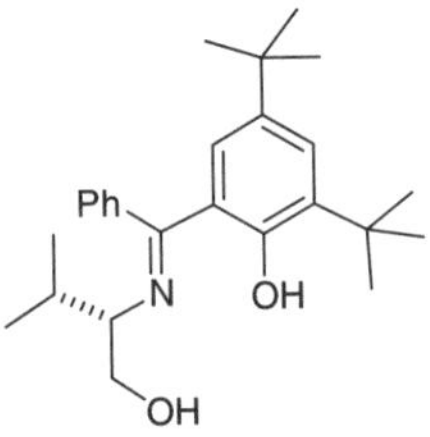
Ph
N
OH
OH

(14)

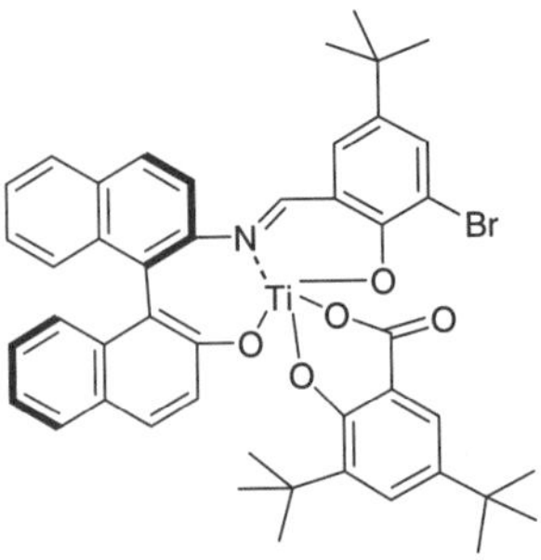
N
Br
Ti
O
O

(15)

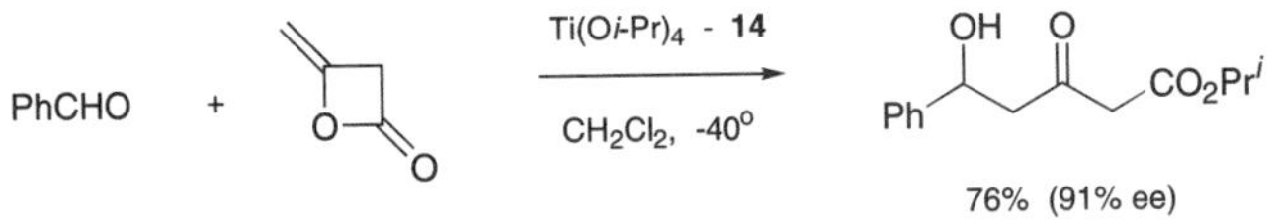
PhCHO
+
Ti(Oi-Pr)4 - 14
CH2Cl2, -40°
OH O
Ph
CO2Pri
76% (91% ee)

A high enantioselectivity is found in the aldol reaction of silylketene acetal derived from ethyl 1,3-dithiolane-2-carboxylate with aldehydes, which is promoted by the oxazaborolidinone **16**. The products are readily desulfurized with the stereogenic center intact.[21]

With trichlorosilyl enolates as donors and a chiral phosphoric triamide **17** as catalyst, asymmetric aldol reactions with aldehydes are realized at low temperatures.[22]

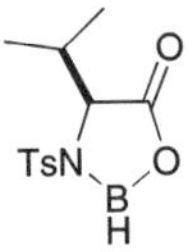
TsN
B
H
O
O

(16)

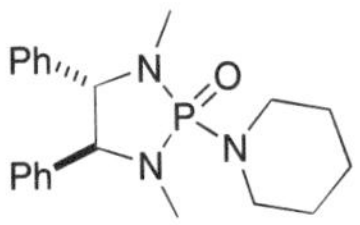
Ph
Ph
N
P
O
N
N

(17)

Aldol donors bearing chiral auxiliaries include *cis*-1-*N*-tosylamino-2-indanyl esters (*anti*-selective with $TiCl_4$/*i*-Pr_2NEt)[23] and (–)-bis(2,2,2-trifluoroethyl) (8-phenylmenthoxycarbonyl)methylphosphonate.[24] A dynamic kinetic resolution of racemic α-amino aldehydes is realized during their reaction with the latter compound.

A chelate derived from glycine is a source of *syn*-β-hydroxy-α-amino acids.[24a] Most intriguingly, kinetic and thermodynamic conditions lead to different stereoisomers.

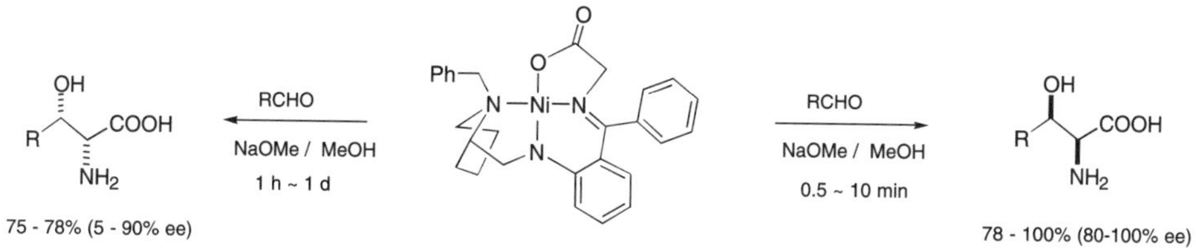
OH
R
COOH
NH2
RCHO
NaOMe / MeOH
1 h ~ 1 d
75 - 78% (5 - 90% ee)
Ph
O
N
Ni
N
RCHO
NaOMe / MeOH
0.5 ~ 10 min
78 - 100% (80-100% ee)

Chiral 2,3-disubstituted DABCOs promote asymmetric Baylis–Hillman reactions under high pressure. While the yields are acceptable, the enantioselectivity is dismal.[25]

The Darzens condensation of (–)-8-phenylmenthyl chloroacetate with ketones has been examined.[26] Good diastereoselectivities are observed.

Substitutions. Radical displacement of chiral α-bromo esters[27] and amides[28] on reaction with allyltributylstannane is under stereocontrol by the 8-phenylmenthyl and 4-diphenylmethyloxazolidin-2-one moieties, respectively.

The asymmetric allylic substitution with Pd catalysts has received much attention regarding chiral ligand development. While ligands **18**[29] and **19**[30] show moderate effects, ligands **11**[31,32], **20A**[33,34], and **21**[35] are excellent for reactions with assorted nucleophiles. Although the substitution of one acyloxy group in *cis*-2,5-diacyloxy-2,5-dihydrofurans with 6-chloropurine using dppe and $(dba)_3Pd_2{\cdot}CHCl_3$ is nonproductive, on changing the ligand to **20B**, the reaction becomes remarkably smooth.[36]

Ligand **11** forms tungsten complexes, which have also been examined successfully for allylic displacement reactions.[37]

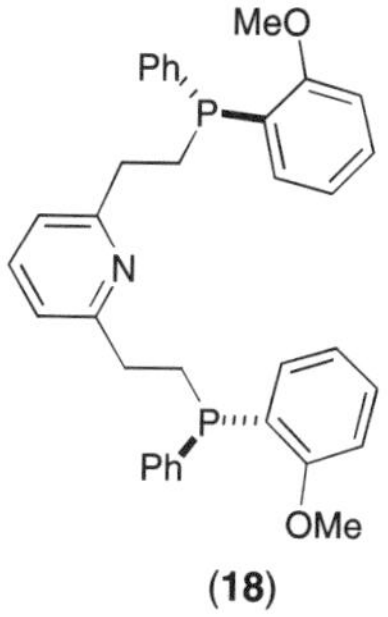
MeO
Ph
P
N
P
Ph
OMe

(18)

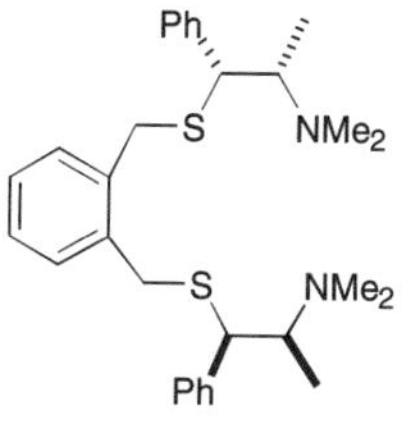
Ph
S
NMe2
S
NMe2
Ph

(19)

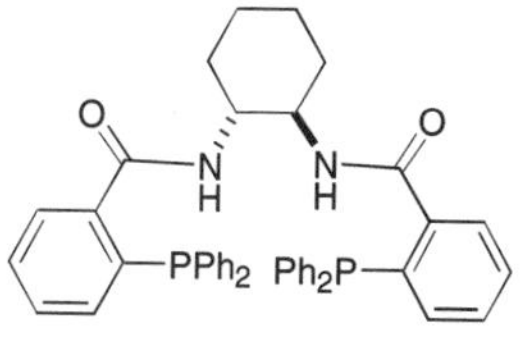
O
N
H
N
H
O
PPh2
Ph2P

(20A)

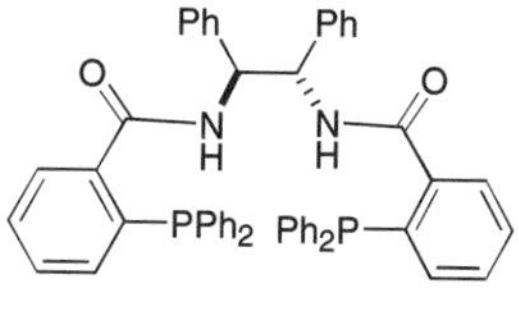
Ph
Ph
O
N
H
N
H
O
PPh2
Ph2P

(20B)

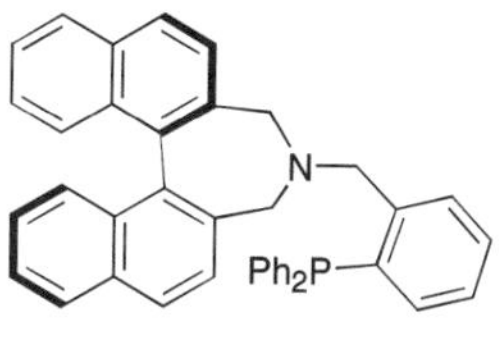
N
Ph2P

(21)

Asymmetric synthesis of 2-vinylmorpholine and 2-vinylpiperazine[38] from (*Z*)-1,4-diacetoxy-2-butene enlists a Pd complex derived from diphosphine ligand **22**.

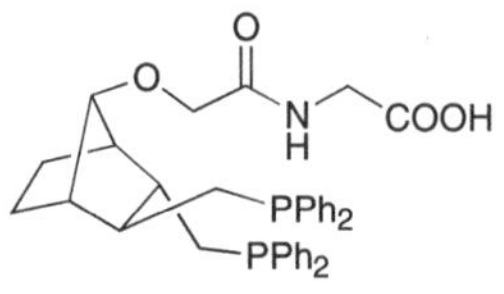

(22)

Openings of *meso*-epoxides to obtain chiral β-hydroxy nitrile derivatives[39] and of *N*-acylaziridines to afford *N*-(β-sulfenylalkyl) amides[40] have enlisted the service of ligand **23** and dicyclohexyl L-(+)-tartrate, respectively. An efficient method for acquiring chiral azido silyl ethers from epoxides and Me_3SiN_3 employs a (salen)Cr–N_3 complex **24**.[41]

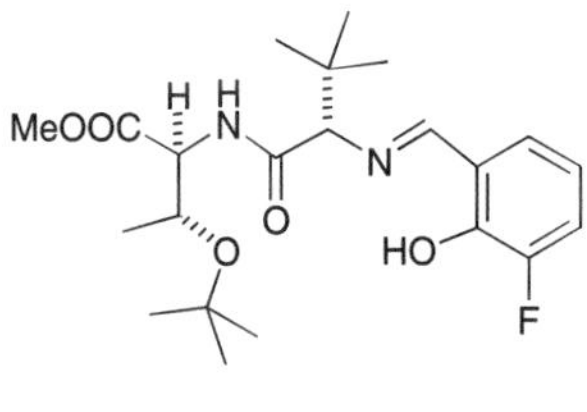

(23)

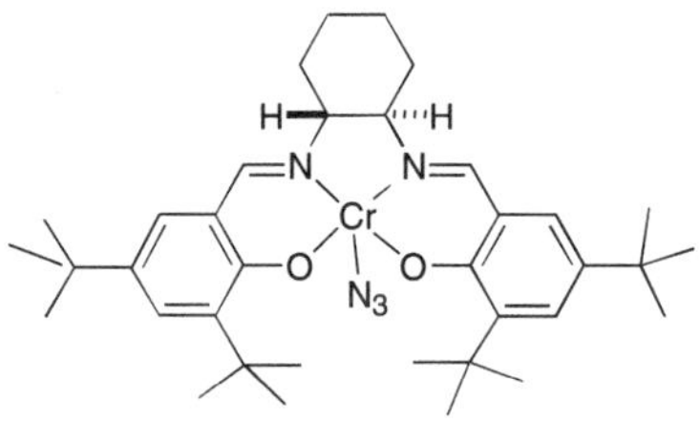

(24)

Alkylations. The alkylation of pseudoephedrine amides with epoxides is a practical method for gaining access to chiral γ-lactones.[42] Reaction of the alkylation products with alkyllithiums leads to γ-hydroxy ketones.

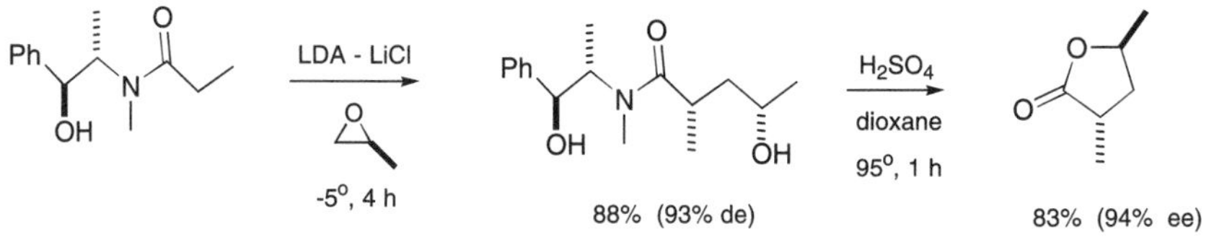

LDA - LiCl; -5°, 4 h — 88% (93% de); H_2SO_4, dioxane, 95°, 1 h — 83% (94% ee)

4-Substituted *N*-acyl-5,5-dimethyloxazolidin-2-ones undergo diastereoselective alkylation,[43] controlled by the ring substituent. The methylation of chiral

2-arylacetyl-1,3-dithiane *S*-oxides (**25**) furnishes precursors of medicinally valuable acids related to Naproxen.[44]

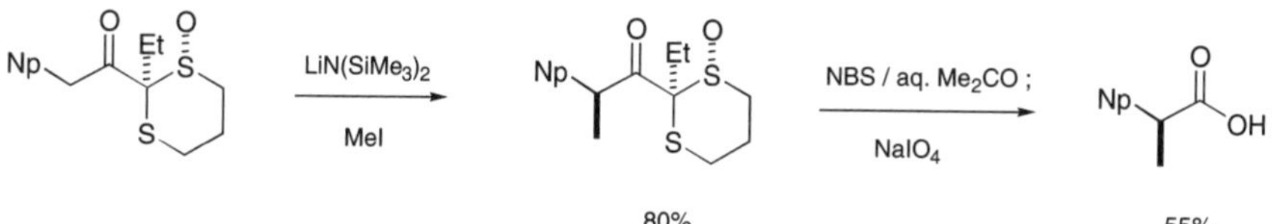

LiN(SiMe$_3$)$_2$ / MeI; NBS / aq. Me$_2$CO ; NaIO$_4$

(**25**) 80% 55%

Allylation of a camphorsultam-modified glycine derivative **26** proceeds on deprotonation and addition of a Pd catalyst and an allyl carbonate.[45] A three-component coupling[46] involving α-allylation of *N*-acryloyloxazolidin-2-one by reactions with free radicals is rendered enantioselective in the presence of a bis(4-phenyloxazoline)methane ligand analogous to **12**. Similarly, a stepwise Baylis–Hillman reaction to furnish chiral α-methylene-β-hydroxy ketones[47] is effective if catalyzed by a tartaric acid derivative **27**.

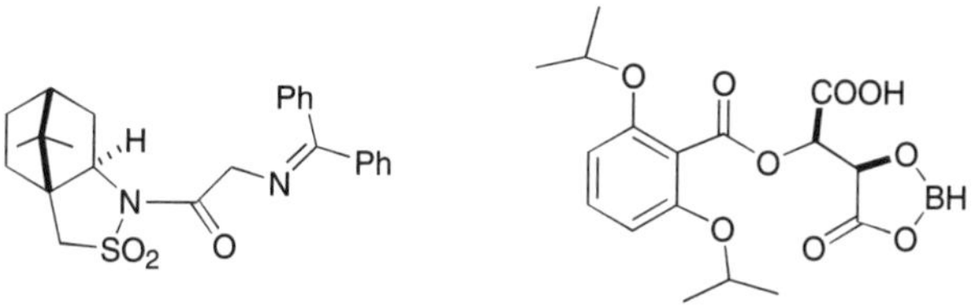

(**26**) (**27**)

Taking advantage of the bulky *t*-butyl group of methyl *N*-(ω-bromoacyl)-2-*t*-butyl-oxazolidine-4-carboxylate to achieve a 1,3-asymmetric induction in the intramolecular alkylation,[48] a series of lactams containing a quaternary chiral center is available. With SAMP/RAMP chiral auxiliaries, the synthesis of 2-substituted (five, six, and seven membered) lactams is readily achieved.[49] Aldehyde hydrazones of this series undergo diastereoselective phosphinylation.[50]

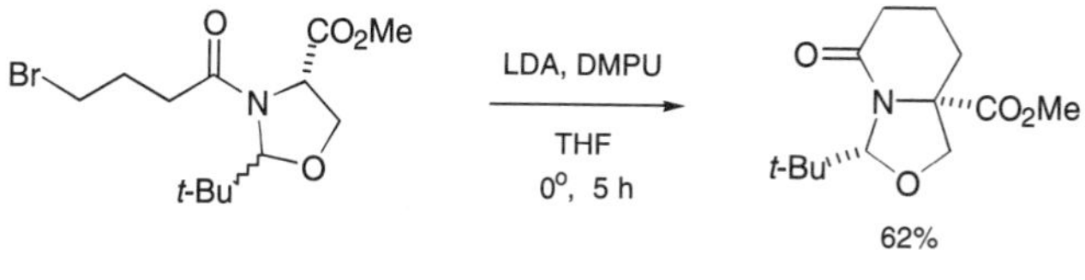

LDA, DMPU / THF 0°, 5 h

62%

The mediation of (–)-sparteine in the deprotonation and alkylation of Boc-protected *N*-allylanilines affords the preparation of either enantiomer of 3-substituted enecarbamates.[51] A synthesis of chiral primary amines[52] relies on *C*-alkylation of imines in the presence of a hydrazonium salt **28**.

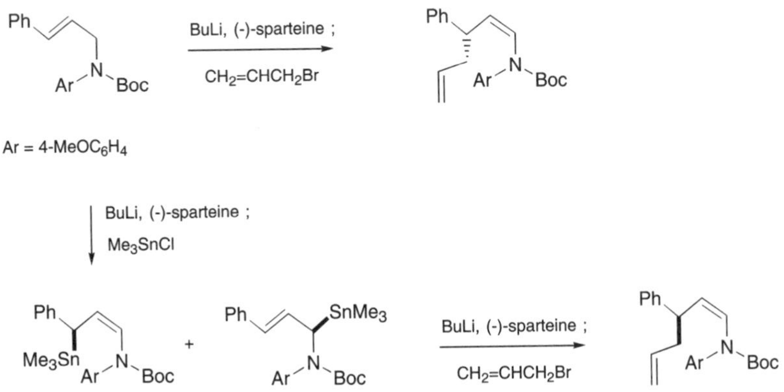

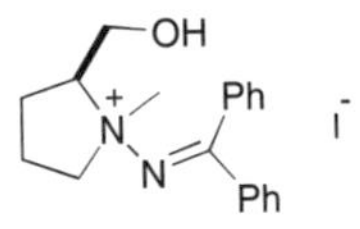

Ar = 4-$MeOC_6H_4$

BuLi, (-)-sparteine ; CH_2=$CHCH_2Br$

BuLi, (-)-sparteine ; Me_3SnCl

BuLi, (-)-sparteine ; CH_2=$CHCH_2Br$

(**28**)

Michael additions. Despite tremendous efforts spent in achieving catalytic asymmetric Michael additions, effective additives of wide applicability are still quite rare. Interestingly, a polyamine ligand **29** promotes the addition of ketone enolates.[53] With *N*-methoxy-*N*-methyl amides of α,β-unsaturated acids as acceptors, the addition of lithium (*S*)-(α-methylbenzyl)benzylamide proceeds in a highly diastereo- and enantioselective manner, ascribable to a six-centered transition state in which the conjugated amide adopts an *s-cis* conformation.[54]

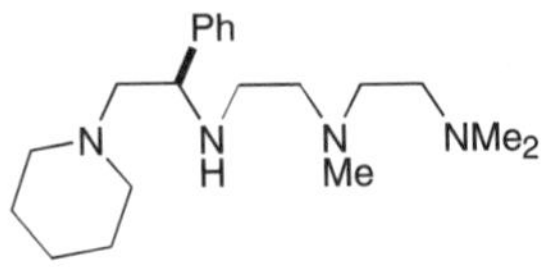

(**29**)

Good results are obtained from reactions of compounds containing chiral auxiliaries. Thus, 1,3-asymmetric induction in the addition of radicals to **30**,[55] 1,4-asymmetric induction by the isopropyl group of the dilactim derived from cyclo(Gly-Val) in the addition to conjugated sulfones,[56] and 1,5-asymmetric induction for SAMP/RAMP hydrazones during the addition to alkenylphosphonate esters[57] are adequate.

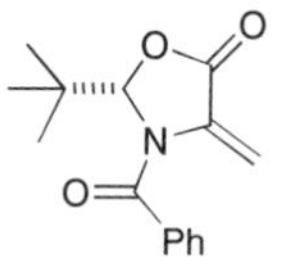

(**30**)

Variants of the Michael addition include the allylation of cyclopropenone acetals[58] and the intramolecular Stetter reaction.[59] So far, only moderate enantioselectivity for the latter reaction has been achieved. (Note that the same chiral catalyst is useful for benzoin condensation.[60])

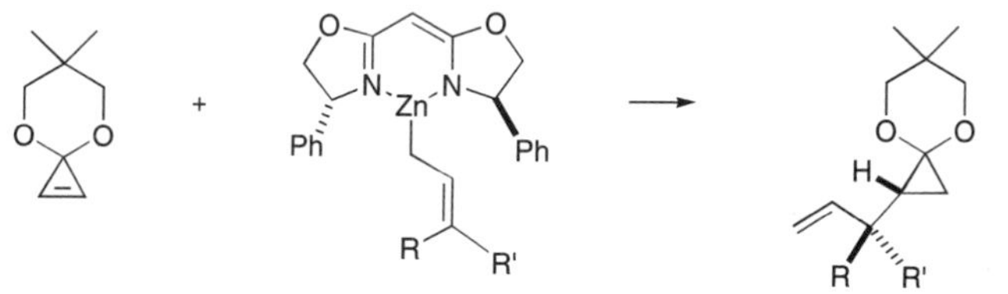

A radical Michael addition whose steric course is dominated by the *t*-butyl substituent at C-2 of perhydro-1-aroyl-2-*t*-butylpyrimidin-4-ones occurs at C-6. After the radical is created from the action of Bu_3SnH on the 1-(*o*-halobenzoyl) derivatives, a 1,5-hydrogen atom translocation takes place.[61] The products are converted to β-amino acids on acid hydrolysis.

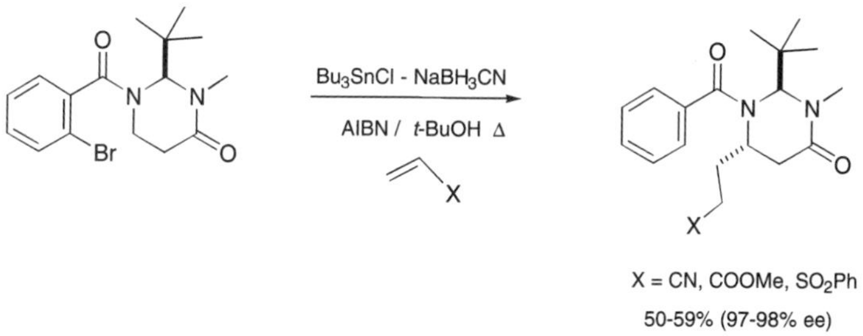

Bu_3SnCl - $NaBH_3CN$

AIBN / *t*-BuOH Δ

X = CN, COOMe, SO_2Ph

50-59% (97-98% ee)

Organometallic reactions. Organocerium reagents (**31**) derived from TADDOL react with aldehydes to give secondary alcohols with moderate to good ee values.[62] For reaction of organozincs, the most reliable chiral ligands seem to be the *trans*-1,2-bis(sulfonylamino)cyclohexanes. Thus, in their presence the titanium(IV)-isopropoxide-promoted addition of functionalized diorganozincs to enals gives allylic alcohols,[63] which are valuable precursors of α-hydroxy aldehydes and 1,2-amino alcohols. This method is the basis of an approach to (−)-*exo*-brevicomin and (−)-*endo*-brevicomin. Propargylic alcohols are also accessible in the same manner.[64] A superior ligand is **32**.[65]

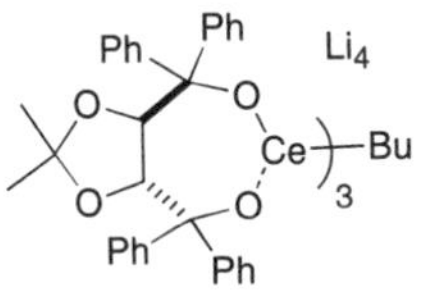

Ph Ph O O Ce Bu 3 Li4 O O Ph Ph

(31)

O2S−N H N−SO2 H Cl OH HO Cl Cl Cl

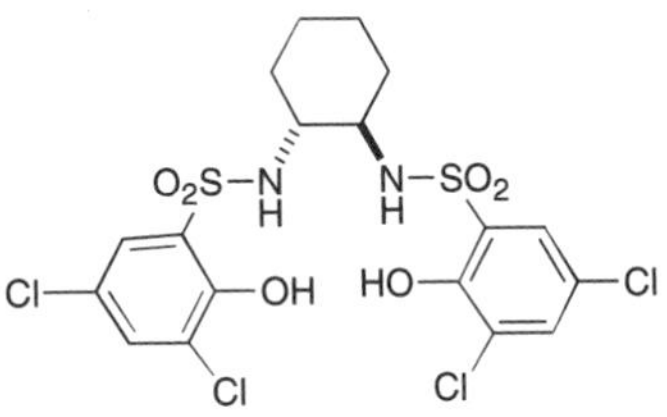

(32)

The reaction of Et_2Zn with aromatic aldehydes is almost a standard for evaluating the ability of chiral ligands to promote enantioselection. The following ligands show promising results: **33**,[66] **34**,[67] **35**,[68] and **36**.[69] Interestingly, the addition catalyzed by pyridylphenol **37** shows a linear free-energy relationship in enantioselectivity with electrophilicity of the aldehyde.[70]

The Reformatsky reaction of alkyl bromodifluoroacetates in the presence of *N*-methylephedrine has been studied.[71,72]

Di-*t*-butyl tartrate and tin(II) catecholate, together with CuI and DBU, promote the allylation of aromatic aldehydes,[73] producing homoallylic alcohols with high ee values. Diallyltin dibromide in the presence of the *N*-butyl-2-piperidinomethylpyrrolidine **38** also gives useful results.[74]

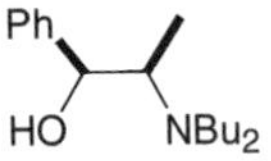

Ph HO NBu2

(33)

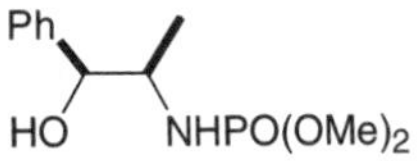

Ph HO NHPO(OMe)2

(34)

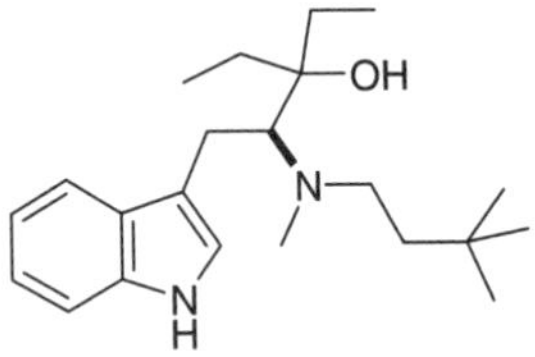

OH N N H

(35)

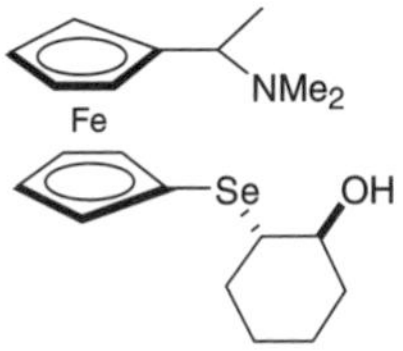

(36) 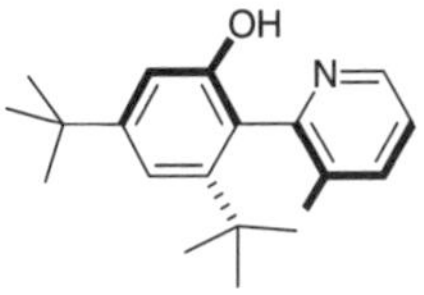 (37) 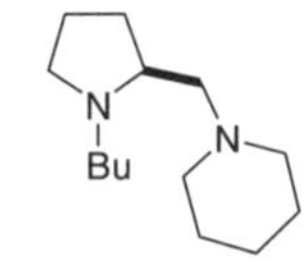 (38)

Optically active *syn*-1,2-diols are obtained from an allyltin reagent attached to a chiral tetrahydropyranyloxy moiety.[75]

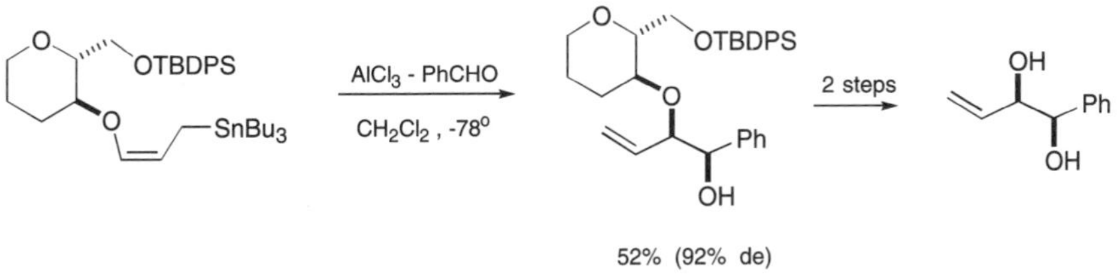

OTBDPS SnBu3 → $AlCl_3$ - PhCHO, CH_2Cl_2, -78° → OTBDPS, Ph, OH → 2 steps → OH, Ph, OH

52% (92% de)

(Z)-3-Chloroallyldiisopinocampheylborane is a valuable allylating agent for aldehydes.[76] The products are *syn*-chlorohydrins, which can be transformed into *cis*-1-alkyl-2-vinyloxiranes. α-Allenic alcohols are prepared from propargylic diisopinocampheylboranes.[77]

A very good stereoselectivity is manifested in the organometallic addition to imines when they are derived from chiral amines such as the SAMP/RAMP block[78] and *erythro*-2-amino-1,2-diphenylethanol.[79]

The conjugate base of *(S,S)*-bis(4-isopropyloxazolin-2-yl)methane **39** complexes to allylzinc halides and then delivers the allyl group to cyclic aldimines[80] stereoselectively. This method is applicable to the synthesis of isoquinoline alkaloids.

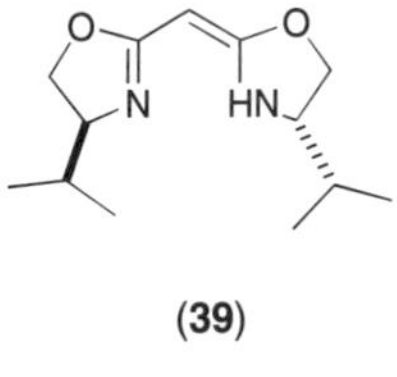

(39)

Cyanohydrination and Strecker synthesis. A high enantioselectivity in the cyanohydrination (with Me_3SiCN as the cyanide source) is achieved by using $(i\text{-PrO})_4Ti$ as catalyst and **40** as ligand.[81] Electronic fine-tuning of diarylphosphinite ligands derived from chiral 1,2-diols is effective for increasing the enantioselectivity of hydrocyanation catalyzed by $Ni(cod)_2$.[82] Cyclo(Phe-Arg) **41** appears to be an excellent catalyst for the Strecker synthesis.[83]

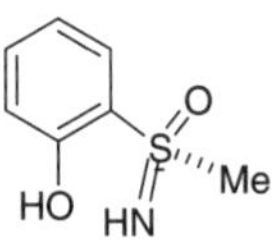

(**40**)

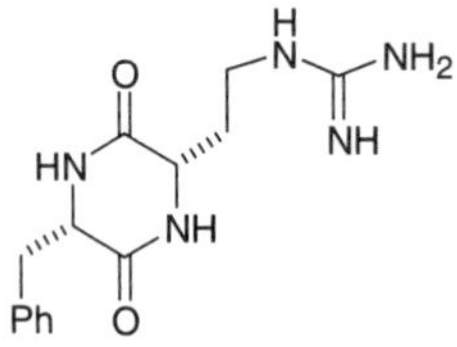

(**41**)

Formation of three-membered rings. Among chiral ligands for the Simmons–Smith reaction are a C_2-symmetric dioxaborolidine prepared from tartaric acid bis(dimethylamide) **42**[84] and the disulfonamide **43**.[85] A thorough survey of a series of ligands indicates that **43A** is the best.[85a]

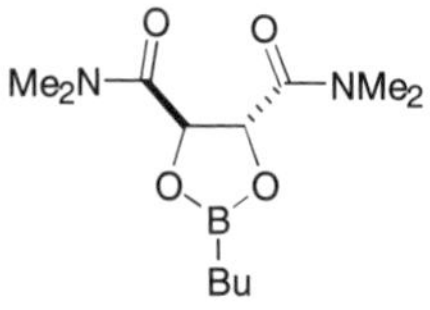

(**42**)

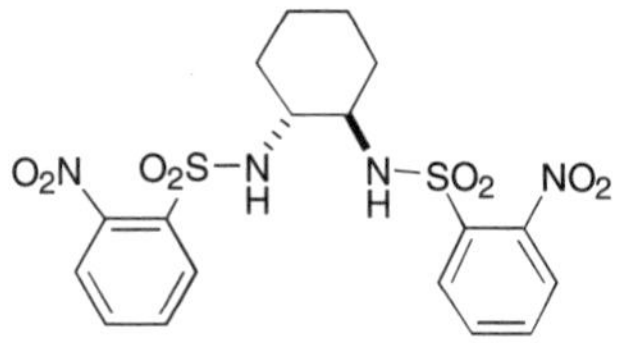

(**43**)

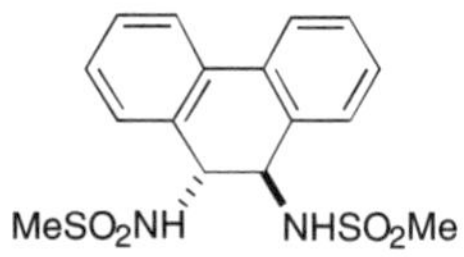

(**43A**)

The cyclopropanation with diazo compounds via decomposition is amenable to asymmetric induction using chiral metal chelates. Rhodium complexes of 3(*S*)-phthalimido-2-piperidinone[86] and *N*-(arenesulfonyl)proline **44**[87] are typical. The latter catalyst is suitable for generating alkenylcarbenoids. For intramolecular reactions, the cognate complex **45**[88] and the semicorrin-copper **46**[89] are effective.

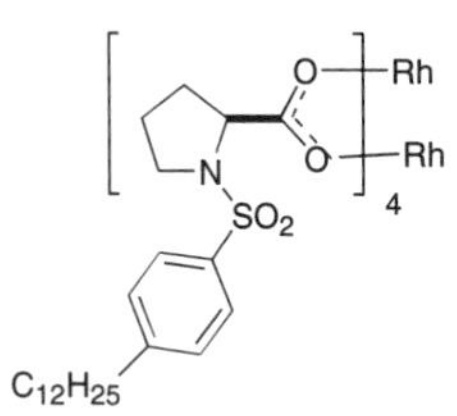

(**44**)

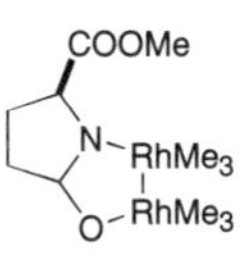

(**45**)

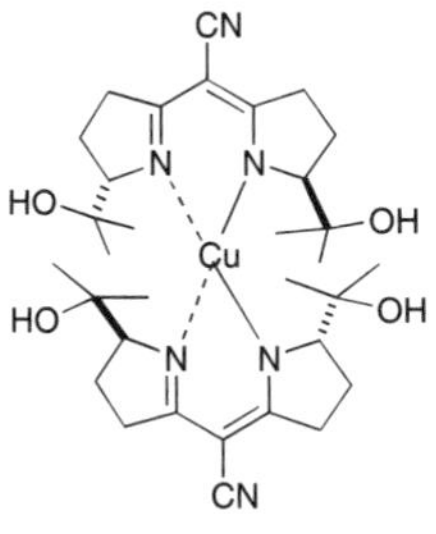

(**46**)

Despite the relatively low yields of aziridines, the carbenoid addition to imines using a copper complex of the bis(oxazolinyl)methane type (cf. **12**) proceeds with high enantioselectivity.[90]

The addition of aldehydes to carbenoids derived from the Cu-catalyzed decomposition of $ArCHN_2$ to form stilbene epoxides is subject to asymmetric induction by 1,3-oxathiane **47** prepared from 10-mercaptoisoborneol and acetaldehyde.[91] The attack of sulfonium ylides derived from **48** on aldehydes also affords epoxides of high optical purity.[92] The same principle underlies a synthesis of chiral aziridines.[93]

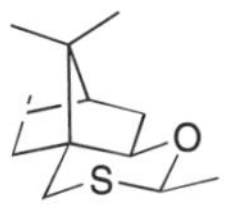

TfO⁻

(**47**) 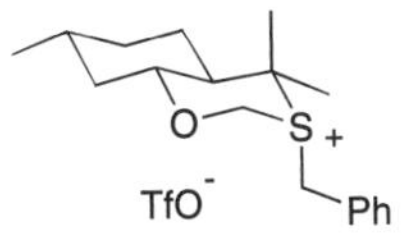 (**48**)

Cycloadditions. Glycine incorporated into Oppolzer's sultam is transformed into the sulfamido ketene that reacts with imines to form β-lactams.[94]

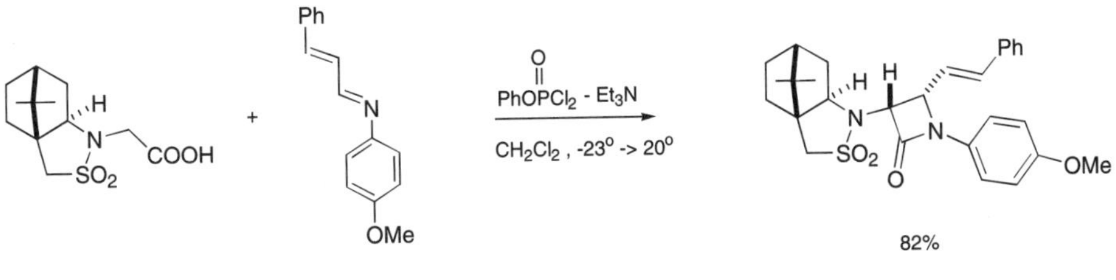

PhOP(O)Cl₂ - Et₃N

CH₂Cl₂, -23° -> 20°

82%

The participation of TADDOL–$Ti(OTs)_2$ in the 1,3-dipolar cycloaddition[95] of nitrones to alkenes ensures high diastereo- and enantioselectivity.

The novel chiral Lewis acid (**49**) is an excellent catalyst for Diels–Alder reactions.[96] The *trans*-4,5-diaryl-1,3-bistriflyl-1,3,2-diazaluminolidines (**50A**) have been exploited in the elaboration of a chiral tetrahydrophthalimide intermediate for the synthesis of gracillin-B and gracillin-C.[97]

A dioxaborepin **51** prepared from a substituted BINOL and 3,5-bis(trifluoromethyl)benzeneboronic acid is a Lewis acid catalyst for enantioselective Diels–Alder reactions.[98]

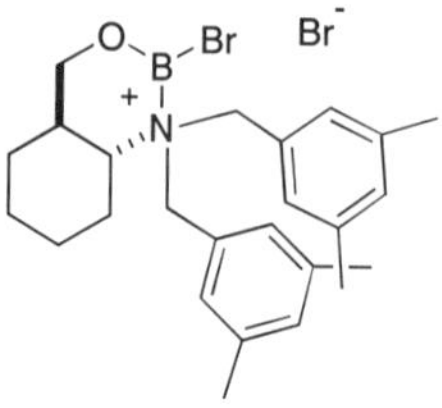

(**49**)

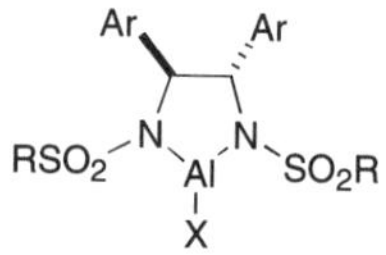

$R = CF_3$ (**50A**)
$R = 3,5\text{-}(F_3C)_2C_6H_3$ (**50B**)

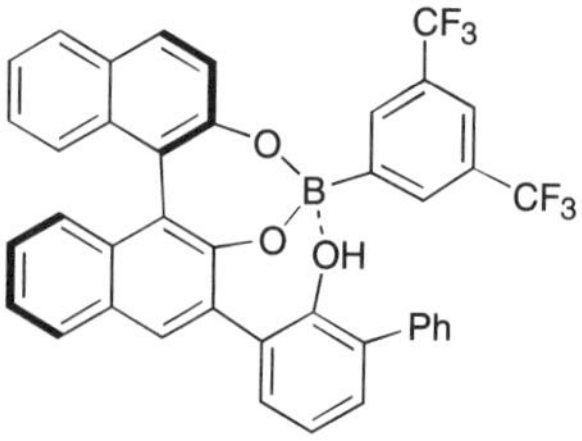

(**51**)

Pauson–Khand reaction. The intramolecular annulation of enynes promoted by a chiral titanocene derivative **52** exhibits high degrees of enantioselectivity.[99]

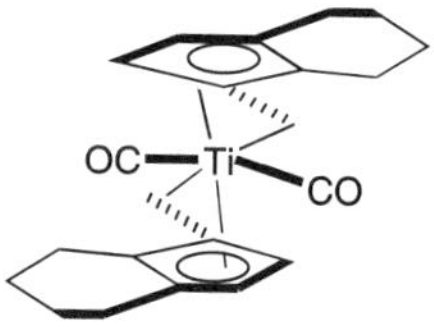

(**52**)

Epoxidation. Asymmetric epoxidation has been an active area of research, and many catalytic systems have been developed. The (salen)Mn series continues to yield catalysts suitable for particular types of alkenes. Thus, **53A** and **53B** are good for tetrasubstituted alkenes[100] and unfunctionalized alkenes,[101] respectively.

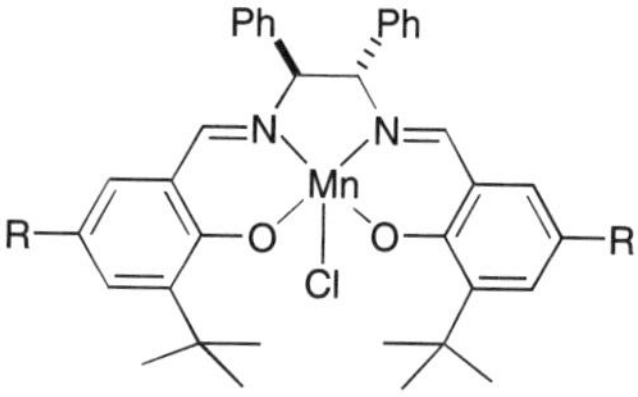

R = Me (**53A**)
R = OTIPS (**53B**)

Other catalysts are a ketodiolide **54**[102] and a ketone **55** derived from fructose.[103] The epoxidation with the latter catalyst using nascent dioxirane indicates that a spiro transition state is favored.

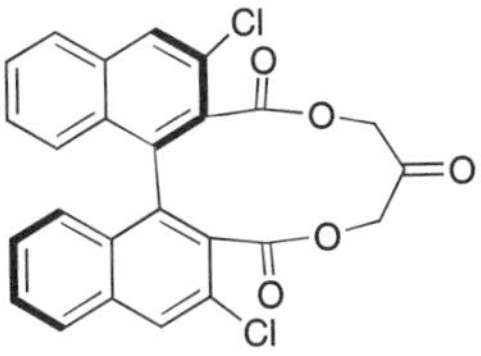

(54) 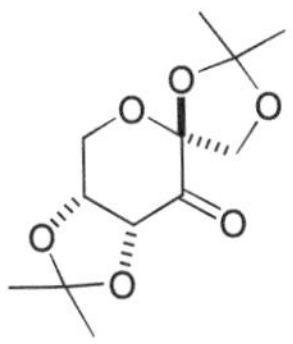 **(55)**

For epoxidation of enones and related substances with alkaline hydrogen peroxide, polyleucine exerts desirable effects, yielding products with greater than 90% ee.[104–106] Enone epoxidation with oxygen in the presence of Et_2Zn and (*R,R*)-*N*-methylpseudoephedrine[107] is also satisfactory.

Oxidation of sulfides. The synthesis of chiral sulfoxides by oxidation of sulfides has also been a subject of intense research. Unfortunately, only a few oxidation systems give consistently good results over a wide range of substrates. In this regard, the combination of $(i\text{-PrO})_4Ti$, cumyl hydroperoxide, and a chiral diethyl tartrate[108] performs extremely well.

The enantioselectivity for sulfide oxidation depends strictly on structural features of the substrates, and therefore, variations can be substantial. Oxaziridines derived from camphor, such as **56A**[109] and **56B**[110], give acceptable, and in some cases, superior results. The sultam precursor of **56B** can also be used as a catalyst in combination with hydrogen peroxide in the presence of DBU.[111]

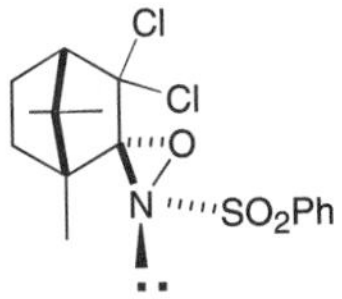

(56A) 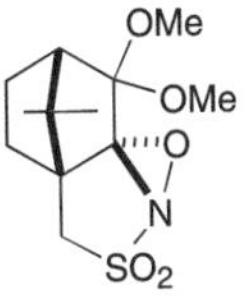 **(56B)**

For metal chelate-mediated oxidation of sulfides, (salen)Mn systems such as **57**,[112] **58**,[113] and **59**[114] have been successfully exploited. A vanadium complex of **60** is also of some utility.[115] Certain multidentate ligands can prolong the lifetimes of catalysts[116] at the expense of enantioselectivity.

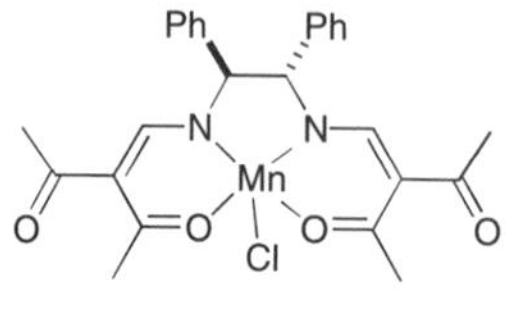

(57)

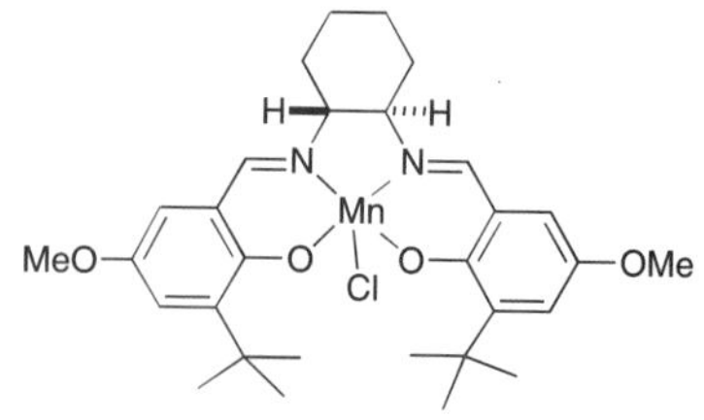

(58)

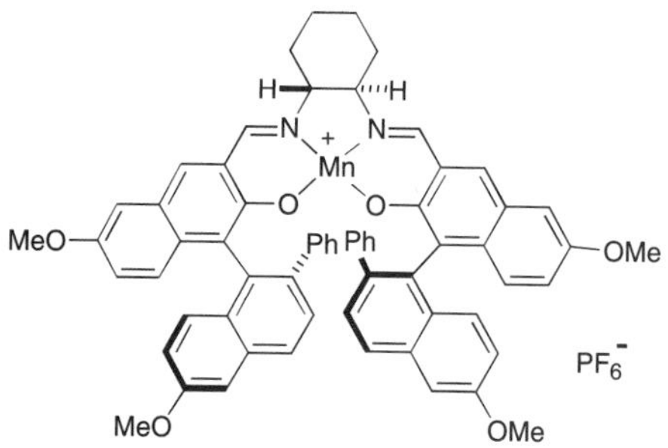

(59)

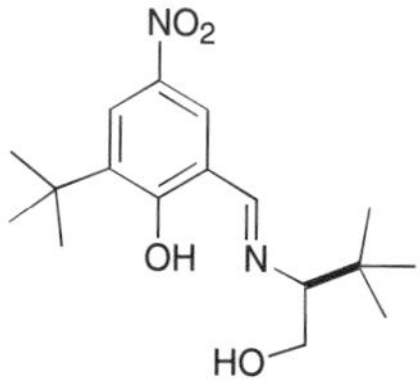

(60)

Allylic oxidations. Allylic esters are obtained in the 50–75% ee range using peroxide, peracid, or perester oxidants in the presence of copper complexes of proline[117,118] or bis(oxazolinyl)methanes.[119,120]

Reduction of ketones with borane. The basic principle for employing chiral 1,2-amino alcohols to mediate borane reduction (with in situ formation of oxazaborolidines) is now well developed. The reduction is accelerated by organoaluminums.[121] New modifications of the ligands afford **61**,[122] **62**,[123] and **63**.[124]

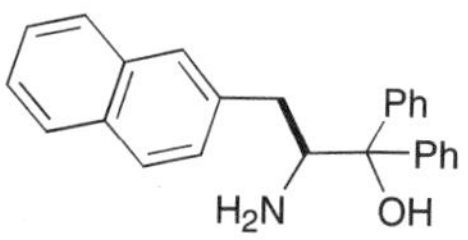

(61)

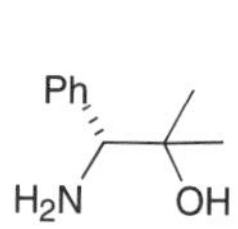

(62)

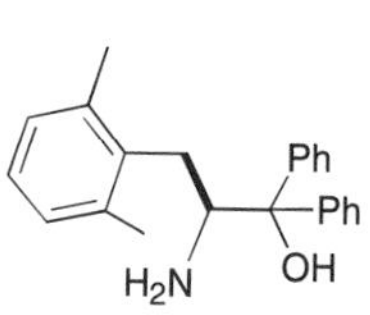

(63)

An optimized procedure (in situ) for the oxazaborolidine-catalyzed reduction is disclosed.[125] Several papers describe the access to propargylic alcohols,[126–128] and it is interesting to note that strong remote steric effects are transmitted across the triple bond.[128]

A modified Corey catalyst in which the boron atom is attached to the benzene ring of polystyrene has been prepared and evaluated.[129]

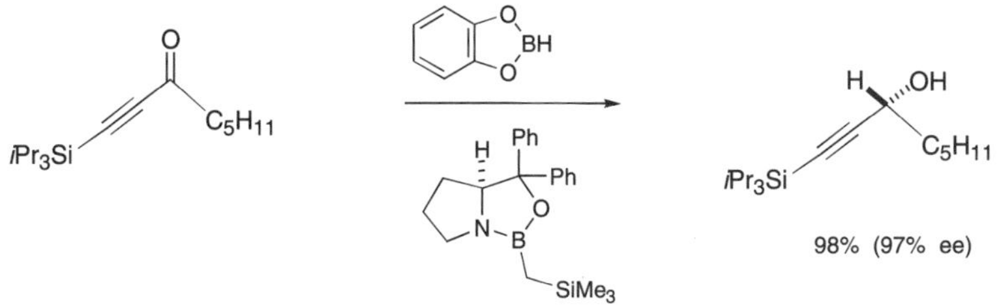

98% (97% ee)

Also interesting are the observations[130] that, with (*R*)-**64**, the reduction of enones branched at both α- and α′-positions gives (*R*)-alcohols, whereas most enones give (*S*)-alcohols. An explanation based on molecular orbital calculations has been presented.

An oxazaborolidine **65** prepared from α-pinene[131] is a new system. More drastic variations are the oxazaphospholidine oxides, such as **66**, which have been examined for their effectiveness in catalyzing the asymmetric reduction.[132] The corresponding oxazaphospholidine–borane complex catalyzes the reduction of imines[133] with diminished stereocontrol.

For ketone reduction with catecholborane, chiral titanium alkoxide species prepared from $(i\text{-PrO})_4Ti$ and diol **67** have been used.[134]

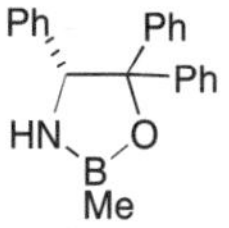

(64) 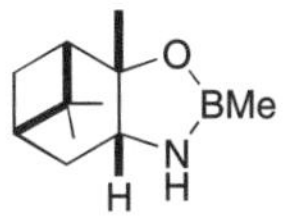 **(65)** 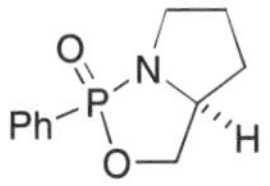 **(66)** 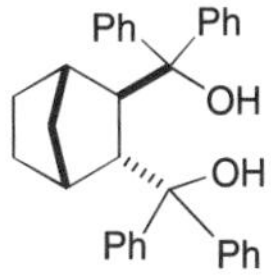 **(67)**

B-Chlorodiisopinocampheylborane is a valuable reagent for the reduction of *o*-substituted benzophenones.[135]

Reduction with complex metal hydrides. Useful reduction systems are those involving $NaBH_4$/ **68**[136] and $LiAlH_4$/**69**.[137] Synthesis of chiral α-hydroxy carboxylic acids can take advantage of forming esters or amides with chiral auxiliaries. Typical examples are depicted by **70**,[138] **71**,[139] and **72**.[140]

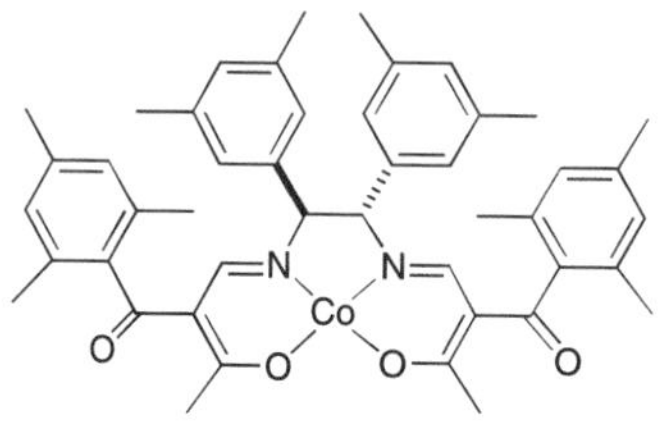

(68) 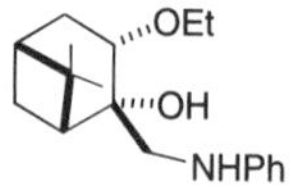 **(69)**

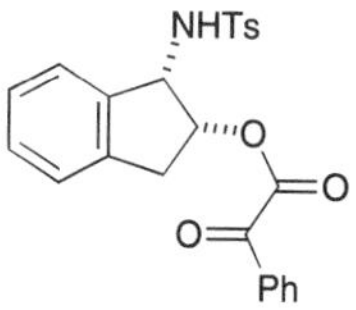

(70) 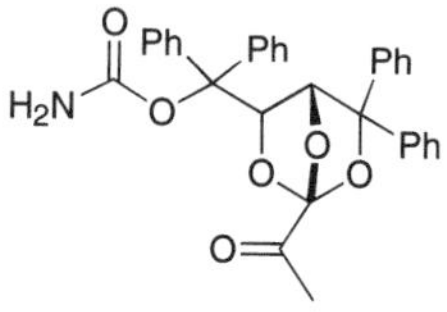 **(71)** 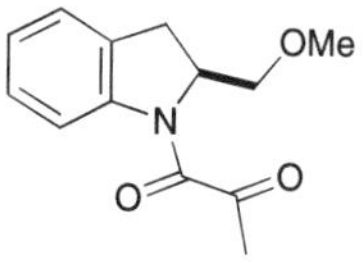 **(72)**

Reduction with hydrosilanes. Ketones undergo enantioselective reduction by Ph_2SiH_2 in the presence of $[Rh(cod)Cl]_2$ and $Ir(cod)Cl_2$. Ligands such as **11**,[141] **73**,[142,143] and **74** (for ketoesters)[144] are effective. Note that it is possible to use isopropanol instead of Ph_2SiH_2 as source of hydrogen by changing the metal catalyst into $(Ph_3P)_3RuCl_2$.[145]

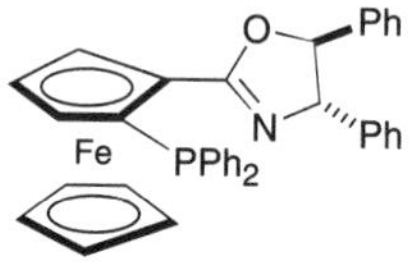

(73) 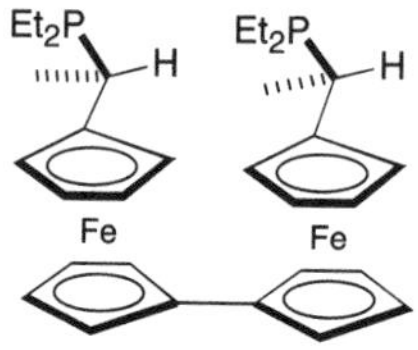 **(74)**

A catalytically active chiral hydridotitanium complex obtained from (*S,S*)-ethylene(η^5-tetrahydroindenyl)titanium difluoride and $PhSiH_3$ hydrosilylates imines under mild conditions with significantly higher substrate:catalyst ratios than known methods yield.[146]

Asymmetric hydrogenation. A practical enantioselective hydrogenation of aromatic ketones consists of a binap-$RuCl_2$ catalyst and the C_2-symmetric α,α'-diaminobibenzyl.[147] The analogous reduction using an Rh catalyst gives inferior results.[148] β-Ketoesters are hydrogenated under mild conditions ($RuBr_2$-**75**/MeOH, 60psi H_2), affording the hydroxy esters with high ee values.[149] BINAP renders the Ru-catalyzed hydrogenation of β-ketophosphonates enantioselective.[150]

The asymmetric hydrogenation of dehydroamino acid derivatives is still a popular research area. For the synthesis of *N*-acetyl D-alanine methyl ester, the hydrogenation is carried out in supercritical carbon dioxide with catalyst **76**.[151] On the other hand, the preparation of protected L-valine and related compounds requires the hydrogenation of a tetrasubstituted double bond. An Rh catalyst with ligand **77** is well suited for this purpose.[152]

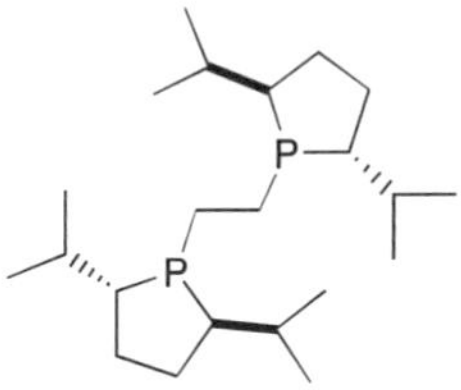

(75) 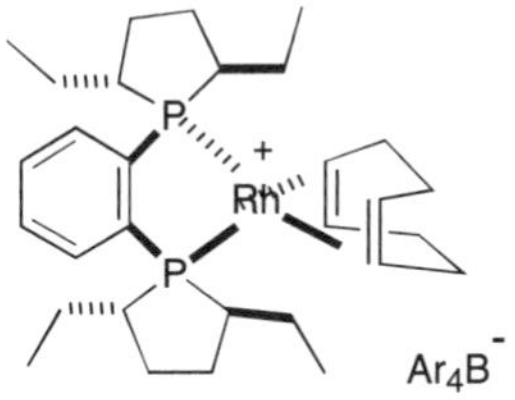 **(76)** 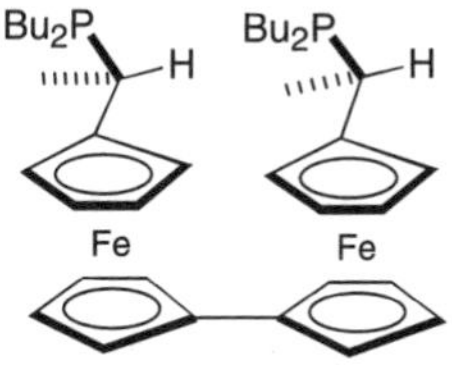 **(77)**

Simpler imines are hydrogenated with great efficiency using catalysts such as **78**,[153] (with HCOOH–Et_3N as hydrogen source) and for enamides, **79**.[154]

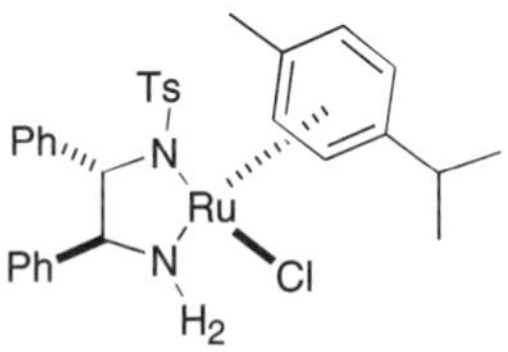

(78) 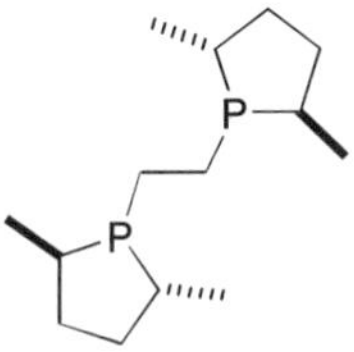 **(79)**

The same HCOOH–Et_3N system as hydrogen source, in conjunction with Ru(II) catalyst **78**, is also quite superior to 2-propanol for the asymmetric reduction of aryl ketones.[155] It allows for a much higher substrate concentration (2–10 M vs. < 0.1 M in 2-propanol).

Ireland–Claisen rearrangement. The effectiveness of the *trans*-4,5-diaryl-1,3-disulfonyl-1,3,2-diazaborolidines (the boron analogue of **50B**) to mediate the rearrangement is shown in its application to a synthesis of (+)-fuscol[156] and dollabellatrienone.[157]

(+)-fuscol

(+)-dolabellatrienone

[1] Takahashi, T., Nakao, N., Koizumi, T. *CL* 207 (1996).
[2] Ishihara, K., Nakamura, S., Kaneeda, M., Yamamoto, H. *JACS* **118**, 12854 (1996).
[3] Amadji, M., Vadecard, J., Plaquevent, J.-C., Duhamel, L., Duhamel, P. *JACS* **118**, 12483 (1996).
[4] Vedejs, E., Daugulis, O., Diver, S.T. *JOC* **61**, 430 (1996).
[5] Vedejs, E., Chen, X. *JACS* **118**, 1809 (1996).
[5a] Seebach, D., Jaeschke, G., Wang, Y.M. *ACIEE* **34**, 2395 (1995).
[6] Uozumi, Y., Kitayama, K., Hayashi, T., Yanagi, K., Fukuyo, E. *BCSJ* **68**, 713 (1995).
[7] Kitayama, K., Uozumi, Y., Hayashi, T. *CC* 1533 (1995).
[8] Brown, J.M., Hulmes, D.I., Layzell, T.P. *CC* 1673 (1993).
[9] Kondakov, D.Y., Negishi, E.-I. *JACS* **118**, 1577 (1996).
[10] Becker, H., Sharpless, K.B. *ACIEE* **35**, 448 (1996).
[11] Han, H., Janda, K.D. *JACS* **118**, 7632 (1996).
[12] Fujita, K., Murata, K., Iwaoka, M., Tomoda, S. *TL* **36**, 5219 (1995).
[13] Deziel, R., Malenfant, E. *JOC* **60**, 4660 (1995).
[14] Nishibayashi, Y., Srivastava, S.K., Takada, H., Fukuzawa, S., Uemura, S. *CC* 2321 (1995).
[15] Back, T.G., Dyck, B.P. *CC* 2567 (1996).
[16] Loiseleur, O., Meier, P., Pfaltz, A. *ACIEE* **35**, 200 (1996).
[17] Larock, R.C., Zenner, J.M. *JOC* **60**, 482 (1995).
[18] Evans, D.A., Murry, J.A., Kozlowski, M.C. *JACS* **118**, 5814 (1996).
[19] Hayashi, M., Tanaka, K., Oguni, N. *TA* **6**, 1833 (1995).
[20] Carreira, E.M., Lee, W., Singer, R.A. *JACS* **117**, 3649 (1995); Singer, R.A., Carreira, E.M. *JACS* **117**, 12360 (1995).
[21] Kiyooka, S.-I., Hena, M.A. *TA* **7**, 2181 (1996).
[22] Denmark, S.E., Winter, S.B.D., Su, X., Wong, K.-T. *JACS* **118**, 7404 (1996).
[23] Ghosh, A.K., Onishi, M. *JACS* **118**, 2527 (1996).
[24] Rein, T., Kreuder, R., von Zezschwitz, P., Wulff, C., Reiser, O. *ACIEE* **34**, 1023 (1995).
[24a] Soloshonok, V.A., Avilov, D.V., Kukhar, V.P., Tararov, V.I., Savel'eva, T.F., Churkina, T.D., Ikonnikov, N.S., Kochetkov, K.A., Orlova, S.A., Pysarevsky, A.P., Struchkov, Y.T., Raevsky, N.I., Belokon, Y.N. *TA* **6**, 1741 (1995).
[25] Oishi, T., Oguri, H., Hirama, M. *TA* **6**, 1241 (1995).
[26] Ohkata, K., Kimura, J., Shinohara, Y., Tagaki, R., Hiraga, Y. *CC* 2411 (1996).
[27] Hamon, D.P.G., Massy-Westropp, R.A., Razzino, P. *T* **51**, 4183 (1995).
[28] Sibi, M.P., Ji, J. *ACIEE* **35**, 190 (1996).
[29] Zhu, G., Terry, M., Zhang, X. *TL* **37**, 4475 (1996).
[30] Koning, B., Hulst, R., Kellogg, R.M. *RTC* **115**, 49 (1996).
[31] Rieck, H., Helmchen, G. *ACIEE* **34**, 2687 (1995).
[32] Eichelmann, H., Gais, H.-J. *TA* **6**, 643 (1995).
[33] Trost, B.M., Lee, C.B., Weiss, J.M. *JACS* **117**, 7247 (1995).
[34] Trost, B.M., Organ, M.G., O'Doherty, G.A. *JACS* **117**, 9662 (1995); Trost, B.M., Krische, M.J., Radinov, R., Zanoni, G. *JACS* **118**, 6297 (1996).

[35] Kubota, H., Koga, K. *H* **42**, 543 (1996).
[36] Trost, B.M., Shi, Z. *JACS* **118**, 3037 (1996).
[37] Llyod-Jones, G.C., Pfaltz, A. *ACIEE* **34**, 462 (1995).
[38] Yamazaki, A., Achiwa, K. *TA* **6**, 1021 (1995).
[39] Cole, B.M., Shimizu, K.D., Krueger, C.A., Harrity, J.P.A., Snapper, M.L., Hoveyda, A.H. *ACIEE* **35**, 1668 (1996).
[40] Hayashi, M., Ono, K., Hoshimi, H., Oguni, N. *T* **52**, 7817 (1996).
[41] Martinez, L.E., Leighton, J.L., Carsten, D.H., Jacobsen, E.N. *JACS* **117**, 5897 (1995).
[42] Myers, A.G., McKinstry, L. *JOC* **61**, 2428 (1996).
[43] Davies, S.G., Sanganee, H.J. *TA* **6**, 671 (1995).
[44] Page, P.C.B., McKenzie, M.J., Buckle, D.R. *JCS(P1)* 2673 (1995).
[45] Voigt, K., Stolle, A., Salaün, J., de Meijere, A. *SL* 226 (1995).
[46] Wu, J.H., Radinov, R., Porter, N.A. *JACS* **117**, 11029 (1995).
[47] Barrett, A.G.M., Kamimura, A. *CC* 1755 (1995).
[48] Andrews, M.D., Brewster, A.G., Moloney, M.G., Owen, K.L. *JCS(P1)* 227 (1996).
[49] Enders, D., Grobner, R., Raabe, G., Runsink, J. *S* 941 (1996).
[50] Enders, D., Berg, T. *SL* 796 (1996).
[51] Weisenburger, G.A., Beak, P. *JACS* **118**, 12218 (1996).
[52] Eddine, J.J., Cherqaoui, M. *TA* **6**, 1225 (1995).
[53] Yasuda, K., Shindo, M., Koga, K. *TL* **37**, 6343 (1996).
[54] Davies, S.G., McCarthy, T.D. *SL* 700 (1995).
[55] Axon, J.R., Beckwith, A.L.J. *CC* 549 (1995).
[56] Shapiro, G., Buechler, D., Marzi, M., Schmidt, K., Gomez-Lor, B. *JOC* **60**, 4978 (1995).
[57] Enders, D., Wahl, H., Papadopoulos, K. *LA* 1177 (1995).
[58] Nakamura, M., Arai, M., Nakamura, E. *JACS* **117**, 1179 (1995).
[59] Enders, D., Breuer, K., Runsink, J., Teles, J.H. *HCA* **79**, 1899 (1996).
[60] Enders, D., Breuer, K., Teles, J.H. *HCA* **79**, 1217 (1996).
[61] Beaulieu, F., Arora, J., Veith, U., Taylor, N.J., Chapell, B.J., Snieckus, V. *JACS* **118**, 8727 (1996).
[62] Greeves, N., Pease, J.E. *TL* **37**, 5821 (1996).
[63] Vettel, S., Lutz, C., Knochel, P. *SL* 731 (1996).
[64] Lutjens, H., Nowotny, S., Knochel, P. *TA* **6**, 2675 (1995).
[65] Zhang, X., Guo, C. *TL* **36**, 4947 (1995).
[66] Soai, K., Inoue, Y., Takahashi, T., Shibata, T. *T* **52**, 13355 (1996).
[67] Soai, K., Ohno, Y., Inoue, Y., Tsuruoka, T., Hirose, Y. *RTC* **114**, 145 (1995).
[68] Dai, W.-M., Zhu, H.J., Hao, X.-J. *TA* **6**, 1857 (1995).
[69] Fukuzawa, S.-I., Tsudzuki, K. *TA* **6**, 1039 (1995).
[70] Zhang, H., Xue, F., Mak, T.C.W., Chan, K.S. *JOC* **61**, 8002 (1996).
[71] Braun, M., Vonderhagen, A., Waldmüller, D. *LA* 1447 (1995).
[72] Andres, J.M., Martinez, M.A., Pedrosa, R., Perez-Encabo, A. *S* 1070 (1996).
[73] Nishida, M., Tozawa, T., Yamada, K., Mukaiyama, T. *CL* 1125 (1996).
[74] Kobayashi, S., Nishio, K. *TL* **36**, 6729 (1995).
[75] Kadota, I., Kobayashi, K., Okano, H., Asao, N., Yamamoto, Y. *BSCF* **132**, 615 (1995).
[76] Jayaraman, S., Hu, S., Oehlschlager, A.C. *TL* **36**, 4765 (1995).
[77] Kulkarni, S.V., Brown, H.C. *TL* **37**, 4125 (1996).
[78] Enders, D., Reinhold, U. *ACIEE* **34**, 1219 (1995).
[79] Hashimoto, Y., Takaoki, K., Sudo, A., Ogasawara, T., Saigo, K. *CL* 235 (1995).
[80] Nakamura, M., Hirai, A., Nakamura, E. *JACS* **118**, 8489 (1996).
[81] Bolm, C., Müller, P. *TL* **36**, 1625 (1995).
[82] RajanBabu, T.V., Casalnuovo, A.L. *JACS* **118**, 6325 (1996).
[83] Iyer, M.S., Gigstad, K.M., Namdev, N.D., Lipton, M. *JACS* **118**, 4910 (1996).

[84] Charette, A.B., Juteau, H., Lebel, H., Deschenes, D. *TL* **37**, 7925 (1996); Barrett, A.G.M., Hamprecht, D., White, A.J.P., Williams, D.J. *JACS* **118**, 7863 (1996).
[85] Takahashi, H., Yoshioka, M., Shibasaki, M., Ohno, M., Imai, N., Kobayashi, S. *T* **51**, 12013 (1995).
[85a] Denmark, S.E., Christenson, B.L., O'Connor, S.P. *TL* **36**, 2219 (1995).
[86] Watanabe, N., Matsuda, H., Kuribayashi, H., Hashimoto, S. *H* **42**, 537 (1996).
[87] Davies, H.M., Bruzinski, P.R., Lake, D.H., Kong, N., Fall, M.J. *JACS* **118**, 6897 (1996).
[88] Doyle, M.P., Kalinin, A.V. *JOC* **61**, 2179 (1996).
[89] Pique, C., Fahndrich, B., Pfaltz, A. *SL* 491 (1995).
[90] Hansen, K.B., Finney, N.S., Jacobsen, E.N. *ACIEE* **34**, 676 (1995).
[91] Aggarwal, V.K., Ford, J.G., Thompson, A., Jones, R.V.H., Standen, M.C.H. *JACS* **118**, 7004 (1996).
[92] Solladie-Cavallo, A., Diep-Vohuule, A., Sunjic, V., Vinkovic, V. *TA* **7**, 1783 (1996).
[93] Aggarwal, V.K., Thompson, A., Jones, R.V.H., Standen, M.C.H. *JOC* **61**, 8368 (1996).
[94] Srirajan, V., Puranik, V.G., Deshmukh, A.R.A.S., Bhawal, B.M. *T* **52**, 5579 (1996).
[95] Gothelf, K.V., Thomsen, I., Jorgensen, K.A. *JACS* **118**, 59 (1996).
[96] Hayashi, Y., Rohde, J.J., Corey, E.J. *JACS* **118**, 5502 (1996).
[97] Corey, E.J., Letavic, M.A. *JACS* **117**, 9616 (1995).
[98] Ishihara, K., Kurihara, H., Yamamoto, H. *JACS* **118**, 3049 (1996).
[99] Hicks, F.A., Buchwald, S.L. *JACS* **118**, 11688 (1996).
[100] Brandes, B.D., Jacobsen, E.N. *TL* **36**, 5123 (1995).
[101] Palucki, M., McCormick, G.J., Jacobsen, E.N. *TL* **36**, 5457 (1995).
[102] Yang, D., Yip, Y.-C., Tang, M.-W., Wong, M.-K., Zheng, J.-H., Cheung, K.-K. *JACS* **118**, 491 (1996); Yang, D., Wang, X.C., Wong, M.-K., Yip, Y.-C., Tang, M.W. *JACS* **118**, 11311 (1996).
[103] Tu, Y., Wang, Z.-X., Shi, Y. *JACS* **118**, 9806 (1996).
[104] Lasterra-Sanchez, M.E., Felfer, U., Mayon, P., Roberts, S.M., Thornton, S.R., Todd, C.J. *JCS(P1)* 343 (1996).
[105] Kroutil, W., Lasterra-Sanchez, M.E., Maddrell, S.J., Mayon, P., Morgan, P., Roberts, S.M., Thornton, S.R., Todd, C.J. *JCS(P1)* 2837 (1996).
[106] Kroutil, W., Mayon, P., Lasterra-Sanchez, M.E., Maddrell, S.J., Roberts, S.M., Thornton, S.R., Todd, C.J., Tüter, M. *CC* 845 (1996).
[107] Enders, D., Zhu, J., Raabe, G. *ACIEE* **35**, 1725 (1996).
[108] Page, P.C.B., Gareh, M.T., Porter, R.A. *TA* **4**, 2139 (1993); Brunel, J.-M., Diter, P., Duetsch, M., Kagan, H.B. *JOC* **60**, 8086 (1995); Scettri, A., Bonadies, F., Lattanzi, A. *TA* **7**, 629 (1996).
[109] Schwan, A.L., Pippert, M.F. *TA* **6**, 131 (1995).
[110] Page, P.C.B., Heer, J.P., Bethell, D., Collington, E.W., Andrews, D.M. *TA* **6**, 2911 (1995).
[111] Page, P.C.B., Heer, J.P., Bethell, D., Collington, E.W., Andrews, D.M. *SL* 773 (1995).
[112] Imagawa, K., Nagata, T., Yamada, T., Mukaiyama, T. *CL* 335 (1995).
[113] Hirano, M., Yakabe, S., Clark, J.H., Kudo, H., Morimoto, T. *SC* **26**, 1875 (1996).
[114] Kokubo, C., Katsuki, T. *T* **52**, 13895 (1996).
[115] Bolm, C., Bienewald, F. *ACIEE* **34**, 2640 (1995).
[116] Di Furia, F., Licini, G., Modena, G., Motterle, R., Nugent, W.A. *JOC* **61**, 5175 (1996).
[117] Rispens, M.T., Zondervan, C., Feringa, B.L. *TA* **6**, 661 (1995).
[118] Levina, A., Muzart, J. *SC* **25**, 1789 (1995).
[119] Gokhale, A.S., Minidis, A.B.E., Pfaltz, A. *TL* **36**, 1831 (1995).
[120] Gupta, A.D., Singh, V.K. *TL* **37**, 2633 (1996).
[121] Zhang, F.-Y., Yip, C.W., Chan, A.S.C. *TA* **7**, 3135 (1996).
[122] Zhang, Y.-W., Shen, Z.-X., Liu, C.-L., Chen, W.-Y. *SC* **25**, 3407 (1995).
[123] Peper, V., Martens, J. *CB* **129**, 691 (1996).
[124] Zhang, Y.-W., Shen, Z.-X., Gu, D.-B., Chen, W.-Y., Fei, Z.-H., Dai, Q.-F. *SC* **26**, 4415 (1996).
[125] Prasad, K.R.K., Joshi, N.N. *TA* **7**, 3147 (1996).

[126] Parker, K.A., Ledeboer, M.W. *JOC* **61**, 3214 (1996).
[127] Bach, J., Berenguer, R., Garcia, J., Loscertales, T., Vilarrasa, J. *JOC* **61**, 9021 (1996).
[128] Helal, C.J., Magriotis, P.A., Corey, E.J. *JACS* **118**, 10938 (1996).
[129] Franot, C., Stone, G.B., Engeli, P., Spöndlin, C., Waldvogel, E. *TA* **6**, 2755 (1995).
[130] Bach, J., Berenguer, R., Farras, J., Garcia, J., Meseguer, J., Vilarrasa, J. *TA* **6**, 2683 (1995).
[131] Masui, M., Shioiri, T. *SL* 49 (1996).
[132] Chiodi, O., Fotiadu, F., Sylvestre, M., Buono, G. *TL* **37**, 39 (1996).
[133] Brunel, J.M., Buono, G. *SL* 177 (1996).
[134] Giffels, G., Dreisbach, C., Kragl, U., Weigerding, M., Waldmann, H., Wandrey, C. *ACIEE* **34**, 2005 (1995).
[135] Shieh, W.-C., Cantrell, W.R., Carlson, J.A. *TL* **36**, 3797 (1995).
[136] Nagata, T., Yorozu, K., Yamada, T., Mukaiyama, T. *ACIEE* **34**, 2145 (1995).
[137] Cherng, Y.-J., Fang, J.-M., Lu, T.-J. *TA* **6**, 89 (1995).
[138] Ghosh, A.K., Chen, Y. *TL* **36**, 6811 (1995).
[139] Dube, D., Deschenes, D., Tweddell, J., Gagnon, H., Carlini, R. *TL* **36**, 1827 (1995).
[140] Byun, I.S., Kim, Y.H. *SC* **25**, 1963 (1995).
[141] Langer, T., Janssen, J., Helmchen, G. *TA* **7**, 1599 (1996).
[142] Nishibayashi, Y., Segawa, K., Ohe, K., Uemura, S. *OM* **14**, 5486 (1995).
[143] Nishibayashi, Y., Segawa, K., Takada, H., Ohe, K., Uemura, S. *CC* 847 (1996).
[144] Sawamura, M., Kuwano, R., Shirai, J., Ito, Y. *SL* 347 (1995).
[145] Langer, T., Helmchen, G. *TL* **37**, 1381 (1996).
[146] Verdaguer, X., Lange, U.E.W., Reding, M.T., Buchwald, S.L. *JACS* **118**, 6784 (1996).
[147] Ohkuma, T., Ooka, H., Hashiguchi, S., Ikariya, T., Noyori, R. *JACS* **117**, 2675 (1995).
[148] Gamez, P., Fache, F., Lemaire, M. *TA* **6**, 705 (1995).
[149] Burk, M.J., Harper, T.G.P., Kalberg, C.S. *JACS* **117**, 4423 (1995).
[150] Gautier, I., Ratouelomanama-Vidal, V., Savignac, P., Genet, J.P. *TL* **37**, 7721 (1996).
[151] Burk, M.J., Feng, S., Gross, M.F., Tumas, W. *JACS* **117**, 8277 (1995).
[152] Sawamura, M., Kuwano, R., Ito, Y. *JACS* **117**, 9602 (1995).
[153] Uematsu, N., Fujii, A., Hashiguchi, S., Ikariya, T., Noyori, R. *JACS* **118**, 4916 (1996).
[154] Burk, M.J., Wang, Y.M., Lee, J.R. *JACS* **118**, 5142 (1996).
[155] Fujii, A., Hashiguchi, S., Uematsu, N., Ikariya, T., Noyori, R. *JACS* **118**, 2521 (1996).
[156] Corey, E.J., Roberts, B.E., Dixon, B.R. *JACS* **117**, 193 (1995).
[157] Corey, E.J., Kania, R.S. *JACS* **118**, 1229 (1996).

Chlorobis(cyclopentadienyl)(dimethylaluminum)methylenetitanium (Tebbe reagent).

Cyclic ethers.[1] Alkenyl esters undergo deoxygenative cyclization, probably via an intramolecular metathesis of the derived enol ethers. This method is applicable to the elaboration of complex systems characteristic of several marine neural toxins that present formidable synthetic challenges.

BnO BnO O O O O → Cp_2Ti Al Cl, THF → BnO BnO O O O

64%

Ph ... Cp_2Ti ... Al / Cl; THF → Ph ... OH 71%

[1]Nicolaou, K.C., Postema, M.H.D., Claiborne, C.F. *JACS* **118**, 1565 (1996).

***m*-Chloroperoxybenzoic acid. 13**, 76–79; **14**, 84–87; **15**, 86; **16**, 80–83; **17**, 76; **18**, 101

Epoxidations. Whereas hydroxyl group direction of epoxidation is well known, the previously unsuspected direction by a ketone side chain in the epoxidation of cycloalkenes has been authenticated.[1]

3-ClC_6H_4COOOH; $NaHCO_3$ / CH_2Cl_2 → 61%

For the epoxidation of alkenes sensitive either to acids alone or to both acids and bases, the presence of 2,6-di-*t*-butylpyridine is advantageous.[2] Liposomized MCPBA is useful for enantioselective epoxidation (ee 62–95%).[3]

The preparation of 3-aryl-*N*-(*t*-butoxycarbonyl)oxaziridines is readily accomplished by reaction of the *N*-Boc-imines with lithium *m*-chloroperbenzoate.[4]

Baeyer–Villiger oxidation. β-Formyl-β-lactams give formate esters instead of being oxidized to acids.[5] Zeolites catalyze the Baeyer–Villiger oxidation of ketones.[6]

CHO ... N ... OMe; 3-ClC_6H_4COOOH, hydrotalcite, $ClCH_2CH_2Cl$ → OCHO ... N ... OMe 91%

Trithiocarbonate oxides.[7] These species (sulfines) are obtained on treatment of dialkyl trithiocarbonates with MCPBA at 0°. They react with organometallic reagents to provide bis(alkylthio)methyl sulfoxides.

S
MeS SMe
3-ClC$_6$H$_4$COOOH
CH$_2$Cl$_2$ 0°
O
S
MeS SMe
73%

1,3-Amino alcohols.[8] A two-step procedure for converting isoxazolidine to 1,3-amino alcohols involves oxidation with MCPBA and hydrogenation (4 examples, 78–88%).

[1]Armstrong, A., Barsanti, P.A., Clarke, P.A., Wood, A. *JCS(P1)* 1373 (1996).
[2]Svensson, A., Lindstrom, U.M., Somfai, P. *SC* **26**, 2875 (1996).
[3]Kumar, A., Bhakuni, V. *TL* **37**, 4751 (1996).
[4]Vidal, J., Damestoy, S., Collet, A. *TL* **36**, 1439 (1995).
[5]Alcaide, B., Aly, M.F., Sierra, M.A. *JOC* **61**, 8819 (1996).
[6]Kaneda, K., Yamashita, T. *TL* **37**, 4555 (1996).
[7]Leriverend, C., Metzner, P., Capperucci, A., Degl'innocenti, A. *T* **53**, 1323 (1997).
[8]Broggini, G., Zecchi, G. *S* 1280 (1996).

N-(5-Chloro-2-pyridyl)triflimide.

6-Substituted 1,2,3,4-tetrahydropyridines.[1] Enol triflates are readily prepared from *N*-alkoxycarbonyl-2-piperidones by successive treatment with LHMDS and reagent (**1**). The triflyloxy group can be replaced on reaction with organocuprates and carbonylation; the product from the latter reaction is a precursor of pipecolic acid.

N O
COOPh
LiN(SiMe$_3$)$_2$ / THF
Cl
N NTf$_2$
(**1**)
N OTf
COOPh
90%

[1]Foti, C.J., Comins, D.L. *JOC* **60**, 2656 (1995).

***N*-Chlorosuccinimide. 13**, 79–80; **15**, 86–88; **18**, 101–102

Lactones.[1] *N*-Haloamides, including NCS, oxidize diols to lactones at room temperature. However,the scope of this reaction has to be established further.

Chlorination of 2-(N-pyridinium)aminopyridines.[2] The ylide substituent increases the nucleophilicity of the pyridine ring so that chlorination occurs at –20°. The charged pyridinium moiety is removed by Zn–HOAc.

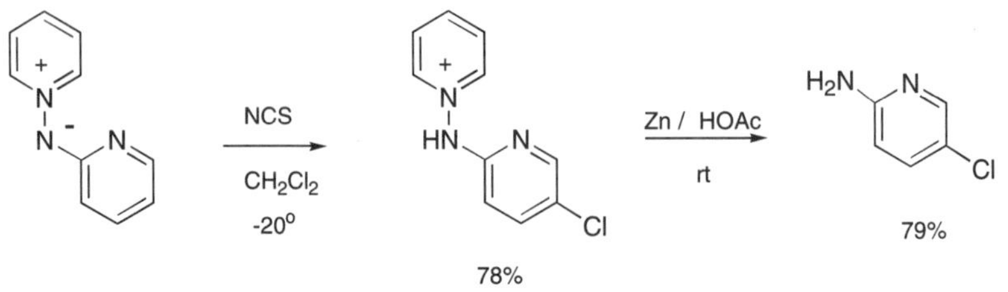

NCS, CH_2Cl_2, -20° → 78% ; Zn / HOAc, rt → 79%

1,5-Diaryl-2,4-pentanediones.[3] These diones are available from isoxazolines via the β-hydroxy ketones. Thus, treatment of the latter compounds with NCS, Me_2S, and Et_3N in dichloromethane at –78° furnishes the stabilized sulfonium ylides, which are readily desulfurized with Zn–HOAc.

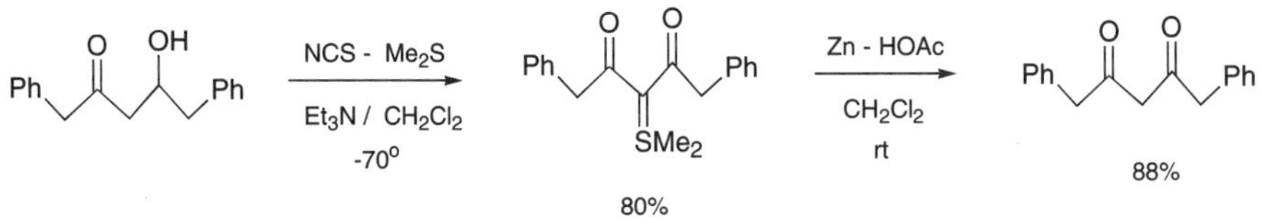

NCS - Me_2S, Et_3N / CH_2Cl_2, -70° → SMe_2 80% ; Zn - HOAc, CH_2Cl_2, rt → 88%

Aldehydes.[4] Oxidation of primary alcohols is readily achieved under phase transfer conditions (Bu_4NCl, H_2O–CH_2Cl_2) using NCS and a catalytic amount of 2,2,6,6-tetramethyl-1-piperidinyloxy (18 examples, 83–100%).

[1]Kondo, S., Kawasoe, S., Kunisada, H., Yuki, Y. *SC* **25**, 719 (1995).
[2]Burgos, C., Delgado, F., Garcia-Navio, J.L., Izquierdo, M.L., Alvarez-Builla, J. *T* **51**, 8649 (1995).
[3]Pulkkinen, J.T., Vepsalainen, J.J. *JOC* **61**, 8604 (1996).
[4]Einhorn, J., Einhorn, C., Ratajczak, F., Pierre, J.-L. *JOC* **61**, 7452 (1996).

Chlorotris(triphenylphosphine)rhodium(I).

Isomerization of allyl ethers.[1] The reaction of Wilkinson catalyst with BuLi gives a catalyst for the isomerization of allyl ethers to enol ethers. Accordingly, this is another option for the deprotection of alcohols from allyloxy derivatives.

Hydroboration. The Rh(I) complex facilitates the reactions of the less reactive catecholborane[2] and pinacolborane.[3] Thus, 1-haloalkylboranes[2] and alkenylboranes[4] become readily accessible.

Alkenylsilanes.[5] In a dehydrogenative silylation reaction, a silyl group is introduced to the terminal carbon atom of 1,5-dienes using various hydrosilanes. The major product has an *(E)*-configuration.

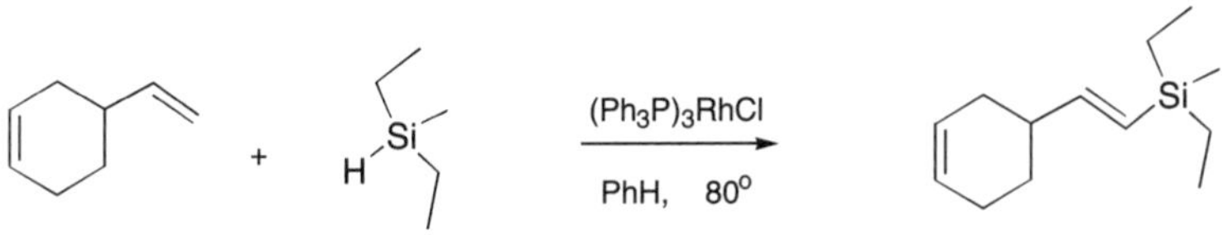

+ H Si $(Ph_3P)_3RhCl$ PhH, 80° Si

90%

Homologous pericyclic reactions. The verisimilitude of a cyclopropane ring to a double bond is manifested in certain reactions. However, some of these reactions need activation by transition metal species. Formal homologous Diels–Alder reactions[6] and electrocyclic reactions[7] leading to seven-membered ring compounds have been discovered.

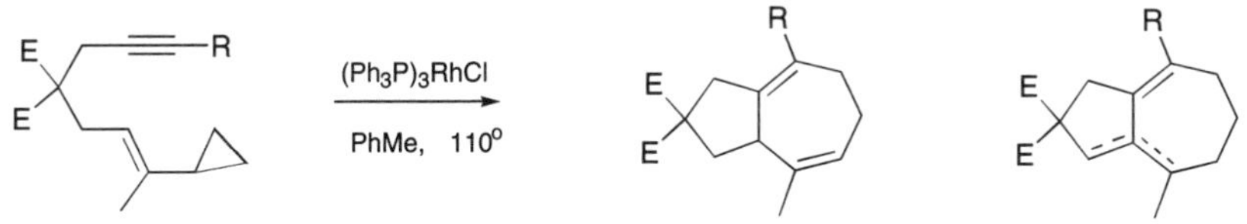

E E R $(Ph_3P)_3RhCl$ PhMe, 110° R E E R E E

E = COOMe

R = H, Me 81 - 82%

R = $SiMe_3$ - 71%

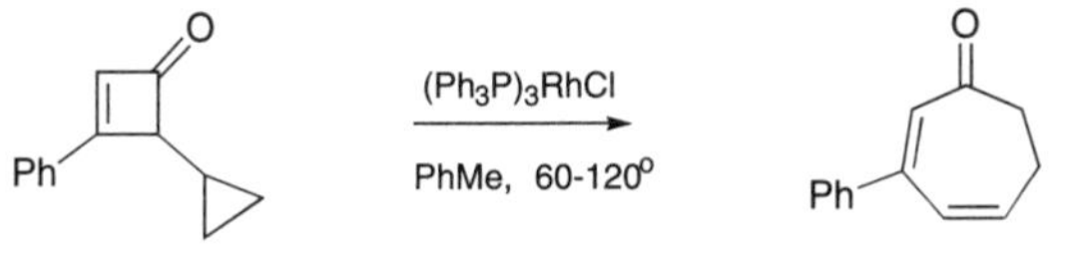

O Ph $(Ph_3P)_3RhCl$ PhMe, 60-120° O Ph

84%

Cycloisomerizations. 2-Substituted pyridines in which the side chain contains two double bonds undergo a catalyzed cyclization.[8] Enediynes give aromatic products in an analogous process involving $(i\text{-}Pr_3P)_2RhCl$.[9]

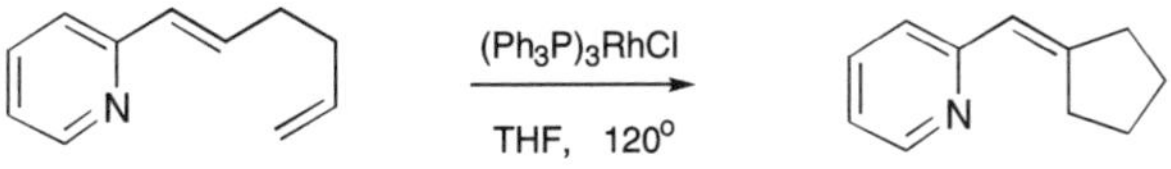

N $(Ph_3P)_3RhCl$ THF, 120° N

93%

S N $(Ph_3P)_3RhCl$ THF, 120° S N

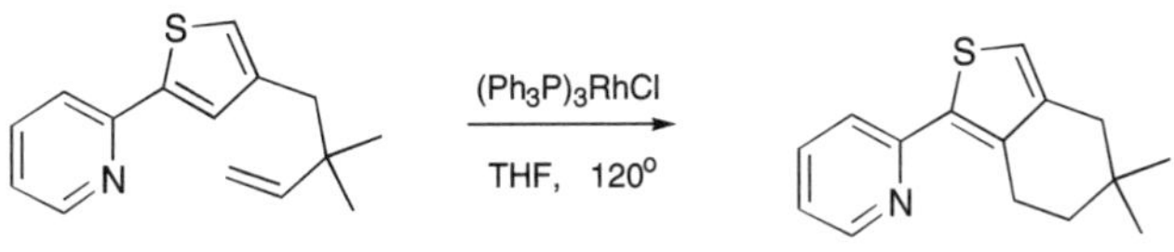

79%

C_6H_{13}
(i-$Pr_3P)_2RhCl$
Et_3N / PhH
50°, 20 h
58% (E:Z 86 : 14)

***C*-Aryl spiroglycosides.**[10] [2+2+2]Cycloaromatization of alkynes catalyzed by $(Ph_3P)_3RhCl$ as applied to anomeric alkynyl propargyl glycosides leads to *C*-aryl spiroglycosides, which have the basic framework of papulacandin-D. A related strategy is apparently serviceable for synthesizing the more common *C*-arylglycosides.

BnO BnO BnO BnO O O
$(Ph_3P)_3RhCl$
≡ / EtOH
0°
BnO BnO BnO BnO O O
89%
HO HO HO RCOO HO O O OH
papulacandin-D

[1]Boons, G.J., Burton, A., Isles, S. *CC* 141 (1996).
[2]Elgendy, S., Patel, G., Kakkar, V.V., Claeson, G., Green, D., Skordalakes, E., Baban, J.A., Beadman, J. *TL* **35**, 2435 (1994).
[3]Pereira, S., Srebnik, M. *JACS* **118**, 909 (1996).
[4]Pereira, S., Srebnik, M. *TL* **37**, 3283 (1996).
[5]Kakiuchi, F., Nogami, K., Chatani, N., Seki, Y., Murai, S. *OM* **12**, 4748 (1993).
[6]Wender, P.A., Takahashi, H., Witulski, B. *JACS* **117**, 4720 (1995).
[7]Huffman, M.A., Liebeskind, L.S. *JACS* **115**, 4895 (1993).
[8]Fujii, N., Kakiuchi, F., Chatani, N., Murai, S. *CL* 939 (1996).
[9]Ohe, K., Kojima, M.-a., Yonehara, K., Uemura, S. *ACIEE* **35**, 1823 (1996).
[10]McDonald, F.E., Zhu, H.Y.H., Holmquist, C.R. *JACS* **115**, 6605 (1993).

Chromium-carbene complexes. 13, 82–83; 14, 91–93; 15, 93–95; 16, 88–92; 17, 80–84; 18, 103–104

Acylations. When Fischer carbene complexes are reacted with organolithiums and then decomposed with iodine, α-alkoxy enones are formed.[1] On the other hand, hydroxy Fischer carbene complexes undergo carbonylation under photochemical conditions, resulting in α-hydroxy esters.[2]

Me_4NO R =$Cr(CO)_5$
hν
CO / HCl - MeOH
-50° -> rt, 20 h
HO R COOMe

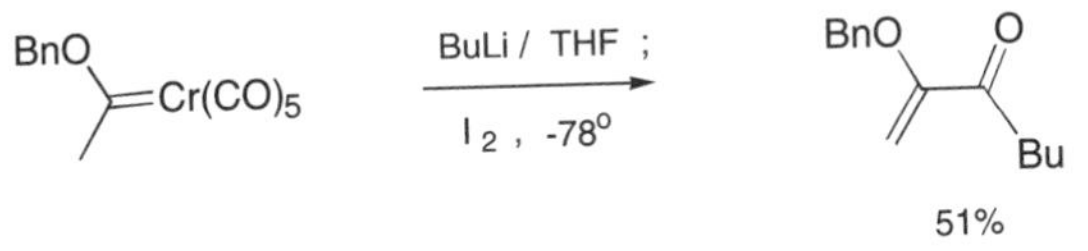

BnO, Cr(CO)5 —BuLi / THF ; I_2, -78°→ BnO, O, Bu 51%

1,3-Dienes. The reaction of Fischer carbene complexes with allenes gives 1,3-dienes into which the nonmetal moiety is incorporated.[3] The transformation is a formal ene-type process. 1-Alkoxy-1,3-dienes arise when the carbene complexes and propargylsilanes are heated together.[4] A 1,2-silicon shift accompanies the carbene insertion into the C–H bond.

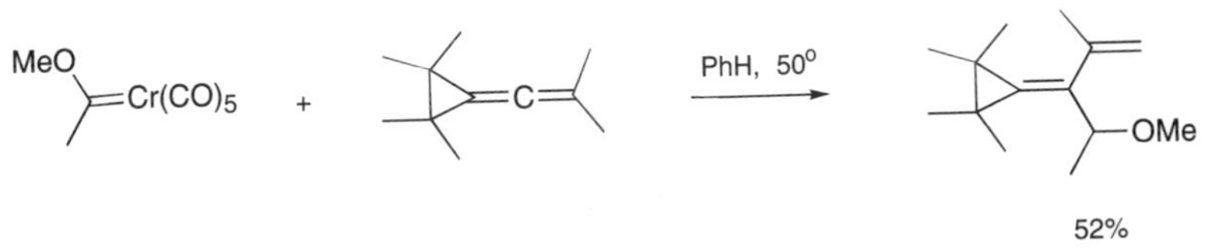

MeO, Cr(CO)5 + C= —PhH, 50°→ OMe 52%

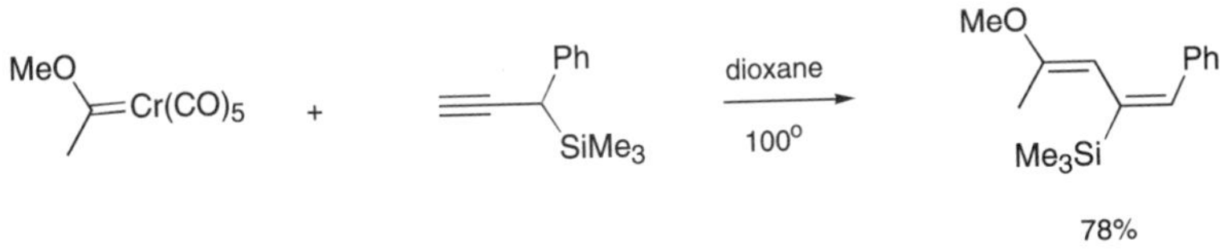

MeO, Cr(CO)5 + Ph, SiMe3 —dioxane, 100°→ MeO, Ph, Me3Si 78%

Carbene relay. The carbene center is translocated upon insertion into an alkyne C–H bond. A remote double bond can then participate in the further reaction, forming a cyclopropane unit.[5]

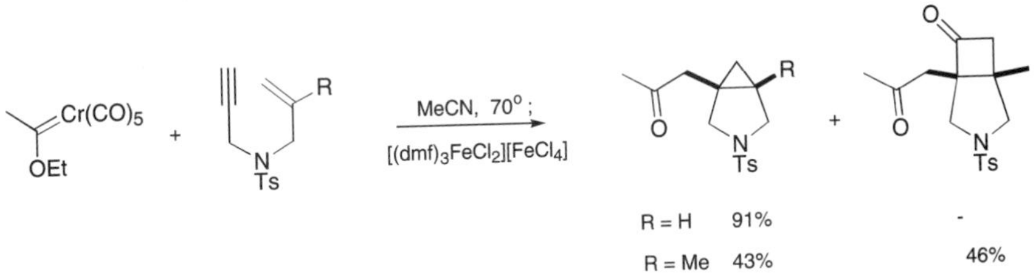

Cr(CO)5, OEt + R, N, Ts —MeCN, 70°; [(dmf)3FeCl2][FeCl4]→ O, R, N, Ts + O, O, N, Ts

R = H 91% -

R = Me 43% 46%

Cycloadditions. Alkynylcarbene complexes furnish three-carbon units for [3+2]-cycloadditions. The C_2 component can be an imine[6] or an enamine.[7]

Ph — C≡C — C(OEt)=Cr(CO)5 + 1-(dimethylamino)cyclopentene (NMe2) —[CH_2Cl_2, 25°]→ bicyclic product (Ph, OEt, NMe2) 73%

With electron-rich dienes, the cycloaddition gives seven-membered ring compounds,[8] including azepines.[9]

Isoindolines and isoquinolines.[10] α,ω-Diynes in which the connecting chains contain a nitrogen atom undergo [2+2+2]-cycloaddition with ketene equivalents derived from the carbene complexes. After the formation of a benzene ring, the products are isoindoline or isoquinoline derivatives.

5-Hydroxyindolines.[11] Bisalkynyl carbene complexes undergo thermolysis in the presence of a hydrogen source (1,4-cyclohexadiene, Et_3SiH), providing 5-hydroxyindolines.

Ph — C≡C — C(=Cr(CO)5) — N(Me) — CH2CH2C≡C — Me —[Et_3SiH, THF, 85°]→ Et_3SiO, Ph, Cr(CO)3 indoline complex 41%

Pyridine synthesis.[12] 1-Aza-1,3-butadienes are stabilized by forming the chromium carbonyl complexes. Thus, they can be used in hetero Diels–Alder reactions. In the reaction shown below the yield can be increased to 77% by using an in situ variation.

(OC)5Cr, EtO, N–H aza-diene complex + 1-pentyne —[hexane, 50°]→ 2,5-dipropylpyridine (N) 21%

[1]Wieber, G.M., Hegedus, L.S., Gale, C. *OM* **14**, 3574 (1995).
[2]Soderberg, B.C., Odens, H.H. *OM* **15**, 5080 (1996).
[3]Hwu, C.-C., Wang, F.-C., Yeh, M.-C.P., Sheu, J.-H. *JOMC* **474**, 123 (1994).
[4]Herndon, J.W., Patel, P.P. *JOC* **61**, 4500 (1996).
[5]Watanuki, S., Mori, M. *OM* **14**, 5054 (1995).

[6]Funke, F., Duetsch, M., Stein, F., Noltemeyer, M., de Meijere, A. *CB* **127**, 911 (1994).
[7]Aumann, R., Meyer, A.G., Fröhlich, R. *OM* **15**, 5018 (1996).
[8]Barluenga, J., Aznar, F., Martin, A., Vazquez, J.T. *JACS* **117**, 9419 (1995).
[9]Barluenga, J., Tomas, M., Ballesteros, A., Santamaria, J., Lopez-Ortiz, F. *CC* 321 (1994).
[10]Mori, M., Kuriyama, K., Ochifuji, N., Watanuki, S. *CL* 615 (1995).
[11]Rahm, A., Wulff, W.D. *JACS* **118**, 1807 (1996).
[12]Duetsch, M., Stein, F., Funke, F., Pohl, E., Herbst-Irmer, R., de Meijere, A. *CB* **126**, 2535 (1993).

Chromium(II) chloride. 13, 84; **14**, 94–97; **15**, 95–96; **16**, 93–94; **17**, 84–85; **18**, 104

Alkenylstannanes and alkenylboranes. $CrCl_2$ mediates the reaction of dibromomethylstannanes[1] and of dichloromethylboranes[2] with aldehydes, each giving functionalized *(E)*-alkenes.

Homoallylic alcohols. The reaction of carbonyl compounds with allylic halides in the presence of $CrCl_2$ shows a difference in stereoselectivity from that with allylic phosphates.[3] Stereoisomers with opposite configuration at the quarternary center can be obtained by using either of the two (*E*)- or (*Z*)- allylic precursors.

OPO(OEt)2 + $C_5H_{11}CHO$ —($CrCl_2$ - LiCl, DMPU, rt)→ OH, C_5H_{11} 94%

OPO(OEt)2 + $C_5H_{11}CHO$ —($CrCl_2$ - LiCl, DMPU, rt)→ OH, C_5H_{11} 93%

[1]Hodgson, D.M., Boulton, L.T., Maw, G.N. *T* **51**, 3713 (1995).
[2]Takai, K., Shinomiya, N., Kaihara, H., Yoshida, N., Moriwake, T. *SL* 963 (1995).
[3]Nowotny, S., Tucker, C.E., Jubert, C., Knochel, P. *JOC* **60**, 2762 (1995).

Chromium(III) chloride.

ArZnX → ArCH(R)OH.[1] The addition of arylzincs to aldehydes is mediated by $CrCl_3$. The reaction is benefited by the presence of Me_3SiCl; various aldehydes react well.

[1]Ogawa, Y., Mori, M., Saiga, A., Takagi, K. *CL* 1069 (1996).

Chromium(III) chloride–lithium aluminum hydride.

Homoallylic alcohols.[1,2] 1,4-Asymmetric induction is observed when the organometallic reagent bears an allylic oxygen. The reaction is also promoted by indium powder.

[1]Maguire, R.J., Mulzer, J., Bats, J.W. *JOC* **61**, 6936 (1996).
[2]Maguire, R.J., Mulzer, J., Bats, J.W. *TL* **37**, 5487 (1996).

Chromium(II) chloride—nickel(II) chloride. 14, 97–98; **15**, 96–97; **17**, 86; **18**, 105

Chiral allylic alcohols.[1] An asymmetric version of the coupling reaction in the presence of a chiral bipyridine (**1**) is demonstrated.

[1]Chen, C., Tagami, K., Kishi, Y. *JOC* **60**, 5386 (1995).

Chromium(VI) oxide–silica gel.

Oxidations.[1] Prepared by grinding anhydrous CrO_3 and SiO_2 together and activating the mixture at 100° for 4.5 h, this reagent is used for the selective oxidation of alcohols.

[1]Khadilkar, B., Chitnavis, A., Khare, A. *SC* **26**, 205 (1996).

Chromium(VI) oxide–sulfuric acid.

2-Aryl-1,3-Diketones.[1] Jones oxidation at –20° is the only useful procedure for oxidation of the 1,3-diols to give the corresponding 1,3-diketones. Other methods, including those of Swern, Collins, Corey (PCC), and Dess-Martin, and by MnO_2 and TEMPO, are not successful.

[1]Khadilkar, B., Chitnavis, A., Khare, A. *SC* **25**, 1907 (1995).

Cobalt–aluminum.

H–D exchange.[1] The Co–Al alloy catalyzes H–D exchange in Na_2CO_3-D_2O with retention of optical purity (7 examples, 59–97%).

Co-Al)))) ; Na_2CO_3 / D_2O ; 60°, 2 h

77% (>99% ee)

[1]Mukumoto, M., Tsuzuki, H., Mataka, S., Tashiro, M., Tsukinoki, T., Nagano, Y *CL* 165 (1996).

Cobalt(II) acetylacetonate–carbon monoxide.

γ-Lactones.[1] Allylic alcohols undergo hydrocarbonylation and cyclization in the presence of pyridine and hydrogen under photosensitization conditions (with xanthone as sensitizer).

Cyclopentenones.[2] Catalytic amounts of $Co(acac)_2$ and $NaBH_4$ promote the Pauson–Khand reaction.

$Co(acac)_2$ - CO ; $NaBH_4$ / CH_2Cl_2 ; 80 - 100°

100%

[1]Chow, Y.L., Huang, Y.-J., Dragojlovic, V. *CJC* **73**, 740 (1995).
[2]Lee, N.Y., Chung, Y.K. *TL* **37**, 3145 (1996).

Cobalt(III) acetylacetonate–oxygen.

Benzylic oxidation.[1] Dihydroisobenzofuran and isochroman are readily converted to the lactones. Solvent effects are manifested in the cleavage of styrenes, since EtOAc favors the formation of aldehydes, whereas in THF, the major products are the benzoic acids.

$Co(acac)_3$ - O_2 ; 70°

	CHO	COOH
in EtOAc	95	5
in THF	14	86

Co(acac)$_3$ - O$_2$, EtOAc, 70°, 1 d; 98%

Oxidation of alcohols and alkanes. An oxidizing system constituting Co(acac)$_3$, oxygen, and *N*-hydroxyphthalimide is effective for converting alcohols to carbonyl compounds in refluxing acetonitrile (17 examples, 50–97%).[2]

Alkanes give ketones and carboxylic acids on heating with this reagent combination in HOAc.[3]

[1]Reetz, M.T., Tollner, K. *TL* **36**, 9461 (1995).
[2]Iwahama, T., Sakaguchi, S., Nishiyama, Y., Ishii, Y. *TL* **36**, 6923 (1995).
[3]Ishii, Y., Iwahama, T., Sakaguchi, S., Nakayama, K., Nishiyama, Y. *JOC* **61**, 4520 (1996).

Cobalt(II) bromide.

Ketones. The reaction of diorganozincs with acid chlorides in *N*-methylpyrrolidinone to give ketones[1] is catalyzed by CoBr$_2$. Symmetrical ketones are obtained when solutions of RZnI containing the catalyst are treated with CO (via bubbling).[2]

Cyclopentenones.[3] The formation of such compounds incorporating two molecules of 1-alkynes and one molecule of CO is effected. Usually, two isomers arise.

$C_{10}H_{21}$—≡ ; CoBr$_2$ - Zn - CO ; CF$_3$COOH, 80° ; 40% + 15%

[1]Reddy, C.K., Knochel, P. *ACIEE* **35**, 1700 (1996).
[2]Devasagayaraj, A., Knochel, P. *TL* **36**, 8411 (1995).
[3]Rao, M.L.N., Periasamy, M. *OM* **15**, 442 (1996).

Cobalt(II) chloride. 14, 99; **15**, 97–98; **18**, 107–108

Ritter reaction.[1] Allylic alcohols and acetates are converted to amides in a cobalt(II)-catalyzed reaction with nitriles.

OR / Ph — $CoCl_2$ - MeCN / AcOR', 80° → NHAc / Ph

R = H, R' = Ac 64%

R = Ac, R' = H 76%

Dialkyl ketones.[2] The reductive coupling of alkenes and carbon monoxide at room temperature is promoted by $CoCl_2$–Ph(Et)$_2$N:BH_3 (7 examples, 50–70%). The generation of $ClBH_2$ as hydroborating agent for the alkenes is implicated, and the subsequent carbonylation is assisted by cobalt carbonyl species formed in situ.

R — $CoCl_2$ - Ph(Et)$_2$N•BH_3 / CO / PhH, 25°; [O] → O / R R

50 - 70%

Trifluoroacetylation.[3] Cobalt(II) chloride is a mild Lewis acid that catalyzes the trifluoroacetylation of methoxyarenes with the use of trifluoroacetic anhydride.

[1]Mukhopadhyay, M., Reddy, M.M., Maikap, G.C., Iqbal, J. *JOC* **60**, 2670 (1995).
[2]Rao, M.L.N., Periasamy, M. *TL* **36**, 9069 (1995).
[3]Ruiz, J., Astruc, D., Gilbert, L. *TL* **37**, 4511 (1996).

Copper. **15**, 99; **16**, 95; **18**, 109

Radical additions.[1] Activated haloalkanes add to alkenes in solvent-free systems in a process initiated by electron transfer from copper.

COOEt / I + $(CH_2)_8COOMe$ — Cu / 130° → O / O / $(CH_2)_8COOMe$

70%

C-Allylation of phenols.[2] This reaction (*vide supra*), promoted by a mixture of Cu and $CuClO_4$, is extended to the *C*-allylation of phenols. The allylation is *ortho*-selective.

Hydrogenation.[3] Selective saturation of conjugated double bonds, such as that present in α-ionone, is accomplished with Cu/SiO_2 as catalyst at atmospheric pressure (11 examples, 69–100%).

[1]Metzger, J.O., Mahler, R. *ACIEE* **34**, 902 (1995).
[2]Baruah, J.B. *TL* **36**, 8509 (1995).
[3]Ravasio, N., Antenori, M., Gargano, M., Mastrorilli, P. *TL* **37**, 3529 (1996).

Copper(II) acetate. 18, 109–110

***N*-Arylation.** Using aryllead triacetates as reagents and $Cu(OAc)_2$ as catalyst, amides[1] and azoles[2] are *N*-arylated in a mixture of CH_2Cl_2 and DMF.

[1]Lopez-Alvarado, P., Avendano, C., Menendez, J.C. *JOC* **61**, 5865 (1996).
[2]Lopez-Alvarado, P., Avendano, C., Menendez, J.C. *JOC* **60**, 5678 (1995).

Copper(II) bromide. 14, 100; **15**, 100; **18**, 111

1,1-Dihalo-1-alkenes.[1] From 1-haloalkynes, through hydroboration and treatment of the resulting 1-boryl-1-haloalkenes with $CuBr_2$, a simple preparation is accomplished.

BH_2 ; $CuBr_2$ / $AcNMe_2$, THF - H_2O → 93%

Bromolactonization.[2] Copper(II) bromide deposited on alumina is suitable for bromolactonization of alkenoic acids in refluxing chloroform (6 examples, 40–98%).

[1]Masuda, Y., Suyama, T., Murata, M., Watanabe, S. *JCS(P1)* 2955 (1995).
[2]Rood, G.A., DeHann, J.M., Zibuck, R. *TL* **37**, 157 (1996).

Copper(II) bromide–lithium *t*-butoxide.

gem-Dihaloalkanes.[1] Hydrazones are converted into *gem*-dibromides on reaction with this reagent in THF at room temperature. The corresponding *gem*–dichlorides are obtained by changing the reagent to $CuCl_2$-Et_3N.

Oxidative decarboxylation.[2] α-Amino acids and α-hydroxy carboxylic acids are degraded, to give nitriles and carbonyl compounds, respectively.

NH_2, Ph, COOH → ($CuBr_2$ - *t*-BuOLi, THF, rt) → Ph-CN 72%

OH, Ph, Ph, COOH → ($CuBr_2$ - *t*-BuOLi, THF, 50°) → Ph, Ph, =O 86%

[1]Takeda, T., Sasaki, R., Nakamura, A., Yamauchi, S., Fujiwara, T. *SL* 273 (1996).
[2]Takeda, T., Yamauchi, S., Fujiwara, T. *S* 600 (1996).

Copper(I) *t*-butoxide–potassium *t*-butoxide.

Vicarious aromatic substitution.[1] The smooth introduction of a carbon chain to C-2 of *m*-dinitrobenzene by reaction with an activated bromo compound is carried out in the presence of the two metal *t*-butoxides at –20° to 0°.

[1]Haglund, O., Nilsson, M. *S* 242 (1994).

Copper(I) chloride. 13, 85; **15**, 101; **18**, 112–113

Destannylation. Alkenylstannanes undergo demetallative homocoupling to form 1,3-dienes,[1] including octahydrobiphenylenes.[2]

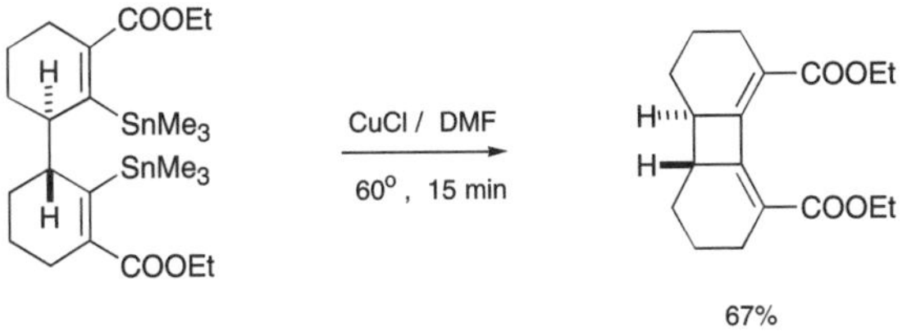

COOEt, H, $SnMe_3$, $SnMe_3$, H, COOEt — CuCl / DMF, 60°, 15 min → H, H, COOEt, COOEt; 67%

Intramolecular Michael additions.[3] Enones bearing an alkenylstannane side chain are cyclized on exposure to CuCl. Thus, bicyclic γ,δ-unsaturated ketones are obtained.

O, $SnMe_3$ — CuCl, DMF, rt, 0.5 h → O, H, H; 96%

Stille coupling. The cross-coupling of alkenylstannanes and enol triflates is a key operation in the construction of complex polyether frameworks.[4]

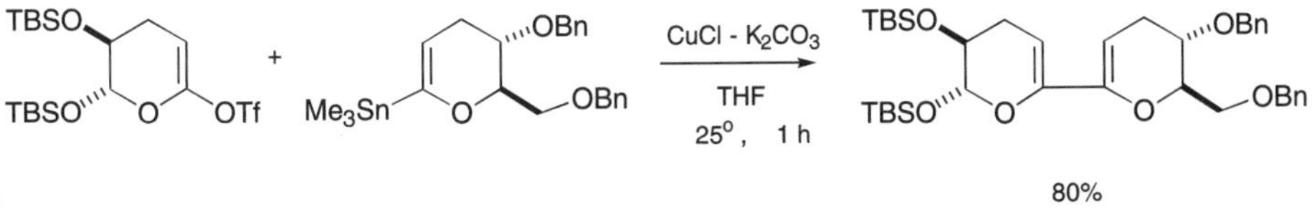

TBSO, TBSO, O, OTf + Me_3Sn, O, OBn, OBn — $CuCl - K_2CO_3$, THF, 25°, 1 h → TBSO, TBSO, O, O, OBn, OBn; 80%

[1]Piers, E., McEachern, E.J., Romero, M.A. *TL* **37**, 1173 (1996).
[2]Piers, E., Romero, M.A. *JACS* **118**, 1215 (1996).
[3]Piers, E., McEachern, E.J., Burns, P.A. *JOC* **60**, 2322 (1995).
[4]Nicolaou, K.C., Sato, M., Miller, N.D., Gunzner, J.L., Renaud, J., Untersteller, E. *ACIEE* **35**, 889 (1996).

Copper(II) chloride. 14, 100; **18**, 113–114

Hydrolysis of nitriles. Heating a nitrile with $CuCl_2 \cdot 2H_2O$ at high temperatures without any solvent accomplishes the hydrolysis to carboxylic acids.[1] The yields range from low to excellent (6 examples, 9–94%).

Protection and deprotection. $CuCl_2$ catalyzes tetrahydropyranylation of alcohols,[2] as well as cleavage of phenacyl esters.[3] In the latter process, oxygen is introduced into the refluxing medium (aqueous DMF).

Dehydrogenation. $CuCl_2$ is an efficient reagent for the preparation of pyridazinones[4] from the dihydro compounds. Dehydrogenation of 1,2,3,4-tetrahydroisoquinolines to give the 3,4-dihydro derivatives is achieved under high pressure of oxygen.[5] Substituted dibenzofurans are formed by an intramolecular oxidative coupling[6] of *o,o′*-dilithiodiaryl ethers with $CuCl_2$.

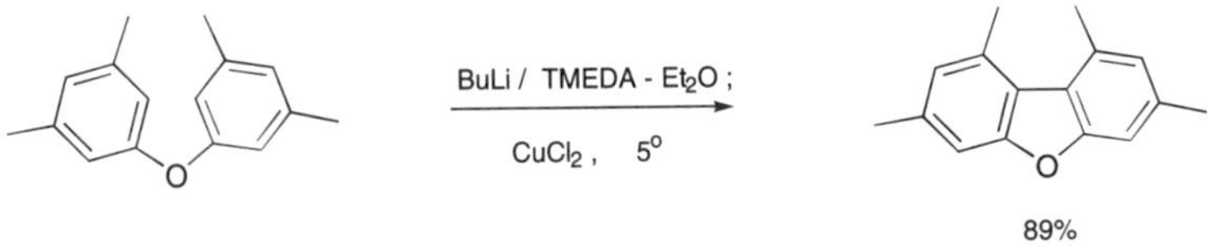

BuLi / TMEDA - Et_2O ; $CuCl_2$, 5°

89%

Chlorohydrination.[7] Allylic amines and sulfides form chlorohydrins with participation of the heteroatom. The reagent system consists of $CuCl_2$ and Li_2PdCl_4 in aqueous THF.

[1]Chemat, F., Poux, M., Berlan, J. *JCS(P1)* 1781 (1995).
[2]Bhalerao, U.T., Davis, K.J., Rao, B.V. *SC* **26**, 3081 (1996).
[3]Ram, R.N., Singh, L. *TL* **36**, 5401 (1995).
[4]Csende, F., Szabo, Z., Bernath, G., Stajer, G. *S* 1240 (1995).
[5]Shimizu, M., Orita, H., Hayakawa, T., Suzuki, K., Takehira, K. *H* **41**, 773 (1995).
[6]Radner, F., Eberson, L. *JCR(S)* 362 (1996).
[7]Lai, J.-Y., Wang, F.-S., Guo, G.-Z., Dai, L.-X. *JOC* **58**, 6944 (1993).

Copper(I) cyanide. 18, 114

Ullmann couplings. The modified method involving a prior halogen–lithium exchange (with *t*-BuLi) and treatment with CuCN proceeds at low temperatures.[1] A synthesis of optically active biaryls is shown.

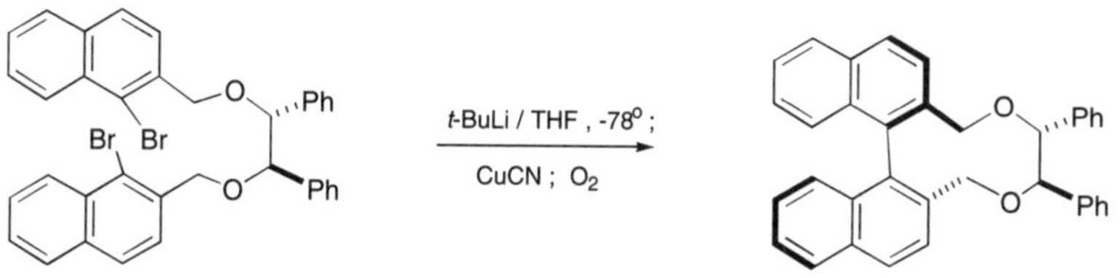

Transmetallations. Very facile Sn → Cu exchange from organostannanes enables new methods to be developed, e.g., for allylic acetates,[2] α-acetoxy ketones,[3] and 3-(1-oxyalkyl)cycloalkanones.[4]

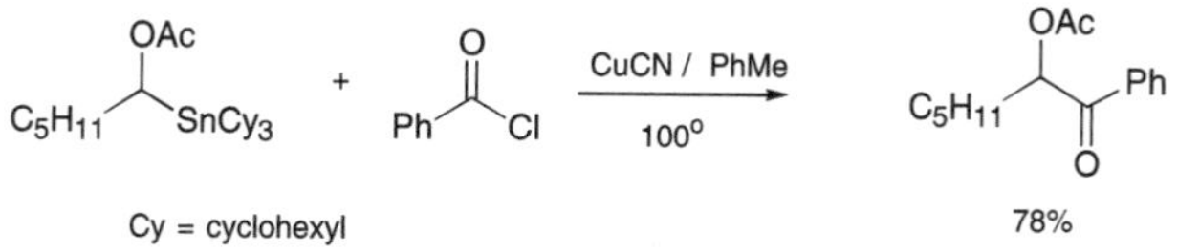

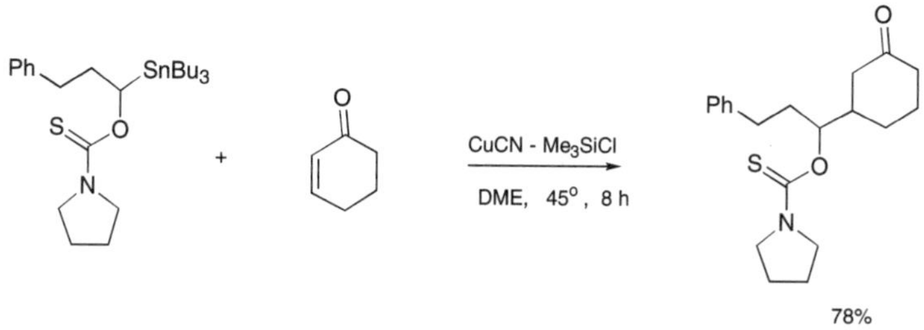

N-Boc allylic amines are readily obtained by CuCN-mediated coupling of enol triflates with the α-lithiated carbamates.[5] This method offers regio- and stereocontrol with respect to the double bond. However, it cannot be extended to the preparation of benzylamine derivatives.

[1]Lipshutz, B.H., Kayser, F., Liu, Z.-P. *ACIEE* **33**, 1842 (1994).
[2]Falck, J.R., Bhatt, R.K., Ye, J. *JACS* **117**, 5973 (1995).
[3]Linderman, R.J., Siedlecki, J.M. *JOC* **61**, 6492 (1996).
[4]Bhatt, R.K., Falck, J.R., Ye, J. *TL* **37**, 3811 (1996).
[5]Dieter, R.K., Dieter, J.W., Alexander, C.W., Bhinderwala, N.S. *JOC* **61**, 2930 (1996).

Copper(I) iodide. 16, 98; **18**, 114–115

ROH → RCF₃.[1] Alcohols react with $ClCF_2COF$ to afford the esters that, on heating with KF and CuI in DMF at 100° under N_2, undergo deoxygenative trifluoromethylation.

Coupling of alkynes with organoiodides. With CuI and pyrrolidine as promoters, the reaction of 1-alkynes with iodoalkynes leads to unsymmetrical 1,3-diynes (17 examples,

61–98%).[2] In the case of 1-bromoalkynes, the addition of $(Ph_3P)_2PdCl_2$ is advantageous. The Pd-catalyzed (also in the presence of CuI) coupling of 1-trimethylsilylethyne with *o*-iodoanilines is a high-yielding reaction; indoles are obtained on subsequent treatment with CuI in DMF.[3]

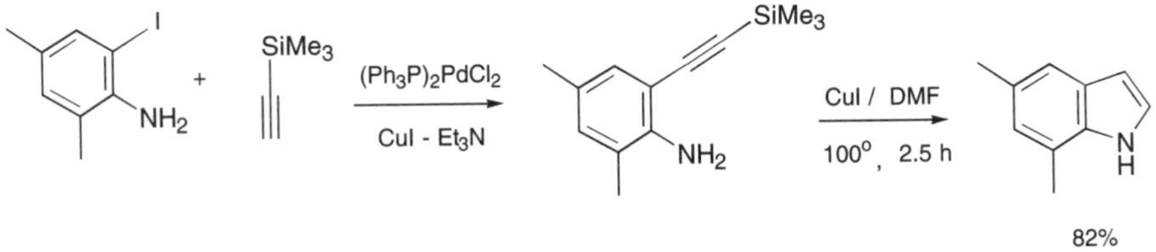

Sulfones.[4] Sodium sulfinates displace aryl iodides in the presence of CuI in DMF.

Cross-coupling of organostannanes. By catalysis of CuI, allylstannanes can react with diaryliodonium tetrafluoroborates to give allylarenes. When CO (1 atm.) is present, allyl aryl ketones are produced.[5] Alkenylstannanes also couple with allyl halides.[6]

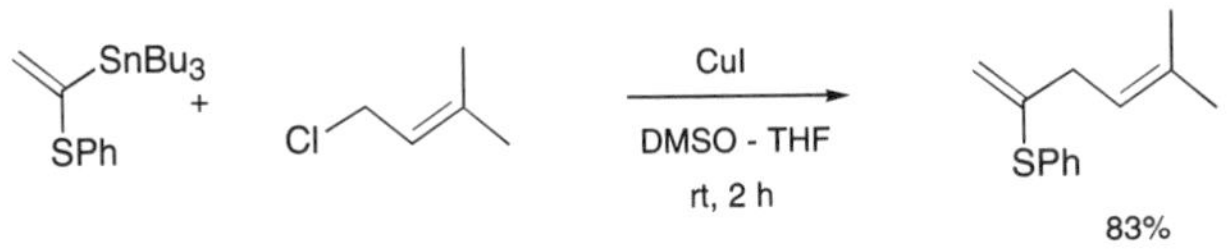

Conjugate additions.[7] Dialkylmagnesiums and Me_3SiCl react with enones to give silyl enol ethers when CuI·2LiI is present.

[1]Duan, J.-X., Chen, Q.-Y. *JCS(P1)* 725 (1995).
[2]Alami, M., Ferri, F. *TL* **37**, 2763 (1996).
[3]Ezquerra, J., Pedregal, C., Lamas, C., Barluenga, J., Perez, M., Garcia-Martin, M.A., Gonzalez, J.M. *JOC* **61**, 5804 (1996).
[4]Suzuki, H., Abe, H. *TL* **36**, 6239 (1995).
[5]Kang, S.-K., Yamaguchi, T., Kim, T.-H., Ho, P.-S. *JOC* **61**, 9084 (1996).
[6]Takeda, T., Matsunaga, K.-I., Kabasawa, Y., Fujiwara, T. *CL* 771 (1995).
[7]Reetz, M.T., Kindler, A. *CC* 2509 (1994).

Copper(II) nitrate–silica gel.

Regeneration of carbonyl compounds.[1] Oximes, tosylhydrazones, and dithioacetals are converted into carbonyl compounds with the supported reagent in CCl_4 at room temperature (42 examples, 85–98%).

[1]Lee, J.G., Hwang, J.P. *CL* 507 (1995).

Copper(II) permanganate.

Oxidation of alcohols.[1] The oxidation in HOAc takes place in a homogeneous medium.

[1]Ansari, M.A., Craig, J.C. *SC* **26**, 1789 (1996).

Copper(II) pivalate.

N-Alkylation of anilines.[1] The copper salt with R_2AlCl forms effective alkylating agents to transfer one R group to aromatic amines.

[1]Barton, D.H.R., Doris, E. *TL* **37**, 3295 (1996).

Copper(II) sulfate.

RX → ROH.[1] A method for the hydrolysis of organohalides is heating with $CuSO_4$ in aqueous DMSO at 110–120° (7 examples, 60–100%).

De(thio)acetalization.[2] Both acetals and dithioacetals are removed with the $CuSO_4/SiO_4$ reagent in refluxing benzene.

$CuSO_4$ / SiO_2, PhH, 80°, 5 h

90%

Retro-Henry reaction. Cyclic β-nitro alcohols undergo fragmentation when heated with $CuSO_4/SiO_4$ in benzene to give nitroalkyl ketones.[3] A suitable intermediate for the synthesis of phoracantholide-I is thus available.

[1]Menchikov, L.G., Vorogushin, A.V., Korneva, O.S., Nefedov, O.M. *MC* 223 (1995).
[2]Caballero, G.M., Gros, E.G. *SC* **25**, 395 (1995).
[3]Saikia, A.K., Hazarika, M.J., Barua, N.C., Bezbarua, M.S., Sharma, R.P., Ghosh, A.C. *S* 981 (1996).

Copper(II) sulfinates.

Alkynyl sulfones.[1] A straightforward preparation of these sulfones consists of the sonochemical induction of coupling of 1-iodoalkynes with copper sulfinates (tosylates, etc.) in THF.

[1]Suzuki, H., Abe, H. *TL* **37**, 3717 (1996).

Copper(I) 2-thienylcarboxylate.

1,3-Dienes and styrene derivatives.[1] With the use of this copper salt (1.5 equiv.), a rapid cross-coupling of aryl, heteroaryl, and alkenylstannanes with iodoalkenes and certain aryl iodides takes place in NMP at or below room temperature (16 examples, 71–97%). The process is economical.

[1]Allred, G.D., Liebeskind, L.S. *JACS* **118**, 2748 (1996).

Copper(II) triflate.

Aziridines.[1] $Cu(OTf)_2$ is a useful catalyst for promoting the reaction between diazoacetic esters and imines.

Ph-N=CHPh + N_2=CHCOOEt —[$Cu(OTf)_2$, THF, -25°]→ cis-aziridine + trans-aziridine (6 : 1) 95%

[1]Rasmussen, K.G., Jorgensen, K.A. *CC* 1401 (1995).

Cyanomethylenetributylphosphorane. 18, 116

Alkylations. $Bu_3P{=}CHCN$ is a useful catalyst for promoting Mitsunobu-type alkylation of *p*-toluenesulfonamide with alcohols at room temperature.[1] Alkylation of active methylene compounds is similarly achieved.[2]

[diol] + $NCCH_2SO_2Ph$ —[$Bu_3P{=}CHCN$, PhH Δ]→ [cyclopentane product] 90%

[amino alcohol] —[$Bu_3P{=}CHCN$, PhH Δ]→ [bicyclic amine] → → (+)-α-skytanthine

[1]Tsunoda, T., Yamamoto, H., Goda, K., Ito, S. *TL* **37**, 2457 (1996).
[2]Tsunoda, T., Yamamoto, H., Goda, K., Ito, S. *TL* **37**, 2459 (1996).

Cyanomethyl formate. 18, 116

***N*-Formylation.**[1] Amino esters are formylated by $HCOOCH_2CN$ by mixing their hydrochloride salt with the reagent and Et_3N (14 examples, 62–97%).

[1]Duczek, W., Deutsch, J., Vieth, S., Niclas, H.-J. *S* 37 (1996).

(1,5-Cyclooctadiene)cyclopentadienylcobalt.

Pyridines from nitriles. 2-Substituted pyridines are readily made by the [2+2+2]-cycloaddition of nitriles with acetylene. A convenient access to analogs of epibatidine by using this method has been reported.[1]

CN (cod)CpCo ≡ hν H_2O ~ 100% epibatidine

[1]Sundermann, B., Scharf, H.-D. *SL* 703 (1996).

(1,5-Cyclooctadiene)rhodium(I) salts.

Hydrogenation. Cationic rhodium complexes catalyze the hydrogenation of imines[1] and carbonyl compounds.[2] When these complexes are coordinated to a chiral phosphine ligand (besides COD), asymmetric hydrogenation by the Rh catalysts is realized.[3]

Tandem hydrosilylation–isomerization.[4] Propargylic alcohols undergo silylation at the sp-carbon with R_3SiH, and the resulting allylic alcohols are converted into β-silyl ketones by the same catalyst, $(cod)_2RhBF_4$.

C_5H_{11} OH Et_3SiH - $(cod)_2RhBF_4$ Ph_3P / Me_2CO 50°, 4 h C_5H_{11} $SiEt_3$ OH 95% $(cod)_2RhBF_4$ Ph_3P / dioxane Δ 16 h C_5H_{11} $SiEt_3$ O 100%

Annulation. Cyclocarbonylation occurs when vinylallenes are exposed to CO in the presence of $[Rh(cod)(MeCN)_2]\ PF_6$. Yields are good (6 examples, 78–99%).[5]

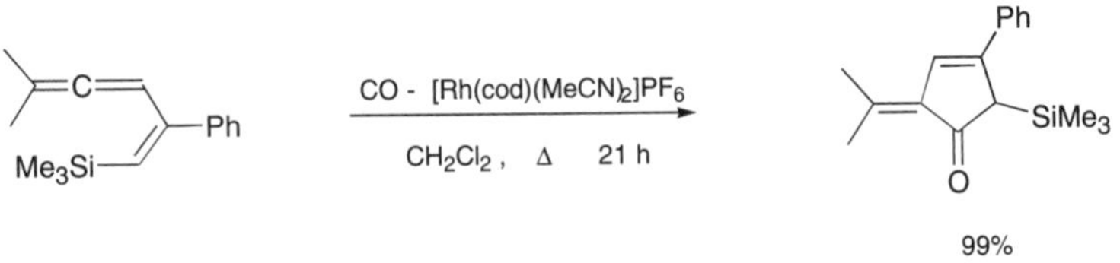

CO - [Rh(cod)(MeCN)$_2$]PF$_6$

CH$_2$Cl$_2$, Δ 21 h

99%

Alkynes bearing a hydrosilane group at a remote site undergo silylformylation.[6] Thus, regio- and stereoselective functionalization of the triple bond is realized.

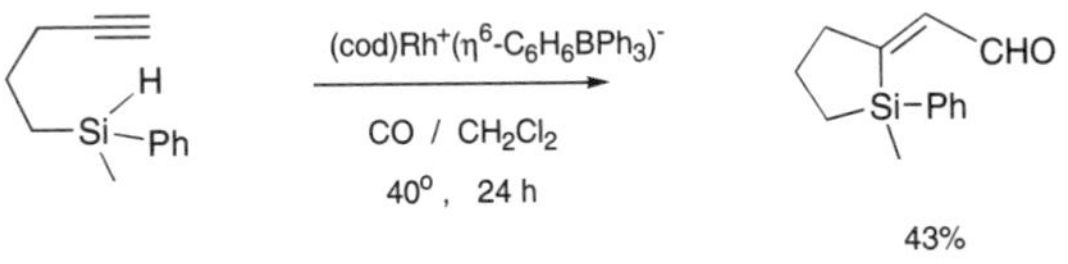

(cod)Rh$^+$(η^6-C$_6$H$_6$BPh$_3$)$^-$

CO / CH$_2$Cl$_2$

40°, 24 h

43%

Cyclopentanones.[7] The (cod)(dppe)RhCl complex promotes the cyclization of 4-alkenals. By heating allyl vinyl ethers with the rhodium complex, a direct transformation to cyclopentanones is accomplished.

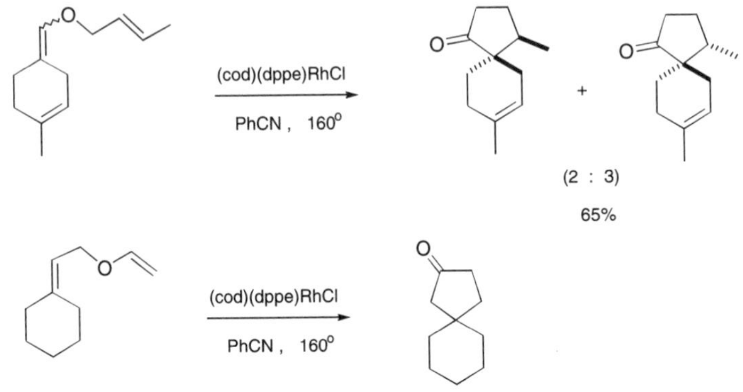

(cod)(dppe)RhCl

PhCN, 160°

(2 : 3)

65%

(cod)(dppe)RhCl

PhCN, 160°

[1]Zhou, Z., James, B.R., Alper, H. *OM* **15**, 4209 (1995).
[2]Burk, M.J., Harper, T.G.P., Lee, J.R., Kalberg, C. *TL* **35**, 4963 (1994).
[3]Burk, M.J., Gross, M.F., Martinez, J.P. *JACS* **117**, 9375 (1995).
[4]Takeuchi, R., Nitta, S., Watanabe, D. *JOC* **60**, 3045 (1995).
[5]Murakami, M., Itami, K., Ito, Y. *ACIEE* **34**, 2691 (1995).
[6]Monteil, F., Matsuda, I., Alper, H. *JACS* **117**, 4419 (1995).
[7]Sattelkau, T., Hollmann, C., Eilbracht, P. *SL* 1221 (1996).

(1,5-Cyclooctadiene)ruthenium chloride.

γ,δ-Unsaturated aldehydes.[1] The catalyst CpRuCl(cod) has been used in the coupling of allyl alcohol with alkynes.

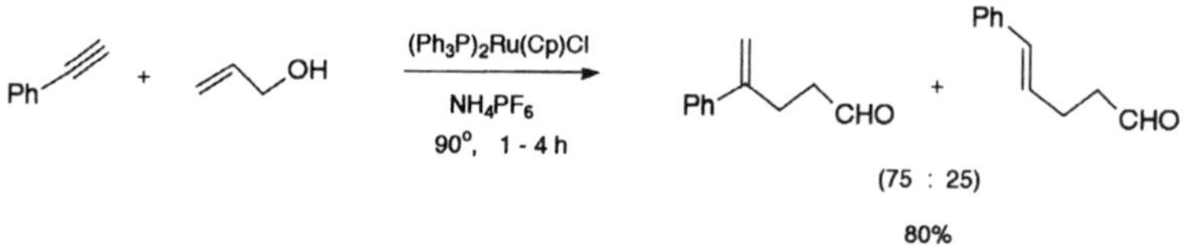

Ph + OH

$(Ph_3P)_2Ru(Cp)Cl$

NH_4PF_6

90°, 1 - 4 h

Ph CHO + Ph CHO

(75 : 25)

80%

[1]Derien, S., Jan, D., Dixneuf, P.H. *T* **52**, 5511 (1996).

D

Diallyltin dibromide.

Allylation of β-keto aldehydes.[1] The reagent attacks the aldehyde group to provide precursors of conjugated dienes and skipped dienes.

[1]Kumaraswamy, S., Nagabrahmanandachari, S., Kumara Swamy, K.C. *SC* **26**, 729 (1996).

1,1-Dianisyl-1,2,2,2-tetrachloroethane.

Hydroxyl group protection.[1] Alcohols form ethers with this reagent in the presence of AgOTs and pyridine. The protecting group is stable to acids and bases, and it is removed by Zn–$ZnBr_2$ or lithium phthalocyaninatocobaltate.

[1]Karl, R.M., Klosel, R., Konig, S., Lehnhoff, S., Ugi, I. *T* **51**, 3759 (1995).

Diarylcadmium.

Aryl C-glycosides.[1] The reaction of glycosyl halides with Ar_2Cd delivers the *C*-glycosides.

[1]Chaudhuri, N.C., Kool, E.T. *TL* **36**, 1795 (1995).

1,4-Diazabicyclo[2.2.2]octane. **13**, 92; **15**, 109; **18**, 120

Baylis–Hillman reaction.[1] DABCO catalyzes the condensation of *N*-tosylimines with acrylic esters at 80° (6 examples, 45–80%).

Benzo-1,3-dioxan-4-ones.[2] Phenyl salicylates form cyclic acetals on reaction with HCHO and DABCO in chloroform. The corresponding methyl esters do not react.

COOPh, OH —(DABCO, HCHO, $CHCl_3$)→ 81%

[1]Campi, E.M., Holmes, A., Perlmutter, P., Teo, C.C. *AJC* **48**, 1535 (1995).
[2]Perlmutter, P., Puniani, E. *TL* **37**, 3755 (1996).

1,8-Diazabicyclo[5.4.0]undec-7-ene. **13**, 92; **14**, 109; **15**, 109–110; **16**, 105–106; **17**, 99–100; **18**, 120–121

Elimination.[1] Tosylhydrazones of α-nitro-[1] and α-benzenesulfonylketones[2] are rapidly converted to the conjugated tosylhydrazones at room temperature on treatment with DBU. *N,N*-Bistosylhydrazones of the unsubstituted ketones also give the same products.[3]

Dehydrochlorination of 2-chloroalkenyl silyl ketones furnishes allenyl silyl ketones (4 examples, 71–86%).[4]

Benzoin condensation.[5] DBU is a suitable base for deprotonation of azolium salts that catalyze the benzoin (and acyloin) condensation.

C-Alkylation of 2-methyl-1,3-diketones.[6] Very little *O*-alkylation interferes when carried out with DBU as base in the presence of LiI.

N-Carboxylation. Amines react with carbon dioxide in the presence of a base (e.g., DBU), and the carbamic acids can be converted to carbamoyl chlorides.[7] Propargylic amines give 5-methyleneoxazol-2-ones.[8]

NHBn —DABCO, CO_2, 20°, 24 h→ O NBn O 93%

[1]Ballini, R., Giantomassi, G. *T* **51**, 4173 (1995).
[2]Ballini, R., Bosica, G., Marcantoni, E. *T* **52**, 10705 (1996).
[3]Magnus, P., Roe, M.B. *TL* **36**, 5479 (1995).
[4]Cunico, R.F., Zhang, C.-P. *SC* **25**, 503 (1995).
[5]Miyashita, A., Suzuki, Y., Iwamoto, K.-I., Higashino, T. *CPB* **42**, 2633 (1994).
[6]Bedekar, A.V., Watanabe, T., Tanaka, K., Fuji, K. *S* 1069 (1995).
[7]McGhee, W.D., Pan, Y., Talley, J.J. *TL* **35**, 839 (1994).
[8]Costa, M., Chiusoli, G.P., Rizzardi, M. *CC* 1699 (1996).

Dibenzyl carbonate.

Benzylation.[1] Active methylene compounds undergo monobenzylation with this electrophile in the presence of K_2CO_3 in refluxing DMF. Phenols are *O*-benzylated.

[1]Selva, M., Marques, C.A., Tundo, P. *JCS(P1)* 1889 (1995).

Di-*t*-butyl pyrocarbonate.

Carbamates. The Boc group is suitable for amine protection in hydroboration reactions.[1] Thus, amino alcohols are synthesized from unsaturated amines via the *N*-Boc derivatives. The cleavage is effected with basic hydrogen peroxide.

Derivatization of sterically hindered amines can be accomplished using tetramethylammonium hydroxide as a base in acetonitrile.[2]

Isocyanates and ureas. In the presence of DMAP as a catalyst, amines form isocyanates at room temperature.[3] On prolonged reaction, ureas are obtained. Unsymmetrical ureas can be synthesized by adding a second amine to the reaction mixture after 10–20 minutes at room temperature.[4]

Carbamates.[5] Instead of the second amine, the addition of an alcohol to the isocyanate formed as described in the previous paragraph gives rise to a carbamate.

NH_2 —(Boc)$_2$O - DMAP; BnOH / MeCN; 82°, 15 h→ NH–C(=O)–OBn 96%

Oxazoles.[6] Cyclodehydration of *N*-acylamino acids occurs on reaction with Boc_2O, DMAP, and pyridine in dioxane. *N*-Acyl anthranilic acids afford acylanthranils under the same conditions.

HN(C(=O)Ph), Ph, COOH —(Boc)$_2$O - DMAP; py - dioxane; rt→ Ph, N, BocO, O, Ph 70%

[1]Kabalka, G.W., Li, N.-S., Pace, R.D. *SC* **25**, 2135 (1995).
[2]Khalli, E.M., Subasinghe, N.L., Johnson, R.L. *TL* **37**, 3441 (1996).
[3]Knölker, H.-J., Braxmeier, T., Schlechtingen, G. *ACIEE* **34**, 2497 (1995).
[4]Knölker, H.-J., Braxmeier, T., Schlechtingen, G. *SL* 502 (1996).
[5]Knölker, H.-J., Braxmeier, T. *TL* **37**, 5861 (1996).
[6]Mohapatra, D.K., Datta, A. *SL* 1129 (1996).

Dicarbonyl(cyclopentadienyl)cobalt. 14, 116; **16**, 112–113; **17**, 102; **18**, 125–126

1-Acyl-2-methylenecycloalkanecarboxylic esters.[1] Under photochemical conditions, 2-(ω-alkynyl)-3-ketoalkanoic esters undergo cyclization in the presence of $CpCo(CO)_2$. This method has been applied to the construction of the phyllocladane skeleton.

O, COOMe, Ph —$CpCo(CO)_2$, hv; PhH Δ→ O, COOMe, Ph 69%

[1]Cruciani, P., Stammler, R., Aubert, C., Malacria, M. *JOC* **61**, 2699 (1996).

Dicarbonyl(cyclopentadienyl)trimethylsiloxymethyliron.

Cyclopropanation.[1] The siloxymethyliron complex (**1**) is prepared from *s*-trioxane, trimethylsilyl iodide, and $CpFe(CO)_2$ anion. Its reaction with alkenes is catalyzed by Me_3SiOTf (5 examples, 52–100%).

(1)
OC–Fe–OSiMe3
OC
Ph, Ph (1,1-diphenylethylene) → Me3SiOTf / CH2Cl2, -78°, 30 min → Ph, Ph (1,1-diphenylcyclopropane) 100%

[1]Du, H., Yang, F., Hossain, M.M. *SC* **26**, 1371 (1996).

Dicarbonylrhodium acetylacetonate.

α,β-Unsaturated aldehydes. Hydroformylation of alkynes with $Rh(CO)_2(acac)$ and CO/H_2 at atmospheric pressure provides enals.[1] Using a silane $PhMe_2SiH$ (instead of hydrogen) and pressurized CO (10 atm.) diverts the reaction to the formation of *(Z)*-β-silyl enals (6 examples, 56–90%). Interestingly, by changing the reagents to $Rh_2Co_2(CO)_{12}$ and Et_3SiH, the reaction of 5-hexynal affords *(Z)*-2-triethylsilylmethylenecyclopentanol.[2]

Ph–Si(Me2) CHO ... CHO 74% ← (acac)Rh(CO)2 / PhMe2SiH / CO / PhMe ← 5-hexynal (CHO) → Rh2Co2(CO)12 / Et3SiH / CO / PhMe → SiEt3, OH 70%

Dehydrobicyclo[3.3.0]octanones. 1,6-Diynes undergo cyclocarbonylation. Different reaction conditions lead to either silylbicyclo[3.3.0]octenones[3] or -dienones.[4]

COOMe, COOMe (diyne) → (acac)Rh(CO)2 / t-BuMe2SiH / CO / PhH, 50° → tBuSiMe2, O=, COOMe, COOMe > 90%

Pipecolic acid derivatives.[5] Under conventional hydroformylation conditions, 2-allylglycine esters undergo cyclohydrocarbonylation. However, the biphenyl phosphite ligand plays a unique role in the reaction, as Ph_3P and Cy_3P are totally ineffective, and in the presence of a bidentate phosphine ligand (dppb), a mixture of pyrrolidine and piperidine products results.

[1]Johnson, J,R., Cuny, G.D., Buchwald, S.L. *ACIEE* **34**, 1760 (1995).
[2]Ojima, I., Tzamarioudaki, M., Tsai, C.-Y. *JACS* **116**, 3643 (1995).
[3]Ojima, I., Fracchiolla, D.A., Donovan, R.J., Banerji, P. *JOC* **59**, 7594 (1994).
[4]Ojima, I., Kaas, D.F., Zhu, J. *OM* **15**, 5191 (1996).
[5]Ojima, I., Tzamarioudaki, M., Eguchi, M. *JOC* **60**, 7078 (1995).

Dicarbonyltitanocene.

Pauson–Khand-type reaction.[1] The commercially available compound (also readily prepared from Cp_2TiCl_2) is the first early transition metal reagent reported for this transformation of enynes into bicyclic cyclopentenones (14 examples, 58–94%).

[1]Hicks, F.A., Kablaoui, N.M., Buchwald, S.L. *JACS* **118**, 9450 (1996).

Dichlorobis(p-cymene)ruthenium(II).

1-Ethoxyvinyl esters.[1] The addition of carboxylic acids to ethoxyacetylene is catalyzed by the Ru catalyst. The products are activated esters that undergo aminolysis very readily.

PhCOOH + (ethoxyacetylene) —[(p-cymene)RuCl2]2, PhMe, 40°, 15 min→ PhCOO(C=CH2)OEt 80% —BnNH2→ PhCONHBn 87%

Furans.[2] The intramolecular addition of an alcohol to the alkyne linkage of an enyne, with the use of (*p*-cymene)$RuCl_2(PPh)_3$ as a catalyst, results in furan formation (8 examples, 50–89%).

(p-cymene)RuCl2 (PPh3), 60 - 110°, 2-20 h; 89%

[1]Kita, Y., Maeda, H., Omori, K., Okuno, T., Tamura, Y. *JCS(P1)* 2999 (1993).
[2]Seiller, B., Bruneau, C., Dixneuf, P.H. *T* **51**, 13089 (1995).

Dichloroborane-dimethyl sulfide.

$RN_3 \rightarrow RNH_2$.[1] The dichloroborane complex is a highly selective reducing agent for converting azides to amines (8 examples, 75–95%).

[1]Salunkhe, A.M., Brown, H.C. *TL* **36**, 7987 (1995).

2,3-Dichloro-5,6-dicyano-1,4-benzoquinone. **13**, 104–105; **14**, 126–127; **15**, 125–126; **16**, 120; **18**, 130

Deprotection of ethers. Both tetrahydropyranyl ethers[1] and allyl ethers[2] are cleaved in aqueous media. For the recovery of alcohols from *p*-methoxylbenzyl ethers[3] (and acetals),[4] a combination of DDQ and $FeCl_3$ is used.

Deacetalization.[5] Selectivity of deacetalization has been observed. The method is compatible with acetyl, benzoyl, benzyl, and tosyl protecting groups.

DDQ, MeCN - H2O, 20°, 4 h; ~ 100%

Solvolysis of allylic alcohols.[6] Exchange of the hydroxyl group of an allylic alcohol with methoxy and acetoxy groups is easily done by refluxing the substrate with the proper solvents for a short time in the presence of DDQ.

Dibenzofurans and hydroxylated biaryls. Several enolboranes derived from cyclohexanones have been converted to dibenzofurans by DDQ at room temperature.[7] Successive treatment of phenols with $AlCl_3$ and DDQ results in biaryls.[8] Interestingly, unsymmetrical products are obtained.

OH OMe + HO — $AlCl_3$ / $MeNO_2$; DDQ / $MeNO_2$ rt, 1 h → OH OMe HO 70%

O — Bu_2OTf - iPr_2NEt ; DDQ → O 34 %

2-Ene-1,4-diones.[9] Furans are cleaved with DDQ (13 examples, 57–98%).

[1]Raina, S., Singh, V.K. *SC* **25**, 2395 (1995).
[2]Yadav, J.S., Chandrasekhar, S., Sumithra, G., Kache, R. *TL* **37**, 6603 (1996).
[3]Yadav, J.S., Chandrasekhar, S., Sumithra, G., Yadav, J.S. *TL* **37**, 1645 (1996).
[4]Zhang, Z., Magnusson, G. *JOC* **61**, 2394 (1996).
[5]Fernandez, J.M.G., Mellet, C.O., Marin, A.M., Fuentes, J. *CR* **274**, 263 (1995).
[6]Iranpoor, N., Mottaghinejad, E. *SC* **25**, 2253 (1995).
[7]Tanemura, K., Yamaguchi, K., Arai, H., Suzuki, T., Hraguchi, T. *H* **41**, 2165 (1995).
[8]Sartori, G., Maggi, R., Bigi, F., Grandi, M. *JOC* **58**, 7271 (1993).
[9]Sayama, S., Inamura, Y. *H* **43**, 1371 (1996).

Dichloro(ethoxy)oxovanadium(V). 18, 131

Allylation of 1,3-dicarbonyl compounds.[1] The allylation with allylsilanes is conducted at 0–25°.

O COOEt + $SiMe_3$ — $VO(OEt)Cl_2$; EtOAc - CH_2Cl_2 0° -> rt → O COOEt 50%

[1]Hirao, T., Sakaguchi, M., Ishikawa, T., Ikeda, I. *SC* **25**, 2579 (1995).

Dichloroketene.

Insertion into acetals.[1] The formation of β-alkoxy-α,α-dichloroalkanoic esters and 4-oxa-lactones from acyclic and cyclic acetals, respectively, in a $ZnCl_2$-catalyzed reaction with dichloroketene has apparent synthetic potentials.

ZnCl2, Et2O
74%

ZnCl2, Et2O

[1]Mulzer, J., Trauner, D., Bats, J.W. *ACIEE* **35**, 1970 (1996).

Dichlorotris(triphenylphosphine)ruthenium(II).

Hydrogenation.[1] The ruthenium chloride complex catalyzes the stereoselective hydrogenation of simple ketones.

H_2 / $(Ph_3P)_3RuCl_2$, $H_2N(CH_2)_2NH_2$ - KOH, *i*-PrOH, 28°, 1 h
(98.4 : 1.6)
96%

Addition to double bonds.[2] Dichlorination of alkenes with $RuCl_2(PPh_3)_3$ and hexachloroethane does not affect other functional groups. However, the reaction is performed in a sealed tube in hot toluene.

X-Alkylation. Heteroaromatic amines are dialkylated by alcohols at 180°. Note that monoalkylation is achieved on using Ru(cod)(cot) under essentially the same conditions.[3]

Related to the process is the formation of indoles.[4]

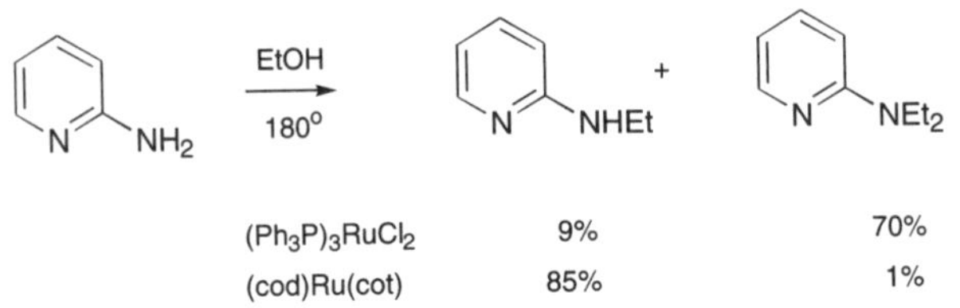

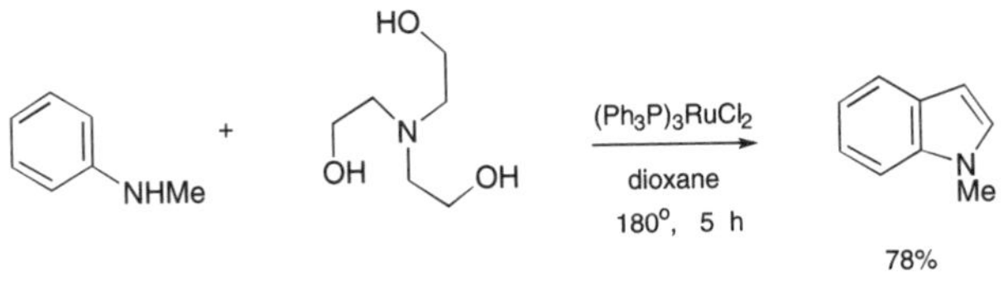

EtOH, 180°

	NHEt	NEt2
$(Ph_3P)_3RuCl_2$	9%	70%
(cod)Ru(cot)	85%	1%

$(Ph_3P)_3RuCl_2$, dioxane, 180°, 5 h — 78%

Insertion of diazocarbonyl compounds into the X–H bonds (S–H, O–H, and N–H) is readily accomplished.[5]

Allylic alcohols. With this catalyst in water, isomerization of homoallylic alcohols to allylic alcohols occurs. The product structures indicate double-bond migration and then transposition of the allylic alcohol system.[6]

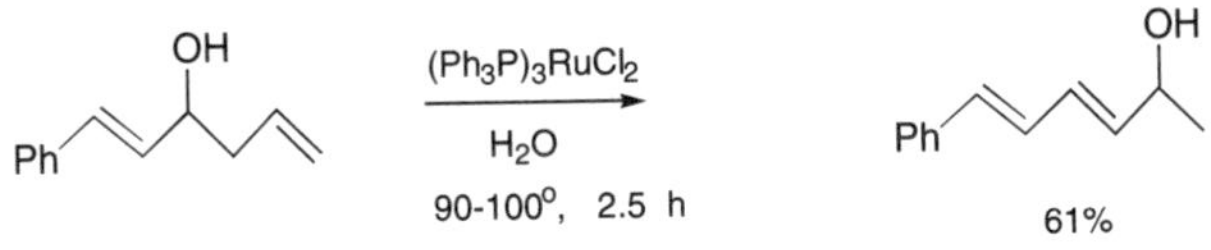

$(Ph_3P)_3RuCl_2$, H_2O, 90-100°, 2.5 h — 61%

Oxidation. Lactone formation from lactols based on the $(PPh_3)_3RuCl_2$-mediated hydrogen transfer to benzalacetone is efficient. The application of this technique to the process of elaborating a dihydropyran unit into two carbon chains played a crucial role in a synthesis of akuammicine.[7]

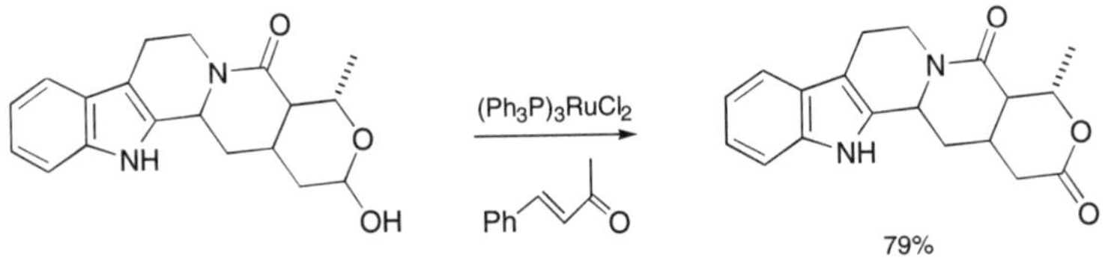

$(Ph_3P)_3RuCl_2$, benzalacetone — 79%

Cyclopropanation.[8] Various ruthenium complexes are effective for catalyzing the decomposition of α-diazocarbonyl compounds. In situ, trapping by an alkene to provide cyclopropane derivatives proceeds as expected.

[1]Ohkuma, T., Ooka, H., Yamakawa, M., Ikariya, T., Noyori, R. *JOC* **61**, 4872 (1996).
[2]Sakai, K., Sugimoto, K., Shigeizumi, S., Kondo, K. *TL* **35**, 737 (1994).
[3]Watanabe, Y., Morisaki, Y., Kondo, T., Mitsudo, T. *JOC* **61**, 4214 (1996).
[4]Shim, S.C., Youn, Y.Z., Lee, D.Y., Kim, T.J., Cho, C.S., Uemura, S., Watanabe, Y. *SC* **26**, 1349 (1996).
[5]Sengupta, S., Das, D., Sarma, D.S. *TL* **37**, 8815 (1996).
[6]Li, C.-J., Wang, D., Chen, D.-L. *JACS* **118**, 12867 (1996).
[7]Martin, S.F., Clark, C.W., Ito, M., Mortimore, M. *JACS* **118**, 9804 (1996).
[8]Demonceau, A., Abreu Dias, E., Lemoine, C.A., Stumpf, A.W., Noels, A.F., Pietraszuk, C., Gulinski, J., Marciniec, B. *TL* **36**, 3519 (1995).

Dicobalt octacarbonyl. 13, 99–101; **14**, 117–119; **15**, 117–118; **16**, 113–115; **17**, 102–105; **18**, 132

Pauson–Khand reaction. The intramolecular reaction can be promoted photochemically.[1] Stereocontrol is demonstrated by attaching the alkynyl group to a chiral auxiliary.[2]

SMe — O — $Co_2(CO)_8$; 80°, 10 h — SMe, O, O

34%

(diastereomer ratio 12 : 1)

Hydroxymethylation.[3] 1,2-Bis(trimethylsiloxy)alkanes are generated when aromatic aldehydes are brought into contact with $Co_2(CO)_8$, CO, and Me_3SiH in benzene at the temperature of ice (6 examples, 50–66%). These silyl ethers are protected diols.

Cyclization.[4] Intramolecular trapping of a complexed propargyl cation by a nucleophilic species (e.g., silyl enol ether) is the key step for the ring formation.

$—Co_2(CO)_6$, OMe, $OSiMe_3$ — $BF_3 \cdot OEt_2$; CH_2Cl_2, rt, 1 h — $(CO)_6Co_2—$, CHO, H

75%

[1]Pagenkopf, B.L., Livinghouse, T. *JACS* **118**, 2285 (1996).
[2]Tormo, J., Verdaguer, X., Moyano, A., Pericas, M.A., Riera, A. *T* **52**, 14021 (1996).

[3]Chatani, N., Tokuhisa, H., Kokubu, I., Fujii, S., Murai, S. *JOMC* **499**, 193 (1995).
[4]Tyrrell, E., Claridge, S., Davis, R., Lebel, J., Berge, J. *SL* 714 (1995).

Dicyclohexylboron triflate–triethylamine.

Aldol reaction of esters.[1] The combination of reagents promotes enolboration, and subsequent reaction under kinetically controlled conditions (–78°) leads predominantly to the *anti* isomers. At higher temperatures (–40°–≈0°) the *syn* isomers are produced with high facial selectivity. The *(E/Z)*-isomerization of the boryl ketene ethers is promoted by Et_3NHOTf that is present in the reaction mixture.

[1]Abiko, A., Liu, J.-F., Masamune, S. *JOC* **61**, 2590 (1996).

2,2-Diethoxyvinyllithium.

α,β-Unsaturated esters.[1] The reagent is formed by lithiation of 2-bromo-1,1-diethoxyethene with BuLi in THF at –70°, and its reaction with carbonyl compounds followed by acid hydrolysis delivers the conjugated esters.

Br–CH=C(OEt)(EtO) —BuLi / THF, - 78°, 30 min→ [Li–CH=C(OEt)(EtO)] —Me3SiCl→ Me3Si–CH=C(OEt)(EtO) 64%; —PhCHO ; H+→ Ph–CH=CH–COOEt 94%

[1]Wei, H., Schlosser, M. *TL* **37**, 2771 (1996).

Diethylaminosulfur trifluoride. (DAST) **13**, 110–112; **16**, 128–129; **18**, 135

Allylic fluorides. Ring cleavage attends the treatment of tertiary cyclopropyl silyl ethers with DAST (11 examples, 45–96%).[1]

OSiMe3 —Et_2NSF_3, CH_2Cl_2, rt→ CH_2F 80%

Fluorination of β-hydroxy sulfides. Fluorination with migration of a thio group is observed.[2] Thus, this reaction can be applied to a preparation of α,α-difluoro-β-(phenylthio)alkanoic esters from α-fluoro-β-hydroxy-α-(phenylthio)esters.[3]

OH SMe MeS SMe Et_2NSF_3 CH_2Cl_2 0° SMe SMe F F 50%

[1]Kirihara, M., Kambayashi, T., Momose, T. *CC* 1103 (1996).
[2]Kuroboshi, M., Furuta, S., Hiyama, T. *TL* **36**, 6121 (1995).
[3]Bildstein, S., Ducep, J.-B., Jacobi, D. *TL* **37**, 8759 (1996).

N,N-Diethylhydroxylamine.

Perfluoroalkene epoxides.[1] The hydroxylamine is able to transfer an oxygen atom to perfluoroalkenes at room temperature (5 examples, 67–98%).

[1]Ono, T., Henderson, P.B. *CC* 763 (1996).

Dihydridotetrakis(triphenylphosphine)ruthenium(II).

Hydration and alcoholysis of nitriles. When heated with water[1] or alcohol[2] in the presence of $RuH_2(PPh_3)_4$, nitriles are converted into amides and esters, respectively.

Aldol and Michael reactions.[3,4] Active methylene compounds such as ethyl cyanoacetate and malonic esters react with aldehydes and electron-deficient alkenes in the aldol and Michael mode, respectively.

Hydrogenolysis.[5] Deoxygenation of 1-alkene-3,4-diols with transposition of the double bond is accomplished via the cyclic carbonates. Heating the latter compounds with $RuH_2(PPh_3)_4$ and HCO_2NH_4 in THF leads mainly to the *(E)*-allylic alcohols. Some saturated alcohols are also formed.

Conjugated enynes. $RuH_2(PPh_3)_4$ catalyzes the addition of alkynes to allenes.[6] A ferrocenylphosphine is also used as ligand.

=C= OH + Ph $(Ph_3P)_4RuH_2$ ferrocenylphosphine cyclohexane Δ OH Ph 58%

[1]Murahashi, S.-I., Sasao, S., Saito, E., Naota, T. *T* **49**, 8805 (1993).

[2]Naota, T., Schichijo, Y., Murahashi, S.-I. *CC* 1359 (1994).
[3]Murahashi, S.-I., Naota, T., Taki, H., Mizuno, M., Takaya, H., Komiya, S., Mizuno, Y., Oyasato, N., Hiraoka, M., Hirano, M., Fukuoka, A. *JACS* **117**, 12436 (1995).
[4]Gomez-Bengoa, E., Cuerva, J.M., Mateo, C., Echavarren, A.M. *JACS* **118**, 8553 (1996).
[5]Kang, S.-K., Kim, D.-Y., Rho, H.-S., Yoon, S.-H., Ho, P.-S. *SC* **26**, 1485 (1996).
[6]Yamaguchi, M., Kido, Y., Omata, K., Hirama, M. *SL* 1181 (1995).

Diiodomethane.

Simmons–Smith reaction. Although *(Z)*-allylic alcohols show very high *syn*-selectivities (> 200:1) in the classical Simmons–Smith reaction, those of the corresponding *(E)*-alcohols are relatively poor. Accordingly, increasing the diastereoselectivity of reactions for the latter substrate is of significant synthetic importance. The findings of a complementary *anti*-selective Sm–Hg/CH_2I_2/THF system and a *syn*-selective $Et_2Zn/CH_2I_2/CH_2Cl_2$ system[1] are pleasing. For optimal results, five equivalents of equimolar Et_2Zn and CH_2I_2 should be present for each unit of the olefin.

Cyclomethylenation of ketones.[2] The combination of CH_2I_2 with MeLi·LiBr generates iodomethyllithium, which converts carbonyl compounds into epoxides (4 examples, 88–90%). The reagent is also used for the one-carbon homologation of a boronic ester.

Homoaldol reaction.[3] The reaction of enolates with $Zn(CH_2I)_2$–BnOLi generates zincate species, which, on further treatment with (*i*-PrO)$TiCl_3$ and then aldehydes, complete the stereoselective carbon-chain elongation.

Bn, O, N, O, H, H; BuLi ; $(ICH_2)_2Zn$ BnOLi, -78°; BnOZn, I, O, N, O, Bn H, H; (*i*PrO)$TiCl_3$, Bn, BocNH CHO; Bn, O, BocNH, N, O, OH Bn H, H; 59% (>99% de)

[1]Charette, A.B., Lebel, H. *JOC* **60**, 2966 (1995).
[2]Wallace, R.H., Battle, W. *SC* **25**, 127 (1995).
[3]McWilliams, J.C., Armstrong, J.D., Zheng, N., Bhupathy, M., Volante, R.P., Reider, P.J. *JACS* **118**, 11970 (1996).

Diisobutylaluminum hydride. 13, 115–116; 15, 137–138; 16, 134–135; 17, 123–125; 18, 140–141

Cyclic ethers from spiroacetals.[1] DIBAH effects a reductive ring opening of spiroacetals.

DIBAH

PhCH3

-12° -> rt, 11 h

92% (main product)

α-Acetoxy ethers.[2] A general procedure for the preparation of these compounds is accomplished by DIBAL-H reduction of esters (lactones) and subsequent acetylation.

Amines from amides.[3] This reduction method is suitable for α,ω-diamides.

Hydroalumination.[4] A stereoselective approach to cycloheptenols is based on the cleavage of 8-oxabicyclo[3.2.1]oct-6-enes. The lack of generality for the fragmentation protocol plagued the method. Now a useful procedure consists of $Ni(cod)_2$-catalyzed hydroalumination and treatment of the alane with *i*-Bu_2AlCl. This method is also appropriate for the asymmetric synthesis of cyclohexenols.

OMe OMe

Ni(cod)2 - (R)-BINAP

i-Bu2AlH

OMe OMe OH

54% (80% ee)

[1]Oikawa, M., Oikawa, H., Ichihara, A. *T* **51**, 6237 (1995).
[2]Dahanukar, V.H., Rychnovsky, S.D. *JOC* **61**, 8317 (1996).
[3]Giboreau, P., Morin, C., Vidal, M. *SC* **26**, 515 (1996).
[4]Lautens, M., Chiu, P., Ma, S., Rovis, T. *JACS* **117**, 532 (1995).

Diisobutyl(phenylselenyl)aluminum.

α-Substituted vinyl selenides.[1] Regioselective hydroselenation of 1-alkynes affords an access to the selenides.

Ph

i-Bu2AlSePh

hexane - PhMe

Δ 24 h

Ph PhSe

85%

[1]Dabdoub, M.J., Cassol, T.M., Batista, A.C.F. *TL* **37**, 9005 (1996).

Diisopropoxytitanium(III) borohydride. 18, 142

Reductions. Aldehydes, ketones, acid chlorides, carboxylic acids, and *N*-Boc amino acids are reduced to the corresponding alcohols, generally in excellent yields.[1] The reduction of α,β-epoxy ketones gives alcohols without affecting the heterocycle.[2]

[1]Ravikumar, K.S., Chandrasekaran, S. *JOC* **61**, 826 (1996).
[2]Ravikumar, K.S., Chandrasekaran, S. *T* **52**, 9137 (1996).

Dilithium tetrachloropalladate.

Annulation with vinylcyclopropanes and -cyclobutanes.[1]

CH(COOEt)2 (o-iodophenyl) + vinylcyclopropane —Li_2PdCl_4, NaOAc, 80°, 3 d→ EtOOC, COOEt indane, 82%

Chlorocarbocyclization of dienes.[2] Dienes tethered to an allylsilane unit undergo Pd(II)-catalyzed cyclization.

MeOOC, COOMe, SiMe2Ph —Li_2PdCl_4 - LiCl, benzoquinone, HOAc - acetone→ MeOOC, COOMe, Cl + MeOOC, COOMe, Cl (3 : 1) 68%

[1]Larock, R.C., Yum, E.K. *T* **52**, 2743 (1996).
[2]Castano, A.M., Bäckvall, J.-E. *JACS* **117**, 560 (1995).

Dimanganesedecacarbonyl.

Unsymmetrical ketones.[1] The successive addition of two different organolithium reagents to $Mn_2(CO)_{10}$ leads to an unsymmetrical ketone. An unsymmetrical α-diketone is produced when the reaction is carried out in the presence of trimethyl phosphite and is quenched with NBS.

[1]Yamamoto, H.M., Sakurai, H., Narasaka, K. *BCSJ* **69**, 157 (1996).

2,4-Dimethoxybenzoylformic acid.

Oxidation of alcohols.[1] The benzoylformate esters are photolabile and decompose to give carbonyl compounds (4 examples, 65–88%).

[1]Pirrung, M.C., Tepper, R.J. *JOC* **60**, 2461 (1995).

2,2-Dimethoxy-5,5-dimethyl-Δ^3-1,3,4-oxadiazoline.

3,3-Dimethoxy hydroindolones.[1] The heterocycle **1** is a convenient precursor of dimethoxycarbene.[2] The cothermolysis of **1** and vinyl isocyanates leads to the formation of hydroindolones via a [1+4]cycloaddition. The development of an approach to alkaloids typified by tazettine is conceivable.

OTHP N=C=O + MeO OMe (**1**) $\xrightarrow{\Delta}$ THPO OMe OMe O N CH(OMe)$_2$ 45%

[1]Rigby, J.H., Cavezza, A., Ahmed, G. *JACS* **118**, 12848 (1996).
[2]Couture, P., Terlouw, J.K., Warkentin, J. *JACS* **118**, 4214 (1996).

2,2-Dimethoxy-3-methyl-3-butanol.

Claisen rearrangement.[1] Heating an allylic alcohol with this reagent in the presence of a mild acid gives a rearranged ketol that can be cleaved by $NaIO_4$ to deliver the γ,δ-unsaturated acid.

OH CN + OH OMe OMe $\xrightarrow[130^{\circ},\ 5\ h]{2,4-(O_2N)_2C_6H_3-OH}$ O OH CN $\xrightarrow{NaIO_4}$ COOH CN

[1] Takayanagi, H., Morinaka, Y. *CL* 565 (1995).

Dimethylaluminum chloride.

β-(Camphor-10-sulfenyl) ketones.[1] The Michael addition of 10-mercapto-isoborneol to enones, mediated by Me_2AlCl, is followed by a stereoselective hydride transfer. Chiral secondary alcohols are readily acquired on desulfurization of the adducts with Raney nickel.

Ph, Ph, O; Me_2AlCl, CH_2Cl_2; OH, H, SH, S, O, Ph; 85%

[1]Nishide, K., Shigeta, Y., Obata, K., Node, M. *JACS* **118**, 13103 (1996).

Dimethylaluminum triflate.

Mukaiyama-type cyclization.[1] A critical step in a synthesis of taxusin involves the formation of the central eight-membered ring by an intramolecular condensation. Various Lewis acids ($TiCl_4$, $SnCl_4$, Me_3SiOTf, and $BF_3 \cdot OEt_2$) are not effective, whereas Me_2AlOTf uniquely induces cyclization in 62% yield.

BnO, $(BnO)_2CH$, TIPSO, H, OPiv; Me_2AlOTf; BnO, OBn, O, H, OPiv; 62%

[1]Hara, R., Furukawa, T., Horiguchi, Y., Kuwajima, I. *JACS* **118**, 9186 (1996).

(8-Dimethylamino-4-azaindenyl)(pentamethylcyclopentadienyl)iron.

Acylation.[1] In the manner of DMAP, the π-complex **1** serves as a catalyst for the acylation of alcohols with diketene. Moreover, since **1** is chiral, it can perform a kinetic resolution of secondary alcohols. The selectivity factor(s) are the highest among nonenzymatic processes.

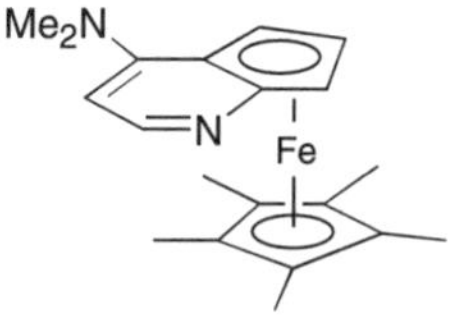

(1)

[1]Ruble, J.C., Fu, G.C. *JOC* **61**, 7230 (1996).

[(3-Dimethylamino)propyl]dimethylaluminum.

Methylation.[1] Carbonyl compounds are attacked by this reagent (**1**), but aryl aldehydes require activation with $AlCl_3$.

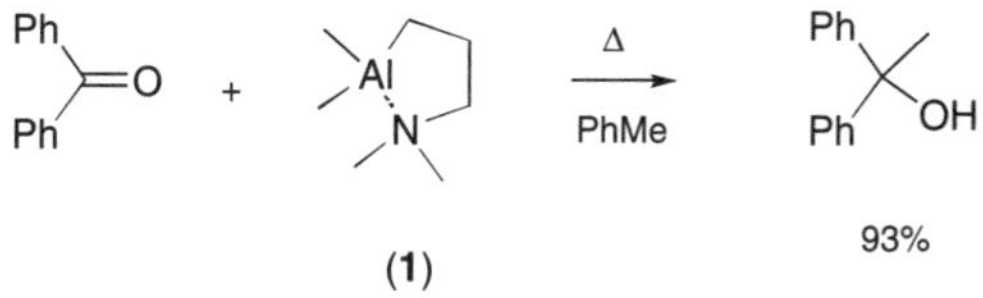

(1)

93%

[1]Baidossi, W., Rosenfeld, A., Wassermann, B.C., Schutte, S., Schumann, H., Blum, J. *S* 1127 (1996).

(2,7-Dimethyl-1,8-biphenylenedioxy)bis(dimethylaluminum).

Bidentate Lewis acid.[1] This useful catalyst (**1**) with a high propensity for double coordination of the carbonyl group is prepared from the corresponding phenol and two equivalents of Me_3Al in CH_2Cl_2 at room temperature. It catalyzes the reduction of 5-nonanone by Bu_3SnH at –78° in 86% yield, whereas a reaction in the presence of the monodentate *O*-dimethylaluminum 2,6-xylenoxide affords 5-nonanol in only 6%. Accordingly, different catalytic efficiencies are also found in the Mukaiyama aldol reaction (e.g., 87% vs. 0% in the reaction between 1-trimethylsiloxy-1-cyclohexene and benzaldehyde) and the Claisen rearrangement of *(E)*-cinnamyl vinyl ether (96% vs. 0%). The contrasting *(E/Z)*-selectivity of the Michael adducts also reflects the different coordination states.

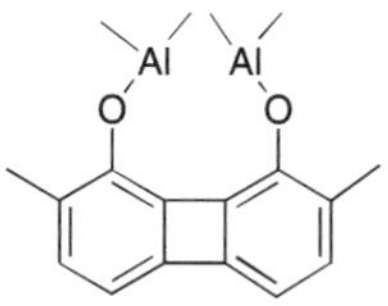

(1)

[1]Ooi, T., Takahashi, M., Maruoka, K., *JACS* **118**, 11307 (1996).

Dimethylboron bromide.

Reductive methylation.[1] Azides are converted to methylamines by Me_2BBr (8 examples, 63–99%). The stereointegrity of chiral azides is retained.

Cleavage of cyclic orthoesters.[2] The regioselective oxidative cleavage of protected pyranosides to give mainly the primary alcohols makes orthoesters derived from phthalide a useful device for protection or differentiation.

Me2BBr
CH2Cl2
-78°, 5 min
(10 : 1)
95%

[1]Dorow, R.L., Gingrich, D.E. *JOC* **60**, 4986 (1995).
[2]Arasappan, A., Fuchs, P.L. *JACS* **117**, 177 (1995).

Dimethyl diazomethylphosphonate.

This Seyferth–Gilbert reagent is conveniently prepared through trifluoroacetylation and diazo transfer with concomitant detrifluoroacetylation.[1]

BuLi / THF ;
CF3COOCH2CF3
-78°
90%
4-AcNHC6H4SO2N3
Et3N / MeCN
0°
50%

[1]Brown, D.G., Velthuisen, E.J., Commerford, J.R., Brisbois, R.G., Hoye, T.R. *JOC* **61**, 2540 (1996).

Dimethyl 2-diazo-3-oxopropylphosphonate.

1-Alkynes.[1] A mild and expedient homologation of aldehydes to 1-alkynes involves the reaction with (**1**) under the influence of K_2CO_3 in methanol at room temperature. Both aliphatic and aromatic aldehydes react nicely. Isolated double bonds are not affected, although enals give homopropargylic methyl ethers, due to the conjugated addition of methanol prior to the deoxygenative homologation.

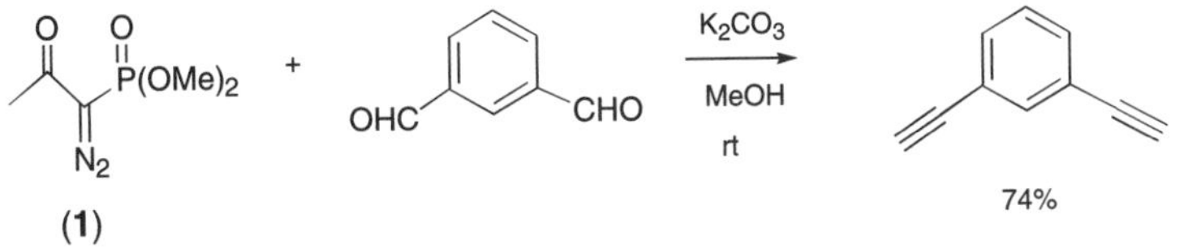

(1) + OHC–C₆H₄–CHO → (K_2CO_3, MeOH, rt) 74%

[1] Müller, S., Liepold, B., Roth, G.J., Bestmann, H.J. *SL* 521 (1996).

Dimethyldioxirane. 13, 120; **14**, 148; **15**, 143–144; **16**, 142–144; **18**, 144–146

Until recently, through state-of-the-art oxidation of acetone, only about 80-mM solutions of DMD were available. More concentrated (a four- to fivefold increase) solutions are now obtained when the normal workup includes washing of organic extracts with phosphate buffer.[1]

Epoxidation. Alkenylsilanes normally undergo epoxidation; however, cyclic substrates give appreciable amounts of allylic oxidation products.[2] In tertiary amines containing a double bond, *N*-oxidation takes precedence,[3] but the basic nitrogen can be protected by precoordination with $BF_3{\cdot}OEt_2$.

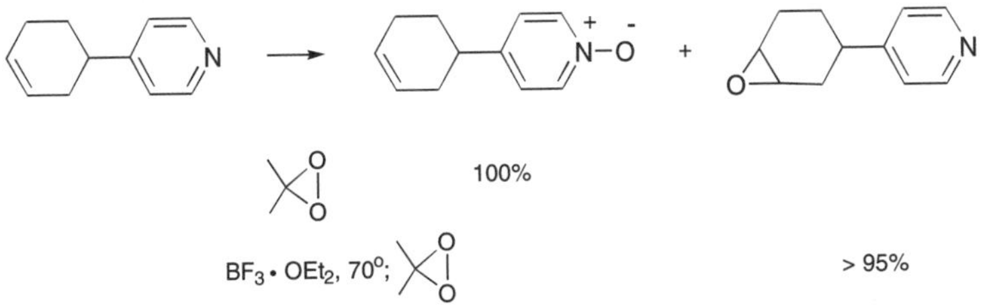

100%

$BF_3 \cdot OEt_2$, 70°; > 95%

vic-Tricarbonyl compounds. The synthesis starting from 1,3-dicarbonyl compounds involves treating the brominated derivatives with dimethyldioxirane in the presence of DMAP.[4] The 2-triphenylphosphoranyl derivatives are also oxidized.[5]

For hydroxylation of 1,3-dicarbonyl compounds at the central carbon (8 examples, > 95%), $Ni(acac)_2$ is employed as a catalyst.[6]

$Ni(acac)_2$, Me_2CO - H_2O; > 95%

Oxidation of alcohols. A neighboring electron-withdrawing group suppresses the oxidation. Thus, the oxidation of 1,2- and 1,3-diols stops at the ketol stage,[7] and β-nitro alcohols also resist oxidation.[8]

Bissulfones.[9] Dithioacetals, including those of ketenes, are rapidly oxidized to the full extent at room temperature by DMD (13 examples, 91–98%).

1-Triflylmethylalkenes.[10] The radical addition of triflylmethyl iodide to 1-alkenes gives γ-iodo triflones. On oxidation with DMD, the allylic triflones are formed (3 examples, 58–93%).

R, I, SO_2CF_3; Me_2CO - Et_2O; R, SO_2CF_3; 58 - 93%

[1] Ferrer, M., Gilbert, M., Sanchez-Baeza, F., Messeguer, A. *TL* **37**, 3585 (1996).
[2] Adam, W., Prechtl, F., Richter, M.J., Smerz, A.K. *TL* **36**, 4991 (1995).
[3] Ferrer, M., Sanchez-Baeza, F., Messeguer, A., Diez, A., Rubiralta, M. *CC* 293 (1995).
[4] Coats, S.J., Wasserman, H.H. *TL* **36**, 7735 (1995).
[5] Wasserman, H.H., Baldino, C.M., Coats, S.J. *JOC* **60**, 8231 (1995).
[6] Adam, W., Smerz, A.K. *T* **52**, 5799 (1996).
[7] Bovicelli, P., Lupattelli, P., Sanetti, A., Mincione, E. *TL* **36**, 3031 (1995).
[8] Ballini, R., Papa, F., Bovicelli, P. *TL* **37**, 3507 (1996).
[9] Curi, D., Pardini, V.L., Viertler, H., Baumstark, A.L., Harden, D.B. *HC* **5**, 555 (1995).
[10] Mahadevan, A., Fuchs, P.L. *JACS* **117**, 3272 (1995).

S,S-Dimethyl dithiocarbonate.

Ureas.[1] Aminolysis in methanol provides either symmetrical or unsymmetrical ureas in 55–92% yields (9 examples) and 40–65% yields (4 examples), respectively.

[1] Leung, M.-K., Lai, J.-L., Lau, K.-H., Yu, H., Hsiao, H.-J. *JOC* **61**, 4175 (1996).

_N,N_-Dimethylformamide.

Allenes.[1] A convenient synthesis of allenes is accomplished by thermolysis of α-alkylidene-β-lactones in DMF. The substrates are available from β-lactones by benzeneselenylation followed by oxidative elimination.

DMF, 105 - 115°, 7.5 h

61%

Formates.[2] Heating tosylates in aqueous DMF gives formates in variable yields (7 examples, 21–74%).

[1]Danheiser, R.L., Choi, Y.M., Menichincheri, M., Stoner, E.J. *JOC* **58**, 322 (1995).
[2]Suri, S.C., Rodgers, S.L., Radhakrishnan, K.V., Nair, V. *SC* **26**, 1031 (1996).

_N,N_-Dimethylformamide–oxalyl chloride.

Inverted esters.[1] Secondary alcohols are converted to esters with inverted configurations by a two-step activation/acylation protocol. First, the alcohols are treated with $[ClCH{=}NMe_2]^+Cl^-$ at 0°. Then they are refluxed with RCOOK in THF.

[1]Barrett, A.G.M., Koike, N., Procopiou, P.A. *CC* 1403 (1995).

Dimethylgallium chloride.

Glycosylation.[1] The reaction of glycosyl fluorides with trimethylsilyl ethers forms *O*-glycosides in the presence of Me_2GaCl-Me_3SiCl.

BnO, BnO, OBn, OBn, F + Me_3SiO — Me_2GaCl, Me_3SiCl / CH_2Cl_2, rt, 3 h — BnO, BnO, OBn, OBn

80% (α : β 29 : 71)

[1]Koide, K., Ohno, M., Kobayashi, S. *S* 1175 (1996).

Dimethylsilyl dichloride.

β-Mannopyranosides.[1] Mixed silyl acetals are formed expediently from a dimethylsilyl-tethered sugar unit in which one is mannosyl phenyl sulfoxide. It simply involves admixture of two equimolar sugars with 1 equiv. of Me_2SiCl_2 in the presence of DMAP and imidazole. The addition of 2,6-di-*t*-butylpyridine and Tf_2O at −100° followed by warming completes the glycosylation.

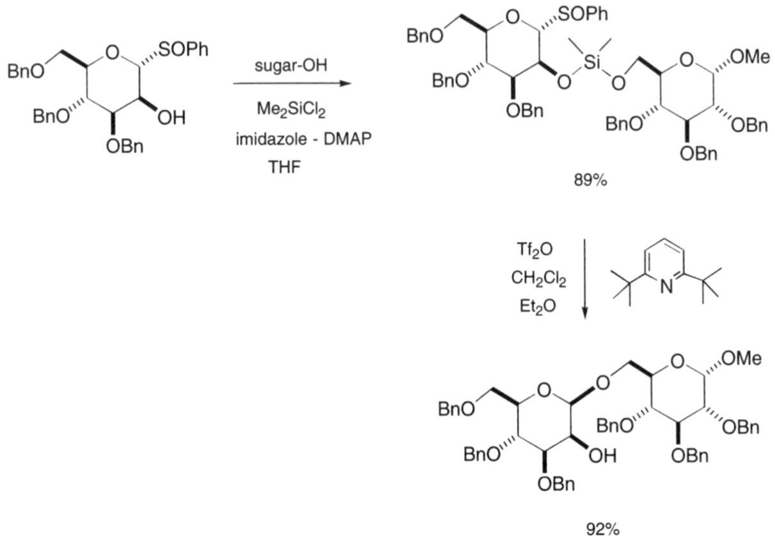

[1]Stork, G., La Clair, J.J. *JACS* **118**, 247 (1996).

Dimethyl sulfoxide–hydrogen bromide.

Bromination.[1] Substituted benzenes undergo site-specific bromination (9 examples, 36–91%).

1,2-Diketones.[2] The oxidation of stilbenes to benzils by DMSO–HBr contrasts with the results of DMSO–I_2 oxidation, which preferentially converts a tolane moiety to α-diketone in the same molecule.

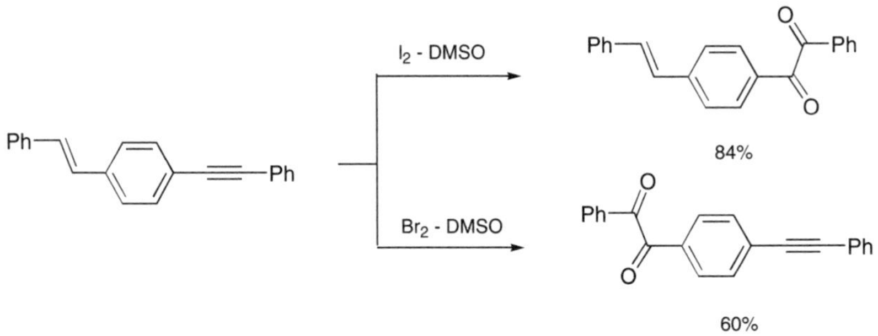

[1]Srivastava, S.K., Chauhan, P.M.S., Bhaduri, A.P. *CC* 2679 (1996).
[2]Yusubov, M.S., Filimonov, V.D., Vasilyeva, V.P., Chi, K.-W. *S* 1234 (1995).

Dimethylsulfoxonium methylide. 14, 152; **15**, 147; **16**, 146; **17**, 126–127; **18**, 148

Cyclopropanation.[1] Under "superbasic" conditions [KOH, $(BnNEt_3)^+Cl^-$, DMSO], the ylide adds to enones efficiently (7 examples, 86–94%).

The ylide can be generated from the salt $(Me_3S{=}O)^+(MeSO_4)^-$, which is obtained from DMSO and Me_2SO_4 at 100°.[2]

[1]Shostakovsky, V.M., Carlson, R.M. *SC* **26**, 1785 (1996).
[2]Forrester, J., Jones, R.V.H., Preston, P.N., Simpson, E.S.C. *JCS(P1)* 2289 (1995).

Dimethyltitanocene.

2-Methyleneoxetanes.[1] These rare compounds are now available by reaction of β-lactones with Cp_2TiMe_2.

[1]Dollinger, L.M., Howell, A.R. *JOC* **61**, 7248 (1996).

2,4-Dimethyl-1,2,4-triazolium iodide.

Benzoin condensation.[1] The azolium salts are effective catalysts for benzoin condensation. The reaction is conducted by heating an aromatic aldehyde with NaH in THF in the presence of the salt.

[1]Miyashita, A., Suzuki, Y., Kobayashi, M., Kuriyama, N., Higashino, T. *H* **43**, 509 (1996).

[(*E*)-γ-(1,3,2-Dioxaborinanyl)allyl]diisopinocampheylborane.

anti-1-Alkene-3,4-diols.[1] The ready availability of *B*-allenyl-(1,3,2-dioxaborinane) from allenylmagnesium bromide and the borinyl bromide makes it possible to prepare the reagent (**1**) by hydroboration. This reagent reacts with aldehydes stereoselectively, yielding products that, on oxidative workup, are transformed into the alkenediols with good de and ee.

O–B–Cl + BrMg–CH=C=CH₂ → (Et_2O, - 78°) → O–B–CH=C=CH₂ → $(Ipc)_2BH$ → (**1**) (Ipc_2B)

(**1**) → RCHO, - 78°; NaOOH →

(de >95%)
(ee 90 ~ >95%)

[1]Brown, H.C., Narla, G. *JOC* **60**, 4686 (1995).

Diphenyl chlorophosphate.

Glycosyl chlorides.[1] Sugars containing acid-labile groups can be converted into glycosyl chlorides by reaction of the corresponding lithium alkoxides with $ClPO(OPh)_2$.

BuLi, $ClPO(OPh)_2$, THF, 0° -> rt, 15 h; 92%

[1]Hung, S.-C., Wong, C.-H. *TL* **37**, 4903 (1996).

Diphenyl diselenide. 13, 125; **18**, 151–152

Very pure PhSeSePh can be prepared via dihydroxy(phenyl)selenonium tosylate $[PhSe(OH)_2]^+OTs^-$ by treatment with 33% H_2O_2 and TsOH.[1]

Benzyl selenides.[2] PhSeNa derived from PhSeSePh and $NaBH_4$ converts benzyl alcohol to BnSePh in the presence of $AlCl_3$. The direct transformation of benzaldehydes into the selenides is achieved by this system. (In situ reduction of ArCHO initiates the process.)

β-Hydroxy selenides.[3] In combination with nitrogen dioxide, PhSeSePh reacts with alkenes to afford β-nitrato selenides, which, on contact with silica gel, are hydrolyzed to the corresponding alcohols.

C_6H_{13}; PhSeSePh, NO_2, 0°, 0.5 h -> rt, 12 h; C_6H_{13}, ONO_2, SePh; SiO_2; C_6H_{13}, OH, SePh; 69% (overall yield)

Peptide synthesis.[4] Peptide-bond formation from *N*-protected amino acids is readily achieved by reaction with α-azidoalkanoic esters, which is promoted by PhSeSePh–Bu_3P in a mixture of DMF and MeCN at room temperature. This is a rapid and self-regulated process.

Bz-Leu + N_3CH_2COOEt $\xrightarrow[\text{Bu}_3\text{P, DMF - MeCN}]{\text{PhSeSePh}}$ Bz-Leu-Gly-OEt 93%

[1]Henriksen, L., Stuhr-Hansen, N. *SC* **26**, 1897 (1996).
[2]Abe, H., Yamasaki, A., Fujii, H., Harayama, T. *CPB* **44**, 2223 (1996).
[3]Han, L.-B., Tanaka, M. *CC* 475 (1996).
[4]Ghosh, S.K., Verma, R., Ghosh, U., Mamdapur, V.R. *BCSJ* **69**, 1705 (1996).

Diphenylsilyl dichloride.

Intramolecular Diels–Alder reaction. For tethering a diene/dienophile pair to control the Diels–Alder reaction, the use of Ph_2SiCl_2 is illustrated in a synthesis of (+)-adrenosterone.[1] The first step is enolsilylation, which is followed immediately by treatment of an alcohol bearing the dienophile.

KN(SiMe$_3$)$_2$; Ph_2SiCl_2 - Et_3N ; DMAP, acid half-ester

PhMe, 200°; K_2CO_3, MeOH

(+)-adrenosterone

[1]Dzierba, C.D., Zandi, K.S., Möllers, T., Shea, K.J. *JACS* **118**, 4711 (1996).

Diphenyltitanocene.

Aldol condensation.[1] α,α′-Dibenzylidenation of cycloalkanones is accomplished on heating the reaction partners and Cp_2TiPh_2 in a sealed tube (without solvent).

Cp$_2$TiPh$_2$

sealed tube

120°, 6 h

86%

[1]Nakano, T., Migita, T. *CL* 2157 (1993).

Disulfur dichloride.

Pentathiepins.[1] *o*-Bis(*t*-butylthio)benzene can be converted to a heterocyclic system containing five sulfur atoms on reaction with S_2Cl_2. This method is well suited to the elaboration of varacin.

S_2Cl_2 - $BaCO_3$ / THF ;

CF_3COOH / CH_2Cl_2

varacin

[1]Toste, F.D., Still, I.W.J. *JACS* **117**, 7261 (1995).

E

[2-(Ethoxycarbonyl)allyl]trihalostannanes.

α-Methylene-γ-lactones.[1] 2-Bromomethylacrylic esters form the functionalized allylating agents that react with aldehydes to afford α-methylene-γ-lactones directly.

$$\text{CH}_2\text{=C(COOEt)CH}_2\text{SnCl}_2\text{Br} + \text{RCHO} \xrightarrow[\text{rt}]{\text{Et}_2\text{O}} \alpha\text{-methylene-}\gamma\text{-lactone (R)}$$

40 - 89%

[1]Fouquet, E., Gabriel, A., Maillard, B., Pereyre, M. *TL* **34**, 7749 (1993).

***(Z)*-2-Ethoxyvinyl(tributyl)stannane.**

α,β-Unsaturated aldehydes.[1] Under the influence of BF_3•OEt_2, the reagent (**1**) is useful for homologation of aldehydes by two carbon atoms.

$$\underset{(\mathbf{1})}{\text{Bu}_3\text{Sn-CH=CH-OEt}} + \text{PhCHO} \xrightarrow[\text{CH}_2\text{Cl}_2\text{, -78}^\circ]{\text{BF}_3\cdot\text{OEt}_2} \text{Ph-CH=CH-CHO}$$

86%

[1]Cabezas, J.A., Oehlschlager, A.C *TL* **36**, 5127 (1995).

Ethylaluminum dichloride.

Homologation of alkenes.[1] The lengthening of the carbon chain of an unsaturated ester on reaction with acetals to give homoallylic ethers is performed with the assistance of $EtAlCl_2$.

$$\text{CH}_2\text{=CH(CH}_2)_8\text{COOMe} + \text{(MeO)}_2\text{CH}_2 \xrightarrow[\text{-78}^\circ\text{ -> rt} \quad \text{24 h}]{\text{EtAlCl}_2\text{ / CH}_2\text{Cl}_2} \text{MeO-CH}_2\text{CH}_2\text{CH=CH(CH}_2)_7\text{COOMe}$$

75%

trans-Carbosilylation of alkynes.[2] Allylsilanes are split and added to alkynes in a trans manner, giving 1-silyl-1,4-dienes. Crotyltrimethylsilane affords γ-adducts only.

[1]Metzger, J.O., Biermann, U. *LA* 1851 (1996).
[2]Asao, N., Yoshikawa, E., Yamamoto, Y. *JOC* **61**, 4874 (1996).

Ethyl diazoacetate. 16, 163–164; **18**, 157

Ethyl tetrahydrofuran-2-carboxylates.[1] β-Alkoxy carbonyl compounds in which the alkoxy group is readily detached in the presence of a Lewis acid undergo condensation with ethyl diazoacetate. Tin(IV) chloride is a suitable catalyst for β-alkoxy aldehydes, but it requires zirconium(IV) chloride to effect a reaction with β-alkoxy ketones.

[1]Angle, S.R., Wei, G.P., Ko, Y.K., Kubo, K. *JACS* **117**, 8041 (1995).

Ethyl *N*-diphenylmethyleneglycinate.

Substituted amino acids.[1] Mono- and disubstitution at the CH_2 group of $Ph_2C{=}N$-CH_2COOEt can be controlled.

[1]Ezquerra, J., Pedregal, C., Moreno-Manas, M., Pleixats, R., Roglans, A. *TL* **34**, 8535 (1993).

Ethyl diphenylphosphonoacetate.

(Z)-α,β-Unsaturated esters.[1] The reagent is available in two steps from the reaction of $(EtO)_2P(O)CH_2COOEt$ with PCl_5 and phenol in 60% yield. It shows a stereoselectivity similar to that of the bis(trifluoroethyl)phosphonoacetate esters.

[1]Ando, K. *TL* **36**, 4105 (1995).

Ethyl trifluoroacetate.

Selective protection of amines. Polyamines undergo trifluoroacetylation only at the primary sites by heating with CF_3COOEt in aqueous acetonitrile.[1] Very highly selective monotrifluoroacetylation of *vic*-diamines has been achieved[2] using one equivalent of the ester at 0°.

[1]O'Sullivan, M.C., Dalrymple, D.M. *TL* **36**, 3451 (1995).
[2]Xu, D., Prasad, K., Repic, O., Blacklock, T.J. *TL* **36**, 7357 (1995).

Europium(III) chloride. 18, 159

Epoxidation.[1] Alkenes are converted into epoxides by oxygen at ambient temperature with a mixture of $EuCl_3$, Zn, and propionic acid in 1,2-dichloroethane.

[1]Yamanaka, I., Nakagaki, K., Akimoto, T., Otsuka, K *CL* 1717 (1994).

Europium tris(1,1,1,2,2,3,3-heptafluoro-7,7-dimethyl-4,6-octanedionate).

Hetero Diels–Alder reactions.[1] The $Eu(fod)_3$ complex is a Lewis acid catalyst that is frequently used to promote the condensation of conjugated carbonyl compounds and enol ethers. High pressure further favors the reaction.[1] Exclusive *endo*-selectivity in the reactions involving *(E)*-1-benzenesulfonyl-3-alken-2-ones has been determined.[2]

Transposition of allylic methoxyacetates.[3] Allylic methoxyacetates undergo rearrangement at room temperature on treatment with $Eu(fod)_3$. Synfacial transfer of the ester group is observed. The rearrangement is efficient and not subject to steric hindrance, because the metal coordinates with the tether atoms.

[1]Vandenput, D.A.L., Scheeren, H.W. *T* **51**, 8383 (1995).
[2]Wada, E., Pei, W., Yasuoka, H., Chin, U., Kanemasa, S. *T* **52**, 1205 (1996).
[3]Shull, B.K., Sakai, T., Koreeda, M. *JACS* **118**, 11690 (1996).

F

Fluorine. 13, 135; **14**, 167; **15**, 160; **18**, 161

Electrophilic halogenation.[1] Fluorine facilitates the introduction of bromine or iodine atoms into aromatic nuclei.

$F_2 - I_2$, H_2SO_4

66%

Fluorodesulfurization.[2] The C–S bonds of sulfides (e.g., thioglycosides) and dithioacetals are readily replaced by the C–F bond on reaction with F_2–I_2.

[1]Chambers, R.D., Skinner, C.J., Atherton, M.J., Moilliet, J.S. *JCS(P1)* 1659 (1996).
[2]Chambers, R.D., Sandford, G., Sparrowhawk, M.C., Atherton, M.J. *JCS(P1)* 1941 (1996).

Fluoroboric acid. 18, 162

Diels–Alder reactions.[1] Enones substituted at the α′-position with an acetal group form cyclic vinyloxocarbenium ions. The enhanced dienophilic reactivity of such species is synthetically useful, and diasteroselective processes have been developed in which the ionization is mediated by fluoroboric acid. Scandium(III) triflate provides lower diastereoselectivity, and $BF_3{\cdot}OEt_3$ is useless, as extensive decomposition of the substrates occurs.

OEt; HBF_4, CH_2Cl_2, - 45°; isoprene ; MeOH; OH

70%

[1]Sammakia, T., Berliner, M.A. *JOC* **60**, 6652 (1995).

1-Fluoro-4-hydroxy-1,3-diazoniabicyclo[2.2.2]octane bis(tetrafluoroborate).

Fluorination of ketones.[1] Direct fluorination with reagent (**1**) often proceeds in high yields, providing α-fluoro ketones.

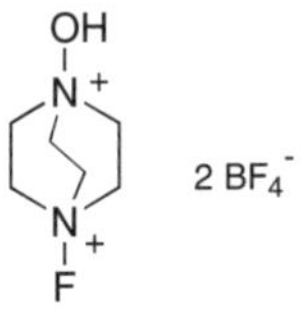

(**1**)

Fluorofunctionalization of alkenes.[2] In the presence of a nucleophile (e.g., MeOH), an addition reaction to alkenes occurs (12 examples, 90–98%).

[1]Stavber, S., Zupan, M. *TL* **37**, 3591 (1996).
[2]Stavber, S., Zupan, M., Poss, A.J., Shia, G.A *TL* **36**, 6769 (1995).

N-Fluoropyridinium sulfonates. 16, 170–171; 18, 162

Fluorination.[1] Internal salts such as (**1**) are excellent fluorinating agents for styrenes (2 examples, 24–51%), anisole (83%), thioanisole (80%), 2,3-dihydrofuran (75%), silyl enol ethers (4 examples, 88–93%), and β-dicarbonyl compounds (5 examples, 46–98%).

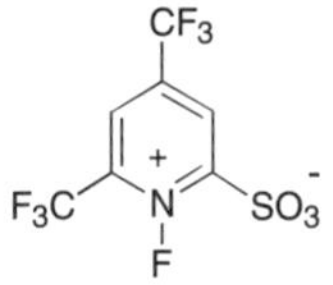

(**1**)

[1]Umemoto, T., Tomizawa, G. *JOC* **60**, 6563 (1995).

N-Fluoro-1,2,3-oxathiazin-4-one.

Fluorides.[1] This stable, crystalline reagent (**1**) is derived from the artificial sweetener Acesulfam. It fluorinates enolates of β-ketoesters and malonic esters, as well as enol acetates and anisoles.

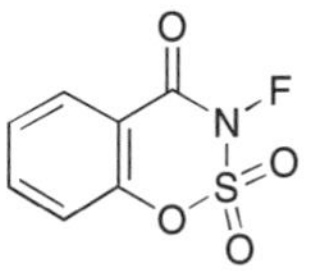

(**1**)

[1]Cabrera, I., Appel, W.K. *T* **51**, 10205 (1995).

Fluorotrichloromethane.

Chlorodecarboxylation.[1] Photochemical decomposition of the Barton esters in $FCCl_3$ leads to chloro compounds.

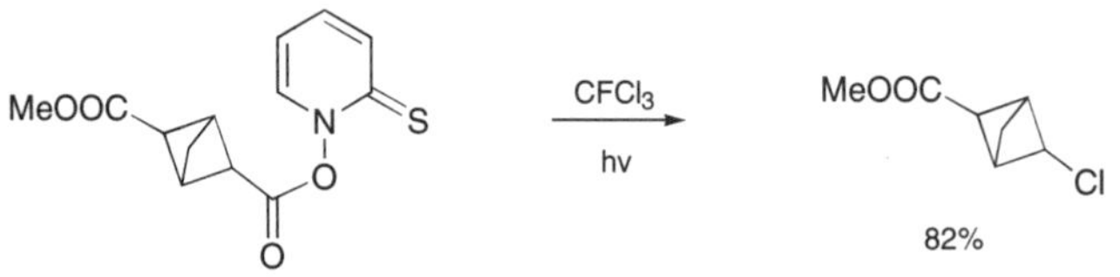

[1]Della, E.W., Taylor, D.K. *JOC* **59**, 2986 (1994).

Formaldehyde dimethylhydrazone.

Hydroformylation and hydrocyanation.[1] The reagent adds to nitroalkenes, giving adducts that are transformed into nitro aldehydes by ozonolysis and into nitro nitriles by treatment with a peracid.

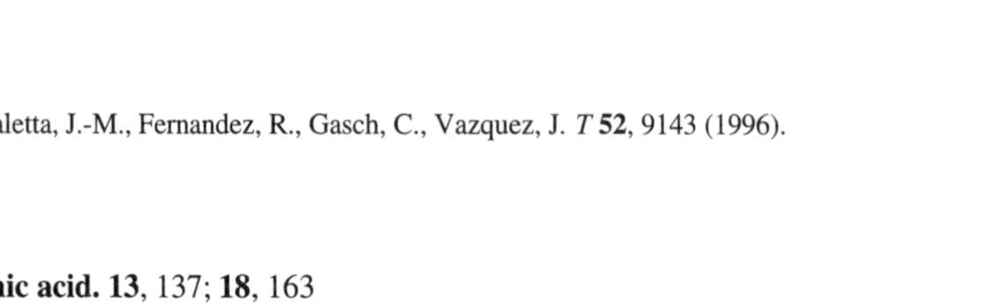

[1]Lassaletta, J.-M., Fernandez, R., Gasch, C., Vazquez, J. *T* **52**, 9143 (1996).

Formic acid. 13, 137; **18**, 163

Reductive cyclization of diynes.[1] The formation of a 1,2-bismethylenecyclohexane unit from the 1,7-diyne is the key feature for an approach to drimane sesquiterpenes. The cyclization conditions using $(dba)_2Pd_2{\cdot}CHCl_3$, Ph_3As, Et_3SiH, and HCOOH in toluene at 80° seem to be optimal. Formic acid is critical to help shift the equilibrium from the Pd(0) catalyst to the active hydridopalladium species.

$(dba)_3Pd_2 \cdot CHCl_3$

Fu_3P - Et_3SiH

HCOOH / PhMe

80°

79%

siccanin

[1]Trost, B.M., Fleitz, F.J., Watkins, W.J. *JACS* **118**, 5146 (1996).

G

Gallium—lead(II) chloride.

Reformatsky-type reagents.[1] The synthesis of β-hydroxy nitriles (9 examples, 55–99%) and β-hydroxy α,α-dichloroalkanoic esters (9 examples, 60–80%) from iodoacetonitrile and trichloroacetic esters, respectively, is mediated by Ga–$PbCl_2$ in refluxing THF.

[1]Zhang, X.-L., Han, Y., Tao, W.-T., Huang, Y.-Z. *JCS(P1)* 189 (1995).

Gallium(III) iodide.

Propargylic alcohols.[1] Condensation of alkynes with carbonyl compounds is simply accomplished by treatment with GaI_3–Bu_3N in THF at room temperature (16 examples, 49–88%).

[1]Han, Y., Huang, Y.-Z. *TL* **36**, 7277 (1995).

Glyoxal. 18, 166

Monoaminals.[1] The chiral monoaminals of glyoxal are valuable building blocks as asymmetric reactions on them become possible. Thus, chiral aldehydes containing an α-acetoxy or α-amino group are readily obtained from organometallic reactions directly or after condensation with tritylamine. The addition of organocuprate to the unsaturated ester derived from an Emmons–Wadsworth reaction generates chiral β-formyl esters.

OHC−CHO → OHC–(imidazolidine, N-R, Ph, Ph) → Ph3CN=CH–(imidazolidine) → R'CH(NHBoc)CHO; R'CH(OAc)–(imidazolidine) → R'CH(OAc)CHO; EtOOC–CH=CH–(imidazolidine) → EtOOC–CH2–CH(R')–CHO

[1]Alexakis, A., Tranchier, J.-P., Lensen, N., Mangeney, P. *JACS* **117**, 10767 (1995).

Grignard reagents. 13, 138–140; **14**, 171–172; **16**, 172–173; **17**, 141–142; **18**, 167–171

As bases. The formation of phosphonium ylides[1] and the transformation of 1,1-dibromoalkenes to acetylenes[2] are two types of applications of Grignard reagents (EtMgBr) as a base.

Ketone syntheses. Acyl derivatives that favor the arrestment of Grignard reactions beyond the first round include *N*-acylpyrazoles,[3] acyl hemiacetals,[4] and acyl tributylphosphonium chlorides (generated in situ from RCOCl and Bu_3P).[5] The protocol involving *N*-methoxy-*N*-methyl carboxamides has been extended to the preparation of α-chloro ketones,[6] α-keto amides, and α-diketones (the last two from the oxalyl diamides).[7] Symmetrical diketones are obtained by the Grignard reaction of bis(benzimidazole) methiodides.[8] Note that an analogous reaction of 1,3-disubstituted benzimidazolium salts furnishes aldehydes.[9]

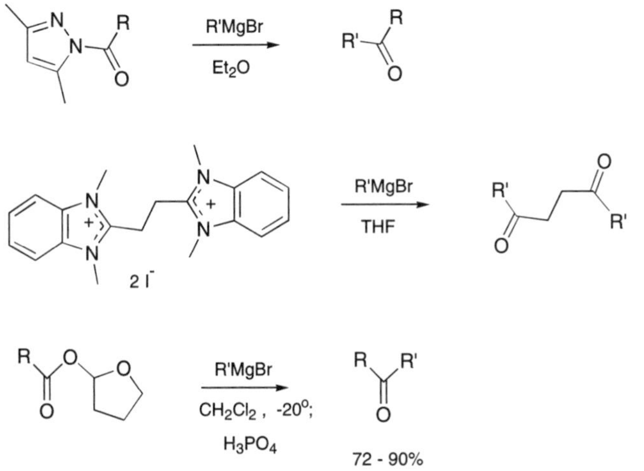

R'MgBr, Et_2O

2 I⁻ R'MgBr, THF

R'MgBr, CH_2Cl_2, -20°; H_3PO_4 72 - 90%

The reaction of 3-(1,1,3,3-tetramethyl-1,3,2-disilaazolidin-2-yl)propylmagnesium bromide (**1**) with aromatic nitriles gives 2-aryl-1-pyrrolines,[10] due to intramolecular cyclodehydration of the released amino ketones.

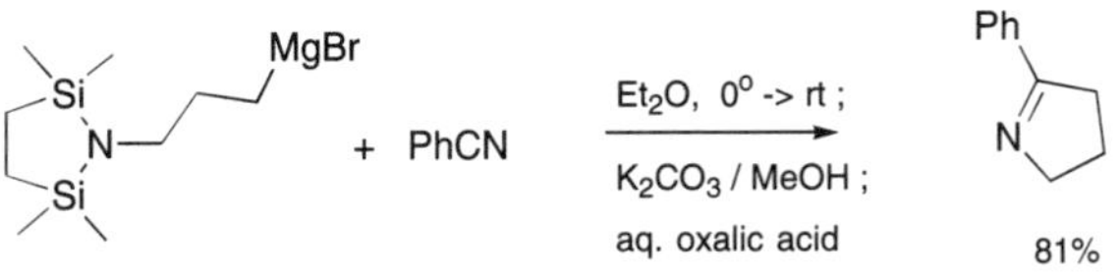

(**1**) + PhCN — Et_2O, 0° -> rt ; K_2CO_3 / MeOH ; aq. oxalic acid → 81%

On reaction with Grignard reagents, β-keto nitriles are transformed into enamino ketones.[11] The carbonyl group is protected as the enolate anion. To retain the nitrogen atom, a mild workup is required.

Attack on high-valent sulfur compounds. Mesylates undergo cleavage on reaction with MeMgBr in THF at room temperature to give alcohols.[12] While the cleavage of α-chloroalkenyl phenyl sulfoxides[13] by EtMgCl provides alkenyl Grignard reagents, in the Grignard reaction of chiral dimethyl α-(p-toluenesulfinyl)methylphosphonate[14] the chemist's interest is in chiral sulfoxide products.

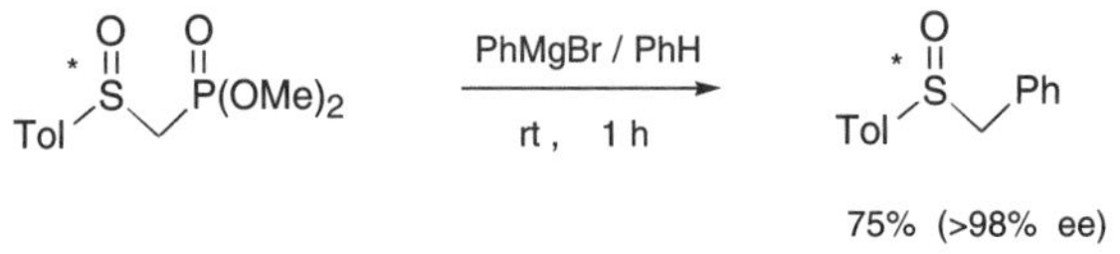
Tol
S
O
P(OMe)2
PhMgBr / PhH
rt, 1 h
Ph
75% (>98% ee)

Conjugate additions. γ-Chloro-α,β-unsaturated sulfoxides give cyclopropane derivatives on Grignard reactions.[15] Apparently, intramolecular displacement of the γ-chloro substituent in the α-sulfinyl anion intermediates is highly favorable.

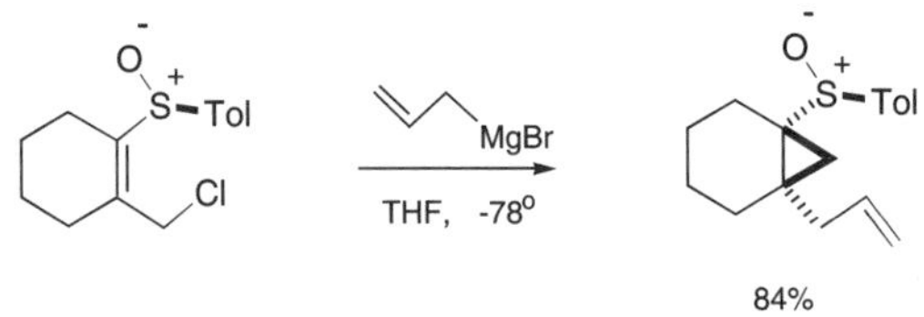
S
Tol
Cl
MgBr
THF, -78°
84%

Substitution of allyloxy derivatives. 2-Alkenyl-1,3-dioxolan-4-ones undergo ring opening with Grignard reagents, with the attack controlled by the substitution pattern at C-5. The products can readily be converted to chiral allylic alcohols and protected α-hydroxy aldehydes (ozonolysis).[16]

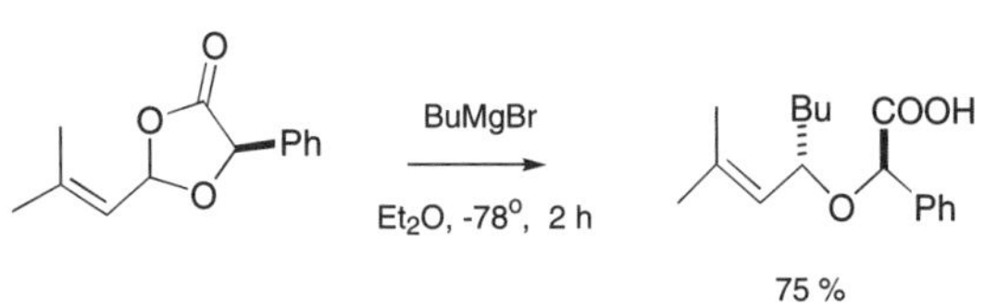
O
Ph
BuMgBr
Et2O, -78°, 2 h
Bu
COOH
75 %

The ring opening of endoperoxides by Grignard reagents and related organometallics is interesting, as monoethers of *cis*-2-cycloalkene-1,4-diols are formed.[17]

OH HO O H MgBr xylene THF, -78° Δ 83% 90%

Reaction with boronates and arylbismuth triflates. Dialkyl 2,3-butadienylboronates, which serve as 1,3-butadiene-2-yl anion equivalents, are formed from allenylmagnesium bromide with dialkyl (halomethyl)boronates.[18]

RCHO PhMe rt (iPrO)$_2$B R OH 59 - 95%

Triarylbismuthanes bearing three different aryl groups are available by succesive Grignard reactions of arylbismuth triflates.[19]

Ar_3Bi Me_3SiOTf -HMPA Ar'MgBr / THF Ar—Bi(Ar)(Ar') Me_3SiOTf -HMPA Ar"MgBr / THF Ar—Bi(Ar")(Ar') 70 - 80%

Vinylmagnesium bromides. These reagents can be employed as alkene equivalents in Diels–Alder reactions. Their reaction with lithium alka-2,4-dienolates followed by heating at 130° leads to 2-cyclohexene-1-methanols.[20] The Mg atom serves as a temporary tether to render the reaction intramolecular, so that highly substituted unactivated dienes and dienophiles are "coaxed" to partake in cycloaddition. Even 4,4-disubstituted alkadienols become reactive. Unlike silicon atom tethers, the cycloadducts undergo protonolysis on workup.

OH BuLi ; MgBr O Mg Δ ; H_2O OH 44%

[1]Shen, Y., Yao, J. *JCR(R)* 394 (1995).
[2]Jiang, B., Ma, P. *SC* **25**, 3641 (1995).
[3]Kashima, C., Kita, I., Takahashi, K., Hosomi, A. *JHC* **32**, 25 (1995).
[4]Mattson, M.N., Rapoport, H. *JOC* **61**, 6071 (1996).
[5]Maeda, H., Okamoto, J., Ohmori, H. *TL* **37**, 5381 (1996).
[6]Tillyer, R., Frey, L.F., Tschaen, D.M., Dolling, U.-H. *SL* 225 (1996).
[7]Sibi, M.P., Marvin, M., Sharma, R. *JOC* **60**, 5016 (1995).
[8]Shi, Z., Gu, H., Xu, L.-L. *SC* **26**, 3175 (1996).
[9]Shi, Z., Gu, H. *SC* **26**, 4175 (1996).
[10]Keppens, M., De Kimpe, N., Fonck, G. *SC* **26**, 3097 (1996).
[11]Sibi, M.P., Marvin, M., Sharma, R. *JOC* **59**, 4040 (1994).
[12]Cossy, J., Ranaivosata, J.-L., Bellosta, V., Wietzke, R. *SC* **25**, 3109 (1995).
[13]Satoh, T., Takano, K., Someya, H., Matsuda, K. *TL* **36**, 7097 (1995).
[14]Cardellicchiu, C., Iacuone, A., Naso, F., Tortorella, P. *TL* **37**, 6017 (1996).
[15]Takemoto, Y., Ohra, T., Sugiyama, K., Imanishi, T., Iwata, C. *CPB* **43**, 571 (1995).
[16]Heckmann, B., Mioskowski, C., Bhatt, R.K., Falck, J.R. *TL* **37**, 1421 (1996).
[17]Schwaebe, M.K., Little, R.D. *TL* **37**, 6635 (1996).
[18]Soundararajan, R., Li, G., Brown, H.C. *JOC* **61**, 100 (1996).
[19]Matano, Y., Miyamatsu, T., Suzuki, H. *OM* **61**, 8 (1996).
[20]Stork, G., Chan, T.Y. *JACS* **117**, 6595 (1995).

Grignard reagents/cerium(III) chloride. 18, 171

Drying is important for improving the efficiency of the addition reactions.[1] Such reactions can be rendered catalytic.

2-Substituted allylic alcohols.[2] The synthesis from esters by a two-step branching homologation process involves nucleophilic attack of an (alkoxysilyl)methylmagnesium chloride (2 equiv.) and oxidative desilylation.

MeOOC–CH(Me)–CH₂–OBn → [iPrO–SiMe₂–CH₂MgCl; $CeCl_3$ / THF; -78° -> rt] → iPrO–SiMe₂–CH₂–C(=CH₂)–CH(Me)–CH₂–OBn (89%) → [KF; H_2O_2] → HO–CH₂–C(=CH₂)–CH(Me)–CH₂–OBn (98%)

[1]Dimitov, V., Kostova, K., Genov, M. *TL* **37**, 6787 (1996).
[2]Mickelson, T.J., Koviach, J.L., Forsyth, C.J. *JOC* **61**, 9617 (1996).

Grignard reagents/copper salts. 18, 171–173

Homoallylic alcohols.[1] When 1,2-diols are treated with *N*-tosylimidazole and then vinylmagnesium bromide in the presence of CuI, homoallylic alcohols are obtained. This one-pot process involves epoxide intermediates that are opened by Grignard reagents. The method can be employed in the synthesis of chiral products.

α-Diketones and α-ketoesters. Oxalyl chloride reacts with two equivalents of Grignard reagents in the presence of CuBr and LiBr to afford α-diketones.[2] A similar reaction of the monochloride/esters of oxalic acid gives the α-ketoesters.[3]

A convenient approach to 4- and 5-oxoacids consists of analogous Cu-catalyzed reactions of cyclic anhydrides (succinic and glutaric anhydrides).[4]

Reaction of aziridines. As aziridines with an activating group on nitrogen (e.g., phosphoamides) undergo electrophilic reactions, primary amines may be prepared by the Grignard reaction and subsequent removal of the activating group.[5] Substitution at the side chain of *N,O*-bis(diphenylphosphinyl)hydroxymethyl aziridine predominates, because a more reactive leaving group is present.[6]

Conjugate additions. The Kharasch reaction using CuI·2LiCl and Me_3SiCl as catalysts has been evaluated.[7] Another report[8] delineates synthetic and spectroscopic aspects of the conjugate allylation.

The replacement of a β-methylthio residue of enones by Cu(I)-catalyzed reaction is both chemo- and stereoselective.[9,10]

S_N2' displacements. A method for the synthesis of 3,3-difluoroalkan-2-ones involves Grignard reaction of $F_2C{=}C(OEt)CH_2OAc$ and hydrolysis of the resulting enol ethers.[11]

F, F, OAc, OEt + PhMgBr —CuCN - LiCl, THF, -20° -> rt→ F, F, Ph, OEt

60%

1,2-Diarylethenyl ethyl ethers.[12] Carbocupration of ethoxyethyne using ArMgBr, CuBr, and LiBr followed by Pd-catalyzed coupling with aryl iodides provides the products (10 examples, 58–84%).

EtO—≡ —PhMgBr, CuBr - LiBr, THF, -20°→ [EtO, Ph, $CuMgBr_2$] —ArI, $(dppf)PdCl_2$, $ZnBr_2$→ EtO, Ph, Ar

[1]Cink, R.D., Forsyth, C.J. *JOC* **60**, 8122 (1995).
[2]Badudri, F., Fiandanese, V., Marchese, G., Punzi, A. *TL* **36**, 7305 (1995).
[3]Badudri, F., Fiandanese, V., Marchese, G., Punzi, A. *T* **52**, 13513 (1996).
[4]Lhommet, G., Freville, S., Thuy, V., Petit, H., Celerier, J.P. *SC* **26**, 2397 (1996).
[5]Osowska-Pacewicka, K., Zwierzak, A. *S* 333 (1996).
[6]Cantrill, A.A., Sweeney, J.B. *SL* 1277 (1995).
[7]Reetz, M.T., Kindler, A. *JOMC* **502**, C5 (1995).
[8]Lipshutz, B.H., Hackmann, C. *JOC* **59**, 7437 (1994).
[9]Mehta, B.K., Ila, H., Junjappa, H. *TL* **36**, 1925 (1995).

[10]Mehta, B.K., Dhar, S., Ila, H., Junjappa, H. *TL* **36**, 9377 (1995).
[11]Shi, G.-Q., Cai, W.-L. *SL* 371 (1996).
[12]Kato, N., Miyaura, N. *T* **52**, 13347 (1996).

Grignard reagents/nickel complexes. 18, 173

S_N2' displacements. 4-Alkenyl-1,3-dioxolan-2-ones undergo ring opening in the Ni(II)-catalyzed Grignard reaction, furnishing allylic alcohols.[1] Allylic ethers are also reactive when there is a phosphine group to coordinate Ni.[2] Note that Grignard reagents which can act as hydride donors give products of reductive cleavage of the allylic ethers without transposition of the double bond.[3]

MeMgBr, $(Ph_3P)_2NiCl_2$, THF, 22°, 18 h

70% (*Z*:*E* > 49 : 1)

The Ni-catalyzed reaction solves the problem in ring opening of recalcitrant oxabicycles.[4] Alkynyl dithioacetals that can be described as allene 1,3-dication synthons give allenes.[5]

MeMgI, (dppe)$NiCl_2$, PhH, 65°, 12 h

90%

C-Aryl-Δ^2-glycopyranosides.[6] Unsaturated glycosides undergo substitution by Grignard reagents with either Ni or Pd catalysts. The opposite stereoselectivity of the two reactions is fascinating.

PhMgBr, THF

+ (dppf)$NiCl_2$	-40°	(70 %)	0	:	100
+ (dppf)$PdCl_2$	25°	(95 %)	100	:	0

Ketones.[7] The Grignard reaction of acyl halides is moderated by Ni complexes. Ketones, including benzils, can be obtained under such conditions.

(dppe)$NiCl_2$

80%

Indolines.[8] Nickel complexes mediate the coupling of phenethylmagnesium halides with organoazides, resulting in indolines.

$(Ph_3P)_2NiCl_2$

ArN_3

R = Ts, Tol

Allylsilanes.[9] Alkenyl selenides are converted to allylsilanes on treatment with Me_3SiCH_2MgCl in DME in the presence of $(Ph_3P)_2NiCl_2$ [or $(Ph_3P)_2PdCl_2$]. Using ether or THF as solvent gives lower yields.

[1]Kang, S.-K., Cho, D.-G., Park, C.-H., Namkoong, E.-Y., Shin, J.-S. *SC* **25**, 1659 (1995).
[2]Didiuk, M.T., Morken, J.P., Hoveyda, A.H. *JACS* **117**, 7273 (1995).
[3]Morken, J.P., Didiuk, M.T., Hoveyda, A.H. *TL* **37**, 3613 (1996).
[4]Lautens, M., Ma, S. *JOC* **61**, 7246 (1996).
[5]Tseng, H.-R., Luh, T.-Y. *JOC* **61**, 8685 (1996).
[6]Moineau, C., Bolitt, V., Sinou, D. *CC* 1103 (1995).
[7]Malanga, C., Aronica, L.A., Lardicci, L. *TL* **36**, 9185 (1995).
[8]Koo, K., Hillhouse, G.L. *OM* **15**, 2669 (1996).
[9]Hevesi, L., Hermans, B., Allard, C. *TL* **35**, 6729 (1994).

Grignard reagents/samarium (II) iodide.

Cycloalkanols.[1] A mixture of RMgX and SmI_2 converts ω-haloalkanoates into cycloalkanols. Thus, cyclopropanols are prepared very readily, while cyclopentanols and cyclohexanols are obtained in lower yields, but cyclobutanols cannot be made by this method, since tetrahydrofuran derivatives are formed.

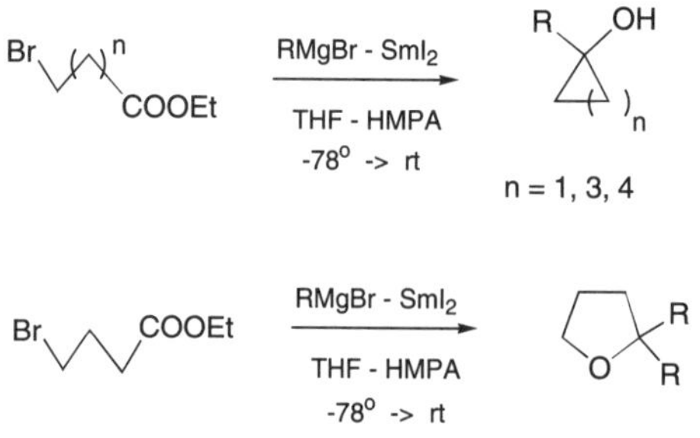

[1]Fukuzawa, S.-I., Furuya, H., Tsuchimoto, T. *T* **52**, 1953 (1996).

Grignard reagents/titanium(IV) compounds. 14, 121–122; **18**, 174

(Z)-Alkenes.[1] The method for generating *(Z)*-alkenes from alkynes can be extended to conjugated and methylene-skipped enynes. On quenching the titanacyclopropene intermediates with D_2O, dideuterio products are obtained.

Pyrroles.[2] Adding imines to the titanacyclopropenes and exposing the adducts to CO lead to 2,3,4-trisubstituted pyrroles.

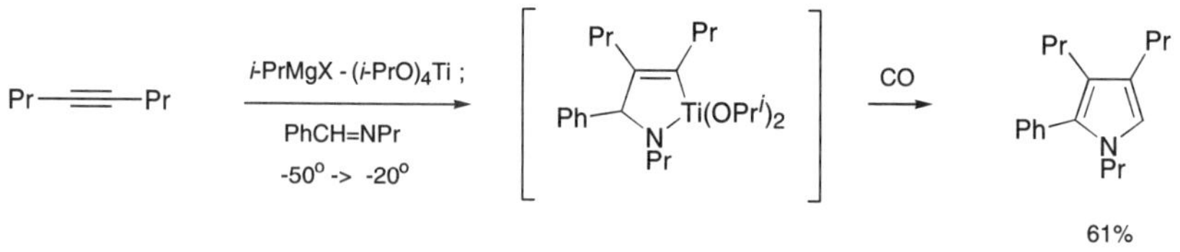

61%

Allylation. Allyl halides are converted to allyltitanium compounds by *i*-PrMgBr/(*i*-PrO)$_4$Ti, which can be used for allylation of carbonyl compounds[3] and aldimines.[4] With imines derived from a chiral α-phenethylamine, the reaction is highly diastereoselective following the Cram pattern exhibiting a 1,3-asymmetric induction.

Allyltitanium species are also obtained from a diene and bis(η^5-indenyl)titanium dichloride in the presence of *i*-PrMgCl.[5]

Cleavage of allyl derivatives. A combination of (*i*-PrO)$_4$Ti and a Grignard reagent is useful for de-*O*-allylation[6] without affecting other protecting groups, such as TIPS. Perhaps more interesting is that an allyl group at the central carbon of a malonic ester can be removed in the same way, indicating a new way to protect acidic hydrogen of malonic esters by C-allylation.[7]

Hydrodehalogenation of gem-dihalocyclopropanes.[8] The rapid reaction is carried out in refluxing ether to give the monobromocyclopropanes.

Aldol reactions.[9] Titanium -ate complexes effect *anti*-selective aldolization (5 examples, 72–81%).

Intramolecular acylation. When a carbonate ester containing an alkyne[10] or allene unit[11] at the proper length is submitted to *i*-PrMgBr/(*i*-PrO)$_4$Ti, the multiple bond is metallated and becomes nucleophilic such that internal attack on the carbonyl group results in O → C transfer of the alkoxycarbonyl group. Diethyl alkynylmalonates also undergo a similar reaction.[12]

Me$_3$Si ... OCOOEt —(*i*-PrMgBr, (*i*-PrO)$_4$Ti)→ Me$_3$Si ... 68%

Bu ... C ... OCOOEt —(*i*-PrMgBr, (*i*-PrO)$_4$Ti)→ EtOOC, Bu ... OH 83 % (*Z* : *E* 77 : 23)

R ... X ... COOEt —(*i*-PrMgBr, (*i*-PrO)$_4$Ti, Et$_2$O, -45°)→ R, EtO$_2$C, CO$_2$Et / R ... O

X = COOEt 42%
R = Me

X = Me 68%
R = Hex

2-Iodo-1,3-dienes.[13] Quenching the reaction of a 2,3-alkadienyl carbonate with iodine furnishes 2-iodo-1,3-alkadienes. Key intermediates are 1,3-alkadien-2-yltitanium species.

Functionalized cyclopropanes. The reductive coupling of esters with 1-alkenes by a Grignard reagent and titanium alkoxide gives rise to cyclopropanols (the Kulinkovich reaction).[14] With the use of ethylene carbonate, the condensation gives rise to 2-substituted cyclopropanone hemiacetals.[15] An intramolecular version delivers bicyclic products.[16]

COOMe —(BuMgBr, (*i*-PrO)$_4$Ti)→ OH

—(c-C$_5$H$_9$MgBr, (*i*-PrO)$_4$Ti, ethylene carbonate)→ O, OH, HO 47%

Similarly, tertiary amides are converted into dialkylaminocyclopropanes.[17]

Note that the diastereoselective synthesis of *cis*-1,2-dialkylcyclopropanol is considerably improved by using $(i\text{-PrO})_3TiCl$ or $(i\text{-PrO})_4Ti$ with RCH_2CH_2MgBr (where R > H).[18] The procedure is further simplified by adding an alkyl bromide to a mixture of the ester, $(i\text{-PrO})_3TiCl$, and Mg in THF at room temperature, thus obviating the preparation of the Grignard reagent. In the presence of a chiral TADDOLate, the reaction becomes enantioselective.

Bicyclo[3.3.0]octanes and bicyclo[3.1.0]hexanes.[19] Heptenynes bearing a conjugated ester group undergo cyclization. However, the result depends on whether they are enoic esters or ynoic esters.

Titanabicycloalkene intermediates derived from diynes or enynes in which the terminal sp-carbon bears a silyl group are alkylating agents. Thus, a subsequent reaction with aldehydes generates allylic alcohols.[20]

[1] Hungerford, N.L., Kitching, W. *CC* 1697 (1995).
[2] Gao, Y., Shirai, M., Sato, F. *TL* **37**, 7787 (1996).
[3] Kasatkin, A., Nakagawa, T., Okamoto, S., Sato, F. *JACS* **117**, 3881 (1995).
[4] Gao, Y., Sato, F. *JOC* **60**, 8136 (1995).
[5] Urabe, H., Yoshikawa, K., Sato, F. *TL* **36**, 5595 (1995).
[6] Lee, J., Cha, J.K. *TL* **37**, 3663 (1996).
[7] Yamazaki, T., Kasatkin, A., Kawanaka, Y., Sato, F. *JOC* **61**, 2266 (1996).

[8]Al Dulayymi, J.R., Baird, M.S., Bolesov, I.G., Tveresovsky, V., Rubin, M. *TL* **37**, 8933 (1996).
[9]Mahrwald, R. *T* **51**, 9015 (1995).
[10]Kasatkin, A., Okamoto, S., Sato, F. *TL* **36**, 6075, 6089 (1995).
[11]Yoshida, Y., Okamoto, S., Sato, F. *JOC* **61**, 7826 (1996).
[12]Kasatkin, A., Yamazaki, T., Sato, F. *ACIEE* **35**, 1966 (1996).
[13]Okamoto, S., Sato, H., Sato, F. *TL* **37**, 8865 (1996).
[14]Lee, J., Kim, H., Cha, J.-K. *JACS* **118**, 4198 (1996).
[15]Lee, J., Kim, Y. G., Bae, J.G., Cha, J.-K. *JOC* **61**, 4878 (1996).
[16]Lee, J., Kang, C.H., Kim, H., Cha, J.-K. *JACS* **118**, 291 (1996).
[17]Chaplinski, V., de Meijere, A. *ACIEE* **35**, 413 (1996).
[18]Corey, E.J., Rao, S.A., Noe, M.C. *JACS* **116**, 9345 (1994).
[19]Suzuki, K., Urabe, H., Sato, F. *JACS* **118**, 8729 (1996).
[20]Urabe, H., Sato, F. *JOC* **61**, 6756 (1996).

Grignard reagents/zirconium complexes. 18, 174

Kinetic resolution.[1] Cyclic allylic ethers react with EtMgBr in the presence of a zirconocene–BINOL complex [*(R)*-(EBTHI)Zr-BINOL], showing enantioselectivity. Thus, unreacted ethers are recovered in the chiral form.

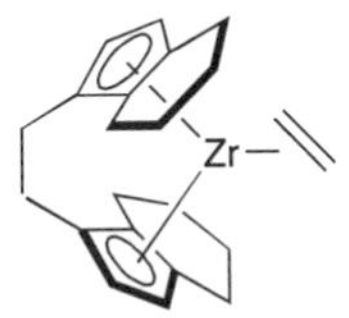

[1]Visser, M.S., Harrity, J.P.A., Hoveyda, A.H. *JACS* **118**, 3779 (1996).

H

Hafnium(IV) triflate. 18, 175

Fries rearrangement.[1] This salt is an efficient catalyst for the rearrangement in a mixture of toluene and nitromethane at 100° (15 examples, 53–97%).

[1]Kobayashi, S., Morikawi, M., Hachiya, I. *TL* **37**, 2053 (1996).

Hafnocene dichloride.

Disaccharides.[1] Glycosyl fluorides are activated by Cp_2HfCl_2–$AgClO_4$ to react with another sugar.

[1]Barrena, M.I., Echarri, R., Castillon, S. *SL* 675 (1996).

Hexaalkylditin. 13, 142; **14**, 173–174; **16**, 174; **17**, 143–144; **18**, 175–176

Allylation.[1] Allyl phenyl sulfide submits the allyl group to a sugar when it is treated with $(Bu_3Sn)_2$ and a glycosyl bromide.

Cyclization with group transfer. Unsaturated organoiodides and tellurides undergo cyclization of free radicals. With a hydrosilane to provide a hydrogen atom the configuration of the alkylidenecycloalkane product is predetermined.[2] There is a distinct preference for cyclization leading to a tetrahydrofuran rather than to a cyclopentane.[3]

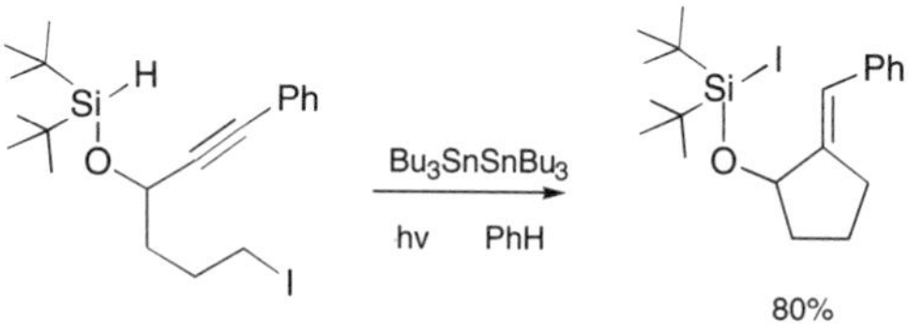

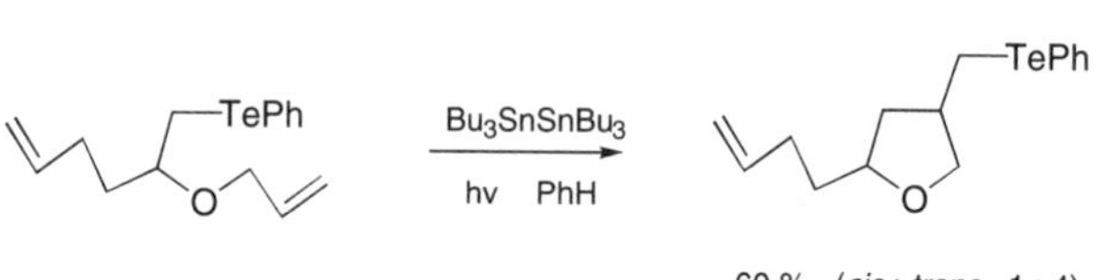

$Bu_3SnSnBu_3$, hv PhH

80%

$Bu_3SnSnBu_3$, hv PhH

60 % (*cis* : *trans* 1 : 4)

Cycloalkanones.[4] ω-Bromo thioesters and selenoesters give cyclic ketones on photolysis in the presence of $(Bu_3Sn)_2$.

EtOOC COOEt, Br, COSPh → $Bu_3SnSnBu_3$, hv (300 nm) → EtOOC COOEt, O; 85%

Synthesis of oxime ethers.[5] Phenylsulfonyloxime ethers accept free radicals generated from organohalides and subsequently expel the $PhSO_2$ unit. Thus, the process constitutes a chain extension by a masked acyl group.

OEt, O, I → $Me_3SnSnMe_3$, hv → [O, OEt] → $PhSO_2$, =N-OBn, R → H, O, OEt, H, =N-OBn, R

R = H 88%
R = Me 78%

vic-Bis(trimethylstannyl)alkenes.[6] With $(Ph_3P)_4Pd$ as a catalyst, hexamethylditin homolytically adds to alkyl 2-alkynoates and *N,N*-dimethyl-2-alkynamides (21 examples, 64–95%).

[1]Ponten, F., Magnusson, G. *JOC* **61**, 7463 (1996).
[2]Martinez-Grau, A., Curran, D.P. *JOC* **60**, 8332 (1995).
[3]Engman, L., Gupta, V. *CC* 2515 (1995).
[4]Kim, S., Jon, S.Y. *CC* 1335 (1996).
[5]Kim, S., Lee, I.Y., Yoon, J.-Y., Oh, D.H. *JACS* **118**, 5138 (1996).
[6]Piers, E., Skerlj, R.T. *CJC* **72**, 2468 (1994).

Hexafluoropropene–diethylamine.

gem-Difluorination.[1] Together with *N*-iodosuccinimide or 1,3-dibromo-5,5-dimethylhydantoin, the combination of reagents converts 1,3-dithiolanes into *gem*-difluorides.

[1]Shimizu, M., Maeda, T., Fujisawa, T. *JFC* **71**, 9 (1995).

Hexamethyldisilazane. 13, 141; **18**, 177–178

Allylsilanes.[1] Allylic acetates and trifluoroacetates are converted into the trimethylsilane derivatives by a Pd(0) catalyst. The reaction of acetates requires higher temperatures and the presence of LiCl.

OAc
C_7H_{15} + $Me_3SiSiMe_3$ → ($Pd(dba)_2$, LiCl; DMF; 100°, 40 h) C_7H_{15} $SiMe_3$

81% (*E* : *Z* 95 : 5)

[1]Tsuji, Y., Funato, M., Ozawa, M., Ogiyama, H., Kajita, S., Kawamura, T. *JOC* **61**, 5779 (1996).

Hexamethylenetetramine. 18, 178

α-Methylenation of aryl ketones.[1] Heating a ketone with the amine in acetic anhydride at 80° furnishes acrylophenones (8 examples, 76–87%), which are precursors of 2-alkylindanones.

[1]Bhattacharya, A., Segmuller, B., Ybarra, A. *SC* **26**, 1775 (1996).

Hydridotetrakis(triphenylphosphine)rhodium(I).

Reduction of enones.[1] This complex catalyzes the reduction of enones with hydrosilanes in excellent yields. Remarkably, the chemoselectivity of the reaction is dependent on the silane (e.g., $PhMe_2SiH$ for 1,4-reduction and Ph_2SiH_2 for 1,2-reduction).

OH ← ($(Ph_3P)_4RhH$; Ph_2SiH_2 / CH_2Cl_2) O → ($(Ph_3P)_4RhH$; $PhMe_2SiH$) $OSiMe_2Ph$

84% 84%

[1]Zheng, G.Z., Chan, T.-H. *OM* **14**, 70 (1995).

Hydrogen fluoride–amine. 16, 286–287; **18**, 181

Fluorocyclization.[1] Unsaturated aldehydes in which the double bond and the carbonyl group are separated by four or five bonds undergo cyclization to give fluoro alcohols.

CHO → (HF - Et_3N; CH_2Cl_2, 25°) OH + OH
F F

78% (89 : 11)

Ring expansion.[2] Cycloalkylideneacetates afford ring expansion esters with the introduction of two geminal fluorine atoms when subjected to electrochemical fluorination (10 examples, 42–83%). Thus, ethyl 2,2-difluorocycloheptanecarboxylate is obtained from the cyclohexylideneacetic ester.

HF - Et_3N; e^-, -20°; 71%

α-Fluoro carbonyl compounds.[3] These substances are readily acquired by the rearrangement of 2-fluorooxiranes on treatment with $3HF{\cdot}NEt_3$.

Furans.[4] γ-Hydroxy-α,β-unsaturated ketones give furans upon reaction with $3HF{\cdot}NEt_3$.

[1]Hayashi, E., Hara, S., Shirato, H., Hatakeyama, T., Fukuhara, T., Yoneda, N. *CL* 205 (1995).
[2]Hara, S., Chen, S.-Q., Hoshio, T., Fukuhara, T., Yoneda, N. *TL* **37**, 8511 (1996).
[3]Michel, D., Schlosser, M. *T* **52**, 2429 (1996).
[4]Sammond, D.M., Sammakia, T. *TL* **37**, 6065 (1996).

Hydrogen fluoride–antimony(V) fluoride–carbon tetrachloride.

Dehydrogenation.[1] The superacid system promotes the ionization of CCl_4 to form Cl_3C^+, which is capable of dehydrogenating polycyclic ketones.

HF - SbF_5 - CCl_4; -30°, 45 min; 63%

Functionalization of nonactivated C–H bonds.[2] Amides and ketones are functionalized at a remote unactivated site after the rearrangement of carbocationic intermediates. *N*-Bromosuccinimide may be used instead of CCl_4.

HF - SbF_5 - CCl_4; Na_2CO_3 / ice H_2O; OH; O; 80%

[1]Martin, A., Jouannetaud, M.-P., Jacquesy, J.-C. *TL* **37**, 7731 (1996).
[2]Martin, A., Jouannetaud, M.-P., Jacquesy, J.-C. *TL* **37**, 2967 (1996).

Hydrogen fluoride–tetrabutylammonium fluoride–*N*-haloimide.

Reaction with sulfur compounds. Dithoic esters [RC(=S)SR′] are transformed into trifluoromethyl derivatives (RCF_3), and thus, *N*-trifluoromethylamines[1] and 3,3,3-trifluoropropenes[2] are readily obtained under very mild conditions (CH_2Cl_2, 0° to room temperature). By changing the reagent to HF-pyridine–HgF_2–KF, the reaction proceeds with retention of one C–S bond, furnishing α,α-difluoroalkyl sulfides.[3] The latter compounds are convenient precursors of 1,1-difluoroalkenes (by way of oxidation and pyrolysis of the resulting sulfoxides).

Sulfides and 1,1,1-tris(methylthio)alkanes behave slightly differently. Fluoro-Pummerer rearrangement of sulfides leads to α-fluoro sulfides,[4] whereas partial replacement of the SMe groups with fluorine atoms and concomitant α-bromination seems to be the option for the orthothioesters.[5]

[1]Furuta, S., Hiyama, T. *SL* 1199 (1996).
[2]Kuroboshi, M., Mizuno, K., Kanie, K., Hiyama, T. *TL* **36**, 563 (1995).
[3]Kim, K.-I., McCarthy, J.R. *TL* **37**, 3223 (1996).
[4]Furuta, S., Kuroboshi, M., Hiyama, T. *TL* **36**, 8243 (1995).
[5]Furuta, S., Hiyama, T. *TL* **37**, 7983 (1996).

Hydrogen peroxide, acidic. 14, 176; **15**, 167, 168; **16**, 177–178; **17**, 145; **18**, 182–183

Epoxidation.[1] Allylic and homoallylic alcohols are epoxidized very effectively with chloral–H_2O_2. *p*-Toluenepersulfonic acid together with hydrogen peroxide in methanol also constitutes a useful epoxidizing agent.[2]

Oxidation of sulfur compounds. With maleic anhydride (forming permaleic acid), sulfides are oxidized to sulfones.[3] The H_2O_2–HOAc system has been used in the desulfurization of cyclic thioureas to yield imidazoles.[4]

[1]Kasch, H. *TL* **37**, 8349 (1996).
[2]Kluge, R., Schulz, M., Liebsch, S. *T* **52**, 2957 (1996).
[3]Drago, R.S., Mateus, A.L.M.L., Patton, D. *JOC* **61**, 5693 (1996).
[4]Grivas, S., Ronne, E. *ACS* **49**, 225 (1995).

Hydrogen peroxide–metal catalysts. 13, 145; **14**, 177; **15**, 294; **17**, 146–148; **18**, 184–185

In addition to those reactions described next, those promoted by the H_2O_2–$MeReO_3$ system are listed under **methylrhenium trioxide**.

Epoxidation. A practical method for the epoxidation of 1-alkenes under halide-free conditions employs 30% H_2O_2–Na_2WO_4 and a phase transfer agent.[1]

Oxidation of 1,2-diols.[2] α-Diketones are produced by the reaction of glycols with H_2O_2, catalyzed by peroxotungstophosphate.

Oxidation of amines. Zeolites (Ti silicalite) serve as catalysts for the selective oxidation of benzylic and allylic amines to oximes,[3] and secondary amines to nitrones.[4]

Bromination. Bromination of arenes (such as anisole) can use HBr or KBr as the bromine source in the presence of H_2O_2 and sodium tungstate[5] or molybdate.[6]

Oxidation of ethers.[7] Ethers give acids or lactones on heating with H_2O_2 over zeolites.

[1]Sato, K., Aoki, M., Ogawa, M., Hashimoto, T., Noyori, R. *JOC* **61**, 8310 (1996).
[2]Iwahama, T., Sakaguchi, S., Nishiyama, Y., Ishii, Y. *TL* **36**, 1523 (1995).
[3]Joseph, R., Ravindranathan, T., Sudalai, A. *TL* **36**, 1903 (1995).
[4]Joseph, R., Ravindranathan, T., Sudalai, A. *SL* 1177 (1995).
[5]Bezodis, P., Hanson, J.R., Petit, P. *JCR(S)* 334 (1996).
[6]Conte, V., Di Furia, F., Moro, S. *TL* **37**, 8609 (1996).
[7]Sasidharan, M., Suresh, S., Sudalai, A. *TL* **36**, 9071 (1995).

Hydrogen peroxide—*N,N'*-dicyclohexylcarbodiimide.

Epoxidation.[1] Alkenes are transformed into epoxides with dilute H_2O_2 in the presence of DCC (12 examples, 19–91%).

[1]Majetich, G., Hicks, R. *SL* 649 (1996).

Hydrogen peroxide—tetracyanoethylene.

Epoxidation.[1] The combination is a mild oxidant that is used in acetonitrile at room temperature.

[1]Masaki, Y., Miura, T., Mukai, I., Iwata, I., Oda, H., Itoh, A. *CPB* **43**, 686 (1995).

Hydrogen peroxide—isoalloxazinium salt.

Baeyer–Villiger oxidation.[1] A 3,5,10-trisubstituted isoalloxazinium salt is capable of catalyzing the conversion of cyclic ketones to lactones in the presence of H_2O_2 at room temperature.

[1]Mazzini, C., Lebreton, J., Furstoss, R. *JOC* **61**, 8 (1996).

Hydrosilanes.

Reductions. Epoxy ketones are reduced by trimethoxysilane, with stereoselectivity dependent on the polarity of the solvent.[1] Thus, *anti*-alcohols are favored in ether, whereas *syn*-alcohols are favored in HMPA.

$(MeO)_3SiH$ / LiOMe

Et_2O / -20°	8	:	92
HMPA / 0°	90	:	10

Amides[2] and lactones[3] are converted to aldehydes and lactols, respectively, by a hydrosilane when a titanium alkoxide is present. Two-level reduction, i.e., of esters to ethers,[4] is achieved by using manganese carbonyl complexes [$(Ph_3P)(CO)_4MnAc$] as catalysts.

An organotin hydride mediates the conjugate reduction of unsaturated carbonyl compounds[5] by $PhSiH_3$ in a free-radical process. The analogous reduction of α,β-unsaturated esters containing a remote oxo group results in the generation of bicyclic lactones after desilylation(6 examples, 66–85%).[6]

Δ; Bu_3SnH - $PhSiH_3$; AIBN, EtOH / PhH ; Bu_4NF, rt

R = H, Me

9,10-Dihydro-9-sila-10-heteroanthracenes are a class of radical-based reducing agents. They effect reductive debromination and deoxygenation of alcohols via *O*-thiocarbonyl derivatives.[7]

Formylation.[8] Aryl and enol triflates are homologated in a Pd-catalyzed reaction involving a hydrosilane and CO (10 examples, 47–94%).

Allyl[tris(trimethylsilyl)]silanes.[9] These valuable synthetic reagents are prepared by free-radical substitution of allyl phenyl sulfides with $(Me_3Si)_3SiH$. They can be used as radical-based allylating agents.

Cyclosilylation of dienes.[10] Phenylsilane with a catalytic amount of $Cp^*_2YMe\cdot THF$ effects the cyclization of 1,5- and 1,6-dienes to give silylmethylcyclopentanes and -cyclohexanes. Substituted piperidines are also available.

$Cp^*_2YMe \cdot THF$; $PhSiMeH_2$; rt, 1 h

SiPhMeH; SiPhMeH; 90%

[1]Hojo, M., Fujii, A., Murakami, C., Aihara, H., Hosomi, A. *TL* **36**, 571 (1995).
[2]Bower, S., Kreutzer, K.A., Buchwald, S.L. *ACIEE* **35**, 1515 (1996).
[3]Verdaguer, X., Berk, S.C., Buchwald, S.L. *JACS* **117**, 12641 (1995).
[4]Mao, Z., Gregg, B.T., Cutler, A.R. *JACS* **117**, 10139 (1995).
[5]Hays, D.S., Scholl, M., Fu, G.C. *JOC* **61**, 6751 (1996).

[6]Hays, D.S., Fu, G.C. *JOC* **61**, 4 (1996).
[7]Oba, M., Kawahara, Y., Yamada, R., Mizuta, H., Nishiyama, K. *JCS(P2)* 1843 (1996).
[8]Kotsuki, H., Datta, P.K., Suenaga, H. *S* 470 (1996).
[9]Chatgilialoglu, C., Ballestri, M., Vecchi, D., Curran, D.P. *TL* **37**, 6383 (1996).
[10]Molander, G.A., Nichols, P.J. *JACS* **117**, 4415 (1995).

N-Hydroxy-*o*-benzenedisulfonimide.

Oxidations.[1] This reagent (**1**) oxidizes benzylic alcohols to aldehydes, thiols to disulfides, and sulfides to sulfoxides. Aromatic aldehydes can be further converted into aroic acids.

(**1**)

[1]Barbero, M., Degani, I., Fochi, R., Perracino, P. *JOC* **61**, 8762 (1996).

1-Hydroxy-2-methylanthraquinone.

Ether formation and oxidation.[1] The ether derivatives that are readily prepared by the Mitsunobu reaction are oxidatively cleaved to afford carbonyl compounds on photolysis. These ethers serve to protect alcohols.

Mitsunobu → ; hv, MeOH → RCHO 76 - 84%

[1]Blankespoor, R.L., Smart, R.P., Batts, E.D., Kiste, A.A., Lew, R.E., Vander Vliet, M.E. *JOC* **60,** 6852 (1995).

N-Hydroxyphthalimide.

Oxidation.[1] This compound has been used to promote the oxidation of alcohols to ketones by oxygen. Benzylic CH_2 is also oxidized.

[1]Ishii, Y., Nakayama, K., Takeno, M., Sakaguchi, S., Iwahama, T., Nishiyama, Y. *JOC* **60**, 3934 (1995).

1-Hydroxypyridine-2-(1*H*)-thione. 18, 187

Amides.[1] Esters derived from the *N*-hydroxy compound are active esters that undergo aminolysis in the dark at room temperature. In sterically demanding cases, the benzenesulfenamides R_2NSPh are more reactive than the free amines. Peptide formation has been demonstrated.

[1]Barton, D.H.R., Ferreira, J.A. *T* **52**, 9347, 9367 (1996).

Hydroxy(tosyloxy)iodobenzene. 14, 179–180; **16**, 179; **17**, 150; **18**, 187

Stille coupling.[1] The Pd-catalyzed coupling of organostannanes can use PhI(OH)OTs as the source of a phenyl group.

Ring expansion.[2] 1-Bromoethynyl-1-cycloalkanols undergo ring expansion, giving the homologous α-(iodobromomethylene)cycloalkanones, on reaction with I_2–PhI(OH)OTs.

Br HO PhI(OTs)OH I_2 O I Br 85%

[1]Kang, S.-K., Lee, H.-W., Kim, J.-S., Choi, S.-C. *TL* **37**, 3723 (1996).
[2]Bovonsombat, P., McNelis, E. *TL* **35**, 6431 (1994).

Hypofluorous acid–acetonitrile. 18, 188

Oxidation. The range of oxidation potential of HOF·MeCN has been extended to the direct preparation of epoxy carboxylic acids from alkenoic acids,[1] oxidative cleavage of methyl ethers to ketones,[2] and conversion of aryl perfluoroalkyl sulfides to the sulfones.[3]

[1]Rozen, S., Bareket, Y., Dayan, S. *TL* **37**, 531 (1996).
[2]Rozen, S., Dayan, S., Bareket, Y. *JOC* **60**, 8267 (1995).
[3]Beckerbauer, R., Smart, B.E., Bareket, Y., Rozen, S. *JOC* **60**, 6186 (1995).

I

Indium. 14, 181; **16**, 181–182; **18**, 189

Allylation in water. Numerous reports exist on indium-mediated allylation using various allylic bromides. The highly *syn*-selective reaction on α-hydroxy aldehydes[1] is synthetically valuable. On the other hand, the $La(OTf)_3$-catalyzed reaction of ethyl γ-bromocrotonate leads mainly to the *anti*-isomer.[2]

PhCHO + EtOOC⁀⁀Br $\xrightarrow[H_2O,\ rt]{In - La(OTf)_3}$ (syn product) + (anti product)

(90 : 10)

99 %

1,3-Dibromopropene acts as a *gem*-allyl dianion synthon.[3] However, the reaction is not as clean as one might desire, due to competing reductive debromination of the 1:1 products. 1,1-Dichloro-2-propene undergoes a γ-selective addition process, affording *syn*-chlorohydrins,[4] but the allylindium species derived from 3-bromo-1-trimethylsilylpropene attacks aldehydes with the α-carbon.[5]

Me_3Si⁀⁀Br + $(CH_3)_2CH$–CHO $\xrightarrow[H_2O,\ rt]{In}$ product (OH, $SiMe_3$)

62% (*E* : *Z* 75 : 25)

Propargylic halides also react similarly, although a small amount of allenyl alcohols is formed.[6]

There is no reactivity problem with enolizable 1,3-dicarbonyl compounds as allyl acceptors. (Tin can be used instead of indium).[7]

Many functionalized allylic halides tolerate the reaction conditions, allowing more straightforward syntheses of useful intermediates, including 2-trimethylsilyl-2-propenyl iodide,[8] 2-bromomethyl-2-propenyl bromide,[9] and 2-bromomethylacrylic acid.[10] The derivative from 2-bromomethylacrylic acid has been applied to carbohydrate synthesis, including *N*-acetylneuraminic acid. Of particular interest is the intramolecular alkylation for ring expansion.[11]

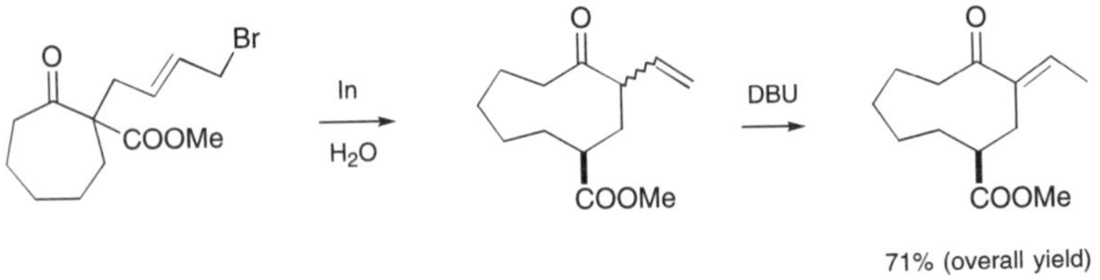

71% (overall yield)

Homoallylic amines.[12] Common methods for the assembly of homoallylic amines by the allylation route employ imines that imply structural limitations. Enamines become useful in the reaction with allylindium reagents to introduce an allyl group α to nitrogen. Thus, tertiary homoallylic amines are formed directly.

Reformatsky-type reaction.[13] Using indium and iodine in DMF, the condensation of phenyl α-bromoalkanoates with carbonyl compounds leads to β-lactones.

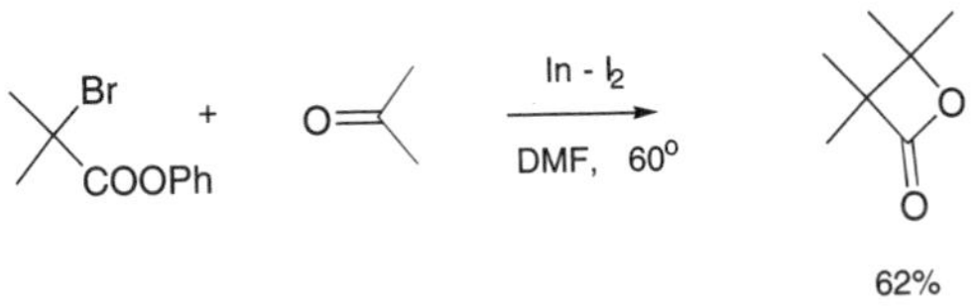

62%

2,6-Dienols.[14] Allylindium reagents add to 2,3-butadienols in a manner regulated by chelation. There are certain structural limitations regarding the γ-selective attack of the reagents, and thus tertiary carbinols and homoallenyl alcohols, as well as allenyl ethers, fail to react.

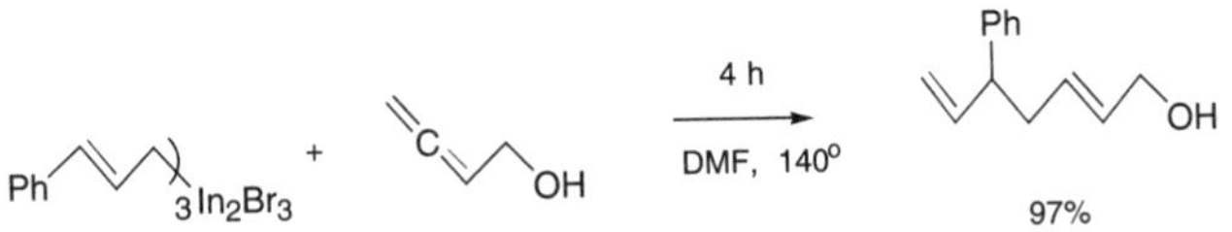

97%

Reaction with diselenides. Organoindiums derived from allyl or propargyl halides and α-halo ketones react with diselenides to afford unsaturated selenides[15] and α-seleno ketones,[16] respectively.

[1]Paquette, L.A., Mitzel, T.M. *TL* **36**, 6863 (1995).
[2]Diana, S.-C.H., Sim, K.-Y., Loh, T.-P. *SL* 263 (1996).
[3]Chen, D.-L., Li, C.-J. *TL* **37**, 295 (1996).
[4]Araki, S., Hirashita, T., Shimizu, K., Ikeda, T., Butsugan, Y. *T* **52**, 2803 (1996).
[5]Isaac, M.B., Chan, T.-H. *TL* **36**, 8957 (1995).
[6]Isaac, M.B., Chan, T.-H. *CC* 1003 (1995).
[7]Li, C.-J., Lu, Y.-Q. *TL* **36**, 2721 (1995).

[8]Bardot, V., Remuson, R., Gelas-Mialhe, Y., Gramain, J.-C. *SL* 37 (1996).
[9]Li, C.-J. *TL* **36**, 517 (1995).
[10]Chan, T.-H., Lee, M.-C. *JOC* **60**, 4228 (1995).
[11]Li, C.-J., Chen, D.-L., Lu, Y.-Q., Haberman, J.X., Mague, J.T. *JACS* **118**, 4216 (1996).
[12]Bossard, F., Dambrin, V., Lintanf, V., Beuchet, P., Mosset, P. *TL* **36**, 6055 (1995).
[13]Schick, H., Ludwig, R., Kleiner, K., Kunath, A. *T* **51**, 2939 (1995).
[14]Araki, S., Usui, H., Kato, M., Butsugan, Y. *JACS* **118**, 4699 (1996).
[15]Bao, W., Zheng, Y., Zhang, Y., Zhou, J. *TL* **37**, 9333 (1996).
[16]Bao, W., Zhang, Y. *SL* 1187 (1996).

Indium(III) chloride.

Redox isomerization.[1] The combination of $InCl_3$ and a ruthenium complex in the presence of $Et_3NH^+PF_6^-$ in refluxing THF promotes the conversion of propargylic alcohols to enals and enones. (For enals, the addition of $NH_4^+PF_6^-$ is beneficial.) The process is highly chemoselective, as functionalities such as isolated carbonyl groups, free alcohol, alkyne, and terminal alkene are not affected.

Allylation and alkynylation. *Anti*-selective allylation of aldehydes with allylstannanes occurs in the presence of $InCl_3$. Thus, premixing RCHO and $InCl_3$ and then adding the stannane at –78° provides the product. The reagent allows *gem*–alkoxylated allylstannanes that are labile to $TiCl_4$ or $SnCl_4$ to undergo the necessary transmetallation. It is noteworthy that crotylstannane and α-MOM crotylstannane show γ- and α-selectivities, respectively, while both reactions are strongly *anti*-selective. Eight diastereomeric hexoses and their enantiomers have been synthesized using this method.[2] α,δ-Dioxygenated allylstannanes undergo similar reactions.[3]

RCHO + OMOM / SnBu3 —InCl3, Me2CO, -78° -> rt→ R, OH, OMOM

74 - 83% (anti : syn 84-98 : 16-2)

Admixture of alkynylstannanes, aldehydes, trimethylsilyl chloride, and $InCl_3$ in acetonitrile at room temperature results in the formation of propargyl silyl ethers.[4] In the synthesis of homoallylic alcohols, simple allylic halides can be used to form the tin halides in situ in water.[5]

Mukaiyama aldol and Diels–Alder reactions. Both reactions[6,7] are catalyzed by $InCl_3$ in water.

[1]Trost, B.M., Livingston, R.C. *JACS* **117**, 9586 (1995).
[2]Marshall, J.A., Hinkle, K.W. *JOC* **61**, 105 (1996).
[3]Marshall, J.A., Garofalo, A.W. *JOC* **61**, 8732 (1996).
[4]Yasuda, M., Miyai, T., Shibata, I., Baba, A. *TL* **36**, 9497 (1995).
[5]Li, X.-R., Loh, T.-P. *TA* **7**, 1535 (1996).
[6]Loh, T.-P., Pei, J., Cao, G.-Q. *CC* 1819 (1996).
[7]Loh, T.-P., Pei, J., Lin, M. *CC* 2315 (1996).

Iodine. 13, 148–149; **14**, 181–182; **15**, 172–173; **16**, 182; **18**, 189–191

Alkyl iodides.[1] The conversion of alcohols to iodides with inverted configuration is accomplished by heating with iodine in an alkane solvent.

Cleavage of ethers. The reaction in refluxing methanol cleaves *p*-methoxybenzyl ethers (10 examples, 75–91%)[2]. Benzyl ethers are not affected, but *t*-butyldimethylsilyl ethers are deprotected.[3]

Dithioacetal/acetal exchange.[4] Iodine in 1,2-ethanediol is effective for this transformation, which is particularly valuable for the synthesis of bissilyl ketones.

$(MeS)_2C(SiMe_2Ph)_2$ —I_2, $HOCH_2CH_2OH$→ dioxolane $(SiMe_2Ph)_2$ (78%) —80% HOAc→ $O{=}C(SiMe_2Ph)_2$

Activation of glycosyl derivatives. Glycosyl methyl sulfides[5] and glycosyl halides[6] are activated by iodine for the glycosylation of alcohols. In the latter cases DDQ is added.

(E)-Alkenyl iodides.[7] (2-Stannylalkenyl)boranes undergo replacement of the boron residue with iodine and protodestannylation with retention of configuration. The bimetallic compounds are readily obtained from the reaction of alkynylborates with organotin chlorides.

$Bu{-}C{\equiv}C{-}B(Hx)_3^-\ Li^+ \xrightarrow{Bu_3SnCl} (Bu)(Bu_3Sn)C{=}C(Hx)(B(Hx)_2) \xrightarrow[HOAc,\ -78^o \rightarrow rt]{I_2 / THF;} (Bu)CH{=}C(Hx)(I)$ 70%

Heterocyclization. Allylmalonic acids[8] and *o*-allylphenols[9] cyclize without incorporating iodine, giving γ-butyrolactones and dihydrobenzofuranes, respectively, when they are treated with I_2 in dichloromethane at room temperature.

The stereoselectivity of pyrrolidine formation by an overall 5-*endo-trig* process is greatly affected by a base.[10]

I_2 / MeCN

--	87%	0	:	100
+ K_2CO_3	83%	> 25	:	1

Methyl carbamates.[11] Amines undergo carbonylation and methanolysis in the presence of I_2, $Pd(OAc)_2$, K_2CO_3, and CO. A strong base is not suitable for carbamate formation from arylamines, because nuclear iodination occurs.

Diels–Alder reactions.[12] The dienophilic reactivity of *N*-allylacrylamides is enhanced by way of oxazolinium ion formation in the presence of iodine.

Disulfides.[13] The oxidative dimerization of thiols is easily achieved with iodine-morpholine.

[1]Joseph, R., Pallan, P.D., Sudalai, A., Ravindranathan, T. *TL* **36**, 609 (1995).
[2]Vaino, A.R., Szarek, W.A. *SL* 1157 (1995).
[3]Vaino, A.R., Szarek, W.A. *CC* 2351 (1996).
[4]Sakurai, H., Yamane, M., Iwata, M., Saito, N., Narasaka, K. *CL* 841 (1996).
[5]Kartha, K.P.R., Aloui, M., Field, R.A. *TL* **37**, 5175 (1996).
[6]Kartha, K.P.R., Aloui, M., Field, R.A. *TL* **37**, 8807 (1996).
[7]Zou, M.-F., Deng, M.-Z. *JOC* **61**, 1857 (1996).
[8]Kim, K.M., Ryu, E.K. *TL* **37**, 1441 (1996).
[9]Kim, K.M., Ryu, E.K. *H* **41**, 219 (1996).
[10]Jones, A.D., Knight, D.W. *CC* 915 (1996).
[11]Pri-Bar, I., Schwartz, J. *JOC* **60**, 8124 (1995).
[12]Kitagawa, O., Aoki, K., Inoue, T., Taguchi, T. *TL* **36**, 593 (1995).
[13]Ramadas, K., Srinivasan, N. *SC* **26**, 4179 (1996).

Iodine–mercury(II) oxide.

Aromatic iodination. Alkyl aryl ethers are iodinated (27 examples, 75–100%).[1] Lead(IV) acetate can be used in place of HgO.[2]

Sulfoxides.[3] Selective oxidation of sulfides to sulfoxides in dichloromethane at room temperature has been reported (11 examples, 38–95%).

[1]Orito, K., Hatakeyama, T., Takeo, M., Suginome, H. *S* 1273 (1995).
[2]Krassowska-Swiebocka, B., Lulinski, P., Skulski, L. *S* 926 (1995).
[3]Orito, K., Hatakeyama, T., Takeo, M., Suginome, H. *S* 1357 (1995).

Iodine–phenyliodine(III) diacetate.

α-Cleavage of carbonyl compounds. A free-radical reaction obtained by using I_2–$PhI(OAc)_2$ is valuable in performing a skeletal simplification of a bridged ketone, which is necessary for a synthesis of (+)-*cis*-lauthisan. Actually, the degradation of the γ-lactone unit also enlists the same combination of reagents.[1]

I_2 - $PhI(OAc)_2$, hv; Bu_3SnH ; *i*-Bu_2AlH; I_2 - $PhI(OAc)_2$

OTBDPS OTBDPS OTBDPS OCHO OTBDPS

(+)-*cis*-lauthisan

[1]Kim, H., Ziani-Cherif, C., Oh, J., Cha, J.K. *JOC* **60**, 792 (1995).

Iodine(I) chloride.

Biaryls.[1] (Bromoaryl)silyl moieties attached to a polymer backbone undergo Suzuki cross-coupling with arylboronic acids. Iodobiaryls are released from the polymer on subsequent treatment with ICl.

Alkenyl iodides.[2] Alkenylboronates derived from Heck reaction of the parent vinylboronate are a source of iodoalkenes. Both *(Z)*- and *(E)*-isomers are accessible.

MeONa / THF ; ICl, -78° -> rt

ICl, -78° ; MeONa, -78° -> rt

Perfluoroalkyl iodides.[3] Fluoroalkenes undergo iodofluorination on reaction with ICl and HF–BF_3.

[1]Han, Y., Walker, S.D., Young, R.N. *TL* **37**, 2703 (1996).
[2]Stewart, S.K., Whiting, A. *TL* **36**, 3929 (1995).
[3]Petrov, V.A., Krespan, C.G. *JOC* **61**, 9605 (1996).

Iodomethyl triflone.

γ-Iodoalkyl triflones.[1] The readily available reagent[2] adds to alkenes in the presence of Bz_2O_2, thus extending an alkene by one methylene unit and introducing two useful functional groups. The adducts can be transformed into cyclopropyl triflones and β,γ-unsaturated triflones on treatment with DBU and dimethyldioxirane, respectively. The analogous addition to alkynes affords the iodoalkenyl triflones.

Ph + I, SO_2CF_3 —(Bz_2O_2; PhH, 100°)→ Ph, I, SO_2CF_3 80%

[1]Mahadevan, A., Fuchs, P.L. *JACS* **117**, 3272 (1995).
[2]Mahadevan, A., Fuchs, P.L. *TL* **35**, 6025 (1994).

***N*-Iodosuccinimide. 16**, 185–186; **18**, 193–194

p-Methoxybenzyl ethers.[1] Using *p*-methoxybenzyl 4-penten-1-yl ether as donor, the group transfer to an alcohol is mediated by NIS. The side product is 2-iodomethyltetrahydrofuran.

Alkenyl allyl ethers.[2] A silyl enol ether, an allylic alcohol, and NIS react to form mixed acetal of iodoacetaldehyde. The elimination of an iodine atom and the siloxy group with BuLi furnishes the alkenyl allyl ether with a defined *(Z)*- or *(E)*-configuration. An important factor is that the elimination pattern in DME is strictly *anti*, while in hexane it is *syn*.

RO, C_8H_{17} + C_3H_7, OH —(NIS; CH_2Cl_2; -78°)→ OR, O, I, C_8H_{17}, H, H_7C_3 94%

R = Si(*t*-Bu)Me_2

BuLi, DME, -78°

H_7C_3, O, C_8H_{17} 76%

Iodoalkenes.[3] Group exchange of alkenylsilanes under mild conditions involves treatment with NIS in acetonitrile at room temperature.

Activation of alkenyl glycosides. Transglycosylation[4] and anomeric phosphorylation[5] are readily achieved by activating alkenyl glycosides with NIS and TfOH or Me_3SiOTf.

[1]Okada, M., Kitagawa, O., Fujita, M., Taguchi, T. *T* **52**, 8135 (1996).
[2]Maeda, K., Shinokubo, H., Oshima, K., Utimoto, K. *JOC* **61**, 2262 (1996).
[3]Stamos, D.P., Taylor, A.G., Kishi, Y. *TL* **37**, 8647 (1996).
[4]Chenault, H.K., Castro, A., Chafin, L.F., Yang, J. *JOC* **61**, 5024 (1996).
[5]Boons, G.-J., Burton, A., Wyatt, P. *SL* 310 (1996).

Iodosylbenzene. 13, 151; **16**, 186; **18**, 194

Epoxidation. With Mn-salen as a catalyst, enol derivatives give 2-hydroxy acetals.[1] Unsaturated acids form lactones during oxidation in the presence of an iron porphyrin.[2]

Glycosylation.[3] The action of a PhIO–Me_3SiOTf combination on thioglycosides consists of oxidation and Lewis acid catalysis, thus allowing the synthesis of disaccharides.

[1]Fukuda, T., Katsuki, T. *TL* **37**, 4389 (1996).
[2]Komuro, M., Higuchi, T., Hirobe, M. *JCS(P1)* 2309 (1996).
[3]Fukase, K., Kinoshita, I., Kanoh, T., Nakai, Y., Hasuoka, A., Kusumoto, S. *T* **52**, 3897 (1996).

Iodosylbenzene–trimethylsilyl azide.

Azidonation. An azido group can be introduced into an allylic position of enol ethers,[1] including silyl enol ethers.[2] In the presence of TEMPO, an addition reaction occurs, resulting in *vic*-diazides.

[1]Magnus, P., Roe, M.B. *TL* **37**, 303 (1996).
[2]Magnus, P., Lacour, J., Evans, P.A., Roe, M.B., Hulme, C. *JACS* **118**, 3406 (1996).

***p*-Iodotoluene difluoride.**

α-Fluoro-β-ketoesters.[1] The selective fluorination of β-ketoesters is readily performed with Tol–IF_2 and a HF–pyridine complex (6 examples, 50–80%).

Xanthates → fluorides.[2] This reaction realizes a new way for the conversion of ROH to RF.

[1]Hara, S., Sekiguchi, M., Ohmori, A., Fukuhara, T., Yoneda, N. *CC* 1899 (1996).
[2]Koen, M.J., Le Guyader, F., Motherwell, W.B. *CC* 1241 (1995).

Iodotrichlorosilane.

Reduction of enones.[1] $ISiCl_3$, generated from $SiCl_4$–NaI in situ, shows chemoselectivity of conjugate reduction. The corresponding saturated ketones are produced.

[1]Elmorsy, S.S., El-Ahl, A.-A.S., Soliman, H., Amer, F.A. *TL* **37**, 2297 (1996).

o-Iodoxybenzoic acid.

Oxidation. The oxidation reactions of primary alcohols to aldehydes, of 1,2-diols to α-ketols or α-diketones without overoxidation or cleavage, and of amino alcohols to amino carbonyl compounds, which do not require protection of the amino group, have been demonstrated.[1] 1,4-Diols give lactols.[2]

DMSO, 23°, 2 h

81%

[1]Frigerio, M., Santagostino, M., Sputore, S., Palmisano, G. *JOC* **60**, 7272 (1995).
[2]Corey, E.J., Palani, A. *TL* **36**, 3485 (1995).

Iridium chloride.

Reductive insertion.[1] Isoxazolidines give perhydrooxazines by $IrCl_3$-catalyzed insertion of CO into the N–O bond. Interestingly, perhydrooxazin-2-ones are obtained by changing the catalyst to $[Rh(cod)Cl]_2$.

Note that the cyclization of hydrazones[2] with an alkyne moiety is an analogous process. It gives heterocycles by methylene-stitching, using $Ir_4(CO)_{12}$, CO, and a hydrosilane.

MeOOC, MeOOC, N-NMe$_2$ → $Ir_4(CO)_{12}$, CO - tBuMe$_2$SiH, MeCN 10 atm, 140°, 20 h → SiMe$_2$Bu^t, MeOOC, MeOOC, N-NMe$_2$

53%

[1]Khumtaveeporn, K., Alper, H. *JOC* **60**, 8142 (1995).
[2]Chatani, N., Yamaguchi, S., Fukumoto, Y., Murai, S. *OM* **14**, 4418 (1995).

Iron.

Reduction of N–O compounds. Nitroxides and nitrones are reduced to secondary amines with Fe–HOAc without affecting other functionalities.[1] In the presence of acetic anhydride, nitroalkenes are converted into enamides at reflux temperatures (11 examples, 60–81%).[2]

Reduction of azides.[3] Primary amines or amides are generated from ArN_3 and $ArCON_3$, respectively, on treatment with Fe and $NiCl_2 \cdot 6H_2O$ in THF at the temperature of ice.

[1]Sar, C.P., Kalai, T., Baracz, N.M., Jerkovich, G., Hideg, K. *SC* **25**, 2929 (1995).
[2]Laso, N.M., Quicklet-Sire, B., Zard, S.Z. *TL* **37**, 1605 (1996).
[3]Baruah, M., Boruah, A., Prajapati, D., Sandhu, J.S., Ghosh, A.C. *TL* **37**, 4559 (1996).

Iron(III) chloride. 13, 133–134; **14**, 164–165; **15**, 158–159; **16**, 167–169, 190–191; **17**, 138–139; **18**, 197

Debenzylation.[1] Benzyl ethers of complex oligosaccharides are selectively cleaved on exposure to $FeCl_3$ in CH_2Cl_2 at 0° without affecting acetates, benzoates, phthalimides, acyl amides, and sensitive glycosidic linkages.

Glycosylation.[2] *O*-Glycosides can be synthesized from unprotected glycosyl donors by using $FeCl_3$ as a catalyst.

Hydration of arylalkynes.[3] $FeCl_3$ with water constitutes a specific system for converting arylalkynes to ketones.

Oxidative cleavage of cyclopropyl trimethylsilyl ethers. 2-Cycloalkenones are obtained from bicyclo[n.1.0]alkan-1-yl trimethylsilyl ethers.[4] When a double bond is present at a suitable distance to interact with the β-radical, cyclization occurs.[5]

2,2-Dichloro-1,3-indanediones.[6] 1,4-Naphthoquinones undergo ring contraction and chlorination on reaction with $FeCl_3$–$HClO_4$ in HOAc at room temperature.

R = H, OH
R' = H, Me, OH

$ArNO_2 \rightarrow ArNH_2$.[7] The $FeCl_3$-catalyzed reduction employs 1,1-dimethylhydrazine.

[1]Rodebaugh, R., Debenham, J.S., Fraser-Reid, B. *TL* **37**, 5477 (1996).
[2]Ferrieres, V., Bertho, J.-N., Plusquellec, D. *TL* **36**, 2749 (1995).
[3]Damiano, J.P., Postel, M. *JOMC* **522**, 303 (1996).
[4]Booker-Milburn, K.I., Thompson, D.F. *T* **51**, 12955 (1995).
[5]Booker-Milburn, K.I., Thompson, D.F. *JCS(P1)* 2315 (1995).
[6]Singh, P.K., Khanna, R.N. *TL* **35**, 3753 (1994).
[7]Boothroyd, S.R., Kerr, M.A. *TL* **36**, 2411 (1995).

(Isocyanocyclohexane)gold(I) chloride.

Imidazolines and isoxazolines. 1,3-Dipolar cycloaddition of isocyanoacetic esters and amides with *N*-tosylimines[1] and aldehydes,[2] respectively, furnishes the five-membered heterocycles. Interestingly, *cis*-imidazolines and *trans*-oxazolines are major products. Since *trans*-imidazolines are obtained on base-catalyzed isomerization of the *cis*-isomers, both *erythro*- and *threo*-α,β-diamino acids are accessible.

Ph, N, Ts + COOMe, NC — (c-HxNC)AuCl, MeCN, 20°, 8 h → Ph, COOMe, TsN, N + Ph, COOMe, TsN, N

> 99% (cis : trans 90 : 10)

[1]Hayashi, T., Kishi, E., Soloshonok, V.A., Uozumi, Y. *TL* **37**, 4969 (1996).
[2]Sawamura, M., Nakayama, Y., Kato, T., Ito, Y. *JOC* **60**, 1727 (1995).

Isocyanuric chloride. 18, 199

Oxidation. The rapid oxidation of aldehydes to methyl esters in the presence of pyridine and methanol is preparatively useful (9 examples, 67–93%).[1] Acetals are similarly converted to esters.[2]

[1]Hiegel, G.A., Bayne, C.D., Donde, Y., Tamashiro, G.S., Hilberath, L.A. *SC* **26**, 2633 (1996).
[2]Benincasa, M., Grandi, R., Ghelfi, F., Pagnoni, U.M. *SC* **25**, 3463 (1995).

Isocyanuric fluoride.

Activation of carboxylic acids. The formation of an acid fluoride and its immediate treatment with $NaBH_4$ accomplishes the reduction of an acid to an alcohol.

γ,δ-Unsaturated ketones.[2] The fluoride is useful for promoting the transfer of the alkenyl group of an alkenylboronic acid to enones.

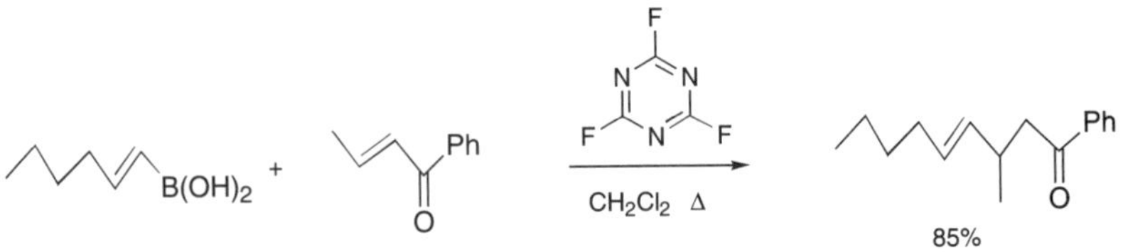

[1]Kokotos, G., Noula, C. *JOC* **61**, 6994 (1996).
[2]Hara, S., Ishimura, S., Suzuki, A. *SL* 993 (1996).

Isopinocampheylchloroborane.

Asymmetric cyclic hydroboration.[1] As illustrated in the following equation, the facile accessibility of this reagent by in situ reduction of the dichloroborane derivative with Me_3SiH and the high optical yield of (+)-*trans*-decalone prepared from the hydroboration route suggest a general approach to such ketones:

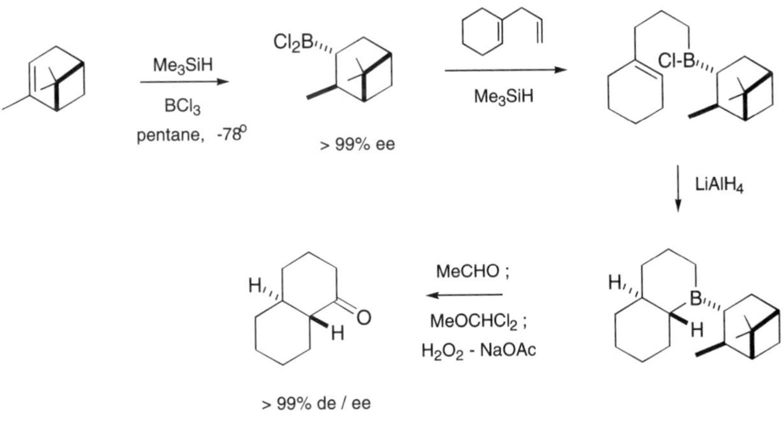

[1]Brown, H.C., Mahindroo, V.K., Dhokte, U.P. *JOC* **61**, 1906 (1996).

4-Isopropyl-2-oxazolin-5-one.

Formyl anion equivalent. Under mildly basic conditions, (**1**) acts as nucleophile toward carbonyl compounds and undergoes conjugate additions as well.

(1) + CHO $\xrightarrow{Et_3N}$ 72% $\xrightarrow[\text{THF}]{\text{5\% HCl}}$ 38%

[1]Barco, A., Benetti, S., De Risi, C., Pollini, G.P., Spalluto, G., Zanirato, V. *TL* **34**, 3907 (1993).

L

Lanthanum(III) isopropoxide. 17, 160; **18**, 201

Transesterification.[1] Heating an ester in another alcohol in the presence of $La(OPr^i)_3$ is a simple method for exchanging the alkoxy group.

[1]Okano, T., Miyamoto, K., Kiji, J. *CL* 246 (1995).

Lanthanum(III) perchlorate hydrate.

β-Glycosylation. Trimethylsilyl ethers are adequate glycosyl acceptors,[1] whereas a combination of $La(ClO)_4$ and $Sn(OTf)_2$ is effective for the glycosylation of α-mannosyl fluorides.[2]

[1]Kim, W.-S., Hosono, S., Sasai, H., Shibasaki, M. *TL* **36**, 4443 (1995).
[2]Kim, W.-S., Sasai, H., Shibasaki, M. *TL* **37**, 7797 (1996).

Lead.

α-Methylene-γ-butyrolactones.[1] Lead is capable of promoting (2-alkoxy-carbonyl)allylation of aldehydes. The products undergo lactonization on contact with CF_3COOH.

RCHO + Br(CH₂C(=CH₂))COOEt —Pb, NH_4Cl, THF - H_2O→ R–CH(OH)–CH₂–C(=CH₂)–COOEt

61 - 97%

Reduction of conjugated carbonyl compounds.[2] The Lindlar catalyst, which contains Pb as a regulator of catalytic activity, promotes the saturation of the conjugated double bonds.

[1]Zhou, J.-Y., Jia, Y., Yao, X.-B., Wu, S.-H. *SC* **26**, 2397 (1996).
[2]Righi, G., Rossi, L. *SC* **26**, 1321 (1996).

Lead(IV) acetate. 13, 155–156; **14**, 188; **16**, 193–194; **18**, 201–202

Cleavage of cyclobutenolones.[1] Squaric acid derivatives undergo oxidative cleavage to give lactones.

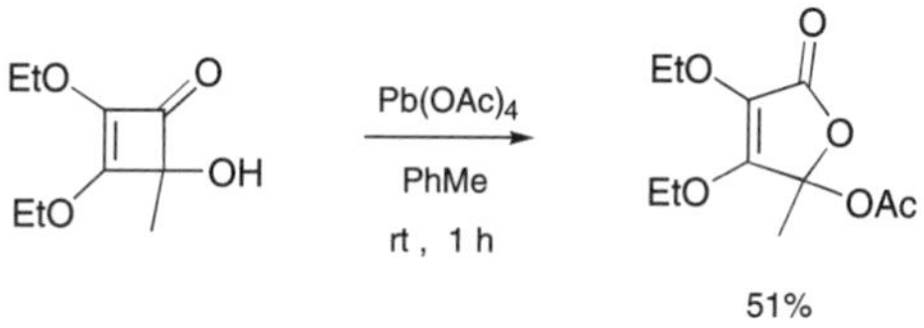

Oxidative cleavage of β-amino alcohols.[2] Adjustment of oxidation conditions can lead to either an aldehyde or a nitrile.

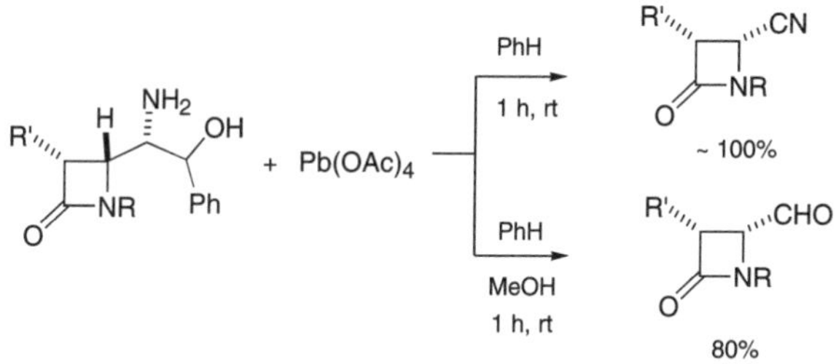

[1]Yamamoto, Y., Ohno, M., Eguchi, S *JACS* **117**, 9653 (1995).
[2]Jayaraman, M., Nandi, M., Sathe, K.M., Deshmukh, A.R.A.S., Bhawal, B.M. *TA* **4**, 609 (1993).

Lipases. 17, 133–134; **18**, 202–204

Hydrolysis. 1,1,1-Trifluoro-2-alkanols are resolved through enzymatic hydrolysis of the corresponding chloroacetates,[1] although the acetates of 1,1,1-trifluoro-3-phenylthio-2-alkanols have also been successfully resolved by a similar procedure.[2] A useful selectivity is shown in the resolution of menthyl α-acetoxy-α-ethylthioacetate, as the menthyl ester group is retained.[3] Selective hydrolysis of 1,5-dibenzoyloxy-3-methyl-3-pentanol provides a chiral diol,[4] which has been used for the synthesis of *(R)*-mevalonolactone and an *(S)*-chromanethanol, which is an important building block for vitamin E. Racemic 6-acetoxymethyldihydropyran is resolved to the chiral isomers by porcine pancreatic lipase (PPL). These can be used to prepare 2,2′-bis(halomethyl)dihydropyrans and 2,2′-bis(phenylthiomethyl)dihydropyran, which are useful protecting and resolving agents for 1,2-diols.[5]

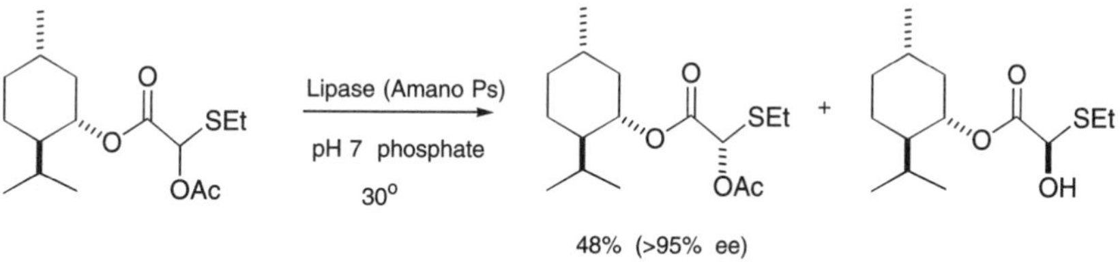

The resolution by enantioselective hydrolysis of cyclic carbonates[6] and 2-oxazolidinones[7] is particularly interesting with respect to the latter class of compounds, in which the ester group is attached to a side chain while the stereogenic center is in the ring.

lipase
i-Pr_2O
rt, 3 h

42 % (75% ee)

46% (70% ee)

Cross-linked crystals of *Candida rugosa* lipase are highly efficient catalysts for resolving racemic esters,[8] whereas PPL immobilized in microemulsion-based gel has been used to hydrolyze one of the polyphenolic perpropanoates.[9] Actually, symmetrical diacetates are readily cleaved to provide the monoesters by PPL.[10]

Resolution by transesterification. A procedure for the large-scale preparation of (1*S*,2*S*)-*trans*-2-methoxycyclohexanol, which is a key intermediate for the synthesis of tricyclic β-lactam antibiotics, has been worked out.[11] Monoprotected 1,2-diols and acetals of α-hydroxy aldehydes afford chiral acetates by the lipase-catalyzed transesterification.[12] α-Ketols are also resolved.[13] Dimethyl *meso*-2,5-dibromoadipate is readily desymmetrized by the transesterification protocol, permitting the synthesis of chiral *cis*-2,5-disubstituted pyrrolidines.[14]

PPL
BnOH

62%

R*NH_2
K_2CO_3
PhH - H_2O

56% (89% de)

PPL = porcine pancreatic lipase

A double-enantioselective synthesis of carbonates and carbamates using *C. antarctica* lipase is eminently successful.[15]

CAL
hexane, 32 h

CAL = *Candida antarctica* lipase

41% conv.
88% de, 98% ee

The enzymatic reaction is promoted by microwave.[16] Significant effects of acyl groups on the enantioselective transesterification have been observed.[17]

Diketene is a recently employed esterification agent in organic media.[18,19] The enantioselective esterification of hemithioacetals,[20] which are formed in situ, is most interesting. Note that the greater-than-50% yield indicates that the unesterified hemithioacetals are recycled by dissociation into the original components.

MeO(C=O)CHO + HS–CH2CH2–OSiEt3 → [PLL - CH_2=CHOAc; t-BuOMe, SiO_2; 30°, 11 days] → MeO(C=O)CH(OAc)–S–CH2CH2–OSiEt3

83% (90% ee)

PLL = *Pseudomonas fluorescens* lipase

Chemoselective transesterification. A survey of enzymatic regioselective acylation (and deacylation) of carbohydrates has been given.[21]

Esterification. The lipase-mediated esterification is promoted by microwave (90°, 15 min.).[22] Four lipases immobilized in microemulsion-based gels retain their activity in catalyzing esterification.[23] Resolution by selective acetylation of ethyl 6-hydroxy-1-cyclohexenecarboxylate gives access to a chiral intermediate for a synthesis of the spirocyclic alkaloids:[24] (+)-nitramine, (+)-isonitramine, and (–)-sibirine.

The resolution of 2-methylalkanoic acid by enzymatic esterification is quite efficient using a long-chained alcohol.[25]

Amidation of esters. Benzyl acetate is a useful acetyl donor.[26] Succinimide formation with simultaneous resolution of α-substituted succinic esters[27] is possible.

COOMe / COOMe → [$BuNH_2$ - CAL; hexane, 7 days] → NBu succinimide

45 % conv. (99% ee)

[1]Yonezawa, T., Sakamoto, Y., Nogawa, K., Yamazaki, T., Kitazume, T. *CL* 855 (1996).
[2]Shimizu, M., Sugiyama, K., Fujisawa, T. *BCSJ* **69**, 2655 (1996).
[3]Milton, J., Brand, S., Jones, M.F., Rayner, C.M. *TA* **6**, 1903 (1995).
[4]Mizuguchi, E., Suzuki, T., Achiwa, K. *SL* 743 (1996).
[5]Ley, S.V., Mio, S., Meseguer, B. *SL* 787, 791 (1996); Ley, S.V., Mio, S. *SL* 789 (1996).
[6]Matsumoto, K., Fuwa, S., Kitajima, H. *TL* **36**, 6499 (1995).
[7]Wakamatsu, H., Terao, Y. *CPB* **44**, 261 (1996).
[8]Lalonde, J.J., Govardhan, C., Khalaf, N., Martinez, A.G., Visuri, K., Margolin, A.L. *JACS* **117**, 6845 (1995).
[9]Parmar, V.S., Pati, H.N., Sharma, S.K., Singh, A., Malhotra, S., Kumar, A., Bisht, K.S. *BMCL* **6**, 2269 (1996).

[10]Houille, O., Schmittberger, T., Uguen, D. *TL* **37**, 625 (1996).
[11]Stead, P., Marley, H., Mahmoudian, M., Webb, G., Noble, D., Ip, Y.T., Piga, E., Rossi, T., Roberts, S., Dawson, M.J. *TA* **7**, 2247 (1996).
[12]Kim, M.-J., Lim, I.T., Choi, G.-B., Whang, S.-Y., Ku, B.-C., Choi, J.-Y. *BMCL* **6**, 71 (1996).
[13]Adam, W., Diaz, M.T., Fell, R.T., Saha-Moller, C.R. *TA* **7**, 2207 (1996).
[14]Ozasa, N., Tokioka, K., Tamamoto, Y. *BBB* **59**, 1905 (1995).
[15]Pozo, M., Gotor, V. *TA* **6**, 2797 (1995).
[16]Parker, M.-C., Besson, T., Lamare, S., Legoy, M.-D. *TL* **37**, 8383 (1996).
[17]Ema, T., Maeno, S., Takaya, Y., Sakai, T., Utaka, M. *TA* **7**, 625 (1996).
[18]Jeromin, G.E., Welsch, V. *TL* **36**, 6663 (1995).
[19]Suginaka, K., Hayashi, Y., Yamamoto, Y. *TA* **7**, 1153 (1996).
[20]Brand, S., Jones, M.F., Rayner, C.M. *TL* **36**, 8493 (1995).
[21]Bashir, N.B., Phythian, S.J., Reason, A.J., Roberts, S.M. *JCS(P1)* 2203 (1995).
[22]Carrillo-Munoz, J.-R., Bouvet, D., Guibe-Jampel, E., Loupy, A., Petit, A. *JOC* **61**, 7746 (1996).
[23]de Jesus, P.C., Rezende, M.C., da Graca Nascimento, M. *TA* **6**, 63 (1995).
[24]Yamane, T., Ogasawara, K. *SL* 925 (1996).
[25]Edlund, H., Berglund, P., Jensen, M., Hedenstrom, E., Hogberg, H.-E. *ACS* **50**, 666 (1996).
[26]Adamczyk, M., Grote, J. *TL* **37**, 7913 (1996).
[27]Puertas, S., Rebolledo, F., Gotor, V. *T* **51**, 1495 (1995).

N-Lithiomethyl-*N,N′,N″N″*-tetramethyldiethylenetriamine.

Allylic alcohols.[1] An epoxide is homologated at the terminus by a methylene group on reaction with the reagent (**1**)(6 examples, 42–72%).

(**1**) + epoxide; C_5H_{12}, rt, 1-3 h; 61%

[1]Schakel, M., Luitjes, H., Dewever, F.L.M., Scheele, J., Klumpp, G.W. *CC* 513 (1995).

Lithiomethyldiphenylphosphonium methylide.

Conjugated amides.[1] The lithiated ylide reacts with isocyanates and then with aldehydes to complete a stereoselective synthesis of conjugated amides.

PhN=C=O, THF Δ; PhCHO; CONHPh; 77%

[1]Cristau, H.J., Taillefer, M., Urbani, J.P., Fruchier, A. *T* **52**, 2005 (1996).

Lithiotrimethylsilyldiazomethane.

Lithium trimethylsilylethynolate.[1] Carbonylation of the lithium compound with CO is accompanied by N_2 elimination and bonding reorganization. The ynolate species can be transformed into bissilyl ketenes and other interesting compounds.

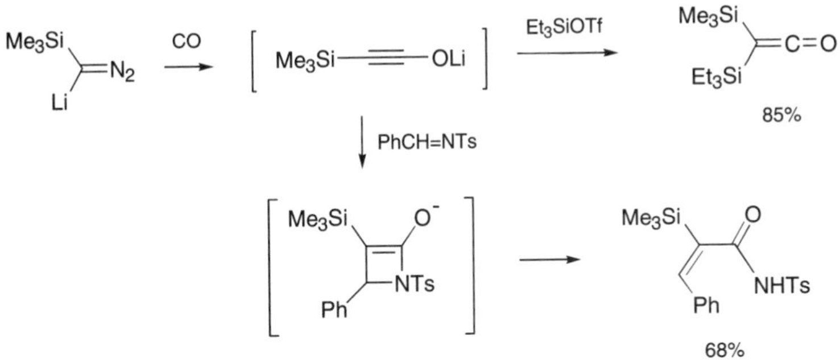

$Me_3SiC(N_2)Li$ → (CO) → [$Me_3Si-C\equiv C-OLi$] → (Et_3SiOTf) → $(Me_3Si)(Et_3Si)C=C=O$ 85%

[$Me_3Si-C\equiv C-OLi$] → (PhCH=NTs) → [4-membered ring intermediate with Me_3Si, O^-, NTs, Ph] → Me3Si-substituted unsaturated amide with Ph and NHTs 68%

Alkylidenation-intramolecular insertion. A stereoselective synthesis of cyclopentenes from acyclic ketones occurs by reaction of the latter with lithiotrimethylsilyldiazomethane. An application of this method to the synthesis of (+)-cassiol[2] and oxo-T-cadinol[3] attests to its usefulness.

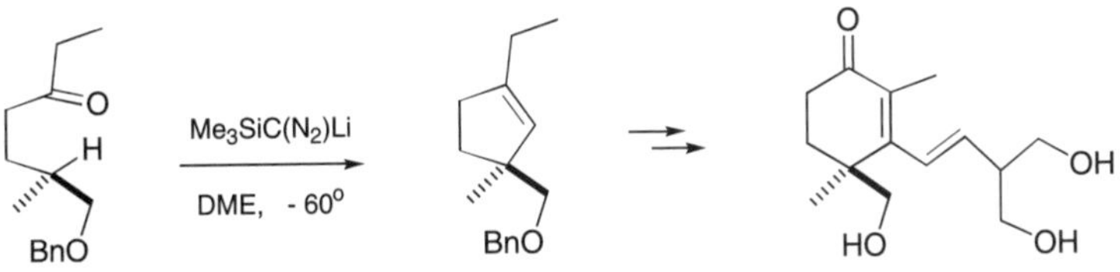

Ketone (with BnO) → ($Me_3SiC(N_2)Li$, DME, - 60°) → cyclopentene (BnO) 85% ⇉ (+)-cassiol

Ketone (with BnO) → ($Me_3SiC(N_2)Li$, DME, -50°-> rt) → bicyclic cyclopentene (BnO) 63% ⇉ oxo-*T*-cadinol

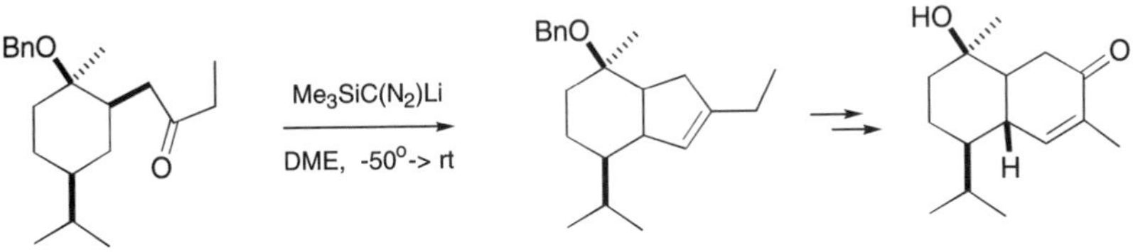

[1]Kai, H., Iwamoto, K., Chatani, N., Murai, S. *JACS* **118**, 7634 (1996).
[2]Taber, D.F., Meagley, R.P., Doren, D.J. *JOC* **61**, 5723 (1996).
[3]Taber, D.F., Christos, T.E., Hodge, C.N. *JOC* **61**, 2081 (1996).

Lithium. **13**, 157–158; **15**, 184; **18**, 205–206

Ketones. Sonochemical Barbier reaction allows direct access to ketones from the condensation of lithium carboxylates and organohalides in the presence of Li.[1] An interesting approach to 2-furanyl ketones[2] involves in situ deprotonation of furan with nascent *t*-BuLi (generated with the assistance of ultrasound) and subsequent reaction with RCOOLi.

3,4-Dilithio-1,2-butadienes.[3] Solutions of these dilithiated species are obtained in high yields by sonicating 1,1-disubstituted butatrienes with Li dust in ether.

α-Disulfones.[4] On exposure to ultrasound, sulfonyl chlorides undergo dechlorinative coupling by lithium (4 examples, 62–69%).

Silyldechlorination of fused gem-dichlorocyclopropanes.[5] The replacement of one or both chlorine atoms with Li–Me_3SiCl occurs by using Li containing 0.01% Na or 1% Na, respectively.

0.01% Na in Li	55%	-
1% Na in Li	-	69%

Dephenylation of phosphine-borane complexes.[6] A *P*-phenylphospholane-borane complex is converted to the *P*-lithio reagent by treatment with lithium. The latter species are used for the synthesis of chiral ligands.

[1]Aurell, M.J., Danhui, Y., Einhorn, J., Einhorn, C., Luche, J.L. *SL* 459 (1995).
[2]Aurell, M.J., Einhorn, C., Einhorn, J., Luche, J.L. *JOC* **60**, 8 (1995).
[3]Maercker, A., Wunderlich, H., Girreser, U. *T* **52**, 6149 (1996).
[4]Prokes, I., Toma, S., Luche, J.L. *TL* **36**, 3849 (1995).
[5]Grelier-Marly, M.-C., Grignon-Dubois, M. *OM* **14**, 4109 (1995).
[6]Morimoto, T., Ando, N., Achiwa, K. *SL* 1211 (1996).

Lithium aluminum hydride. 14, 190–191; **18**, 207

Dephosphonylation.[1] The phosphorus-containing residue from β-keto phosphonates can be removed by reduction of the enolates with $LiAlH_4$ (10 examples, 57–98%).

Allylic alcohols.[2] A method for the conversion of α,β-dihydroxy aldehydes to 1-alken-3-ols via their tosylhydrazones has been found. It requires only that the sodium salts be treated with $LiAlH_4$ (6 examples, 65–75%).

BnO ... N-NHTs → (NaHMDS; $LiAlH_4$, THF; rt, 8 h) → BnO ... OH 69%

[1]Hong, J.E., Shin, W.S., Jang, W.B., Oh, D.Y. *JOC* **61**, 2199 (1996).
[2]Chandrasekhar, S., Mohapatra, S., Takhi, M. *SL* 759 (1996).

Lithium aminoborohydride. 17, 170; **18**, 208–209

$RCONR_2' \rightarrow RCH_2OH$.[1] Tertiary amides are reduced to primary alcohols with $LiBH_3(NH_2)$ at room temperature. The results are complementary to those of other metal hydride reagents that lead to tertiary amines.

[1]Myers, A.G., Yang, B.H., Kopecky, D.J. *TL* **37**, 3623 (1996).

Lithium azide.

Glycosyl azides.[1] These compounds are prepared in a one-pot reaction in DMF via 1,2-cyclic sulfites obtained from the (partially protected) sugars and *N,N*-thionyldiimidazole.

HO, MeO, OH, HO, OH → (N,N-thionyldiimidazole; LiN_3 / DMF; -30° -> rt, 24 h) → HO, MeO, N_3, HO, OH 64%

[1]Meslouti, A.E., Beaupere, D., Demailly, G., Uzan, R. *TL* **35**, 3913 (1994).

Lithium bis(trifluoromethanesulfonyl)amide.

Diels–Alder reactions.[1] For reactions catalyzed by $LiClO_4 \cdot OEt_2$, the catalyst can be replaced by the safer $LiNTf_2$ in acetone.

[1]Handy, S.T., Grieco, P.A., Mineur, C., Ghosez, L. *SL (sp. issue)* 565 (1995).

Lithium borohydride.

Reduction.[1] With $TiCl_4$ to form chelates, β-keto phosphine oxides undergo stereoselective reduction to give the *anti*-alcohols as major products.

[1]Bartoli, G., Bosco, M., Sambri, L. Marcantoni, E. *TL* **37**, 7421 (1996).

Lithium bromide. 18, 209–210

Aldol reactions.[1] The Knoevenagel condensation between aromatic aldehydes and cyanoacetic acid or ethyl ester or malononitrile in the solid state can be promoted by LiBr as a catalyst (13 examples, 25–98%).

[1]Prajapati, D., Lekhok, K.C., Sandhu, J.S., Ghosh, A.C. *JCS(P1)* 959 (1996).

Lithium cobalt-bis(dicarbollide).

Allylic substitutions.[1] A typical reaction is shown in the formation of dehydro *C*-glycosides.

AcO, AcO, O + MeO, OTBS → $LiCo(B_9C_2H_{11})_2$ / $ClCH_2CH_2Cl$ → AcO, O, COOMe 84%

[1]Grieco, P.A., DuBay, W.J., Todd, L.J. *TL* **37**, 8707 (1996).

Lithium 4,4′-di-*t*-butylbiphenylide. (LDTBB). 13, 162–163; **16**, 195–196; **17**, 164; **18**, 210–211

Organolithiums. Allylic and benzylic alcohols undergo deoxygenative lithiation by treatment of their lithium alkoxides or phenyldimethylsilyl ethers with LDTBB.[1] Alkyl phenyl selenides are also cleaved to give organolithium species that can react with aldehydes and allyl bromide.[2] Some special alkyllithiums have been prepared from (2-pyridylthio)alkanes, which are available from carboxylic acids.[3]

hv
CH_2Cl_2
4 h, 0°
65 - 73%
- 78°
LDTBB ;
Me_2CO
- 78°
60 - 65%

Functionalized organolithiums that are prepared by the cleavage protocol include α-oxidoalkyllithium derived from epoxides,[4] ethoxymethyllithium from chloromethyl ethyl ether,[5] and a masked hydroxymethyllithium from *N*-(chloromethyloxycarbonyl)-pyrrolidine.[6] Acetals of tricarbonylchromium-complexed benzaldehydes undergo reductive cleavage.[7]

LDTBB
THF, -78°
Me_3SiCl
81%

Cyclopropanes.[8] 1,3-Diols form cyclic sulfates readily. These sulfates are converted into substituted cyclopropanes on treatment with lithium powder and LDTBB.

[1]Alonso, E., Guijarro, D., Yus, M. *T* **51**, 11457 (1995).
[2]Krief, A., Nazih, A., Hobe, M. *TL* **36**, 8111, 8115 (1995).
[3]Rychnovsky, S.D., Plzak, K., Pickering, D. *TL* **35**, 6799 (1994).
[4]Bachki, A., Foubelo, F., Yus, M. *TA* **7**, 2997 (1996).
[5]Guijarro, A., Mancheno, B., Ortiz, J., Yus, M. *T* **52**, 1643 (1996).
[6]Guijarro, D., Yus, M., Guijarro, A., Ortiz, J. *T* **52**, 5593 (1996).
[7]Siwek, M.J., Green, J.R. *SL* 560 (1996).
[8]Guijarro, D., Yus, M. *T* **51**, 11445 (1995).

Lithium diisopropylamide. 13, 163–164; **15**, 188–189; **16**, 196–197; **17**, 165–167; **18**, 212–214

Alkylations. The use of LDA as a base in alkylations is almost universal. Listed here are some new examples: perfluoroalkylation of *N*-acyloxazolidin-2-ones,[1] alkylation of 2-cyanocycloalkanones via the SAMP derivatives,[2] and alkylation of α-(*p*-toluenesulfinyl)alkanoic esters.[3] The last of these reacts with *N*-tosylimines to generate adducts that are precursors of β-amino esters.

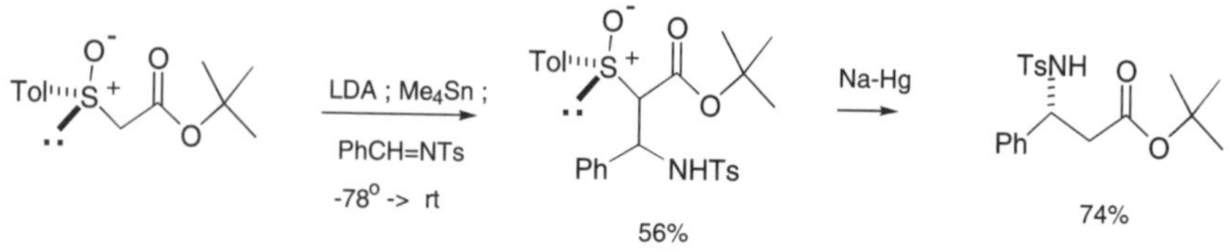

Rearrangements. Anionic rearrangements are readily promoted by LDA. The [2,3]-Wittig rearrangement of *N*-(allyloxyalkyl)benzotriazoles to give β,γ-unsaturated ketones proceeds in high yields (4 examples, 86–92%).[4] Benzyl and allyl hydroximates undergo imino-Wittig rearrangement, resulting in oxime ethers of α-ketols.[5]

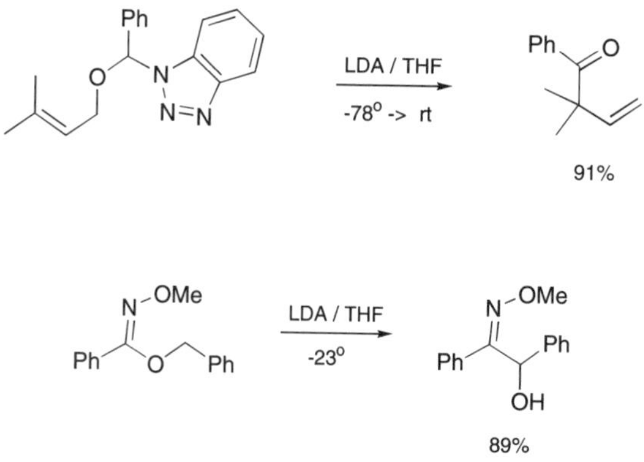

The aza-[2,3]-Wittig rearrangement route[6] to tetrahydropyridines can be coupled to a Wittig olefination of 2-acylaziridines. BuLi was actually used as the base for the demonstration.[7]

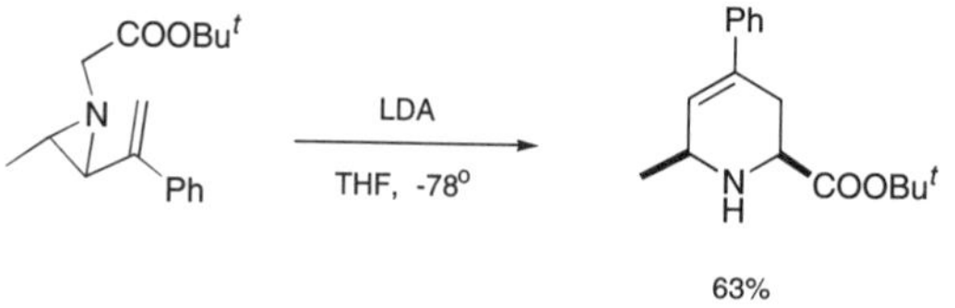

Cyclopentenes.[8] Dihydrothiopyrans prepared by Diels–Alder reactions undergo ring contraction (via thiolated allyl anions) in the presence of LDA/HMPA. The reaction is diastereoselective.

COOEt
LDA / THF-HMPA ;
MeI
COOEt
SMe
+
COOEt
SMe
(8 : 1)
87%

Allenes. Substituted allenes are prepared from ketones via elimination of the enol phosphates.[9] Interestingly, enol triflates tend to give alkynes. The method can be applied to protected aminoalkenyl phosphates.[10] Silyl enol ethers[11] also undergo elimination.

LDA / THF
-78°
OPO(OEt)2
=C=
62%

(EtO)2OPO
N(SiMe3)2
LDA / THF
25 °
C
N(SiMe3)2
70%

Ph
Ph
OSiMe2Bu^t
LDA / THF
25 °
Ph
Ph
Li
Li
=C=
Me2Si(H)Cl
Ph
Ph
=C=
SiMe2H
SiMe2H

Alkenylsilanes and alkenyl sulfones. 1,1-Disilylepoxides undergo an unusual rearrangement,[12] whereas β-silyl sulfones lose the sulfonyl group on treatment with LDA to give alkenylsilanes.[13]

C8H17
O
SiMe3
SiMe3
LDA
Et2O Δ
C8H17
SiMe3
SiMe2OH
90% (*E* : *Z* 1 : 1)

SiMe3
Ph
SO2Ph
LDA
THF Δ
Ph
SiMe3
79%

The base-mediated transformation of episulfones into alkenyl sulfones is stereoselective, giving predominantly the *(E)*-isomers.[14]

Catalytic Shapiro reaction.[15] Small amounts of LDA in ether successfully promote the decomposition of hydrazones derived from *N*-amino-2-phenylaziridine. *(Z)*-Alkenes are the major products.

Heterocycles. New syntheses of pyrroles[16] and β-thiolactams[17] have been reported. A synthesis of the antibiotic (−)-dysidazirine features an elimination of an *N*-toluenesulfinylaziridine.[18] Remarkably, the deprotonation is not at the α-carbon of the ester group.

[1] Iseki, K., Takahashi, M., Asada, D., Nagai, T., Kobayashi, Y. *JFC* **74**, 269 (1995).
[2] Enders, D., Zamponi, A., Raabe, G., Runsink, J. *S* 725 (1993).
[3] Shimizu, M., Kooriyama, Y., Fujisawa, T. *CL* 2419 (1994).
[4] Katritzky, A.R., Wu, H., Xie, L. *JOC* **61**, 4035 (1996).
[5] Miyata, O., Koizumi, T., Ninomiya, I., Naito, T. *JOC* **61**, 9078 (1996).
[6] Ahman, J., Jarevang, T., Somfai, P. *JOC* **61**, 8148 (1996).
[7] Coldham, I., Collis, A.J., Mould, R.J., Rathmell, R.E. *JCS(P1)* 2739 (1995).
[8] Larsen, S.D., Fisher, P.V., Libby, B.E., Jensen, R.M., Mizsak, S.A., Watt, W., Ronk, W.R., Hill, S.T. *JOC* **61**, 4725 (1996).
[9] Brummond, K.M., Dingess, E.A., Kent, J.L. *JOC* **61**, 6096 (1996).
[10] Cunico, R.F., Kuan, C.P. *JOMC* **487**, 89 (1995).
[11] Seyferth, D., Langer, P., Doring, M. *OM* **14**, 4457 (1995).
[12] Hodgson, D.M., Comina, P.J. *TL* **37**, 5613 (1996).
[13] Menichetti, S., Stirling, C.J.M. *JCS(P1)* 1511 (1996).
[14] Dishington, A.P., Muccioli, A.B., Simpkins, N.S. *SL* 27 (1996).
[15] Maruoka, K., Oishi, M., Yamamoto, H. *JACS* **118**, 2289 (1996).
[16] Nagafuji, P., Cushman, M. *JOC* **61**, 4999 (1996).

[17]Creary, X., Zhu, C. *JACS* **117**, 5859 (1995).
[18]Davis, F.A., Reddy, G.V., Liu, H. *JACS* **118**, 3651 (1996).

Lithium hexamethyldisilazide. 13, 165; **14**, 194; **18**, 215–216

Julia condensation.[1] Application of the method to form a benzooxacyclooctenone is crucial to a synthesis of heliannuol A.

MeO … SO_2Ph, COOMe —$LiN(SiMe_3)_2$, THF→ MeO … SO_2Ph (87%) → HO … OH (heliannuol-A)

1,3-Elimination.[2] The formation of alkynylcyclopropanes by exposing 4-alkynyl tosylates to lithium hexamethyldisilazide is instrumental to the acquisition of chiral 1,2-disubstituted cyclopropanes required for the synthesis of bicyclo[3.3.0]oct-1-en-3-ones.

TsO, OEt, $SiMe_3$ —$LiN(SiMe_3)_2$, THF→ OEt, $SiMe_3$ (74%)

N-Trimethylsilylimines. The preparation of these derivatives from carbonyl compounds in hexane paves the way to *N*-sulfonyl imines.[3] The *N*-trimethylsilylimines are also useful for the synthesis of imidazoles.[4]

RCHO —$LiN(SiMe_3)_2$→ RCH=N–$SiMe_3$ —R'CH₂NC, -78° -> rt→ imidazole (R, R', HN, N) 23 - 66%

[1]Grimm, E.L., Levac, S., Trimble, L.A. *TL* **35**, 6847 (1994).
[2]Bräse, S., Schömenauer, S., McGaffin, G., Stolle, A., de Meijere, A. *CEJ* **2**, 545 (1996).

[3]Georg, G.I., Harriman, G.C.B., Peterson, S.A. *JOC* **60**, 7366 (1995).
[4]Shih, N.-Y. *TL* **34**, 595 (1993).

Lithium hydride. 13, 165–166; **18**, 217

The activation of LiH by *t*-BuOH in the presence of $Ni(OAc)_2$ endows LiH with the power to reduce alkyl halides, ketones, alkenes, and sulfurated compounds. The reagent can also effect coupling of aryl halides.[1]

[1]Fort, Y. *TL* **36**, 6051 (1995).

Lithium hydroxide. 18, 217

Michael addition.[1] A very facile reaction between β-carbonyl compounds and conjugated compounds occurs in the presence of $LiOH{\cdot}H_2O$ in DME at room temperature (8 examples, 79–96%).

Emmons–Wadsworth olefination.[2] Condensation with α-ketols leads directly to butenolides (7 examples, 25–70%).

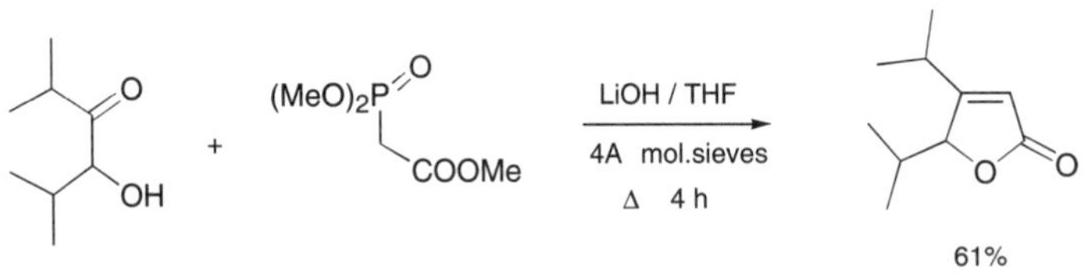

[1]Bonadies, F., Forcellese, M.L., Locati, L., Scettri, A., Scolamiero, C. *G* **124**, 467 (1994).
[2]Bonadies, F., Cardilli, A., Lattanzi, A., Pesci, S., Scettri, A. *TL* **36**, 2839 (1995).

Lithium (α-methylbenzyl)allylamide.

β-Amino esters.[1] The reagent is useful for the synthesis of chiral amines. Michael addition of this compound to conjugated esters affords *N,N*-diprotected β-amino esters in a stereoselective manner. The allyl group is easily removed with $(Ph_3P)_3RhCl$ in aqueous MeCN, and the products can then be converted into β-lactams.

[1]Davies, S.G., Fenwick, D.R. *CC* 1109 (1995).

Lithium (α-methylbenzyl)amide.

a-Hydroxy-β-amino esters.[1] The sequential addition of this compound [(*R*)-isomer)] and (+)-(camphorsulfonyl)oxaziridine to conjugated esters gives direct access to the chiral *anti*-α-hydroxy-β-amino esters.

86% (92% de)

[1]Bunnage, M.E., Davies, S.G., Goodwin, C.J. *JCS (P1)* 1375 (1993).

Lithium naphthalenide (LN). 15, 190–191; **18**, 217–218

Lithium–halogen exchange. The direct synthesis of ketones[1] from carboxylic acids, organic chlorides and LN is quite useful. The cyclization of 1,4-diiodobutane derivatives is the basis of an approach to cyclobutyl ketones.[2]

Li-naphth. THF, 0° ; H_3O^+

45% (over two steps)

Functionalized silyllithiums such as (t-BuO)Ph_2SiLi are formed by a similar exchange reaction. 1-Dimethylaminonaphthalene appears to be superior to naphthalene as an electron carrier.[3]

Cleavage of C–S bonds. The preparation of bis(lithiomethyl)silanes[4] from the corresponding bis(phenylthiomethyl)silanes with LN is quite straightforward. 1,2-Diaryl-1,2-bisphenylthioethenes undergo elimination to furnish diarylalkynes (5 examples, 67–89%).[5] Such a bisphenylthioethene is prepared by alkylation of a bis(phenylthio)acetal of an aromatic aldehyde with another aromatic aldehyde and treatment of the adduct with MsCl–EtN_3.

MsCl - Et_3N ; Li-naphth. THF, -100°

R = 2-Naphthyl

82%

67%

[1]Alonso, F., Lorenzo, E., Yus, M. *JOC* **61**, 6058 (1996).
[2]Ramig, K., Dong, Y., Van Arnum, S.D. *TL* **37**, 443 (1996).
[3]Tamao, K., Kawachi, A. *OM* **14**, 3108 (1995).

[4]Strohmann, C., Ludtke, S., Wack, E. *CB* **129**, 799 (1996).
[5]Sato, T., Tsuchiya, H., Otera, J. *SL* 628 (1995).

Lithium perchlorate–diethyl ether. 18, 218–219

Isomerization of epoxides.[1] In 5 M $LiClO_4$–Et_2O, epoxides are converted to carbonyl compounds.

Glycosylation.[2] Protected glycosyl fluorides and trichloroimidates undergo glycosylation, whereas the corresponding glycosyl bromides and phosphates are inert.

Organometallic reactions. The reactions of ketones with alkynyllithiums and alkylmagnesium halides in the presence of $LiClO_4{\cdot}OEt_2$ have been compared.[3] The catalyzed Grignard reaction effects a substitution of the benzenesulfonyl group of β-benzenesulfonyl-γ-oxo-γ-arylbutanenitriles.[4]

Mukaiyama aldol and Michael reactions. Aldol reactions catalyzed by $LiClO_4{\cdot}OEt_2$ proceed only with the combination of silyl enol ethers and dimethoxyacetals of aldehydes.[5] Free aldehydes or acetals of ketones do not react.

The Michael addition using silyl enol ethers and nitroalkenes involves silyl group transfer.[6]

Diels–Alder reactions. Compounds that give an oxygen-stabilized allyl cation are super-dienophiles in the presence of $LiClO_4.OEt_2$. The method obviates the difficulties experienced with the low reactivities of 2-cyclohexenones.[7] An intramolecular variant[8] serves to construct tricyclic ketones expediently.

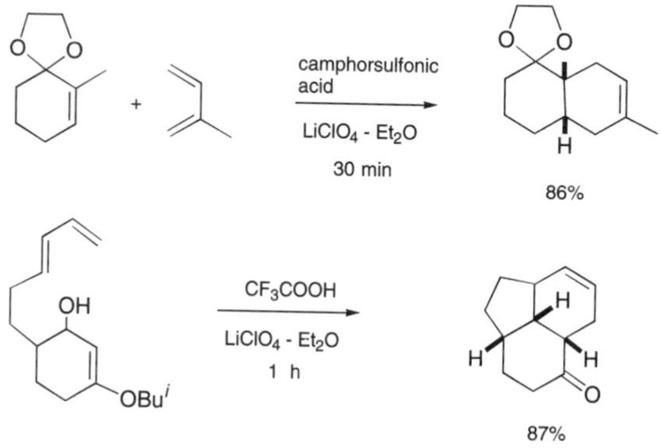

camphorsulfonic acid / $LiClO_4$ - Et_2O, 30 min — 86%

CF_3COOH / $LiClO_4$ - Et_2O, 1 h — 87%

Other cycloadditions. A [3+2]-cycloaddition involving a benzoquinone and an alkene to give 2,3-dihydrobenzofuran derivatives,[9] and an intramolecular [4+3]-cycloaddition to provide functionalized polycyclic compounds,[10] are further demonstrations of the utility of $LiClO_4{\cdot}OEt_2$. The reaction of aromatic or α,β-unsaturated aldehydes with acid chlorides proceeds via ketenes and then 2-oxetanones.[11]

Rearrangements and ene reactions. The [1.3]-rearrangement of hindered allyl vinyl ethers as catalyzed by $LiClO_4.OEt_2$ has previously been reported. Allyl ethers that ionize to give stabilized carbocations such as 2-furylalkyl and 2-thienylalkyl ions are found to undergo this type of reaction.[12]

Intramolecular ene reactions are readily catalyzed by $LiClO_4$ dispersed in silica gel.[13]

[1]Sudha, R., Narasimhan, K.M., Saraswathy, V.G., Sankararaman, S. *JOC* **61**, 1877 (1996).
[2]Bohm, G., Waldmann, H. *LA* 613 (1996).
[3]Ipaktschi, J., Eckert, T. *CB* **128**, 1171 (1995).
[4]Giovannini, R., Petrini, M. *SL* 1001 (1996).
[5]Saraswathy, V.G., Sankararaman, S. *JCS(P2)* 29 (1996).
[6]Saraswathy, V.G., Sankararaman, S. *JOC* **60**, 5024 (1995).
[7]Grieco, P.A., Collins, J.L., Handy, S.T. *SL* 1155 (1995).
[8]Grieco, P.A., Kaufman, M.D., Daeuble, J.F., Saito, N. *JACS* **118**, 2095 (1996).
[9]Asano, J., Ryu, Y., Chiba, K., Tada, M. *JCR(S)* 124 (1995).
[10]Harmata, M., Elomari, S., Barnes, C.L. *JACS* **118**, 2860 (1996).
[11]Arrastia, I., Cossio, F.P. *TL* **37**, 7143 (1996).
[12]Palani, N., Balasubramanian, K.K. *TL* **36**, 9527 (1995).
[13]Sarkar, T.K., Nandy, S.K., Ghorai, B.K., Mukherjee, B. *SL* 97 (1996).

Lithium telluride.

vic-Diols → alkenes.[1] After derivatization to the dimesylate, a *vic*-diol is amenable to elimination by Li_2Te at room temperature.

[1]Clive, D.L.J., Wickens, P.L., Sgarbi, P.W.M. *JOC* **61**, 7426 (1996).

Lithium 2,2,6,6-tetramethylpiperidide (LTMP). 13, 167; **14**, 194–195; **17**, 171–172; **18**, 220–221

1-Hydroxyalkylmaleic esters.[1] LTMP can initiate a Michael addition to maleic esters to give adducts that are reactive toward carbonyl compounds. The expulsion of LTMP completes a reaction cycle similar to the Baylis–Hillman reaction.

Ph(C=O)CH3 + EtOOC–CH=CH–COOEt → (LTMP, -78° -> rt) Ph–C(OH)(CH3)–C(COOEt)=CH–COOEt

51%

Interring-directed lithiation.[2] With LTMP at −40° or −70°, 2,4′-bipyridyl is lithiated at the C-3 of the 4-substituted pyridine ring, owing to the directing effect of the second nitrogen atom. Thus, selective functionalization (such as by stannylation and Heck reaction) at that position is readily achieved.

[1]Harrowven, D.C., Poon, H.S. *T* **52**, 1389 (1996).
[2]Zoltewicz, J.A., Dill, C.D. *T* **52**, 14469 (1996).

Lithium tetraorganoindate.

Allylation.[1]Coupling of the organic residue of $Li(InR_4)$ with allylic halides occurs at room temperature. Products that include 1,5-dienes are readily formed, and thus, the method is applicable to the synthesis of rosefuran and sesquirosefuran.

[1]Araki, S., Jin, S.-J., Butsugan, Y. *JCS(P1)* 549 (1995).

Lithium tri-s-butylborohydride (L-Selectride).

Reduction of the C=N bond.[1] The highly stereoselective reduction of *N*-diphenylphosphinoyl imines[1] and *O*-methyl α-(*p*-toluenesulfinyl)ketoximes[2] leading to amino derivatives are reported.

2-methylcyclohexylidene=N–P(O)Ph2 → ($LiBH(sBu)_3$, THF, rt) cis-NH–P(O)Ph2 + trans-NH–P(O)Ph2

(>97 : <3)

64%

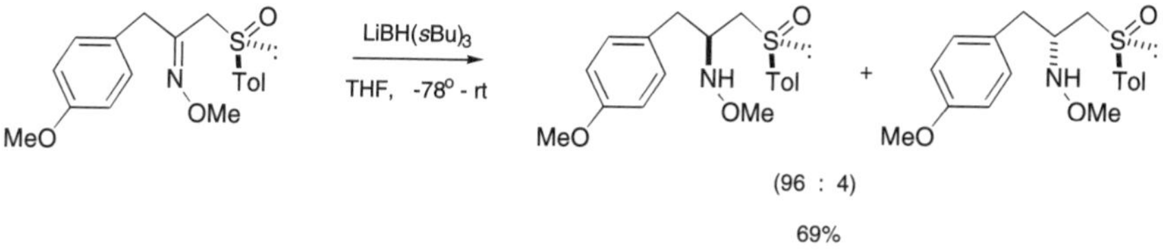

LiBH(sBu)$_3$
THF, -78° - rt

(96 : 4)
69%

Chiral 2-hydroxyalkylaziridines.[3] The diastereoselectivity in the reduction of certain 2-acylaziridines is high.

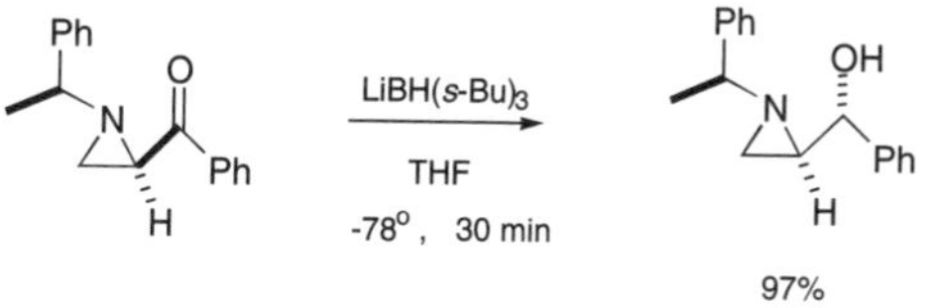

LiBH(s-Bu)$_3$
THF
-78°, 30 min

97%

Reductive cyclization.[4] 2,7-Nonadienedioic diesters undergo cyclization to cyclohexane derivatives. With proper placement of substituents, a building block for terpene synthesis is readily prepared by this method.

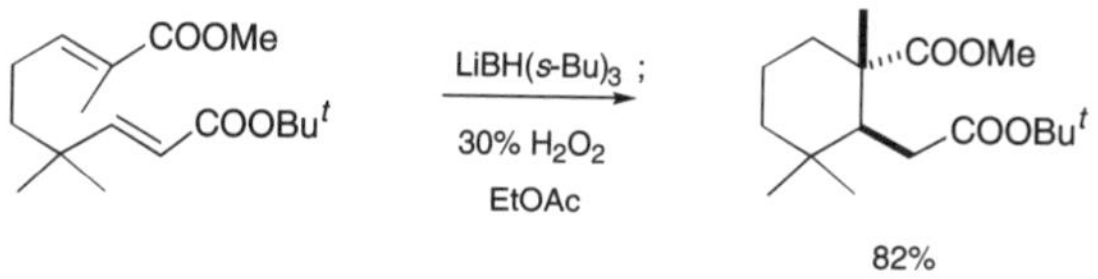

LiBH(s-Bu)$_3$;
30% H_2O_2
EtOAc

82%

[1]Hutchins, R.O., Adams, J., Rutledge, M.C. *JOC* **60**, 7396 (1995).
[2]Miyashita, K., Toyoda, T., Miyabe, H., Imanishi, T. *SL* 1229 (1995).
[3]Kim, B.C., Lee, W.K. *T* **52**, 12117 (1996).
[4]Yoshii, E., Hori, K., Nomura, K., Yamaguchi, K. *SL (Sp. Issue)* 568 (1995).

Lithium trifluoromethanesulfonate.

2-Amino alcohols.[1] This salt is an excellent substitute for the potentially hazardous lithium perchlorate in many catalyzed reactions. Thus, the epoxide ring opening by an amine is promoted by LiOTf.

[1]Auge, J., Leroy, F. *TL* **37**, 7715 (1996).

Lithium tris(methylthio)methide. 18, 222–223

α-Keto (trithio)orthoesters and dimethyl α-keto dithioacetals.[1] The reaction of the salt with acid chlorides gives either the (trithio)ortho esters or the dithioacetals, depending on the presence or absence of *N*-methylthiophthalimide in the reaction mixture.

$PhCOCl + (MeS)_3CLi \xrightarrow{THF,\ -78^o} PhCOCH(SMe)_2$ 98%

$PhCOCl + (MeS)_3CLi \xrightarrow{THF,\ -78^o,\ N\text{-methyldithiophthalimide}} PhCOC(SMe)_3$ 94%

[1]Degani, I., Dughera, S., Fochi, R., Serra, E. *JOC* **61**, 9572 (1996).

M

Magnesium. 13, 170; **15**, 194; **16**, 198–199; **18**, 224–225

Reduction of nitroarenes.[1] With iodine in dry methanol, Mg causes reductive coupling of $ArNO_2$ to give diazoarenes (ArN=NAr).[1] In the presence of $(NH_4)_2SO_4$, anilines are the products.[2]

Reductive amination.[3] Imines formed in situ from amines and carbonyl compounds are subject to reduction. This method constitutes a convenient synthesis of secondary amines.

Reductive coupling.[4] Using Mg as a sacrificial anode to perform electrolysis esters are reductively coupled to either α-diketones or 1,2-bistrimethylsilylalkenes, depending on the nature of the additives.

Enolsilylation.[5] Magnesium hexamethyldisilazide, which is prepared electrochemically with a Mg anode, deprotonates ketones. An ensuing silylation delivers the silyl enol ethers.

Reductive silylation of carbonyl compounds.[6] α-Silylalkyl silyl ethers [e.g., $RR'C(SiMe_3)OSiMe_3$] are readily obtained from a mixture of RR′C=O, Me_3SiCl, and Mg in DMF.

Cycloalkylphosphonates.[7] An electrochemically generated α-phosphono carbanion from a dialkyl trichloromethylphosphonate (Mg anode) undergoes alkylation with an α,ω-dihaloalkane to furnish the 1-chlorocycloalkylphosphonate. Further reduction removes the remaining chlorine atom.

(*i*-PrO)$_2$P(O)-CCl$_3$ + Br(CH$_2$)$_3$Br —e$^-$, Mg anode (1.5 F/mol)→ (*i*-PrO)$_2$P(O)-C(Cl)(cyclobutyl) 64% —e$^-$, Mg anode (2 F/mol)→ (*i*-PrO)$_2$P(O)-CH(cyclobutyl) 52%

[1]Khurana, J.M., Ray, A. *BCSJ* **69**, 407 (1996).
[2]Prajapati, D., Borah, H.N., Sandhu, J.S., Ghosh, A.C. *SC* **25**, 4025 (1995).
[3]Micovic, I.V., Ivanovic, M.D., Roglic, G.M., Kiricojevic, V.D., Popovic, J.B. *JCS(P1)* 265 (1996).
[4]Kashimura, S., Murai, Y., Ishifune, M., Masuda, H., Murase, H., Shono, T. *TL* **36**, 4805 (1995).
[5]Bonafoux, D., Bordeau, M., Biran, C., Cazeau, P., Dunogues, J. *JOC* **61**, 5532 (1996).
[6]Ishino, Y., Maekawa, H., Takeuchi, H., Sukata, K., Nishiguchi, I. *CL* 829 (1995).
[7]Jubault, P., Feasson, C., Collignon, N. *TL* **37**, 3679 (1996).

Magnesium–copper chloride.

Allylation and reduction.[1] The Mg–$CuCl_2 \cdot 2H_2O$ couple reduces carbonyl compounds to alcohols (7 examples, 70–95%). In the presence of an allyl halide, homoallyl alcohols are produced.

Coupling of organohalides.[2] Various kinds of halides (alkyl, aryl, benzyl, etc.) RX give dimers R–R on reaction with Mg and CuCl–2LiCl.

[1]Sarangi, C., Nayak, A., Nanda, B., Das, N.B., Sharma, R.P. *TL* **36**, 7119 (1995).
[2]Johnson, D.K., Ciavarri, J.P., Ishmael, F.T., Schillinger, K.J., van Geel, T.A.P., Stratton, S.M. *TL* **36**, 8565 (1995).

Magnesium–tin(II) chloride dihydrate.

Nitroalkenes → carbonyl compounds.[1] The transformation is effected by Mg–$SnCl_2 \cdot 2H_2O$ in aqueous THF.

[1]Das, N.B., Sarangi, C., Nanda, B., Nayak, A., Sharma, R.P. *JCR(S)* 28 (1996).

Magnesium bromide. **15**, 194–196; **16**, 199; **17**, 174; **18**, 226–227

Debenzylation.[1] $MgBr_2$ effects a selective cleavage of benzyl ethers of *o*-hydroxy phenones.

Esterification.[2] The combination of $MgBr_2$ and a tertiary amine (such as Et_3N) activates hindered alcohols to undergo esterification (6 examples, 80–99%).

α-Amino acids.[3] A route to α-amino acids from aziridine-2-carboxylic esters involves a reaction with $MgBr_2$ and dehalogenation. Interestingly, the regioselectivity for the ring opening is completely reversed on changing the metal halide to NaI (with Amberlyst-15).

EtOOC N COOMe — $MgBr_2 \cdot OEt_2$, Et_2O, rt, 2 h → NHCOOEt COOMe Br

~ 100%

Allylation.[4] $MgBr_2$ mediates the attack of allyltributylstannane on α-alkoxy carbonyls through chelation.

O O CHO OBn — $MgBr_2 \cdot OEt_2$, allyl-$SnBu_3$, CH_2Cl_2, -78° -> rt → O O OH OBn

98% (ds >50 : 1)

Dieckmann and Claisen condensations. Dieckmann cyclization[5] and Claisen condensation of *N*-acylpyrazoles[6] are catalyzed by $MgBr_2$-R_3N. Note that a practical synthesis of α-acylamino-β-ketoesters employs $MgCl_2$-Et_3N to induce the reaction between alkyl hydrogen (acylamino)malonates and acyl chlorides.[7]

[1]Haraldsson, G.G., Baldwin, J.E. *T* **53**, 215 (1997).
[2]Vedejs, E., Daugulis, O. *JOC* **61**, 5702 (1996).
[3]Righi, G., D'Achille, R., Bonini, C. *TL* **37**, 6893 (1996).
[4]Charette, A.B., Benslimane, A.F., Mellon, C. *TL* **36**, 8557 (1995).
[5]Tamai, S., Ushirogochi, H., Sano, S., Nagao, Y. *CL* 295 (1995).
[6]Kashima, C., Takahashi, K., Fukusaka, K. *JHC* **32**, 1775 (1995).
[7]Krysan, D.J. *TL* **37**, 3303 (1996).

Magnesium methoxide. 18, 227

Selective deacylation. By using $Mg(OMe)_2$ in methanol, it is possible to remove primary acetates in the presence of tertiary acetates.[1] For different esters, the order of reactivity to this reagent is *p*-nitrobenzoate > acetate > benzoate > pivalate.[2]

BzO ... OPiv → [$Mg(OMe)_2$ / MeOH / rt, 24 h] → HO ... OPiv (96%)

Ethers.[3] Benzylic and allylic halides undergo alcoholysis to give alkyl ethers.

Directed aldol reaction.[4] Unsymmetrical γ-diketones are cyclized via the magnesium chelates.

COOMe → [$Mg(OMe)_2$ / MeOH ; NaOH / H_2O] → (~ 80%)

[1]Xu, Y.-C., Bizuneh, A., Walker, C. *JOC* **61**, 9086 (1996).
[2]Xu, Y.-C., Bizuneh, A., Walker, C. *TL* **37**, 455 (1996).
[3]Lin, J.-M., Li, H.-H., Zhou, A.-M. *TL* **37**, 5159 (1996).
[4]Cadman, M.L.F., Crombie, L., Freeman, S., Mistry, J. *JCS(P1)* 1397 (1995).

Magnesium monoperoxyphthalate (MMPP). 14, 197; 16, 199–200; 18, 228

Baeyer–Villiger oxidation. The preparation of lactones from cyclic ketones using MMPP as reagent is mediated by acetonitrile,[1] as well as in the presence of bentonite clay

(6 examples, 65–95%),[2] although other solvent systems, such as aqueous DMF (for the synthesis of phthalides from benzocyclobutenones),[3] have been employed.

OMe O OMe
CO_3H $COO)_2Mg$
DMF - H_2O
40°, 0.5 h
OMe O O OMe
73%

2,5-Dimethoxy-2,5-dihydrofurans.[4] 2,5-Disubstituted furans give the dimethoxylated products on treatment at room temperature with MMPP in methanol (5 examples, 83–95%). The original substituents in 2,5-positions are preserved.

Sulfoxides.[5] Bentonite assists the MMPP oxidation of sulfides (15 examples, 68–96%).

2,2-Disubstituted β,γ-unsaturated nitriles.[6] After deconjugative alkylation of enal *N,N*-dimethylhydrazones, MMPP oxidation (MeOH, 0°) furnishes the nitriles.

[1]Hirano, M., Yakabe, S., Satoh, A., Clark, J.H., Morimoto, T. *SC* **26**, 4591 (1996).
[2]Hirano, M., Ueno, Y., Morimoto, T. *SC* **25**, 3765 (1995).
[3]Hosoya, T., Kuriyama, Y., Suzuki, K. *SL* 635 (1995).
[4]D'Annibale, A., Scettri, A. *TL* **36**, 4659 (1995).
[5]Hirano, M., Ueno, Y., Morimoto, T. *SC* **25**, 3125 (1995).
[6]Mino, T., Yamashita, M. *JOC* **61**, 1159 (1996).

Magnesium perchlorate. 18, 228

Allyl ethers.[1] In the presence of $Mg(ClO_4)_2$, the admixture of an allylic alcohol and another alcohol prone to ionize results in the formation of the mixed allyl ether.

[1]De Mico, A., Margarita, R., Piancatelli, G. *TL* **36**, 2679 (1995).

Manganese–lead(II) chloride–trimethylsilyl chloride.

Allylation.[1] Manganese is activated by a catalytic amount of $PbCl_2$ and Me_3SiCl by way of removing the tightly bound metal oxide that coats the surface. The system is useful for allylation of carbonyl compounds by allyl halides.

Tandem reactions initiated by conjugate addition. A three-component condensation[2] of alkyl iodides, electron-deficient olefins, and carbonyl compounds is an expedient synthetic procedure for skeletal construction. Exposure of allyl or propargyl acrylates and alkyl iodides to the Mn–$PbCl_2$–Me_3SiCl system gives γ,δ-unsaturated or allenic acids, respectively, through a conjugate addition and Ireland–Claisen reaction sequence.[3]

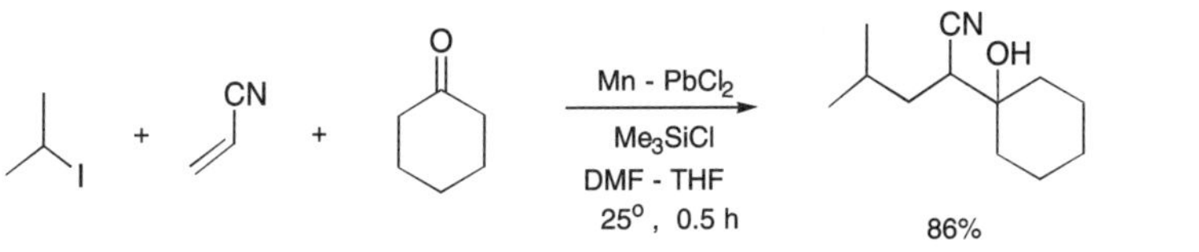

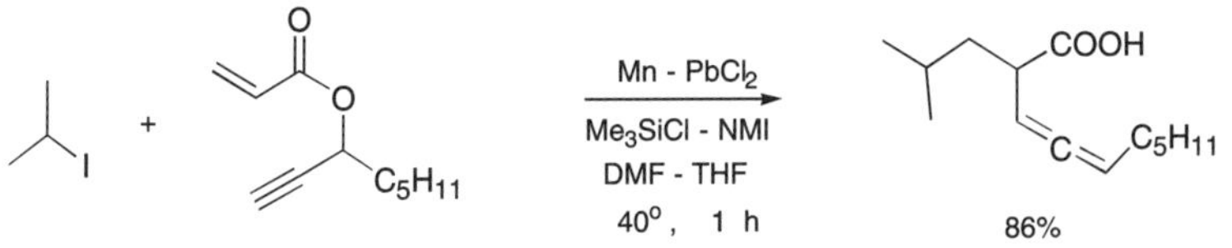

^{1}Takai, K., Ueda, T., Hayashi, T., Moriwake, T. *TL* **37**, 7049 (1996).
^{2}Takai, K., Ueda, T., Ikeda, N., Moriwake, T. *JOC* **61**, 7990 (1996).
^{3}Takai, K., Ueda, T., Kaihara, H., Sunami, Y., Moriwake, T. *JOC* **61**, 8728 (1996).

Manganese(III) acetate. 13, 171; **14**, 197–199; **16**, 200; **17**, 175–176; **18**, 229–230

Alkylation of malonate esters. Special procedures can be applied to form either homologous diesters or lactones in the free-radical reaction with olefins.[1] An intramolecular reaction[2] involving simultaneous addition to a triple bond and carboxylation of the latter has synthetic potentials.

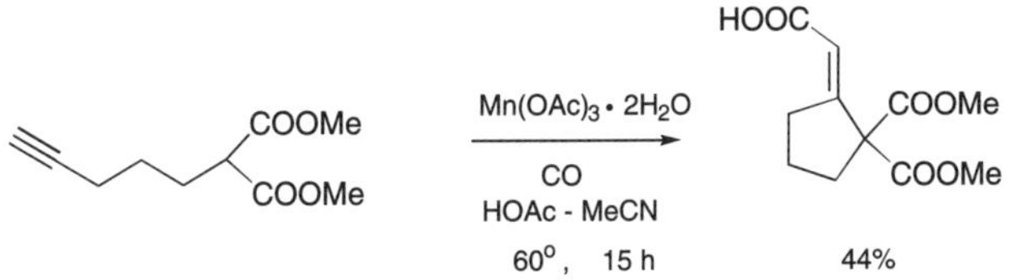

Benzyl acetates.[3] Arylacetic acids undergo oxidative decarboxylation with $Mn(OAc)_3$ in HOAc. The intermediate benzyl cations are quenched by HOAc. An electron-donating *p*-substituent in the aromatic ring and higher α-substitution favor the reaction.

β-Lactams.[4] *N*-Malonylenamines cyclize in the presence of $Mn(OAc)_3$ in HOAc to afford β-lactams (7 examples, 39–73%).

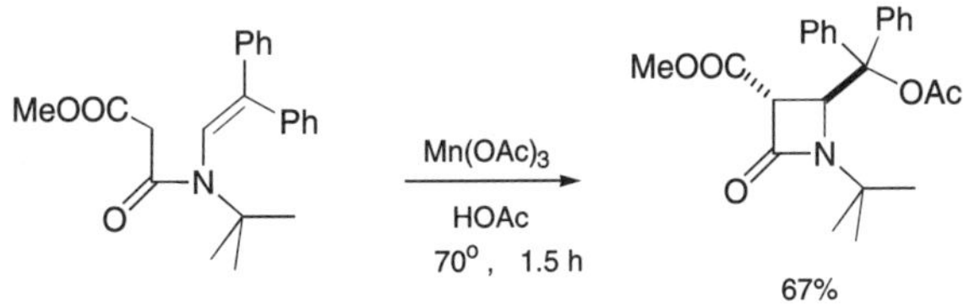

Chlorination and selenenylation. Aromatic chlorination using $Mn(OAc)_3$–AcCl is promoted by ultrasound,[5] and the $Mn(OAc)_3$–PhSeSePh combination is useful for the introduction of a phenylselenyl group to C-5 of pyrimidine nucleosides.[6]

[1]Linker, T., Kersten, B., Linker, U., Peters, K., Peters, E.-M., von Schnering, H.G. *SL* 468 (1996).
[2]Okuro, K., Alper, H. *JOC* **61**, 5312 (1996).
[3]Mohri, K., Mamiya, J., Kasahara, Y., Isobe, K., Tsuda, Y. *CPB* **44**, 2218 (1996).
[4]D'Annibale, A., Resta, S., Trogolo, C. *TL* **36**, 9039 (1995).
[5]Prokes, I., Toma, S., Luche, J.-L. *JCR(S)* 164 (1996).
[6]Lee, D.H., Kim, Y.H. *SL* 349 (1995).

Manganese dioxide. **14**, 200–201; **15**, 197–198; **18**, 230–231

Oxime hydrolysis.[1] Aldoximes and ketoximes are readily cleaved by stirring with activated MnO_2 in hexane at room temperature.

Aromatization.[2] Rapid dehydrogenation of Hantzsch 1,4-dihydropyridines occurs on exposure to manganese dioxide under sonochemical conditions.

Chlorination.[3] Steroidal olefins and ketones can be chlorinated with AcCl–MnO_2 to give dichlorides and α-chloroketones (e.g., 17α-chloro-20-ones), respectively.

Carbazoles.[4] *o*-Alkylation of *p*-substituted anilines with a cationic (1,3-cyclohexadienyl)molybdenum complex proceeds in moderate yields. Oxidation of the products with MnO_2 leads directly to carbazoles.

OC Mo CO H H_2N OMe — MnO_2, PhH, 80° → OMe N H 53%

[1]Shinada, T., Yoshihara, K. *TL* **36**, 6701 (1995).
[2]Vanden Eynde, J.J., Delfosse, F., Mayence, A., Van Haverbeke, Y. *T* **51**, 6511 (1995).
[3]Borah, P., Chowdhury, P. *JCR(S)* 502 (1996).
[4]Knölker, H.-J., Goesmann, H., Hofmann, C. *SL* 737 (1996).

Manganese(II) iodide.

Allylmagnesiation of alkynes.[1] Allylmagnesium bromide adds to alkynes in the presence of MnI_2. The alkenylmagnesium bromides may be quenched with electrophilic agents. Propargyl ethers are converted to 3-substituted 1,2,5-hexatrienes.

Note that the complex $MeC_5H_4Mn(CO)_3$ promotes *syn*-1,2-diallylation by the Grignard reagent.

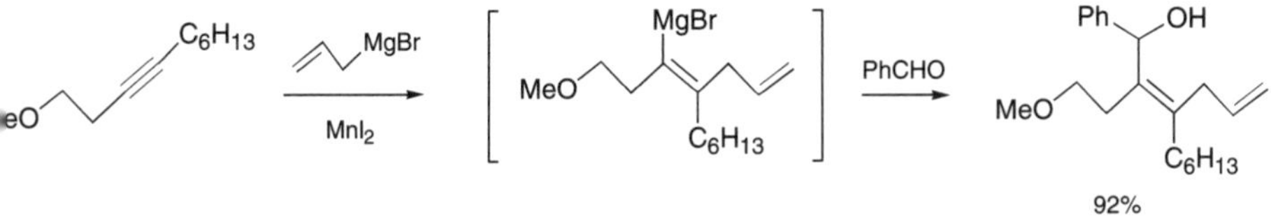

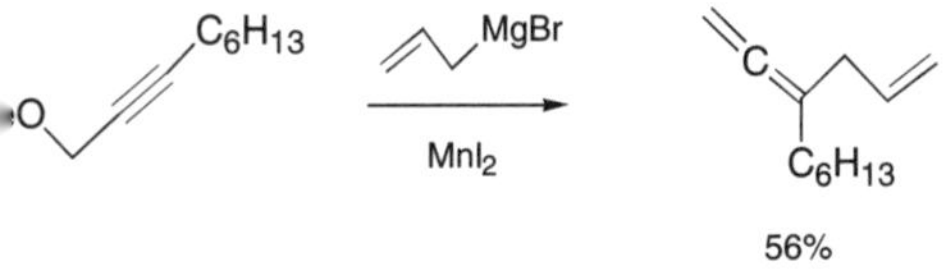

[1]Okada, K., Oshima, K., Utimoto, K. *JACS* **118**, 6076 (1996).

Mercury(II) acetate. 15, 198–199; **17**, 176–177; **18**, 232

RNCO → RNH2.[1] With the use of the common protocol for hydroxymercuration–demercuration [$Hg(OAc)_2$, $THF–H_2O$; $NaBH_4$, OH^-], aliphatic isocyanates are converted to primary amines. However, the yields are variable (6 examples, 10–100%).

[1]Malanga, C., Urso, A., Lardicci, L. *TL* **36**, 8859 (1995).

Mercury(II) chloride. 13, 175; **15**, 200; **18**, 232

Cyclization.[1] By using $HgCl_2$-HMDS in CH_2Cl_2 for intramolecular carbomercuration of 5,6-dimethyl-6-(3-butynyl)trimethylsiloxycyclohexene, a bridged ring system is formed. Carbonylation of the alkenylmercury chloride results in a key intermediate (76% for two steps) for a synthesis of trifarienol-A and -B. This method has great potential for the elaboration of cyclic systems.

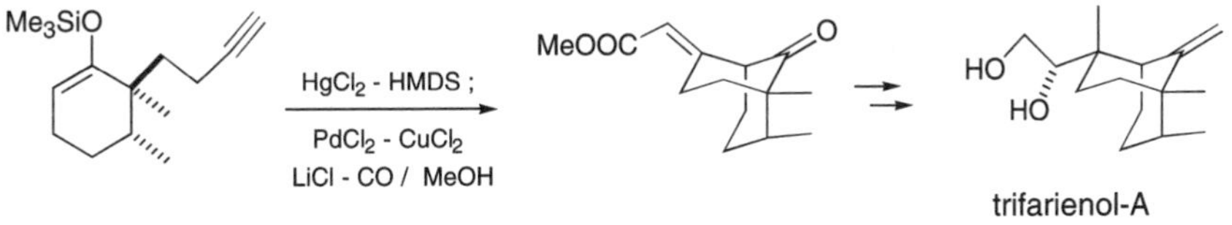

[1]Huang, H., Forsyth, C.J. *JOC* **60**, 5746 (1995).

Methanesulfonyl chloride. 13, 176; **18**, 233

Carbodiimides.[1] Thioureas eliminate H_2S rapidly under mild conditions (MsCl, Et_3N, DMAP, CH_2Cl_2, room temperature) to furnish RN=C=NR′ (7 examples, 85–100%).

[1]Fell, J.B., Coppola, G.M. *SC* **25**, 43 (1995).

_B_-Methoxy-9-borabicyclo[3.3.1]nonane.

Coupling.[1] Arylation of organometallics using 9-MeO-BBN as a mediator extends the scope of the Suzuki coupling.

OHC, Br, OMe; B, OMe; (dppf)$PdCl_2$; THF; Δ, 1 h; OHC, OMe; 60%; O; O; dehydrotremetone

[1]Fürstner, A., Seidel, G. *T* **51**, 11165 (1995).

_B_-Methoxydiisopinocampheylborane.

syn-α-Vinyl chlorohydrins.[1] After allylic boration of allyl chloride with Ipc_2BOMe in the presence of $LiN(c\text{-}Hx)_2$, the reaction of the chloroallylborane with carbonyl compounds proceeds in a diastereo- and enantioselective manner to give the chlorohydrins (15 examples, 68–85%). The products are cyclized to *cis*-vinyl epoxides (85–99%).

Cl; $(Ipc)_2BOMe$; $LiN(cHx)_2$; $BF_3 \cdot OEt_2$; Cl, $B(Ipc)_2$; CHO; -95° ; $HOCH_2CH_2NH_2$; OH, Cl; 72% (95% ee)

[1]Hu, S., Jayaraman, S., Oehlschlager, A.C. *JOC* **61**, 7513 (1996).

2-(*p*-Methoxyphenylamino)ethyl phenyl sulfone.

Allylamines.[1] Condensation of the *N,C*-dianion of this reagent with carbonyl compounds leads to oxazin-2-ones. On treatment with sodium amalgam, the heterocycles undergo desulfonylative ring opening.

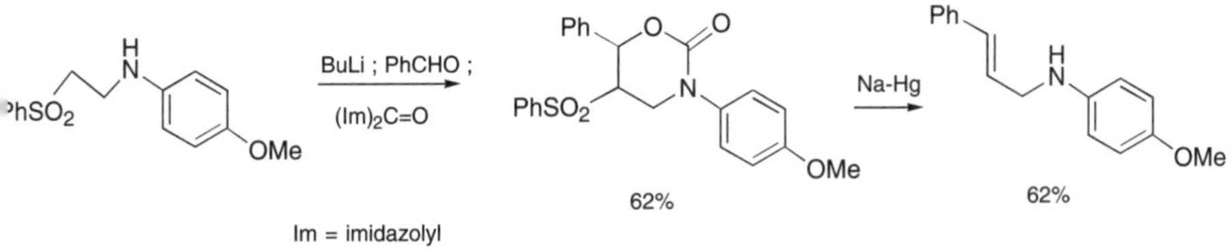

[1]Breuilles, P., Kaspar, K., Uguen, D. *TL* **36**, 8011 (1995).

Methoxytributylstannane.

Lactone opening.[1] A crucial operation in a synthesis of 12a-deoxytetracycline requires a mild alcoholysis of a lactone under conditions precluding closure of the B-ring, because intermediates with the linear tricarbocyclic array resist further cyclization. After successful lactone opening by treatment with Bu_3SnOMe, the α-tetralone unit can be masked as a lactol silyl ether to permit a Dieckmann cyclization to form the A-ring.

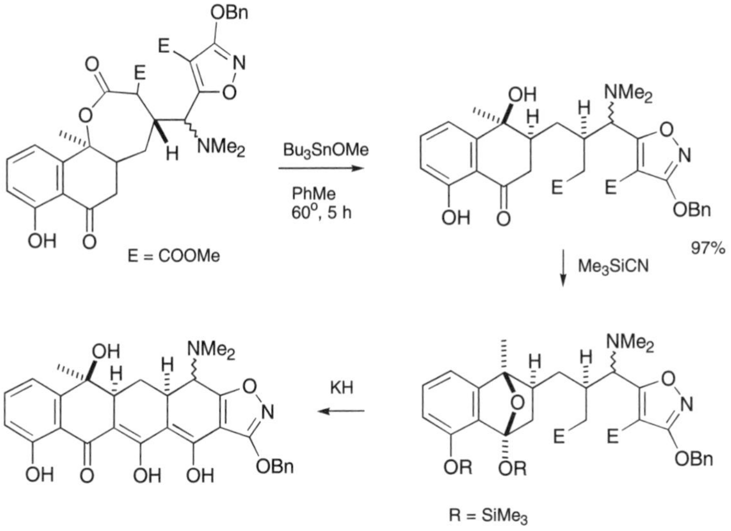

[1]Stork, G., La Clair, J.J., Spargo, P., Nargund, R.P., Totah, N. *JACS* **118**, 5304 (1996).

Methylaluminum bis(2,6-di-*t*-butyl-4-methylphenoxide) MAD. 13, 203; **14**, 206–207; **15**, 204–205; **16**, 212; **17**, 187–188, **18**, 237

Dihydroisoxazine N-oxides.[1] The [4+2]cycloaddition of β-nitrostyrene derivatives with 2-butoxy-1,4-pentadiene gives stereochemically complementary products under the influence of $SnCl_4$ and MAD. Each series is transformed into diastereomeric cyclohexylamines.

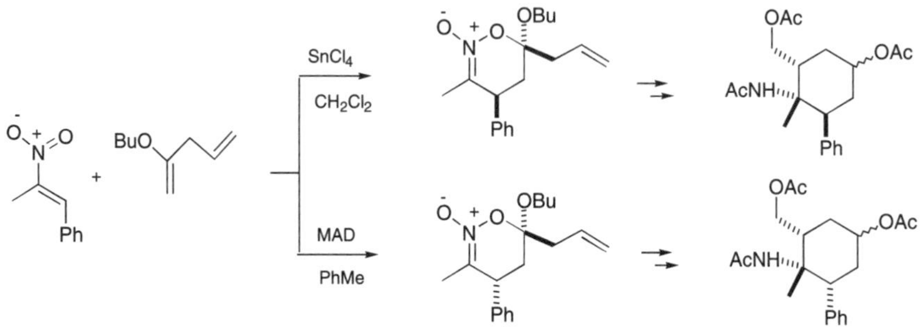

[1]Denmark, S.E., Stolle, A., Dixon, J.A., Guagnano, V. *JACS* **117**, 2100 (1995).

Methyl β-(benzotriazol-1-yl)vinyl ketone.

3-Buten-2-one 4-cation equivalent.[1] The reagent is available by an exchange reaction of 3-oxobutyraldehyde 1,1-dimethylacetal with benzotriazole and partial elimination with NaOH under phase transfer conditions. The reagent undergoes addition–elimination with various nucleophiles (piperidine, thiophenol, 2-nitropropane, malonic esters, etc.).

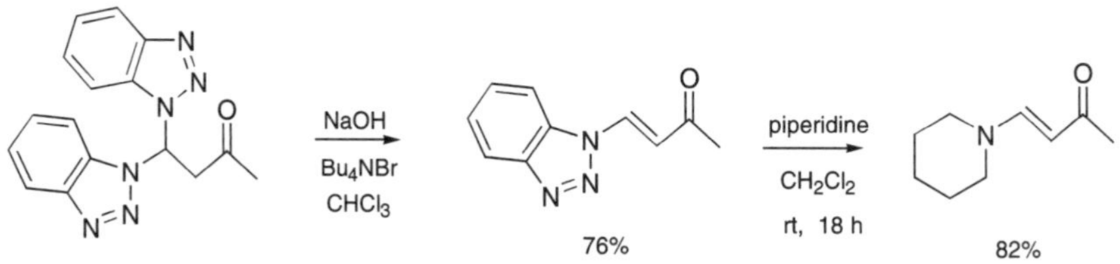

[1]Katritzky, A.R., Blitzke, T., Li, J. *SC* **26**, 3773 (1996).

Methyl chloroaluminum amide.

Amidines.[1] Nitriles are converted to amidines on reaction with $MeAl(Cl)NH_2$ in toluene at slightly elevated temperatures.

$RCOCF_3 \rightarrow RCN$. The degradative formation of nitriles (4 examples, 76–92%)[2] involves successive treatment of trifluoromethyl ketones with $MeAl(Cl)NH_2$ and *t*-BuOK. Conjugated nitriles can be prepared in this manner.

Ph COCF3 MeAl(NH2)Cl ; t-BuOK / THF N Me → Ph CN N Me 91%

[1]Moss, R.A., Ma, W., Merrer, D.C., Xue, S. *TL* **36**, 8761 (1995).
[2]Kende, A.S., Liu, K. *TL* **36**, 4035 (1995).

(1*R*,3*R*)-2-Methylene-1,3-dithiolane 1,3-dioxide.

Chiral ketene equivalent.[1] The reagent prepared from 2-benzyl-oxymethyl-1,3-dithiolane by an asymmetric oxidation and elimination sequence undergoes Diels–Alder reactions. Thus, the adduct with cyclopentadiene affords (+)-norbornenone in high yield.

Me2NH ; MeI - iPr2NEt 90% BF3 • OEt2 -78° 20 min 74%

[1] Aggarwal, V.K., Drabowicz, J., Grainger, R.S., Gültekin, Z., Lightowler, M., Spargo, P.L. *JOC* **60**, 4962 (1995).

Methyl iodide.

Bis(trimethylsilyl)amines.[1] Methyl iodide can be used to activate the silyl donor Me_3SiNEt_2 during the transformation of RNH_2 into $RN(SiMe_3)_2$.

[1]Hamada, Y., Yamamoto, Y., Shimizu, H. *JOMC* **510**, 1 (1996).

Methyllithium. **13**, 188–189; **14**, 211; **15**, 208; **18**, 240

Debenzylation.[1] A new method for *N*-debenzylation of substituted indoles consists of treatment with MeLi.

3-Alkylpyrrolidines.[2] The anionic cyclization of *N*-trimethylstannylmethyl-3-butenylamines is promoted by MeLi. During the reaction, the trimethylstannyl group is transferred to a new position for further elaboration.

[1]Suzuki, H., Tsukuda, A., Kondo, M., Aizawa, M., Senoo, Y., Nakajima, M., Watanabe, T., Yokoyama, Y., Murakami, Y. *TL* **36**, 1671 (1995).
[2]Coldham, I., Hufton, R. *T* **52**, 12541 (1996).

N-Methylmorpholine *N*-oxide (NMO).

ArBr → ArOH.[1] The conversion of bromoarenes to phenols involves lithiation, and reaction with trimethyl borate and with NMO in THF (3 examples, 75–80%).

Epoxidation.[2] Unsaturated lactams give epoxy derivatives without a metal catalyst.

Oxidation of nitrile oxides.[3] A simple access to O=N–COR from nitrile oxides is the addition of NMO in a dichloromethane solution at room temperature. The reactive acyl nitroso compounds are trapped (usually with dienes).

Pauson–Khand reaction.[4] An enantioselective synthesis of cyclopentenones employs alkyne complexes in which a chiral phosphine ligand is attached to the dicobalt moiety. NMO serves as the oxidant to complete the process.

[1]Gotteland, J.-P., Halazy, S. *SL* 931 (1995).
[2]Andres, C.J., Spetseris, N., Norton, J.R., Meyers, A.I. *TL* **36**, 1613 (1995).

[3]Quadrelli, P., Invernizzi, A.G., Caramella, P. *TL* **37**, 1909 (1996).
[4]Hay, A.M., Kerr, W.J., Kirk, G.G., Middlemiss, D. *OM* **14**, 4986 (1995).

Methyltrioxorhenium(VII) (MTO). 17, 192–193

Alkylations.[1] The *O*-alkylation of an alcohol by another alcohol is catalyzed by MTO, and thus, both symmetrical and unsymmetrical ethers are readily prepared. Amines are also alkylated by alcohols.

Epoxides and aziridines.[2] The rhenium carbenoid generated from the decomposition of ethyl diazoacetate undergoes cycloaddition with aromatic aldimines and with carbonyl compounds. The yields of the ethyl 3-arylaziridine-2-carboxylates are excellent.

[1]Zhu, Z., Espenson, J.H. *JOC* **61**, 324 (1996).
[2]Zhu, Z., Espenson, J.H. *JOC* **60**, 7090 (1995).

Methyltrioxorhenium(VII): hydrogen peroxide.

Oxidation of C–H bonds.[1] Carbonyl compounds are obtained from alcohols by oxidation with aq.H_2O_2-MTO. Some hydrocarbons are also oxidized: adamantane to 1-adamantanol (88%), *cis*-decalin to *cis*-4a-decalol (90%), and *trans*-decalin to *trans*-4a-decalol (20%).

Oxidation of amines. Secondary amines give nitrones,[2] whereas 1-adamantanamine is converted to 1-nitroadamantane, and anilines are converted to nitroarenes.[3]

Reactions of ethyl diazoacetates.[4] Diazo compounds form metal carbenoids, which can insert into O–H, S–H, and N–H bonds. With compounds containing a C=X bond (alkenes, carbonyl compounds, and imines), cycloaddition takes place to afford cyclopropanes, epoxides, and aziridines. (Note that RN_3 and ArCHO form ArCH=NR in the presence of MTO and Ph_3P).

α-Diketones.[5] Alkynes are oxidized to α-diketones at room temperature.

Cleavage of β-diketones.[6] 1,3-Cycloalkanediones lose the central carbon to give diacids.

[1]Murray, R.W., Iyanar, K., Chen, J., Wearing, J.T. *TL* **36**, 6415 (1995).
[2]Murray, R.W., Iyanar, K., Chen, J., Wearing, J.T. *JOC* **61**, 8099 (1996).
[3]Murray, R.W., Iyanar, K., Chen, J., Wearing, J.T. *TL* **37**, 805 (1996).
[4]Zhu, Z., Espenson, J.H. *JACS* **118**, 9901 (1996).
[5]Zhu, Z., Espenson, J.H. *JOC* **60**, 7728 (1995).
[6]Abu-Omar, M.M., Espenson, J.H. *OM* **15**, 3543 (1996).

Methyltrioxorhenium(VII)–urea–hydrogen peroxide.

Epoxidation.[1,2] Allylic alcohols undergo epoxidation in a *threo*-selective manner analogous to that of peracid, but contrary to the $VO(acac)_2$-mediated process.

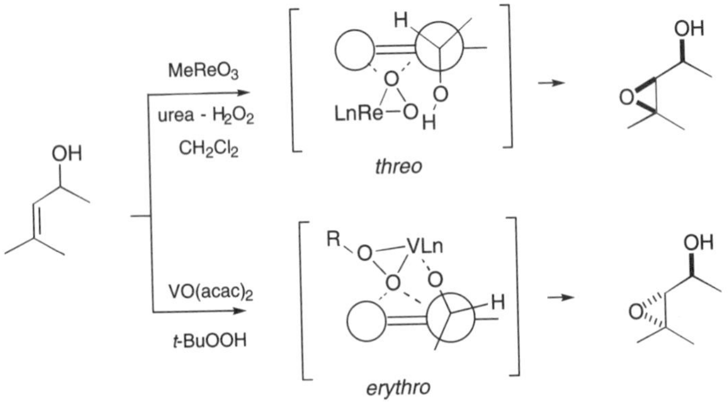

Nitrones.[3] The oxidation of secondary amines to nitrones with this oxidant mixture shows about the same efficiency as aq.H_2O_2-MTO.

[1]Adam, W., Mitchell, C.M. *ACIEE* **35**, 533 (1996).
[2]Boehlow, T.R., Spilling, C.D. *TL* **37**, 2717 (1996).
[3]Goti, A., Nannelli, L. *TL* **37**, 6025 (1996).

Methyl triflate.

Activation of imidazolides.[1] *N*-Acylimidazoles are further activated by *N′*-methylation, making them susceptible to attack by hindered alcohols and amines.

[1]Ulibarri, G., Choret, N., Bigg, D.C.H. *S* 1286 (1996).

Methyl(trifluoromethyl)dioxirane. 15, 212; **16**, 224; **18**, 242

The reagent can be prepared[1] in situ from 1,1,1-trifluoroacetone and Oxone®.

Epoxidation.[1] Fast reaction in homogeneous system for all kinds of alkenes are observed.

α-Ketols.[2] Acetonides undergo oxidative ring cleavage.

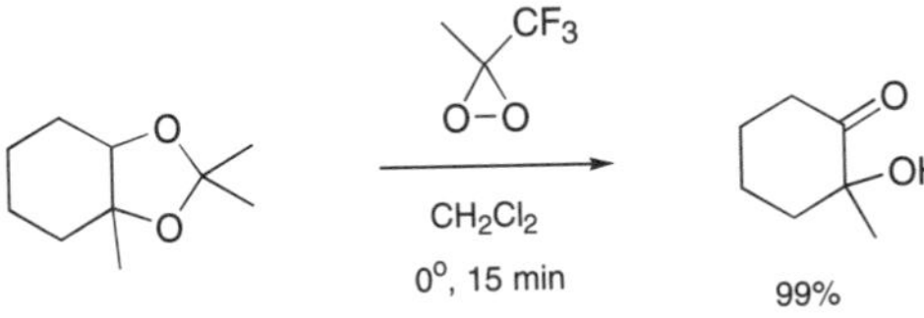

Oxyfunctionalization of unactivated C–H bonds. 2,2,3,3-Tetramethylbutane affords a primary alkyl trifluoroacetate[3] in 99% yield when it is oxidized with $Me(CF_3)CO_2$ in the presence of $(CF_3CO)_2O$ in dichloromethane at 0°. Esters show a remarkable regioselectivity in their oxidation.[4]

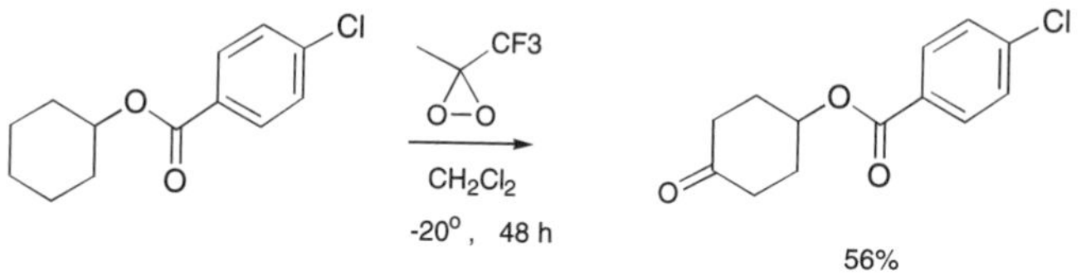

[1]Yang, D., Wong, M.-K., Yip, Y.-C. *JOC* **60**, 3887 (1995).
[2]Curci, R., D'Accolti, L., Dinoi, A., Fusco, C., Rosa, A. *TL* **37**, 115 (1996).
[3]Asensio, G., Mello, R., Gonzalez-Nunez, M.E., Castellano, G., Corral, J. *ACIEE* **35**, 217 (1996).
[4]Asensio, G., Castellano, G., Mello, R., Gonzalez-Nunez, M.E. *JOC* **61**, 5564 (1996).

Methyltriphenylphosphonium borohydride.

Reduction.[1] The borohydride reagent is highly selective, reducing aldehydes in the presence of ketones, conjugated carbonyl compounds to allylic alcohols, and acid chlorides to alcohols.

[1]Firouazabadi, H., Adibi, M. *SC* **26**, 2429 (1996).

Methyl triphenylphosphoranylacetate.

Olefination of lactones.[1] At high temperatures in a nonpolar solvent, lactones react with Ph_3P=CHCOOMe to give the methyl 2-oxacycloalkylideneacetates. However, the stereoselectivity for this reaction is low.

Methyl donor.[2] This Wittig reagent is able to donate a methyl group to amines, phenols, acids, and phthalimide.

[1]Lakhrissi, M., Chapleur, Y. *ACIEE* **35**, 750 (1996).
[2]Desmaele, D. *TL* **37**, 1233 (1996).

Molybdenum carbene complexes. **17**, 194–195; **18**, 242–243

Olefin metathesis. Cross metathesis permits a facile access to allylsilanes[1] and (*Z*)-α,β-unsaturated nitriles.[2] Furthermore, employing an alkene metathesis in tandem with asymmetric allylboration provides functionalized alkenes.[3]

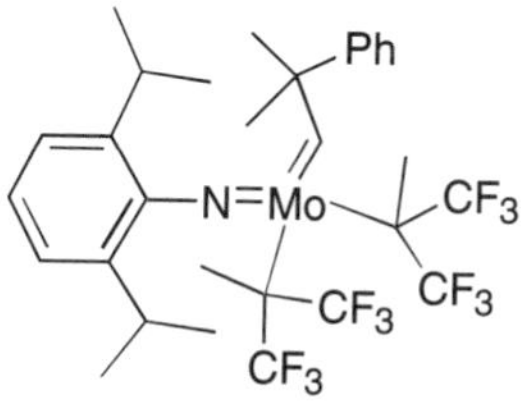

(1)

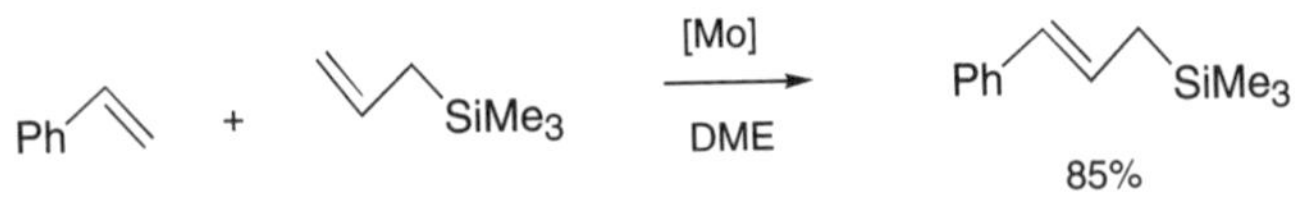

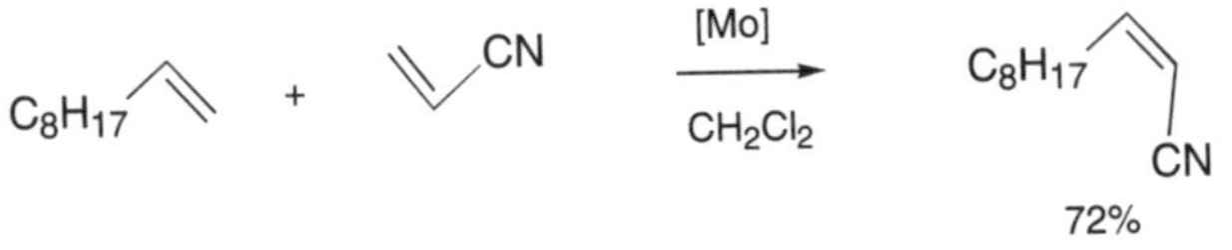

Ph + SiMe3 —[Mo], DME→ Ph–CH=CH–CH2–SiMe3 85%

C8H17 + CN —[Mo], CH_2Cl_2→ C8H17 CN 72%

There are several important applications of the Mo–carbene complex (**1**) in the synthesis of natural products, illustrating the power of the ring-closing olefin metathesis. Thus, elegant syntheses of coronafacic acid,[4] dactylol,[5] and fluvirucin[6] have been reported.

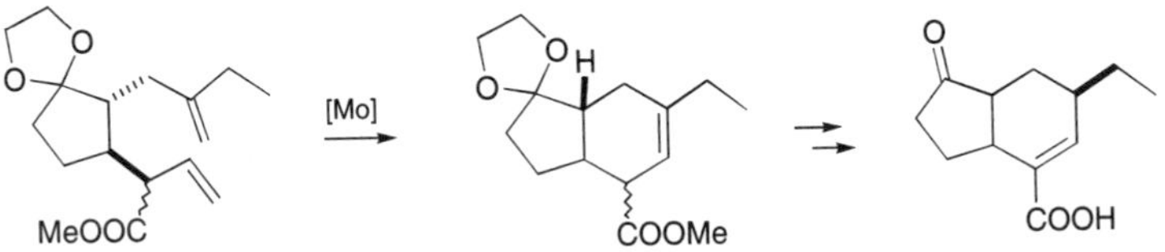

MeOOC —[Mo]→ COOMe → → COOH coronafacic acid

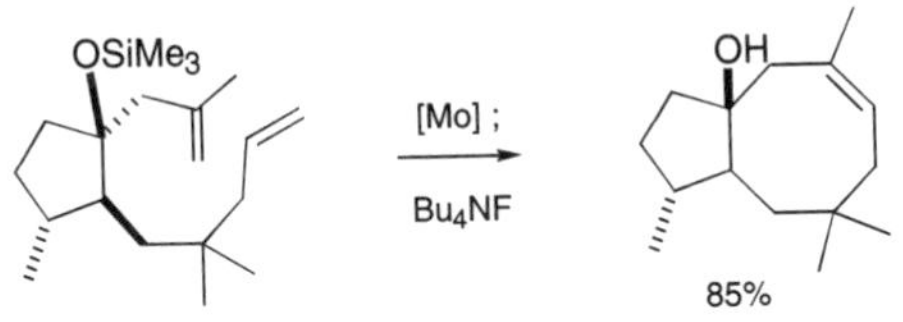

OSiMe3 —[Mo] ; Bu_4NF→ OH 85% dactylol

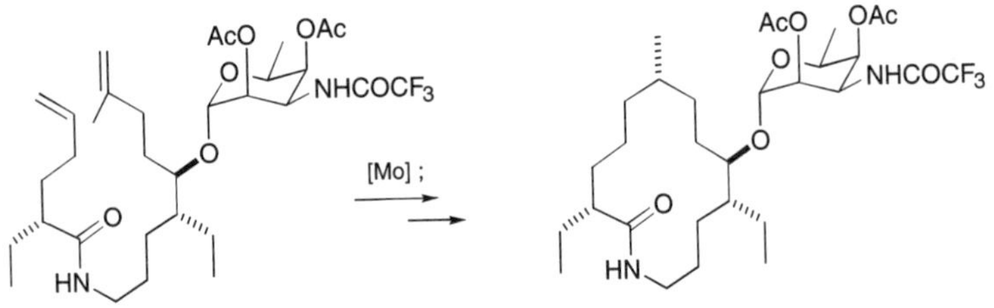

AcO OAc NHCOCF3 HN O —[Mo] ; →→ fluvirucin

It is interesting that a Mo–carbene complex and a Ru–carbene complex can show different stereoselective patterns in ring-closing olefin metathesis.[7]

An asymmetric version of olefin metathesis utilizes a Mo–carbene complex (**2**) with (*R,R*)-1,2-bis(2-hydroxy-2,2-bistrifluoromethyl)ethylcyclopentane as a ligand.[8]

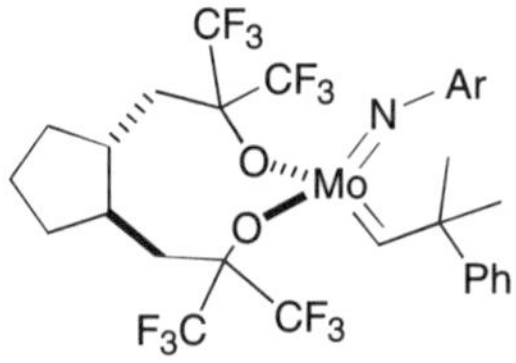

(2)

[1]Crowe, W.E., Goldberg, D.R., Zhang, Z.J. *TL* **37**, 2117 (1996).
[2]Crowe, W.E., Goldberg, D.R. *JACS* **117**, 5162 (1995).
[3]Barrett, A.G.M., Beall, J.C., Gibson, V.G., Giles, M.R., Walker, G.L.P. *CC* 2229 (1996).
[4]Hölder, S., Blechert, S. *SL* 505 (1996).
[5]Fürstner, A., Langemann, K. *JOC* **61**, 8746 (1996).
[6]Xu, Z., Johannes, C.W., Salman, S.S., Hoveyda, A.H. *JACS* **118**, 10926 (1996).
[7]Huwe, C.M., Velder, J., Blechert, S. *ACIEE* **35**, 2376 (1996).
[8]Fujimura, O., Grubbs, R.H. *JACS* **118**, 2499 (1996).

Molybdenum hexacarbonyl. **13**, 194–195; **15**, 212–213; **16**, 225–226; **18**, 243–244

Cyclopentenones.[1] The Pauson–Khand reaction catalyzed by $Mo(CO)_6$ in DMSO has been extended to the preparation of α-methylenecyclopentenones.

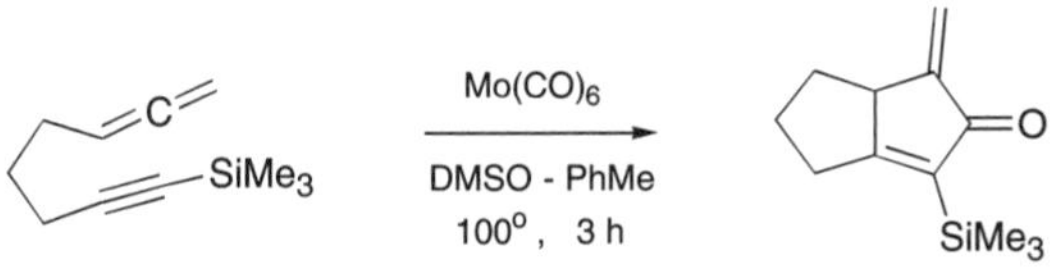

$Mo(CO)_6$; DMSO - PhMe; 100°, 3 h

68%

Alkyne metathesis. Substituted tolanes are formed from alkyl aryl acetylenes.[2] Cross-alkyne coupling provides a new route to unsymmetrical disubstituted alkynes from two symmetrical alkynes.[3]

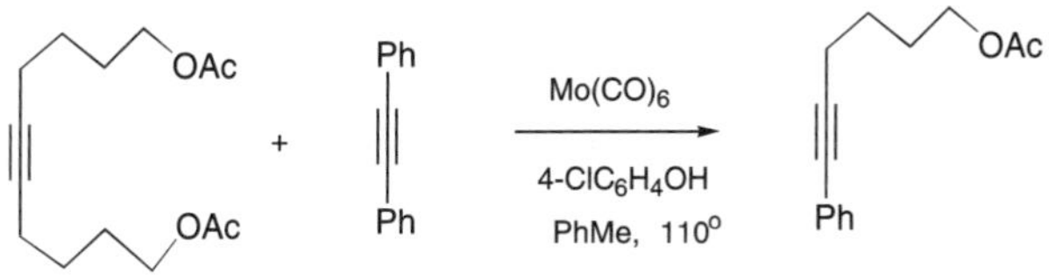

$Mo(CO)_6$; 4-ClC_6H_4OH; PhMe, 110°

74%

Allylic substitutions. A new Mo complex (**1**) derived from $Mo(CO)_6$ is an active catalyst for allylic substitution.[4]

Mo(CO)$_6$ $\xrightarrow[\text{AgOTf}]{[\text{BnNEt}_3]^+\text{Cl}^-;}$ $[(\text{OC})_5\text{MoOTf}]^+\text{OTf}^-$ (**1**) $\xrightarrow[\text{MeOH, DME - CH}_2\text{Cl}_2\text{, rt, 4 h}]{}$ 93%

[1]Kent, J.L., Wan, H., Brummond, K.M. *TL* **36**, 2407 (1995).
[2]Kaneta, N., Hirai, T., Mori, M. *CL* 627 (1995).
[3]Kaneta, N., Hikichi, K., Asaka, S.-I., Uemura, N., Mori, M. *CL* 1055 (1995).
[4]Dvorakova, H., Dvorak, D., Srogl, J., Kocovsky, P. *TL* **36**, 6351 (1995).

Molybdenyl acetylacetonate. 18, 244

Deacetalization.[1] The deprotection of acetals is performed in high yields (9 examples, 70–95%).

Oxidation.[2] $MoO_2(acac)_2$ catalyzes the oxidation of alcohols by sulfoxides. Aldehydes are obtained from primary alcohols.

Meyer–Schuster rearrangement.[3] The conversion of propargylic alcohols to conjugated carbonyl compounds is effected by heating with $MoO_2(acac)_2$ and $Bu_2S{=}O$ in *o*-dichlorobenzene at 100°.

OH $\xrightarrow[\text{Bu}_2\text{S=O, 1,2-Cl}_2\text{C}_6\text{H}_4\text{, 100}^\circ\text{, 5 h}]{\text{MoO}_2(\text{acac})_2}$ CHO 73%

[1]Kantam, M.L., Swapna, V., Santhi, P.L. *SC* **25**, 2529 (1995).
[2]Lorber, C.Y., Pauls, I., Osborn, J.A. *BSCF* **133**, 755 (1996).
[3]Lorber, C.Y., Osborn, J.A. *TL* **37**, 853 (1996).

Montmorillonite clays. 15, 213–214; **18**

Acid-catalyzed organic reactions by using clays continue to attract attention, and many reports on acetalization have appeared. The formation of C-glycosides by the reaction of glycals with enol silyl ethers, acetates, and allylsilanes is efficient (53–97%).[1]

Catalyzed reactions that are further assisted by microwave heating include aldolization[2] and Beckmann rearrangement.[3]

[1]Toshima, K., Miyamoto, N., Matsuo, G., Nakata, M., Matsumura, S. *CC* 1379 (1996).
[2]Abdullah-El ayoubi, S., Texier-Boullet, F. *JCR(S)* 208 (1995).
[3]Bosch, A.I., de la Cruz, P., Diez-Barra, E., Loupy, A., Langa, F. *SL* 1259 (1995).

Montmorillonite clays, metal ion-doped. 15, 101, 178–179; **18**, 244–245

Methylthiolation of thiophenes.[1] When autoclaved with $ZnCl_2$-doped-montmorillonite K 10, thiophene undergoes permethylthiolation.

thiophene + MeS-SMe → (Montmorillonite K 10; $ZnCl_2$, 150°, 5 h; autoclave) → tetrakis(methylthio)thiophene (MeS, MeS, SMe, SMe) 50%

Nitration.[2] Aromatic nitration using fuming nitric acid is catalyzed by $Cu(NO_3)_2$-impregnated acidic montmorillonite clay.

Oxidations. Potassium ferrate(VI) supported on clay has been used in the oxidation of alcohols to carbonyl compounds, of thiols to disulfides, and of organonitrogen compounds.[3] Another method for oxidizing alcohols involves acyl nitrates absorbed in clays.[4]

Homologation. The hydroesterification of alkenes using $Pd(OAc)_2$, CO, Ph_3P, and an alcohol under pressure also works when the clay-supported catalyst is used.[5]

[1]Clark, P.D., Mesher, S.T.E., Primak, A. *PSS* **114**, 99 (1996).
[2]Gigante, B., Prazeres, A.O., Marcelo-Curto, M.J., Cornelis, A., Laszlo, P. *JOC* **60**, 3445 (1995).
[3]Delaude, L., Laszlo, P. *JOC* **61**, 6360 (1996).
[4]de Oliveira Filho, A.P., Moreira, B.G., Moran, P.J.S., Rodrigues, J.A.R. *TL* **37**, 5029 (1996).
[5]Lee, C.W., Alper, H. *JOC* **60**, 250 (1995).

N

Nickel. 12, 355; **13**, 197; **14**, 213; **18**, 246

β-Lactams.[1] Free radicals are generated from trichloroacetamides on reaction with nickel powder–acetic acid. One report describes four examples of intramolecular addition of the radical, leading to β-lactams in 24–65%.

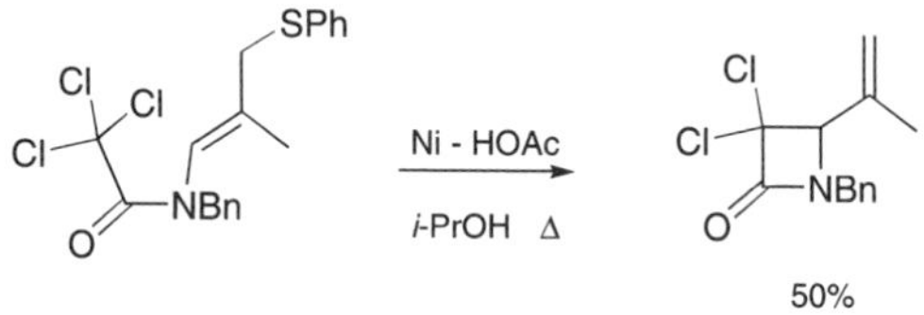

Cleavage of perfluoro epoxides.[2] In combination with iodine, nickel powder promotes the cleavage (C–O and C–C) to form acyl fluorides and *gem*-diiodoperfluoroalkanes.

[1]Quiclet-Sire, B., Saunier, J.-B., Zard, S.Z. *TL* **37**, 1397 (1996).
[2]Yang, Z.-Y. *JACS* **118**, 8140 (1996).

Nickel, Raney–sodium hypophosphite.

Chiral alcohols.[1] The reagent system in a buffered medium (pH 5.2) accomplishes the desulfurization of thioethers without causing racemization of a secondary alcohol (11 examples, 54–89%).

[1]Nishide, K., Shigeta, Y., Obata, K., Inoue, T., Node, M. *TL* **37**, 2271 (1996).

Nickel(0)–phosphine complexes.

Displacements.[1] Nickel complexes can serve as an alternative to the Pd catalysts, e.g., in the synthesis of allylamines[1] and α-allylmalonic esters.[2] A report on the latter reaction states that bis(aminophosphine) ligands are more efficient than dppb and other usual phosphines.

1,4-Dienes.[3] Coupling of allylamines with alkenylboronic acids with $(Ph_3P)_4Ni$ in benzene leads to 1,4-dienes.

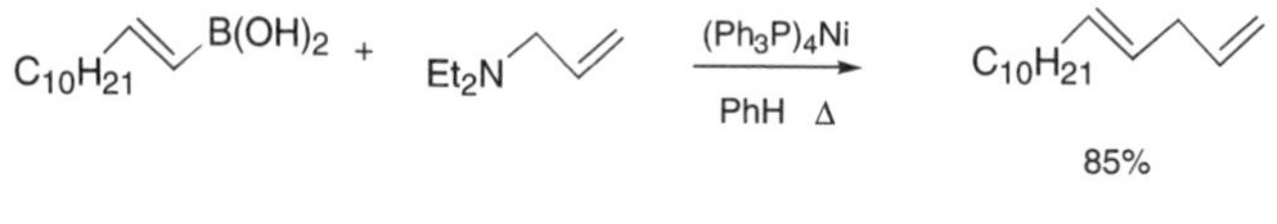

[1]Bricout, H., Carpentier, J.-F., Mortreux, A. *CC* 1863 (1995).
[2]Bricout, H., Carpentier, J.-F., Mortreux, A. *TL* **37**, 6105 (1996).
[3]Trost, B.M., Spagnol, M.D. *JCS(P1)* 2083 (1995).

Nickel(II) acetylacetonate. 17, 201; **18**, 247–248

Homocouplings.[1] Metallated sulfones undergo coupling to give symmetrical alkenes under Ni catalysis.

Cross-couplings. The enynes are formed from allyl halides and alkynylstannanes.[2] The Ni(acac)$_2$-catalyzed cross-coupling[3] of alkyl iodides and dialkylzincs is particularly useful because many functionalities are tolerated. An intramolecular addition to an alkene can intervene in this coupling thereby creating functionalized cyclic compounds.[4]

COOEt; Zn[(CH$_2$)$_n$OAc]$_2$; Ni(acac)$_2$; THF - NMP -78° -> -35°; COOEt; OAc; 78%

BuO; O; I; Et$_2$Zn - Ni(acac)$_2$; CuCN - 2 LiCl / THF ; Br; COOEt; COOEt; BuO; O; 70%

Efficient coupling of benzylic chlorides with alkenylaluminums is catalyzed by Ni(0) species formed by the reduction of Ni(acac)$_2$ in THF with *i*-Bu$_2$AlH in the presence of four equivalents of Ph$_3$P.[5] By contrast, Pd(0) catalysts are much inferior in this cross coupling, particularly in the cases of hindered benzylic chlorides, such as those required in the synthesis of coenzyme Q*n*.

MeO; MeO; Cl; + Me$_2$Al; 3; H; Ni(0)-cat.; MeO; MeO; 3; H; 87%

Conjugate additions. Michael additions involving cyclic β-keto esters[6] and alkenylboranes[7] as donors proceed in the presence of Ni(acac)$_2$. By using a mixture of dimethylzinc, a 1-alkyne, chlorotrimethylsilane, and Ph_3P, in addition to $Ni(acac)_2$, the introduction of a 2,2-disubstituted alkenyl group to enones[8] is accomplished in one step (5 examples, 58–89%).

C_6H_{13}—≡ + cyclopentenone → [$Ni(acac)_2$ - PPh_3 / Me_2Zn - Me_3SiCl / THF, 25°, 2 h; / HCl - H_2O] → 89%

The regioselectivity in the conjugate addition of alkyl aluminums and methyltitanium ate complexes to androsta-1,4-diene-3,17-dione is reversed in the presence of $Ni(acac)_2$.[9]

[4+2]Cycloaddition.[10] The cycloaddition between dienes and unactivated π-systems often proceeds at room temperature in the presence of transition metal catalysts. The catalytic system [$Ni(acac)_2$ and Et_2AlOEt] for such a stereoselective process can be replaced by $Ni(cod)_2$, as applied to a synthesis of a precursor to A-aromatic steroids.

OMOM, RO, MeO, R = $SiMe_3$ → [$Ni(cod)_2$ / $[(i\text{-}C_3HF_6)O]_3P$] → OR, H, MeO, OMOM, 90%

[1]Gai, Y., Julia, M., Verpeaux, J.-N. *BSCF* **133**, 805 (1996).
[2]Cui, D.-M., Hashimoto, N., Ikeda, S.-I., Sato, Y. *JOC* **60**, 5752 (1995).
[3]Devasagayaraj, A., Studemann, T., Knochel, P. *ACIEE* **34**, 2723 (1995).
[4]Vaupel, A., Knochel, P. *JOC* **61**, 5743 (1996).
[5]Lipshutz, B.H., Bulow, G., Lowe, R.F., Stevens, K.L. *JACS* **118**, 5512 (1996).
[6]Rao, H.S.P, Reddy, K.S., Jeyalakshmi, K. *IJC(B)* **34B**, 809 (1995).
[7]Yanagi, T., Sasaki, H., Suzuki, A., Miyaura, N. *SC* **26**, 2503 (1996).
[8]Ikeda, S., Yamamoto, H., Kondo, K., Sato, Y. *OM* **14**, 5015 (1995).
[9]Westermann, J., Neh, H., Nickisch, K. *CB* **129**, 963 (1996).
[10]Wender, P.A., Smith, T.E. *JOC* **60**, 2962 (1995).

Nickel bromide.

Arylations. $NiBr_2$ is a useful catalyst for the electrochemical arylation of activated alkenes.[1] Cross coupling of aryl halides with activated alkyl halides (e.g., α-halo ketones)[2] occurs under similar conditions.

Note that $(bipy)_3Ni(BF_4)_2$ has been employed as the catalyst for the electrochemical cleavage of allyl ethers (Mg anode).[3]

[1]Condon-Gueugnot, S., Leonel, E., Nedelec, J.-Y., Perichon, J. *JOC* **60**, 7684 (1995).
[2]Durandetti, M., Nedelec, J.-Y., Perichon, J. *JOC* **61**, 1748 (1996).
[3]Olivero, S., Dunach, E. *CC* 2497 (1995).

Nickel carbonyl. 13, 198–199; **15**, 216–217; **18**, 249

Cyclopentenones. The cyclocarbonylation of allyl halides and alkynes furnishes polyfunctional products in one step. Stereoselectivity is also observed.

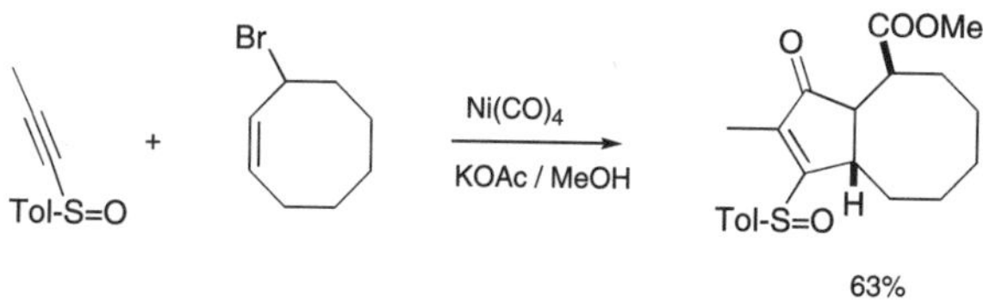

[1]Villar, J.M., Delgado, A., Llebaria, A., Moreto, J. M., Molins, E., Miravitlles, C. *T* **52**, 10525 (1996).

Nickel chloride dihydrate.

Hydrogenation.[1] Hydrated $NiCl_2$ and Li with a catalytic amount of naphthalene form a reducing agent for alkenes[1] and alkynes.[2]

[1]Alonso, F., Yus, M. *TL* **37**, 6925 (1996).
[2]Alonso, F., Yus, M. *TL* **38**, 149 (1997).

Nickel chloride–phosphine complexes. 14, 125; **15**, 122; **16**, 124; **18**, 250

Transfer hydrogenation.[1] The $(Ph_3P)_2NiCl_2$ complex catalyzes the reduction of carbonyl compounds by isopropanol in the presence of NaOH.

Addition to activated imines.[2] Nucleophiles such as malonate esters add to imines in the presence of $(Ph_3P)_2NiCl_2$.

Coupling with organoborates. Coupling partners can be aryl halides[3] and sulfonates,[4] as well as allylic carbonates.[5,6]

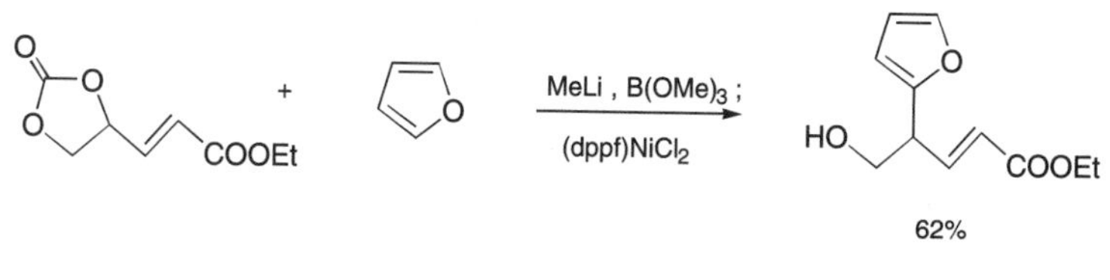

Alkenes.[7] *vic*-Dibromides undergo debromination with EtMgBr–(dppe)$NiCl_2$ in THF at 0°. Alkynes are obtained from dibromoalkenes.

Alkylidenation of dithioacetals.[8] With (dppe)$NiCl_2$, *gem*-bimetallic reagents (RCH[MgBr]ZnBr) react with dithioacetals to form trisubstituted alkenes.

[1]Iyer, S., Varghese, J.P. *CC* 465 (1995).
[2]Shida, N., Kubota, Y., Fukui, H., Asao, N., Kadota, I., Yamamoto, Y. *TL* **36**, 5023 (1995).
[3]Saito, S., Sakai, M., Miyaura, N. *TL* **37**, 2993 (1996).
[4]Kobayashi, Y., Mizojiri, R. *TL* **37**, 8531 (1996).
[5]Mizojiri, R., Kobayashi, Y. *JCS(P1)* 2073 (1995).
[6]Kobayashi, Y., Mizojiri, R., Ikeda, E. *JOC* **61**, 5391 (1996).
[7]Malanga, C., Aronica, L.A., Lardicci, L. *TL* **36**, 9189 (1995).
[8]Tseng, H.-R., Luh, T.-Y. *OM* **15**, 3099 (1996).

Niobium(V) chloride.

Homologation.[1] α-Trimethylstannylmethyl-β-ketoesters are converted to γ-ketoesters in the presence of $NbCl_5$, presumably via cyclopropanol intermediates (6 examples, 34–88%). The tributylstannylmethyl analogues give inferior results.

O Ph COOEt SnMe3 —— $NbCl_5$ / CH_2Cl_2, - 17° → O Ph COOEt 51%

Diels–Alder reactions.[2] At –78° in ether, the $NbCl_5$-catalyzed cycloaddition between cyclopentadiene and crotonaldehyde is complete in 1 h. Interestingly, the *endo*-to-*exo* ratio is 1:9.

[1]Yamamoto, M., Nakazawa, M., Kishikawa, K., Kohmoto, S *CC* 2353 (1996).
[2]Howarth, J., Gillespie, K. *TL* **37**, 6011 (1996).

Nitric acid. 18, 251–252

Nitrations. Sulfuric acid on silica gel is an inexpensive catalyst for nitration.[1] For nitration of strongly deactivated arenes, the mixture of HNO_3/$2CF_3SO_3H$–$B(OTf)_3$ is effective.[2]

Oxidation of sulfides.[3] The oxidation to sulfoxides is accomplished by HNO_3 catalyzed by $FeBr_3$ and $(FeBr)_2(dmso)_3$.

[1]Riego, J.-M., Sedin, Z., Zaldivar, J.M., Marziano, N.C., Tortato, C. *TL* **37**, 513 (1996).
[2]Olah, G.A., Orlinkov, A., Oxyzoglou, A.B., Prakash, G.K.S. *JOC* **60**, 7348 (1995).
[3]Suarez, A.R., Rossi, L.I., Martin, S.E. *TL* **36**, 1201 (1995).

Nitric oxide.

Nitrations.[1] Nitration of alkenes (e.g., styrene to β-nitrostyrene) is readily performed with NO in 1,2-dichloroethane at room temperature. Nitro alcohol side products are dehydrated by heating with acidic alumina.

Anisole and its derivatives are nitrated using an NO–O_2 system.[2]

Reductive deamination.[3] Aromatic amines are converted to deaminated aromatic compounds with NO in THF under argon. The yields range from trace amounts to 92%.

[1]Hata, E., Yamada, T., Mukaiyama, T. *BCSJ* **68**, 3629 (1995).
[2]Mizuno, K., Tamai, T., Hashida, I., Otsuji, Y., Inoue, H. *JCR(S)* 284 (1995).
[3]Itoh, T., Matsuya, Y., Nagata, K., Ohsawa, A. *TL* **37**, 4165 (1996).

Nitridomanganese(V) salen complexes.

Amination.[1] The complexes (**1**) are very effective for aminating electron-rich silyl enol ethers, providing α-amino ketones. These are the first members of nitrogen transfer agents other than porphyrin derivatives. Usually, the reaction is conducted in the presence of $(CF_3CO)_2O$ to acylate the amino group.

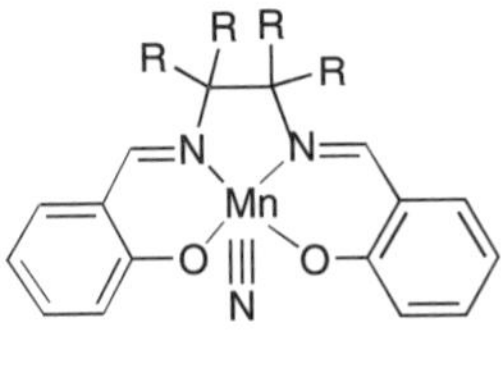

(**1**) R = H, Me

[1]Du Bois, J., Hong, J., Carreira, E.M., Day, M.W. *JACS* **118**, 915 (1996).

Nitrobenzene.

α-Substituted lactic acids.[1] At high temperatures under basic conditions (NaOH), ketones are transformed into lactic acids with $PhNO_2$ serving as an oxidant.

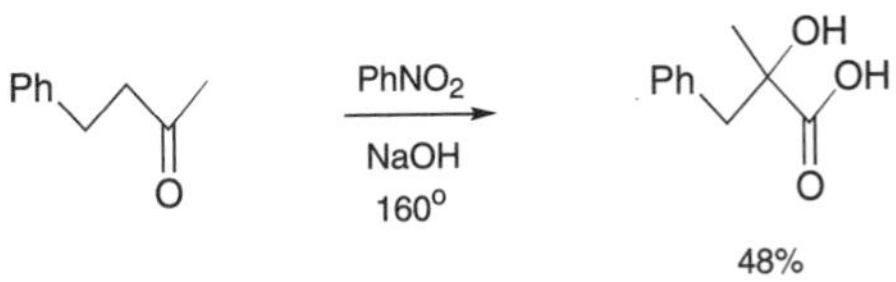

$PhNO_2$, NaOH, 160°

48%

[1]Srinivasan, P.S., Mahesh, R., Rao, G.V., Kalyanam, N. *SC* **26**, 2161 (1996).

4-Nitrobenzenesulfonyl azide.

α-Diazo carbonyl compounds. β-Diketones and β-ketoesters are converted to α-diazo ketones[1] and diazo esters,[2] respectively, on reaction with the sulfonyl azide in CH_2Cl_2 at room temperature, using DBU as a base.

[1]Taber, D.F., Gleave, D.M., Herr, R.J., Moody, K., Hennessy, M.J. *JOC* **60**, 2283 (1995).
[2]Taber, D.F., You, K., Song, Y. *JOC* **60**, 1093 (1995).

Nitrogen dioxide. 15, 219; **18**, 252–253

Deacetalization and dethioacetalization. Cyclic acetals of ketones are cleaved by NO_2 in the presence of silica gel (5 examples, 88–100%).[1] Carbonyl compounds are similarly recovered from dithioacetals[2] by treatment with "nitrogen oxides," which are prepared from arsenous oxide with concentrated HNO_3.

Selenoxides. Autoxidation in the presence of NO_2 is a mild method for the preparation.[3]

[1]Nishiguchi, T., Ohosima, T., Nishida, A., Fujisaki, S. *CC* 1121 (1995).
[2]Mehta, G., Uma, R. *TL* **37**, 1897 (1996).
[3]Bosch, E., Kochi, J.K. *JCS(P1)* 2731 (1996).

Nitronium tetrafluoroborate. 14, 215

Nitration.[1] Allylsilanes undergo desilylative nitration with transposition.

[1]Beresis, R.T., Masse, C.E., Panek, J.S *JOC* **60**, 7714 (1995).

4-Nitropyridine *N*-oxide.

α-Siloxy aldehydes.[1] α-Trimethylsilyl epoxides are converted to the siloxy aldehydes with 4-nitropyridine *N*-oxide with TMS-OTf as a catalyst.

Me3Si, O, O, NO_2 — Me_3SiOTf / CH_2Cl_2, O_2N–C5H4N(+)–O(−), Δ 0.5 h → OHC, OSiMe3, O, O, NO_2 80%

[1]Raubo, P., Wicha, J. *TL* **35**, 3387 (1994).

Nitryl iodide.

Nitration. Cyclic conjugated alkenes give nitroalkenes on reaction with nitryl iodide, which is prepared in situ from iodine and KNO_2 (18-crown-6).

[1]Ghosh, D., Nichols, D.E. *S* 195 (1996).

O

Organoantimony reagents. **17**, 204; **18**, 255

Knoevenagel reactions.[1] Alkylidenemalonic esters are prepared from the dibromomalonic esters and carbonyl compounds with promotion by Bu_3Sb in THF (6 examples, 70–95%).

[1]Davis, A.P., Bhattarai, K.M. *T* **51**, 8033 (1995).

Organocerium reagents. **13**, 206; **14**, 217–218; **15**, 221; **16**, 232; **17**, 205–207; **18**, 256

Reaction with carbonyl compounds. $RCeCl_3$ in which R is a cyclopropenyl residue[1] behaves normally as a nucleophile. Lithium enolates derived from tertiary amides have been converted to the corresponding cerium species, and their reactions with aldehydes have been studied.[2]

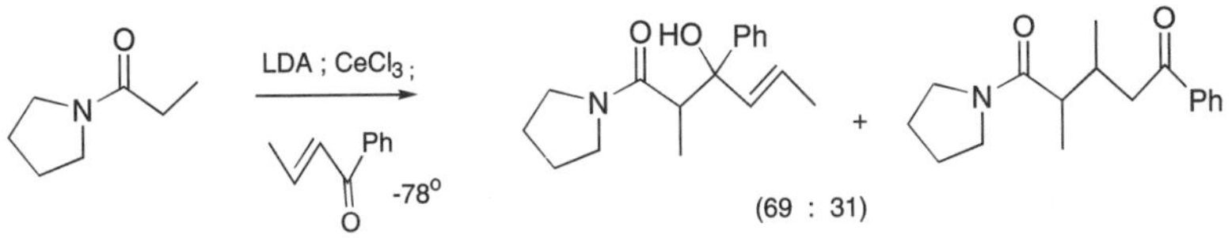

LDA ; $CeCl_3$; -78° (69 : 31)

β-Imino alcohols.[3] Alkaneimidoylcerium chlorides are generated from isonitriles and $RLi/CeCl_3$. On quenching with carbonyl compounds, imino alcohols are obtained.

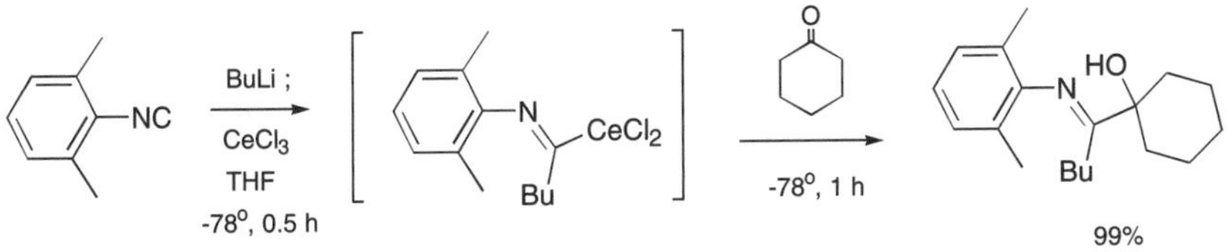

BuLi ; $CeCl_3$ THF -78°, 0.5 h; -78°, 1 h; 99%

[1]Tokuyama, H., Isaka, M., Nakamura, E. *SC* **25**, 2005 (1995).
[2]Shang, X., Liu, H.-J. *SC* **25**, 2155 (1995).
[3]Murakami, M., Ito, H., Ito, Y. *BCSJ* **69**, 25 (1996).

Organocopper reagents. 13, 207–209; **14**, 218–229; **15**, 221–227; **16**, 232–238; **17**, 207–218; **18**, 257–262

Substitutions. A methoxy group at C-4 of oxazolidin-2-ones is more reactive toward cuprates than a primary chloride is.[1] The displacement of a methoxy group from tin-containing mixed acetals shows diastereoselectivity.[2]

Ph_3Cu_2Li; $BF_3 \cdot OEt_2$, THF — 62% (trans : cis > 12 : 1)

β-Amino acid derivatives of the *(R)*-series have been obtained from the β-tosylamino-γ-butyrolactone derived from L-aspartic acid (two steps).[3]

Me_2CuLi; THF, -40° — 78%

Organocopper reagents prepared from the more readily available organozincs have found much use in the synthesis of highly functionalized molecules. For example, Ni-catalyzed hydrozincation of allylic alcohols initiates the preparation of α-(4-hydroxyalkyl)acrylates when the cuprate intermediates are used in the coupling with the α-bromomethylacrylic esters.[4]

Et_2Zn, $Ni(acac)_2$; TMSCl — CuCN - LiCl; CO_2Et, Br — 62%

Substitution at an sp^2 carbon can involve vinylic triflates[5] and tellurides.[6]

Heterocycle openings. α,β-Epoxy silanes furnish β-hydroxy silanes due to the attack of cuprate reagents at the α-carbon.[7] The regioselectivity for the reaction of methyl *(E)*-4,5-epoxypent-2-enoate with arylcopper is dependent on the presence or absence of BF_3. The S_N2' pathway giving the α-aryl esters becomes significant with the added Lewis acid.[8] The S_N2' products are observed in the reaction of epoxy vinyl sulfoxides with cyanocuprates.[9] Note that the ring opening of *N*-diphenylphosphinyl-2-vinylaziridines also favors a similar course of reaction.[10]

MeCu(CN)Li
Et$_2$O
-78° -> rt
91%

Dioxolanes derived from α,β;γ,δ-dienals can react with organocopper reagents at both the α- and γ-position. Interestingly, chiral dioxolanes of such a substitution pattern give optically active products,[11] due to remote asymmetric induction.

PhCu - BF$_3$• OEt$_2$
Et$_2$O, -30°, 30 min
(70 : 30)

Addition to C=X bonds. A preparation of C_2-symmetric di-*s*-phenethylamine is performed by the addition of Me_2CuLi to the benzal derivative of *(S)*-1-phenethylamine in the presence of BF_3.[12]

Conjugate additions. A remarkable contrast in stereoselectivity of the copper and lithium reagents has been unveiled.[13]

RM / Et$_2$O
-100°, 2 h

M = Cu 80%	5	:	1
M = Li 60%	1	:	36

Novel organocuprates used in conjugate additions are 2-(trimethylgermyl)allylcopper(I) [Me_2S complex],[14] dilithium bis[2-trimethylstannyl)vinyl]cyanocuprate,[15] and the α-azoalkylcuprates.[16]

Long-chained alkylphosphonates are formed by the addition of cuprate to a vinylphosphonate ester. The addition-trapping protocol gives rise to the less accessible branched compounds.[17]

Alkenyl triflones have unusual reactivities. A stereoselective synthesis[18] of trisubstituted alkenyl triflones involves the organocoppers and alkynyl triflones.

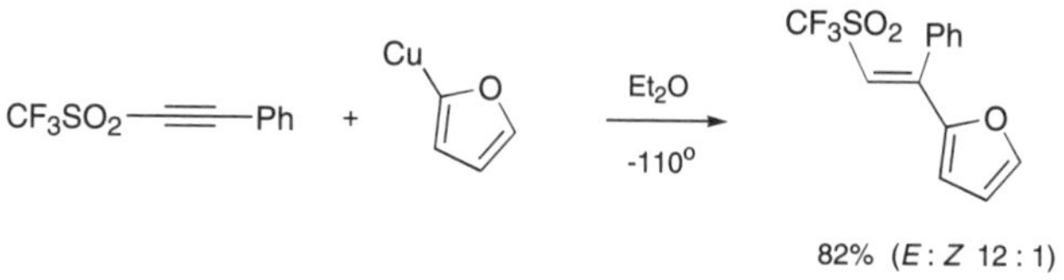

CF_3SO_2—≡—Ph + (2-furyl)Cu —(Et_2O, -110°)→ CF_3SO_2CH=C(Ph)(2-furyl)

82% (*E* : *Z* 12 : 1)

Allenes are obtained from the 1,6-addition to sulfonyl enynes.[19] Actually, acetylene itself is receptive to attack by silylcuprate reagents, giving rise to *(Z)*-alkenylsilanes after trapping with electrophiles.[20]

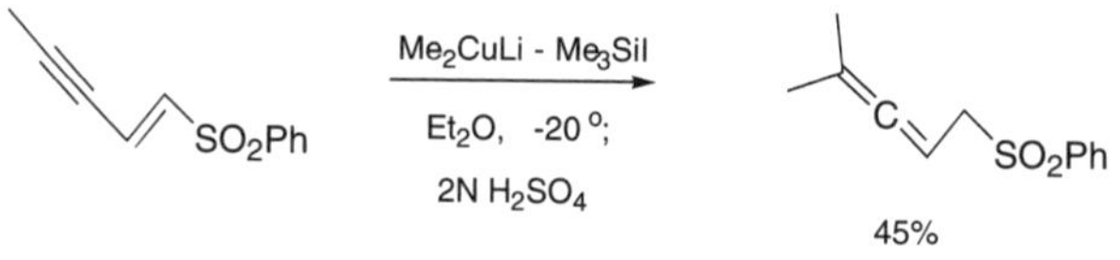

—(Me_2CuLi - Me_3SiI; Et_2O, -20 °; 2N H_2SO_4)→ SO_2Ph

45%

3-Methoxy-2-acylmethylation of 2-cycloalkenols.[21] Fused-ring cyclopropyl carbinols derived from 2-cycloalkenols undergo ring opening in methanol with Hg(II) salts; the ensuing intermediates are readily converted to the alkyl(methyl)mercury derivatives. *O*–Acylation followed by reaction with Me_3CuLi_2 gives the functionalized cycloalkanols.

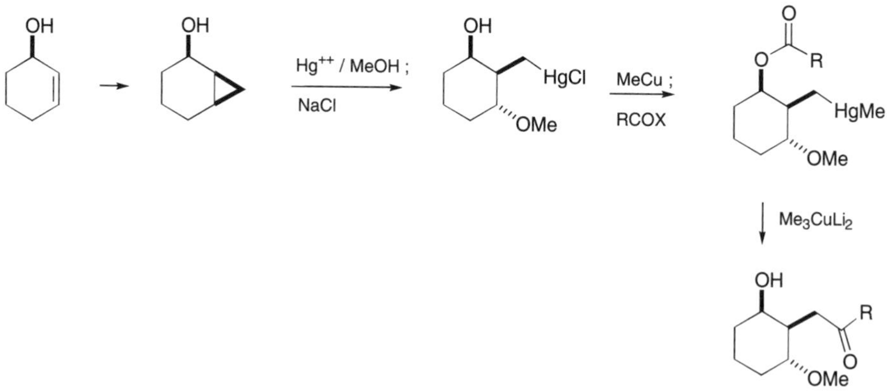

OH → OH —(Hg^{++} / MeOH ; NaCl)→ OH, HgCl, OMe —(MeCu ; RCOX)→ O(CO)R, HgMe, OMe —(Me_3CuLi_2)→ OH, R, O, OMe

Reductive umpolung of aldehydes.[22] A three-step reaction sequence accomplishes this transformation. Thus, the reaction of an aldehyde with tricyclohexylstannyllithium, followed by acetylation and treatment with an active electrophile (acid chloride, allylic bromide) in the presence of CuCN effects a C–C bond formation with the aldehyde in an *umpolung* manner. It is important that, during the Sn → Cu transmetallation, the other substituents remain on the tin atom. Therefore, the particular intermediate that bears secondary cyclohexyl groups has an advantage over the methyl or butyl derivatives.

OAc
C_5H_{11} SnCy3
Cy = cyclohexyl
CuCN
RCOCl
OAc
C_5H_{11} R
O
73 - 92%

[1]Danielmeier, K., Schierle, K., Steckhan, E. *ACIEE* **35**, 2247 (1996).
[2]Linderman, R.J., Chen, S. *TL* **36**, 7799 (1995).
[3]Jefford, C.W., McNulty, J., Lu, Z.-H., Wang, J.B. *HCA* **79**, 1203 (1996).
[4]Vettel, S., Vaupel, A., Knochel, P. *TL* **36**, 1023 (1995).
[5]Tsushima, K., Hirade, T., Hasegawa, H., Murai, A. *CL* 801 (1995).
[6]Chieffi, A., Comasseto, J.V. *SL* 671 (1995).
[7]Hudrlik, P.F., Ma, D., Bhamidipati, R.S., Hudrlik, A.M. *JOC* **61**, 8655 (1996).
[8]Nagumo, S., Irie, S., Akita, H. *CC* 2001 (1995).
[9]Marino, J.P., Anna, L.J., de la Pradilla, R.F., Martinez, M.V., Montero, C., Visa, A. *TL* **37**, *8031* (1996).
[10]Cantrill, A.A., Jarvis, A.N., Osborn, H.M.I., Ouadi, A., Sweeney, J.B. *SL* 847 (1996).
[11]Rakotoarisoa, H., Perez, R.G., Mangeney, P., Alexakis, A. *OM* **15**, 1957 (1996).
[12]Alvaro, G., Savoia, D., Valentinetti, M.R. *T* **52**, 12571 (1996).
[13]Leonard, J., Mohialdin, S., Reed, D., Ryan, G., Swain, P.A. *T* **51**, 12843 (1995).
[14]Piers, E., Kaller, A.M. *SL* 549 (1996).
[15]Pereira, O.Z., Chan, T.-H. *JOC* **61**, 5406 (1996).
[16]Alexander, C.W., Lin, S.-Y., Dieter, R.K. *JOMC* **503**, 213 (1995).
[17]Baldwin, I.C., Beckett, R.P., Williams, J.M.J. *S* 34 (1996).
[18]Xiang, J., Fuchs, P.L. *JACS* **118**, 11986 (1996).
[19]Hohmann, M., Krause, N. *CB* **128**, 851 (1995).
[20]Barbero, A., Cuadrado, P., Fleming, I., Gonzalez, A.M., Pulido, F.J., Sanchez, A. *JCS(P1)* 1525 (1995).
[21]Kocovsky, P., Grech, J.M., Mitchell, W.L. *JOC* **60**, 1482 (1995).
[22]Linderman, R.J., Siedlecki, J.M. *JOC* **61**, 6492 (1996).

Organocopper/zinc reagents. **18**, 262–263

Methylenation.[1] After the reaction of aldehydes with (dialkoxyboryl)-methylcopper–zinc reagents in the presence of $BF_3{\cdot}OEt_2$, subsequent heating gives monosubstituted alkenes. If the initial products are treated with buffered H_2O_2, 1,2-diols result.

Conjugate addition. The Michael reaction using functionalized organozinc reagents as donors requires only catalytic amounts of Cu(I) species.[2] Thus, the addition of MeLi to an RZnMe followed by exposure to catalytic amounts of $Me_2Cu(CN)Li_2$ in the presence of Me_3SiCl effects the transmetallation (from organozinc to lithiocuprate), and the reagent is ready to deliver the functionalized R group to an enone.

(E)-Alkenyl trifluoromethyl ketones are prepared by the reaction of the *(Z)*-β-butyltelluro enone.[3]

BuTe–CH=CH–C(=O)CF₃ → Ph₂CuCN(ZnCl)₂, THF, -60° -> -30°, 3 h → Ph–CH=CH–C(=O)CF₃ (90%)

[1]Sakai, M., Saito, S., Kanai, G., Suzuki, A., Miyaura, N. *T* **52**, 915 (1996).
[2]Lipshutz, B.H., Wood, M.R., Tirado, R. *JACS* **117**, 6126 (1995).
[3]Mo, X.-S., Huang, Y.-Z. *SL* 180 (1995).

Organogallium reagents.

Epoxide opening.[1] Tertiary alcohols are obtained from the reaction of lithium [β-*(E)*-trimethylsilylvinyl]trimethylgallate with 1,1-disubstituted epoxides. The MEM-ethers of these alcohols fragment to give ketones on contact with $TiCl_4$. The behavior is different from that of the *(Z)*-isomers, which form dihydropyrans.

R(R')-epoxide + Me_3Ga^-–CH=CH–SiMe₃ Li^+ → THF; MEM-Cl, *i*-Pr₂NEt → R(R')C(OMEM)–CH₂–CH=CH–SiMe₃ (52 - 55%) → TiCl₄ → R(R')C=O (95%)

(R = $C_{10}H_{21}$, R' = Me)

[1]Horiuchi, Y., Taniguchi, M., Oshima, K., Utimoto, K. *TL* **35**, 7977 (1994).

Organomanganese reagents.

Couplings. Organomanganese species prepared in situ from MnX_2 and a Grignard reagent catalyzes the coupling of alkyl halides, including *gem*-dibromocyclopropanes.[1] RMnCl couples with alkenyl halides in the presence of $Fe(acac)_3$.[2] Selective manganation of *o*-bromofluorobenzene using manganese–graphite in THF followed by the coupling enables the synthesis of *o*-fluorostyrenes (10 examples, 64–75%).[3]

Br, Br / cyclopropane–C_6H_{13} → ($Bu_3MnMgBr$ / THF 0°; MeI) → Me, Bu / cyclopropane–C_6H_{13} + Bu, Me / cyclopropane–C_6H_{13} (94 : 6) 65%

Addition to multiple bonds. In the presence of MnI_2, Grignard reagents add to alkynes. Therefore, 1,4-dienes are formed by using allylmagnesium halides.[4] Cyclization of ω-alkenyl bromides by treatment with a mixture of Et_2Zn and $MnBr_2$/CuCl has been observed.[5] Iodine quench of the reaction provides iodomethylcycloalkanes.

RCOCl → RCOR′. A ketone synthesis[6] by the reaction of acid chlorides with R′MnCl is mild enough that chiral α-acyloxy ketones can be prepared. Such organomanganese halides can be directly made from the halides and Rieke manganese.[7]

Enolization.[8] Phenylmanganese chloride deprotonates ketones in the presence of a catalytic amount of an amine. Quenching with an anhydride gives enol esters.

Et–C(O)–Et (3-pentanone) → (PhMnCl - PhNHMe; THF - Sulfolane, 0°; 4 h) → [Et–C(OMnCl)=CHEt] → ($(EtCO)_2O$; 0° -> rt; 2 h) → Et–C(OCOEt)=CHEt 94% (Z: E 87 : 13)

[1]Inoue, R., Shinokubo, H., Oshima, K. *TL* **37**, 5377 (1996).
[2]Cahiez, G., Marquais, S. *TL* **37**, 1773 (1996).
[3]Fürstner, A., Brunner, H. *TL* **37**, 7009 (1996).
[4]Okada, K., Oshima, K., Utimoto, K. *JACS* **118**, 6076 (1996).
[5]Riguet, E., Klement, I., Reddy, C.K., Cahiez, G., Knochel, P. *TL* **37**, 5865 (1996).
[6]Cahiez, G., Metais, E. *TL* **36**, 6449 (1995).
[7]Kim, S.-H., Hanson, M.V., Rieke, R.D. *TL* **37**, 2197 (1996).
[8]Cahiez, G., Kanaan, M., Clery, P. *SL* 191 (1995).

Organonickel reagents.

Couplings.[1] Alkyl cyanonickelates (also cyanocobaltates and cyanoferrates) prepared in situ from $M(CN)_2$ and RLi at −78° couple with organic halides.

[1]Kauffmann, T., Nienaber, H. *S* 207 (1995).

Organotellurium reagents.

Diorganotellurides are useful precursors of organolithium compounds. The latter species are obtained by treating the tellurides with BuLi. This indirect protocol obviates the difficulty in preparing RLi from RX.

Alkylations. β-Lithiocarbonyl synthons[1] and propargyl anions[2] are generated from the bromides by reacting with BuTeLi and then BuLi. *(Z)*-1,3-Butadien-1-yllithium[3] is available from β-(organotelluro)acroleins.

BuTeTeBu —(NaBH$_4$; ≡—CHO)→ BuTe\=/CHO (84%) —($Ph_3P=CH_2$)→ BuTe\=/= (92%) —(BuLi / THF ; RCHO)→ R—CH(OH)—CH=CH—CH=CH$_2$ (85 - 92%)

The mild conditions for the Te/Li exchange allow the generation of RLi containing a cyano group.[4] A route to homopropargylic alcohols involves the conversion of propargylic bromides to allenyl butyl tellurides (via Grignard reactions with BuTeBr or reactions with Bu_2Te in the presence of $NaBH_4$ [however, the latter reaction is limited to 3-bromo-1-alkynes]) and then to propargyllithium reagents.[5]

Enyne and arylalkene synthesis from alkenyl tellurides can proceed with transmetallation to give Zn species.[6,7] Alternatively, the partial reduction of 1,3-diynes via hydrotelluration, Te/Li exchange, and protonation also provides *(E)*-enynes.[8]

—≡—≡—Ph —(BuTeTeBu, NaBH$_4$, EtOH Δ)→ (TeBu)enyne—≡—Ph (80%) —(BuLi ; H$_2$O)→ enyne—≡—Ph (73%)

α,β-Unsaturated acids and esters. Alkenyl tellurides of defined constitution and configuration can be prepared from alkynes.[9] Treating such tellurides with BuLi and CO_2 or ClCOOR leads to the enoic acids or esters.[10]

β-(Z)-Alkenyl[cyclo]alkanones.[11] Di-*(Z)*-alkenyl tellurides obtained from the hydrotelluration of alkynes undergo Te/Cu exchange [e.g., with $Bu_2Cu(CN)Li_2$]. Conjugate addition to alkenones (including cycloalkenones) gives the substituted ketones. Reaction with epoxides furnishes *(Z)*-homoallylic alcohols.

[1]Inoue, T., Atarashi, Y., Kambe, N., Ogawa, A., Sonoda, N. *SL* 209 (1995).
[2]Kanda, T., Ando, Y., Kato, S., Kambe, N., Sonoda, N. *SL* 745 (1995).
[3]Mo, X.-S., Huang, Y.-Z. *TL* **36**, 3539 (1995).
[4]Kanda, T., Kato, S., Sugino, T., Kambe, N., Ogawa, A., Sonoda, N. *S* 1102 (1995).
[5]Dabdoub, M.J., Rotta, J.C.G. *SL* 526 (1996).
[6]de Araujo, M.A., Comasseto, J.V. *SL* 1145 (1995).
[7]Terao, J., Kambe, N., Sonoda, N. *TL* **37**, 4741 (1996).
[8]Dabdoub, M.J., Dabdoub, V.B. *T* **51**, 9839 (1995).

[9]Dabdoub, M.J., Cassol, T.M. *T* **51**, 12971 (1995).
[10]Dabdoub, M.J., Begnini, M.L., Cassol, T.M. Guerrero, P.G., Silveira, C.C. *TL* **36**, 7623 (1995).
[11]Tucci, F.C., Chieffi, A., Comasseto, J.V., Marino, J.P. *JOC* **61**, 4975 (1996).

Organotin reagents.

β-Hydroxy esters.[1] *anti*-Selective reduction of α-methyl β-alkylketo esters by R_3SnH in the presence of $TiCl_4$ is unusual.

β-Hydroxy ketones.[2] α,β-Epoxy ketones undergo reductive ring cleavage with a mixture of Bu_2SnH_2 and Bu_2SnI_2. The actual reagent is Bu_2SnHI.

Desulfurization.[3] Thionoesters and thionolactones are converted to ethers andcyclic ethers, respectively, with Ph_3SnH-AIBN in refluxing toluene. (10 examples, 72–99%).

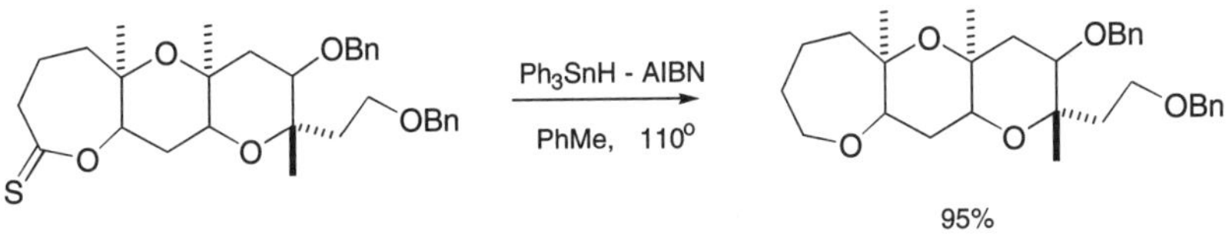

Ph_3SnH - AIBN, PhMe, 110°

95%

Fluorous tin reagents.[4] A general method for the syntheis of the reagents is shown below. These compounds have successfully been applied in Stille coupling using $(Ph_3P)PdCl_2$ with LiCl as an additive in DMF–THF at 80°.

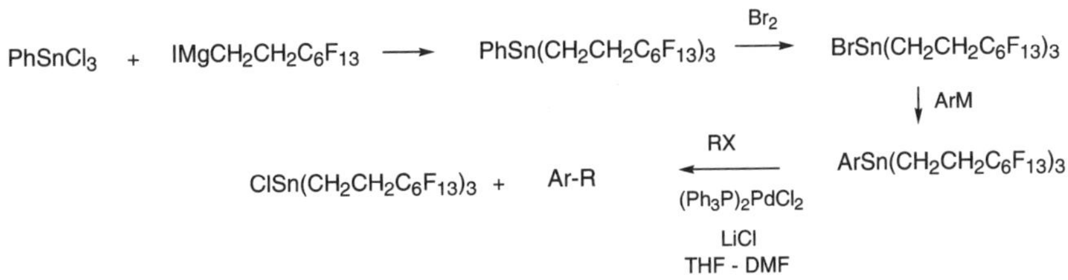

$PhSnCl_3$ + $IMgCH_2CH_2C_6F_{13}$ ⟶ $PhSn(CH_2CH_2C_6F_{13})_3$ —(Br_2)⟶ $BrSn(CH_2CH_2C_6F_{13})_3$ —(ArM)⟶ $ArSn(CH_2CH_2C_6F_{13})_3$ —(RX, $(Ph_3P)_2PdCl_2$, LiCl, THF - DMF)⟶ $ClSn(CH_2CH_2C_6F_{13})_3$ + Ar-R

Polystyrene-supported tin hydride.[5] With the use of this reagent, tropones are prepared from dichloromethylcyclohexadienones.

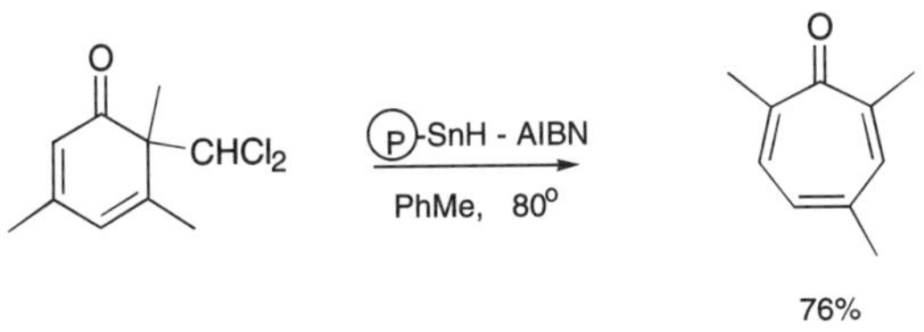

CHCl$_2$ —(Ⓟ-SnH - AIBN, PhMe, 80°)⟶

76%

[1]Sato, T., Nishio, M., Otera, J. *SL* 965 (1995).
[2]Kawakami, T., Shibata, I., Baba, A. *JOC* **61**, 82 (1996).
[3]Nicolaou, K.C., Sato, M., Theodorakis, E.A., Miller, N.D. *CC* 1583 (1995).
[4]Curran, D.P., Hoshino, M. *JOC* **61**, 6480 (1996).
[5]Dygutsch, D.P., Neumann, W.P., Peterseim, M. *SL* 363 (1994).

Organovanadium reagents.

Allylation.[1] The combination of an allyl halide, VCl_2 and Zn can be used to allylate carbonyl compounds. Sequential Grignard reaction of a ketone and treatment with allyl bromide and $VCl_2(tmeda)_2$ together complete the deoxygenative alkylation.

Pinacolization.[2] Aldehydes undergo reductive coupling with Zn, Me_3SiCl, and catalytic amounts of $CpV(CO)_4$ (3 examples, 70–90%). Such a diol product is actually tied up as dioxolane (with one equivalent of the original aldehyde).

[1]Kataoka, Y., Makihira, I., Akiyama, H., Tani, K. *TL* **36**, 6495 (1995).
[2]Hirao, T., Hasegawa, T., Muguruma, Y., Ikeda, I. *JOC* **61**, 366 (1996).

Organozinc reagents. 13, 220–222; 14, 233–235; 15, 238–240; 16, 246–248; 17, 228–234; 18, 264–265

1,3-Dizincs.[1] The preparation of one such species is illustrated below.

BEt2 —HBEt2→ (BEt2, BEt2) 85% —Et2Zn→ (Zn, Zn) 90%

Alcohols and hydroperoxides. The exposure of organozinc halides to air–HMPA in THF affords alcohols (10 examples, 56–98%).[2] On the other hand, aeration in perfluorohexanes at –78° leads to alkyl hydroperoxides.[3]

The zinc carbenoid derived from $CFBr_3$ and Et_2Zn adds to aldehydes readily to form $RCH(OH)CFBr_2$.[4] The remarkable chemoselectivity favoring aldehydes to ketones (> 99:1) is in stark contrast to that of $LiCFBr_2$, which reacts almost indiscriminately.

Nitriles. The reaction of organozincs with TsCN to give nitriles[5] is particularly interesting in cases that involve homopropargylic sulfonates, because 1-zincoalkylidenecyclopropanes are formed, and these can be derivatized.[6]

S ZnI —TsCN, THF, -78° -> rt→ S CN 69%

E = $SiMe_3$, COOEt, COR, CN, CH(OH)R, I

Conjugate additions. A Cu(I)-sulfonamide system[7] is useful for catalyzing the addition of diorganozincs. However, in *N*-methylpyrrolidinone and in the presence of TMSCl, uncatalyzed reactions are quite efficient.[8] Trialkylsilyl(dialkyl)zincates undergo conjugate addition[9] quite well.

[1]Eick, H., Knochel, P. *ACIEE* **35**, 218 (1996).
[2]Chemla, F., Normant, J. *TL* **36**, 3157 (1995).
[3]Klement, I., Knochel, P. *SL* 1113 (1995).
[4]Hata, T., Shimizu, M., Hiyama, T. *SL* 831 (1996).
[5]Klement, I., Lennick, K., Tucker, C.E., Knochel, P. *TL* **34**, 4623 (1993).
[6]Harada, T., Wada, H., Oku, A. *JOC* **60**, 5370 (1995).
[7]Kitamura, M., Miki, T., Nakano, K., Noyori, R. *TL* **37**, 5141 (1996).
[8]Reddy, C.K., Devasagayaraj, A., Knochel, P. *TL* **37**, 4495 (1996).
[9]Vaughan, A., Singer, R.D. *TL* **36**, 5683 (1995).

Osmium tetroxide. **13**, 222–225; **14**, 235–239; **15**, 240–241; **16**, 249–253; **17**, 236–240; **18**, 265–267

Asymmetric dihydroxylation. Efficient and practical polymeric catalysts for heterogeneous dihydroxylation of olefins have been developed.[1] An electrochemical method[2] using Pt electrodes in undivided cells enables a synthesis of chiral 1,2-diols that requires much-reduced quantities of potassium osmate and $K_3Fe(CN)_6$.

Homoallylic alcohol derivatives undergo functionalization.[3] 3-Diphenylphosphinoyl-1,2-alkanediols are now available by this method.[4] New ligands (**1**, **2**) that show equal or better enantioselectivities have been developed.[5]

K_2OsO_4 - $K_3Fe(CN)_6$
(DHQD)$_2$PYZD
K_2CO_3 / aq. *t*-BuOH
0°, 4 h

99% (enantios. 95.5 : 4.5)

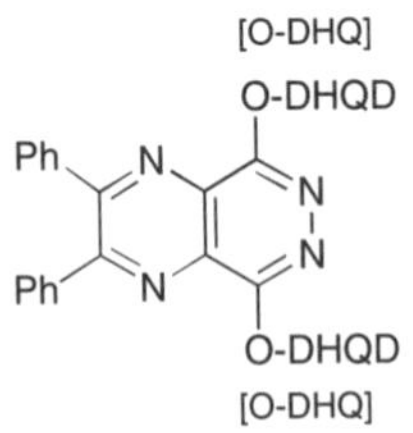

(1)

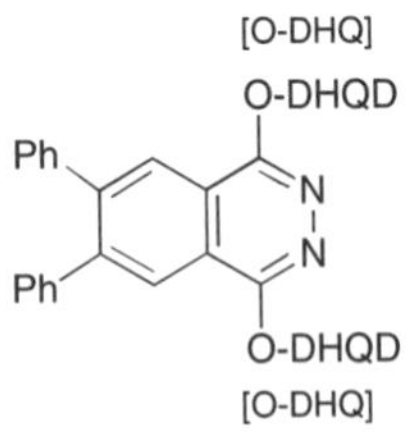

(2)

DHQD = dihydroquinidine
DHQ = dihydroquinine

Aminohydroxylation. Chloramine-T is used in the reaction,[6] but smaller organic substituents on the sulfur atom of the chloramine salts have higher selectivities. Thus, chloramine-M [$MeSO_2N(Na)Cl$] is better.[7] *N*–Halocarbamate salts are more efficient reagents.[8]

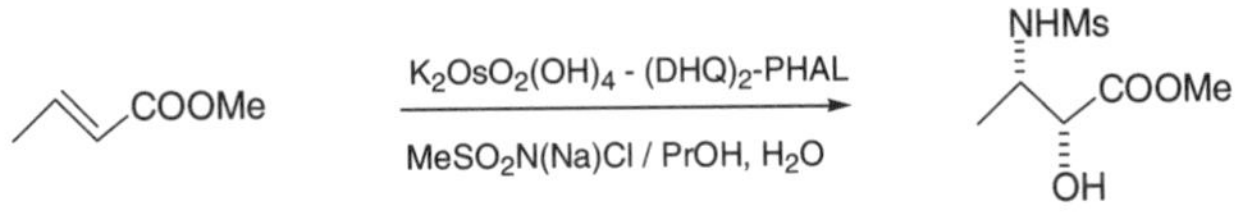

$K_2OsO_2(OH)_4$ - $(DHQ)_2$-PHAL
$MeSO_2N(Na)Cl$ / PrOH, H_2O

65% (95% ee)

[1]Song, C.E., Yang, J.W., Ha, H.J., Lee, S.-G. *TA* **7**, 645 (1996).
[2]Torii, S., Liu, P., Tanaka, H. *CL* 319 (1995); Torii, S., Liu, P., Bhuvaneswari, N., Amatore, C., Jutand, A. *JOC* **61**, 3055 (1996).
[3]Corey, E.J., Guzman-Perez, A., Noe, M.C. *TL* **36**, 3481 (1995).
[4]O'Brien, P., Warren, S. *JCS(P1)* 2129 (1996).
[5]Becker, H., King, S.B., Taniguchi, M., Vanhessche, K.P.M., Sharpless, K.B. *JOC* **60**, 3940 (1995).
[6]Li, G., Chang, H.-T., Sharpless, K.B. *ACIEE* **35**, 451 (1996).
[7]Rudolph, J., Sennhenn, P.C., Vlaar, C.P., Sharpless, K.B. *ACIEE* **35**, 2810 (1996).
[8]Li, G., Angert, H.H., Sharpless, K.B. *ACIEE* **35**, 2813 (1996).

Osmium tetroxide–*N*-methylmorpholine oxide.

Dihydroxylation.[1] Baylis–Hillman adducts undergo *syn*-selective dihydroxylation.

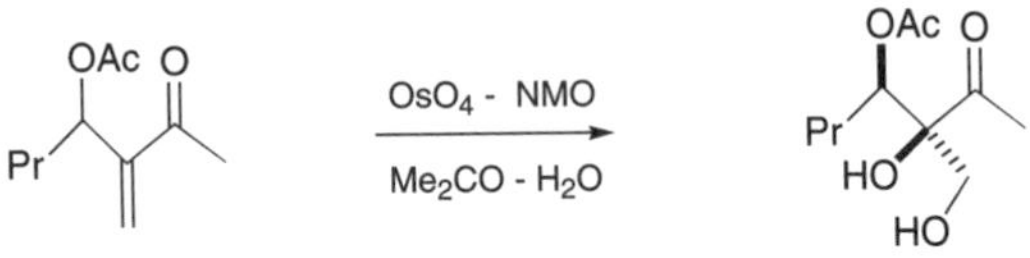

OsO_4 - NMO
Me_2CO - H_2O

89% (syn : anti 19 : 1)

[1]Marko, I.E., Giles, P.R., Janousek, Z., Hindley, N.J., Declercq, J.-P., Tinant, B., Feneau-Dupont, J., Svendsen, J.S. *RTCP* **114**, 239 (1995).

Oxalyl chloride. 17, 241–242; **18**, 267–268

Formamides.[1] An improved procedure for the formylation of amines involves formic acid in combination with oxalyl chloride and imidazole.

2H-Azirine-2-carboxylic esters.[2] Dehydrogenation of aziridine-2-carboxylic esters occurs under the Swern oxidation conditions.

[1]Kitagawa, T., Ito, J., Tsutsui, C. *CPB* **42**, 1931 (1994).
[2]Gentilucci, L., Grijzen, Y., Thijs, L., Zwanenburg, B. *TL* **36**, 4665 (1995).

Oxygen. 18, 268–269

Oxidation of secondary alcohols. The oxidation in the presence of PhCHO does not require a metal catalyst.[1]

Baeyer–Villiger oxidation. The O_2/*i*-PrCHO/metal catalyst system transforms methoxyaryl aldehydes mainly to aryl formates.[2]

Oxidation of sulfides. The conversion of sulfides to either sulfoxides or sulfones by oxygen can be achieved by using a Pd complex[3] or isobutyraldehyde.[4]

Alcohols from organometallics. Direct oxidation of organoboranes[5] and organozincs is useful.[6] The former reaction in perfluoroalkanes (e.g., $C_8F_{17}Br$) provides alcohols with retention of configuration.

N-Alkylation.[7] Lithiated amines form amidocuprates with organocopper reagents. Treatment with oxygen accomplishes the transfer of an organic residue from copper to nitrogen.

Oxidation of hydrocarbons. Several variants of the aldehyde–metal salt combination effect the oxidation of alkanes and alkenes, among them salen–cobalt,[8] $CuCl_2$/18-crown-6,[9] and vanadium-substituted zeolites.[10] A change of ligand for the cobalt ion has dramatic effects on the course of oxidation [allylic oxidation vs. epoxidation of cyclohexene by (**1**) and (**2**), respectively].

Ph COOMe N Co O O N MeOOC Ph

(**1**)

N N ++ Co N N 2 Cl⁻

(**2**)

Nitroalkanes to carbonyl compounds.[11] Copper complexes (TMEDA, 2,2-bipyridine, 1,10-phenanthroline) promote the conversion in quantitative yields.

Interrupted Pauson–Khand reaction.[12] Admitting oxygen to the system while heating the enyne complex yields 1-acyl-2-methylcyclopentene as the product of an intramolecular reaction.

$Co_2(CO)_6$ R $\xrightarrow[\text{PhMe } (O_2)]{\Delta}$ R

R = Bn, Et, CN... 54 - 75%

[1]Choudary, B.M., Sudha, Y. *SC* **26**, 1651 (1996).
[2]Anoune, N., Lanteri, P., Longeray, R., Arnaud, C. *TL* **36**, 6679 (1995).
[3]Aldea, R., Alper, H. *JOC* **60**, 8365 (1995).
[4]Khanna, V., Maikap, G.C., Iqbal, J. *TL* **37**, 3367 (1996).
[5]Klement, I., Knochel, P. *SL* 1004 (1996).
[6]Klement, I., Lutjens, H., Knochel, P. *TL* **36**, 3161 (1995).
[7]Alberti, A., Cane, F., Dembech, P., Lazzari, D., Ricci, A., Seconi, G. *JOC* **61**, 1677 (1996).
[8]Reddy, M.M., Punniyamurthy, T., Iqbal, J. *TL* **36**, 159 (1995).
[9]Komiya, N., Naota, T., Murahashi, S.-I. *TL* **37**, 1633 (1996).
[10]Neumann, R., Khenkin, A.M. *CC* 2643 (1996).
[11]Balogh-Hetgovich, E., Kaizer, J., Speier, G. *CL* 573 (1996).
[12]Krafft, M.E., Wilson, A.M., Dasse, O.A., Shao, B., Cheung, Y.Y., Fu, Z., Bonaga, L.V.R., Mollman, M.K. *JACS* **118**, 6080 (1996).

Oxygen, singlet. 13, 228–229; **14**, 247; **15**, 243; **16**, 257–258; **17**, 251–253; **18**, 269–270

Tetrakis(2,6-dichlorophenyl)porphyrin is a superior sensitizer for singlet oxygen generation.[1] Photooxidation can be carried out in perfluorocarbons with excellent results.[2]

Oxidation of ethers. With benzil as sensitizer, the photooxidation of ethers results in the formation of esters.[3] This reaction may be compared with the process mediated by Cu(II).[4]

[1]Quast, H., Dietz, T., Witzel, A. *LA* 1495 (1995).
[2]Chambers, R.D., Sandford, G., Shah, A. *SC* **26**, 1861 (1996).
[3]Seto, H., Yoshida, K., Yoshida, S., Shimizu, T., Seki, H., Hoshino, M. *TL* **37**, 4179 (1996).
[4]Minakata, S., Imai, E., Ohshima, Y., Inaki, K., Ryu, I., Komatsu, M., Ohshiro, Y. *CL* 19 (1996).

Ozone. 13, 229; **15**, 243–244; **17**, 253–254; **18**, 270–272

Aldehydes. An effective new reagent for the reductive quenching of ozonolysis is the sodium salt of 3,3′-thiobis(propanoic acid).[1]

α-Keto acid derivatives. Ozonides of 1-alkenes are transformed into α-alkylacroleins on treatment with a preheated mixture of CH_2Br_2–Et_3N. Oxidation gives α-keto acid derivatives.[2]

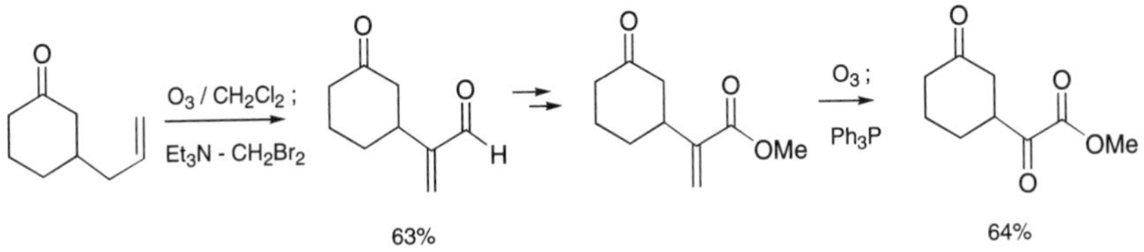
O3 / CH2Cl2 ;
Et3N - CH2Br2
63%
O3 ;
Ph3P
OMe
64%

α-Keto acid derivatives are also formed by a homologation procedure.[3]

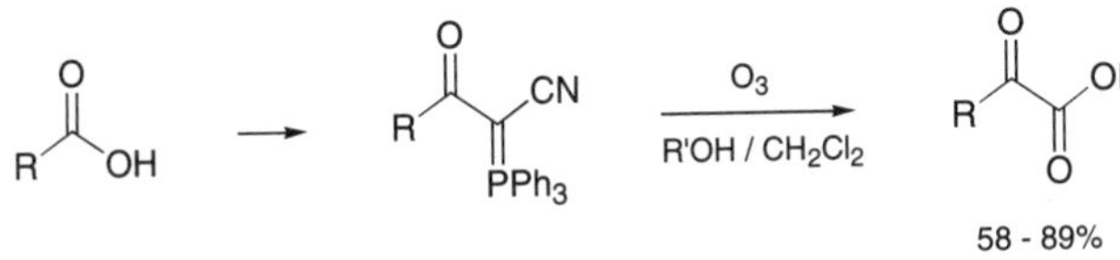
R
OH
CN
PPh3
O3
R'OH / CH2Cl2
OR'
58 - 89%

α,β-Unsaturated esters.[4] Ozonides can be used directly in the Wittig reaction.

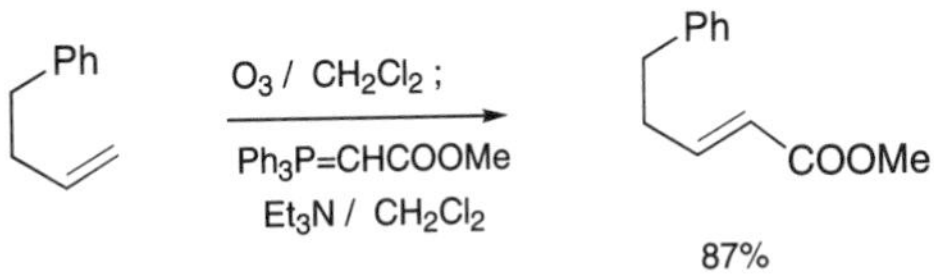
Ph
O3 / CH2Cl2 ;
Ph3P=CHCOOMe
Et3N / CH2Cl2
COOMe
87%

2-Cycloalkenones.[5] Ozonolysis of five- and six-membered cyclic allylic tertiary alcohols with alkaline workup leads directly to cycloalkenones, due to aldolization of the products from ring cleavage.

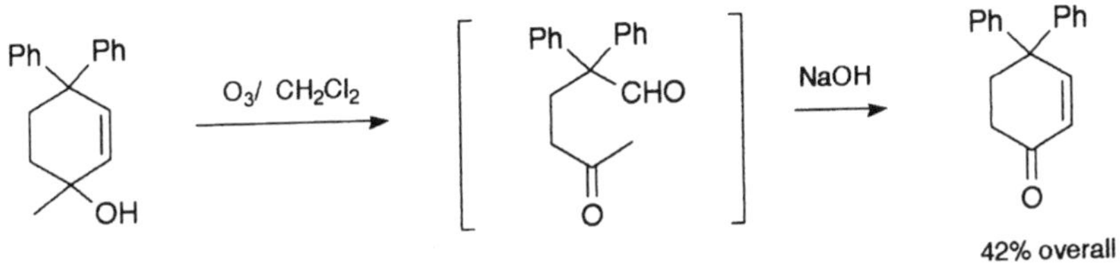
Ph
Ph
OH
O3/ CH2Cl2
CHO
NaOH
42% overall

[1]Appell, R.B., Tomlinson, I.A., Hill, I. *SC* **25**, 3589 (1995).
[2]Hon, Y.-S., Lin, W.-C. *TL* **36**, 7693 (1995).
[3]Wasserman, H.H., Ho, W.B. *JOC* **59**, 4364 (1994).
[4]Hon, Y.-S., Lu, L. *T* **51**, 7937 (1995).
[5]DeNinno, M.P. *JACS* **117**, 9927 (1995).

P

Palladium, colloidal. 18, 273

Heck reactions. The catalyst system consists of colloidal Pd and NaOAc in $AcNMe_2$.

[1]Beller, M., Fischer, H., Kuhlein, K., Reisinger, C.-P., Herrmann, W.A. *JOMC* **520**, 257 (1996).

Palladium/carbon. 13, 230–232; **15**, 245; **18**, 273

Hydrogenolysis. Benzyl phenyl ethers and benzyl benzoates are stable to Pd/C-cyclohexene in refluxing benzene.[1] This is a limitation as well as an exploitable selectivity. 1-Alkene oxides undergo C–O bond scission to give secondary alcohols[2] by treatment with Pd/C and $HCOONH_4$ in ethanol at room temperature.

Amidoximes [$RC(NH_2)=NOH$], which are obtained from nitriles on reaction with hydroxylamine, are hydrogenolyzed (H_2, Pd/C, HOAc–Ac_2O) to give amidines.[3]

Ether synthesis. Hydrogenation of a mixture of an alcohol and an aliphatic aldehyde affords ether.[4] A hemiacetal is the intermediate. Aryl aldehydes cannot be used because the carbonyl group is reduced too rapidly.

Alkynylarenes.[5] Coupling of an alkyne with a haloarene in aqueous DME occurs in the presence of Pd/C, PPh_3, CuI, and K_2CO_3.

Cross-couplings. The coupling between functionalized alkenyl halides and organozinc chlorides can be promoted by Pd/C together with Ph_3As.[6]

Br, Br, COOEt → (PhZnCl; Pd/C, Ph_3As; THF, 20°) → Ph, Br, COOEt

89%

[1]Sansanwal, V., Krishnamurty, H.G. *SC* **25**, 1901 (1995).
[2]Dragovich, P.S., Prins, T.J., Zhou, R. *JOC* **60**, 4922 (1995).
[3]Judkins, B.D., Allen, D.G., Cook, T.A., Evans, B., Sardharwala, T.E. *SC* **26**, 4351 (1996).
[4]Bethmont, V., Fache, F., Lemaire, M. *TL* **36**, 4235 (1995).
[5]Bleicher, L., Cosford, N.D.P. *SL* 1115 (1995).
[6]Rossi, R., Bellina, F., Carpita, A., Gori, R. *SL* 344 (1995).

Palladium clusters.

Heck and Suzuki reactions. The Heck reaction is achieved using Pd clusters stabilized by propylene carbonate.[1] Chloroarenes can be used. The stabilized Pd clusters are prepared either electrochemically (using a sacrificial Pd anode) or by thermolysis of $Pd(OAc)_2$.

For Suzuki coupling that involves arylboronic acids, the stabilization of Pd with a quaternary ammonium bromide [e.g., $(C_8H_{17})_4NBr$] is adequate.[2]

[1]Reetz, M.T., Lohmer, G. *CC* 1921 (1996).
[2]Reetz, M.T., Breinbauer, R., Wanninger, K. *TL* **37**, 4499 (1996).

Palladium(II) acetate. 13, 232–233; **14**, 248; **15**, 245–247; **16**, 259–263; **17**, 255–259; **18**, 274–277

Heck reaction. The Heck reaction has been conducted in superheated steam (260°).[1] *(E/Z)*-Isomers of 3-aryl-3-phenylpropenoates can be synthesized stereospecifically[2] by the Heck method.

In a particular system, the Heck conditions can be manipulated to form either fused or spirocyclic frameworks.[3]

An interesting approach to arylacetaldehydes involves vinylene carbonate as a reactant. With *N*-acyl-2-iodoanilines, the products are the *N*-protected 2-hydroxyindolines,[4] which are readily dehydrated to give the corresponding indole derivatives.

3-Acyloxy-3-buten-2-ones undergo *(Z)*-selective arylation at C-4.[5]

NO_2 + PhI —[$Pd(OAc)_2$ - Et_3N; MeCN, 50°; 48 h]→ Ph ... NO_2 64%

The coupling of enol triflates with γ-hydroxycrotonic esters gives butenolides.[6] The method is suitable for cardenolide synthesis. Nonactivated alkenes also undergo coupling.[7]

OTf, H, AcO + COOMe, OH —[$Pd(OAc)_2$ - KOAc; DMF, 80°]→ O, O, H, AcO 66%

Arylation with onium salts. Diaryliodonium[8,9] and arenediazonium tetrafluoroborates[10] are suitable arylating agents. An important difference with conventional coupling is that cinnamyl alcohols are obtained from the reaction of allylic alcohols with aryliodonium salts[8] instead of β-aryl ketones, due to internal chelation of the positive Pd by the OH group, preventing *syn*-alignment of the hydrogen atom at the carbinyl center. Interestingly, benzylic quaternary ammonium salts also undergo coupling, which, with ethyl acrylate, furnishes ethyl 4-aryl-3-butenoate.[11]

Suzuki couplings. In analogy to the Heck reaction, the use of arenediazonium salts in the Suzuki coupling is successful.[12] The preparation of 1,2-dicyclopropylethenes by such a method gives quite satisfactory results.[13]

Ph, O, B, O + I, OBn —[$Pd(OAc)_2$ - PPh_3; K_2CO_3, Bu_4NCl; DMF - H_2O, 90°]→ Ph, OBn > 95%

Carboxylation. Homologation of allylic halides to give β,γ-unsaturated esters in a two-phase system [$Pd(OAc)_2$, K_2CO_3, CO, EtOH],[14] lactonization of long-chained ω-hydroxyalkynes (e.g., for the synthesis of exaltolide),[15] and carboxylation of arenes[16] proceeds under very mild conditions.

$Pd(OAc)_2$ - CO
Et_3N / THF
rt, 12 h

27%

Addition to multiple bonds. The addition of benzenethiol to allenes to give 2-phenylthio-1-alkenes,[17] and the conjugate transfer of an aryl group from Ar_2SbCl to enones and enals[18] are typical reactions using $Pd(OAc)_2$ as a catalyst. Under reductive conditions (HCOOK present), the delivery of an aryl or alkenyl group from the corresponding halide to 3,3-dialkoxy-1-(*o*-acetaminoaryl)-1-propynes provides precursors of 3-substituted quinolines.[19]

$CH(OEt)_2$ NHAc + I OMe

$Pd(OAc)_2$
HCOOK
DMF, 40°

OMe $CH(OEt)_2$ NHAc

58%

H^+

OMe N

Oxidation. The conversion of silyl enol ethers to enones can be effected by $Pd(OAc)_2$ in DMSO with molecular oxygen to recycle the Pd.[20] This system is also useful for heterocyclization of δ,ε-unsaturated alcohols and amides to form 2-vinylfurans[21] and -pyrrolidines.[22]

Alcohols are oxidized in a nonpolar solvent with $Pd(OAc)_2$–K_2CO_3. Allylic alcohols containing a bromine substituent on the double bond undergo oxidation and debromination.[23]

Br
OH
$Pd(OAc)_2$
K_2CO_3 / PhH
CHO
85%

[1]Reardon, P., Metts, S., Crittendon, C., Daugherity, P., Parsons, E.J., Diminnie, J. *OM* **14**, 3810 (1995).
[2]Moreno-Manas, M., Perez, M., Pleixats, R. *TL* **37**, 7449 (1996).
[3]Rigby, J.H., Hughes, R.C., Heeg, M.J. *JACS* **117**, 7834 (1995).
[4]Samizu, K., Ogasawara, K. *H* **41**, 1627 (1995).
[5]Villar, L., Bullock, J.P., Khan, M.M., Nagarajan, A., Bates, R.W., Bott, S.G., Zepeda, G., Delgado, F., Tamariz, J. *JOMC* **517**, 9 (1996).
[6]Cacchi, S., Ciattini, P.G., Morera, E., Pace, P. *SL* 545 (1996).
[7]Crisp, G.T., Gebauer, M.G. *T* **52**, 12465 (1996).
[8]Kang, S.-K., Lee, H.-W., Jang, S.-B., Kim, T.-H., Pyun, S.-J. *JOC* **61**, 2604 (1996).
[9]Kang, S.-K., Lee, H.-W., Jang, S.-B., Ho, P.-S. *CC* 835 (1996).
[10]Sengupta, S., Bhattacharyya, S. *SC* **26**, 231 (1996).
[11]Pan, Y., Zhang, Z., Hu, H. *S* 245 (1995).
[12]Darses, S., Jeffery, T., Genet, J.-P., Brayer, J.-L., Demoute, J.-P.. *TL* **37**, 3857 (1996).
[13]Charette, A.B., Giroux, A. *JOC* **61**, 8718 (1996).
[14]Kiji, J., Okano, T., Higashimae, Y., Fukui, Y. *BCSJ* **69**, 1029 (1996).
[15]Setoh, M., Yamada, O., Ogasawara, K. *H* **40**, 539 (1995).
[16]Taniguchi, Y., Yamaoka, Y., Nakata, K., Takaki, K., Fujiwara, Y. *CL* 345 (1995).
[17]Ogawa, A., Kawakami, J., Sonoda, N., Hirao, T. *JOC* **61**, 4161 (1996).
[18]Cho, C.S., Motofusa, S.-I., Ohe, K., Uemura, S. *BCSJ* **69**, 2341 (1996).
[19]Cacchi, S., Fabrizi, G., Marinelli, F., Moro, L., Pace, P. *T* **52**, 10225 (1996).
[20]Larock, R.C., Hightower, T.R., Kraus, G.A., Hahn, P., Zheng, D. *TL* **36**, 2423 (1995).
[21]Rönn, M., Bäckvall, J.-E., Andersson, P.G. *TL* **36**, 7749 (1995).
[22]Larock, R.C., Hightower, T.R., Hasvold, L.A., Peterson, K.P. *JOC* **61**, 3584 (1996).
[23]Pitre, S.V., Vankar, P.S., Vankar, Y.D. *T* **52**, 12291 (1996).

Palladium(II) acetate–lithium bromide.

γ,δ-Unsaturated carbonyl compounds. Bromopalladation of alkynes by $Pd(OAc)_2$–LiBr in the presence of conjugated carbonyl compounds is followed by a conjugate addition.[1] The reaction is not confined to activated alkynes. The formation of γ-lactones[2] from alkynoic acids with the simultaneous incorporation of the enal/enone component is certainly a useful synthetic process.

R
( )n
COOH
+
O
R'
$Pd(OAc)_2$
LiBr
O
R'
O
O
R
n = 0
/
O
O
O
R'
R
n = 1

Bromination of allenes.[3] Lithium bromide is the bromine source for addition to 1,2-alkadienes, giving 1,2-dibromides. *p*-Benzoquinone is the oxidant to recycle $Pd(OAc)_2$.

[1]Wang, Z., Lu, X. *CC* 535 (1996).
[2]Wang, Z., Lu, X. *JOC* **61**, 2254 (1996).
[3]Bäckvall, J.-E., Jonasson, C. *TL* **38**, 291 (1997).

Palladium(II) acetate-tertiary phosphine. 13, 91, 233–234, **14**, 249, 250–253; **15**, 247–248; **16**, 264–268; **17**, 259–269; **18**, 277–281

Heck reactions. Arylation of allylic alcohols can also lead to 3-aryl ketones under proper conditions.[1]

MPMO (diol alkene) + $Pd(OAc)_2$ + Bu_3P + PhI → (K_2CO_3) MPMO diol styryl product, 93%; → (Et_3N) MPMO hydroxy ketone Ph product, 85%

Useful preparation reactions include the formation of 3-cyanomethylphthalide from methyl *o*-bromobenzoate and acrylonitrile.[2] An intramolecular Heck reaction involving a nitroalkene gives both saturated and unsaturated products.[3] Alkenylsilanes are arylated, with a concomitant loss of the silyl group.[4] Fully substituted alkylidenecyclopropanes undergo a most unusual Heck reaction, leading to 2-aryl-1,3-dienes.[5]

Me_3Si–CH=CH–CH$_2$–NHBoc + 4-iodoanisole (I, OMe) → [$Pd(OAc)_2$, dppe, Et_3N / MeCN, 50°, 48 h] 2-(4-methoxyphenyl)allyl-NHBoc (82%) + MeO-C$_6$H$_4$-CH=CH-CH$_2$-NHBoc (8%)

Bicyclopropylidene + PhI → [$Pd(OAc)_2$ - PPh_3, Et_3N / DMF, 80°, 4 h] Ph(cyclopropylidene)diene product

The regioselectivity of heterocyclization can be directed by changing the catalytic system.[6]

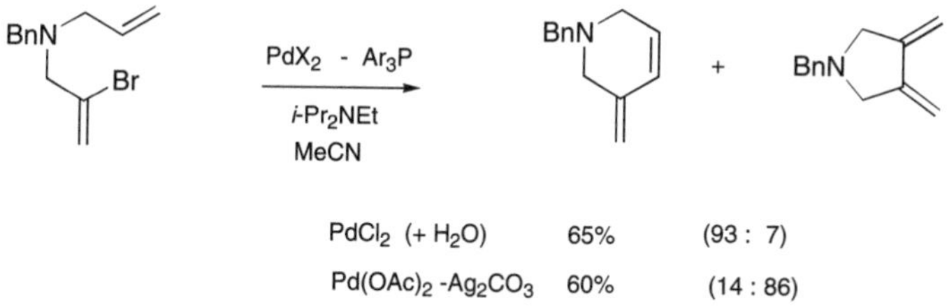

The Heck reaction intermediate from an alkenyl bromide and an alkene is an alkylating agent.[7] Thus, a three-component assembly is feasible, and a bicyclic skeleton can be established by using an intramolecular version of the method.

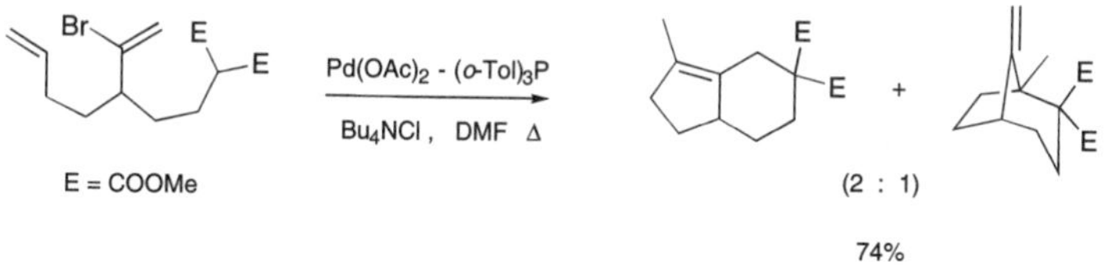

Suzuki couplings. These couplings serve to establish (2*Z*, 4*E*)-alkadienoic esters,[8] making use of a water-soluble catalyst.

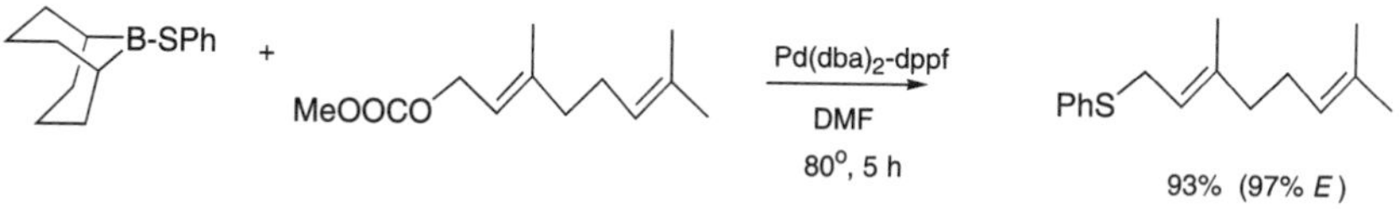

Skipped dienes. Allylic carbonates react with alkenylfluorosilanes in a Pd(II)-catalyzed reaction that results in 1,4-dienes.[9] The C–C bond formation takes place at the sites that originally carry the heteroatoms. A synthesis of 2-alkenyl-2,5-dihydrofurans[10] is initiated by the formation of analogous π-allenylpalladium intermediates.

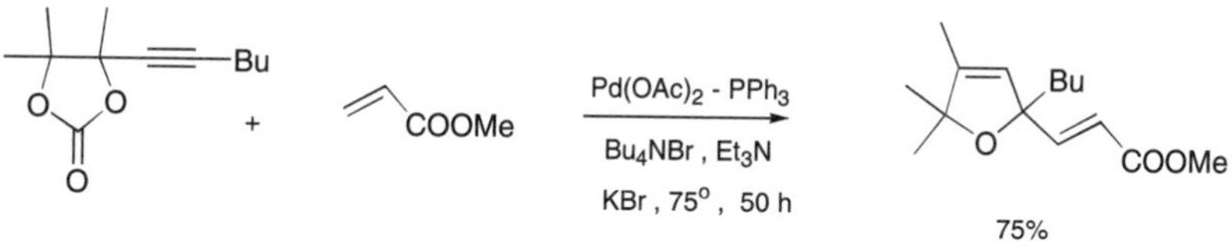

Deallyloxycarbonylation.[11] The mixed carbonate is cleaved in a Pd(II)-catalyzed reaction with NaN_3 in MeCN at room temperature (10 examples, 71–98%).

C-Allylation of naphthols.[12] Using allylic alcohols and neutral conditions is advantageous in accomplishing the *C*-allylation of naphthols. This Pd-catalyzed reaction can be controlled to provide mono-, di-, or triallylated products.

Reactions of alkynes. When 5-hydroxy-2-alkynyl methyl carbonates are subjected to low-pressure carbonylation in the presence of $Pd(OAc)_2$ and $Ph_2P(CH_2)_3PPh_2$ at room temperature, α-allenylidene-γ-butyrolactones are obtained.[13]

OH O O C_6H_{13} $Pd(OAc)_2$ - CO dppp / PhMe rt 1 hr O O C C_6H_{13} 70%

An addition reaction between an alkyne and a propargylic alcohol can lead to either furans or butenolides depending on the reaction conditions.[14]

COOEt Ph O 85% $Pd(OAc)_2$ - PAr_3; DBU Ph + COOEt OH $Pd(OAc)_2$ - PAr_3 Bu_3SnOAc O O Ph 58%

→ O O cleviolide PAr_3 = P MeO MeO 3

α,β-Unsaturated aldehydes. Monosubstituted epoxides undergo isomerization, but the resulting aldehydes are susceptible to aldol condensation.[15] Thus, an enal derived from two molecules of the epoxide can be obtained directly. However, when a large amount of another aldehyde is present, phenylacetaldehyde derived from the isomerization of styrene oxide undergoes a cross-aldol reaction.

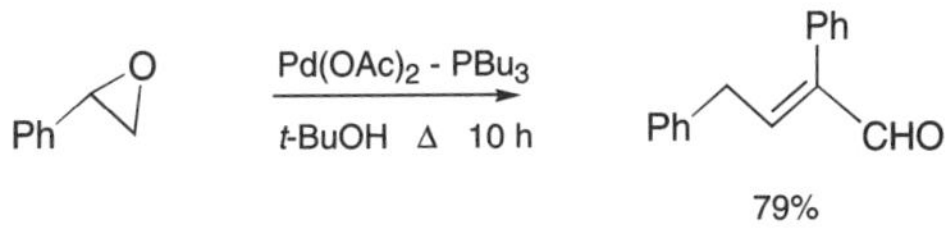

High pressure reactions.[16] A dramatic effect of high pressure on the regioselectivity (i.e., inversion) of [2+3]- and [6+4]-cycloadditions has been observed. The simplest explanation is that the increased rate of bimolecular reaction, in comparison to ligand isomerization within the metal complexes, generates more products from the kinetic complexes.

Cycloisomerization of 1,6-enynes. In sterically difficult situations, such as those reactions proceeding via a transition state disfavored by developing 1,3-diaxial interactions, the catalyst system must be well designed. A case in point is the annulation of a bridged ring system en route to picrotoxinin.[17] Thus, a failed reaction is revitalized when $Pd(OAc)_2$ is combined with a tied-back ligand and an internal source of protons.

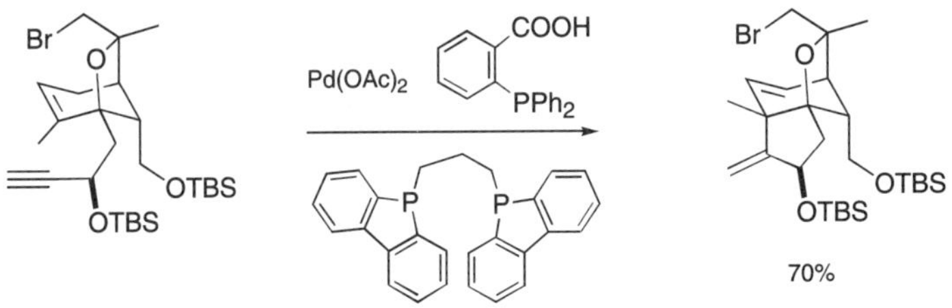

Benzannulated cyclic ethers.[18] (*o*-Haloaryl)alkanols form five- to seven-membered cyclic ethers on reaction with $Pd(OAc)_2$-phosphine in the presence of a base. This transformation is supposed to proceed by C–X bond insertion and the formation of oxapalladacycles. Reductive elimination regenerates the catalytically active Pd(0) species.

[1]Kang, S.-K., Jung, K.-Y., Park, C.-H., Namkoong, E.-Y., Kim, T.-H. *TL* **36**, 6287 (1995).
[2]Parker, J.A., Stanforth, S.P. *JHC* **32**, 1587 (1995).
[3]Denmark, S.E., Schnute, M.E. *JOC* **60**, 1013 (1995).
[4]Alvisi, D., Blart, E., Bonini, B.F., Mazzanti, G., Ricci, A., Zani, P. *JOC* **61**, 7139 (1996).
[5]Brase, S., de Meijere, A. *ACIEE* **34**, 2545 (1995).
[6]Lemaire-Audoire, S., Savignac, M., Dupuis, C., Genet, J.-P. *TL* **37**, 2003 (1996).
[7]Nylund, C.S., Smith, D.T., Klopp, J.M., Weinreb, S.M. *T* **51**, 9301 (1995).
[8]Genet, J.P., Linquest, A., Blart, E., Mouries, V., Savignac, M., Vaultier, M. *TL* **36**, 1443 (1995).
[9]Matsuhashi, H., Hatanaka, Y., Kuroboshi, M., Hiyama, T. *TL* **36**, 1539 (1995).
[10]Darcel, C., Bruneau, C., Albert, M., Dixneuf, P.H. *CC* 919 (1996).
[11]Sigismondi, S., Sinou, D. *JCR(S)* 46 (1996).
[12]Tada, Y., Satake, A., Shimizu, I., Yamamoto, A. *CL* 1021 (1996).
[13]Mandai, T., Tsujiguchi, Y., Matsuoka, S., Saito, S., Tsuji, J. *JOMC* **488**, 127 (1995).
[14]Trost, B.M., McIntosh, M.C. *JACS* **117**, 7255 (1995).

[15]Kim, J.-H., Kulawiec, R.J. *JOC* **61**, 7656 (1996).
[16]Trost, B.M., Parquette, J.R., Marquart, A.L. *JACS* **117**, 3284 (1995).
[17]Trost, B.M., Krische, M.J. *JACS* **118**, 233 (1996).
[18]Palucki, M., Wolfe, J.P., Buchwald, S.L. *JACS* **118**, 10333 (1996).

Palladium(II) acetate–tin(II) chloride. 18, 281

Hydrolysis of dimethylhydrazones.[1] Regeneration of carbonyl compounds from the hydrazones, usually in greater than 94% yield, is accomplished in aqueous DMF at 70° in the presence of $Pd(OAc)_2$ and $SnCl_2$.

[1]Mino, T., Hirota, T., Yamashita, M. *SL* 999 (1996).

Palladium(II) acetylacetonate–1,1,2,2-tetramethylpropyl isocyanide.

(E)-Allylsilanes and allenylsilanes. The preparations from allylic and propargylic alcohols via the (silyl)silyl ethers involve bis(organosilyl)Pd(II) complexes to insert into the multiple C–C bonds and a subsequent *syn*-elimination of the ensuing oxasilacyclobutanes to afford the products and $Ph_2Si{=}O$.

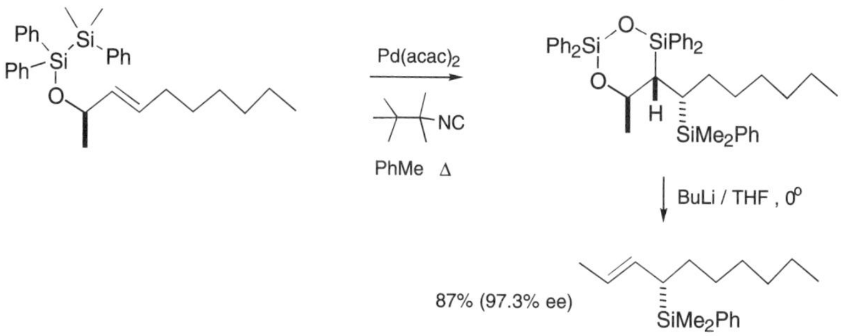

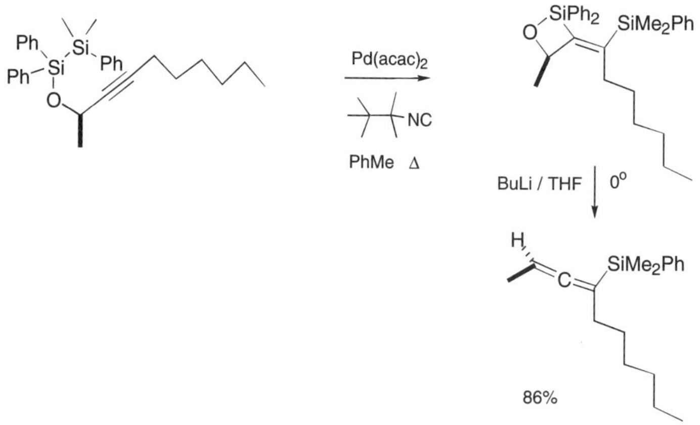

Pd(acac)$_2$ / t-alkyl-NC / PhMe Δ; BuLi / THF, 0°

87% (97.3% ee)

Pd(acac)$_2$ / t-alkyl-NC / PhMe Δ; BuLi / THF 0°

86%

[1]Suginome, M., Matsumoto, A., Ito, Y. *JACS* **118**, 3061 (1996).
[2]Suginome, M., Matsumoto, A., Ito, Y. *JOC* **61**, 4884 (1996).

Palladium(II) chloride. 13, 234–235; **15**, 248–249; **16**, 268–269; **18**, 282

Reduction.[1] The selective reduction of aromatic nitro compounds to the amines by using $PdCl_2$ and a water-soluble phosphine ligand with CO in aqueous NaOH and xylene does not affect other functional groups, such as double bond, ketone, nitrile, and halide groups.

Acetals.[2] *N*-Acryloyloxazolidin-2-ones are oxidatively functionalized to give β,β-dialkoxypropanoic acid derivatives. Asymmetric induction at the α-position by a bulky substituent in the heterocycle is observed.

$PdCl_2$ - CuCl; O_2 / MeOH; DME, rt, 96 h

CH(OMe)$_2$

89% (95% de)

o-Hydroxybenzophenones.[3] $PdCl_2$–LiCl promotes the reaction of salicylaldehydes with aryl iodides in a manner such that the carbonyl group is formally oxidized.

CHO, OH, OMe + I–Ph; $PdCl_2$ - LiCl; Na_2CO_3 / DMF; 100°, 8 h; Ph, OH, OMe

81%

Cross-couplings. The Stille coupling of organostannanes with hypervalent iodine compounds[4] has a broad scope. Diaryl and dialkenyl tellurides are also active toward alkenes if the catalytic system contains AgOAc.[5]

Te, R, R + W; $PdCl_2$ -AgOAc; Et_3N / MeOH; 25°, 20 h; R, W

64 - 99%

[1]Tafesh, A.M., Beller, M. *TL* **36**, 9305 (1995).
[2]Hosokawa, T., Yamanaka, T., Itotani, M., Murahashi, S.-I. *JOC* **60**, 6159 (1995).
[3]Satoh, T., Itaya, T., Miura, M., Nomura, M. *CL* 823 (1996).
[4]Kang, S.-K., Lee, H.-W., Jang, S.-B., Kim, T.-H., Kim, J.-S. *SC* **26**, 4311 (1996).
[5]Nishibayashi, Y., Cho, C.S., Uemura, S. *JOMC* **507**, 197 (1996).

Palladium(II) chloride, bis(triphenylphosphine) complex.

Heck reactions. The catalyst $(Ph_3P)_2PdCl_2$ is inactive under normal conditions. However, at 10^4 atm a large portion of starting materials is consumed to give coupling products.[1] Another reaction condition that is effective consists of using a molten salt (e.g., $C_{16}H_{33}PBu_3Br$ at 100°) as the reaction medium.[2]

A general route to methylenecycloalkenes of various sizes is by an intramolecular cyclization of ω-haloallenes.[3]

E E I E E
$(Ph_3P)_2PdCl_2$
Bu_4NCl - K_2CO_3
DMF
120°, 12 h
E E E E
66%

Br C
$(Ph_3P)_2PdCl_2$
Bu_4NCl - K_2CO_3
DMF
120°, 12 h
55%

Couplings. Partially fluorinated allyl ketones and 1-alkenes are formed by the Stille coupling of β-perfluoroalkyl-substitued alkyl iodides with organostannanes with or without the presence of CO.[4]

C_8F_{17} I + Bu_3Sn
$(Ph_3P)_2PdCl_2$
120°, 16 h
C_8F_{17} / C_8F_{17} O
51%
+ CO - 97%

Functionalized ketones are readily made from acid chlorides and organozinc halides that carry functional groups.[5] The coupling reactions are generally performed in DME, but for functionalized acid chlorides, a mixture of DMA and benzene or toluene is preferred. (Here, $Pd_2(dba)_3$–PPh_3 are used.)

The Suzuki coupling of bis(boryl)alkenes (**1**) is easily controlled. Accordingly, a variety of tetrasubstituted alkenes can be synthesized in a stepwise manner.[6]

(**1**)

Carbonylative cyclization. A synthesis of 2-substituted 1-indanones from 1-(*o*-iodoaryl)-2-propenes[7] is shown in the equations that follow. The solvent and reaction temperature control the types of products.

$(Ph_3P)_2PdCl_2$, CO - Et_3N, Me_2CO - THF, 80°: 58%

$(Ph_3P)_2PdCl_2$, CO - Et_3N, Me_2CO - DMF, MeOH, 100°: COOMe, 90%

$(Ph_3P)_2PdCl_2$, CO - Et_3N, MeCN - PhH, MeOH: COOMe, 84%

Internal acetals of phthalimides are also formed by the carbonylation of 2-(2-bromoaryl)-2-oxazolines in methanol.[8] The catalyst system consists of $(Ph_3P)_2PdCl_2$ and $NiCl_2 \cdot 6H_2O$.

Allylation.[9] $(Ph_3P)_2PdCl_2$ promotes the transfer of the allyl group from allyltributylstannane to aldehydes and imines. Imines are much more reactive under these conditions.

[1]Sugihara, T., Takebayashi, M., Kaneko, C. *TL* **36**, 5547 (1995).
[2]Kaufmann, D.E., Nouroozian, M., Henze, H. *SL* 1091 (1996).

[3]Ma, S., Negishi, E. *JACS* **117**, 6345 (1995).
[4]Shimizu, R., Fuchikami, T. *TL* **37**, 8405 (1996).
[5]Fraser, J.L., Jackson, R.F.W., Porter, B. *SL* 819 (1995).
[6]Brown, S.D., Armstrong, R.W. *JACS* **118**, 6331 (1996).
[7]Negishi, E., Coperet, C., Ma, S., Mita, T., Sugihara, T., Tour, J.M. *JACS* **118**, 5904 (1996).
[8]Cho, C.S., Lee, J.W., Lee, D.Y., Shim, S.C., Kim, T.J. *CC* 2115 (1996).
[9]Nakamura, H., Iwama, H., Yamamoto, Y. *JACS* **118**, 6641 (1996).

Palladium(II) chloride, bis(triphenylphosphine) complex–copper(I) iodide.

Alkynylarenes. The catalyst system, usually in combination with a mild base (trialkylamine or K_2CO_3), is used to couple terminal alkynes with aryl iodides and triflates. A large rate enhancement is observed when an iodide salt is added, as shown in a synthesis of 1,2-bis(trimethylsilylethynyl)benzene.[1]

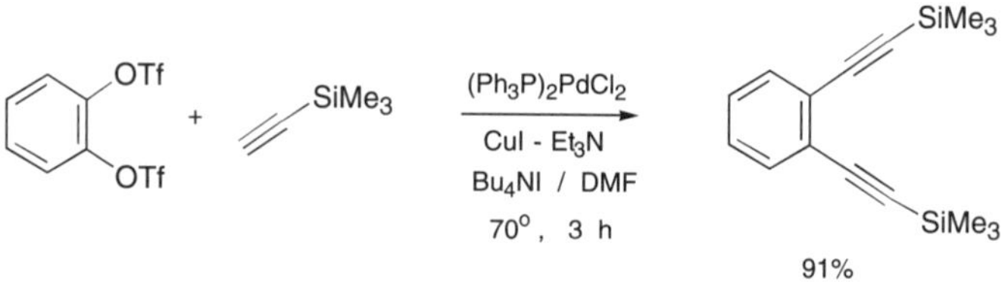

When a nucleophilic functionality is present at an *o*-position with respect to the leaving group, cyclized products (e.g., benzofurans) are obtained.[2] Another variant is that starting from alkynes represented by *o*-hydroxyphenyl propargyl ether, whereby the formation of *(Z)*-2-arylidene-1,4-benzodioxanes is accessible.[3]

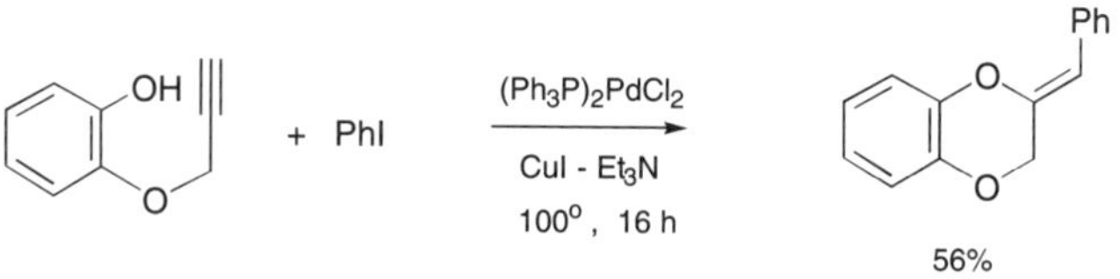

Enynes. Alkyne–bromoalkene couplings are useful for synthesizing enynes. Thus, β-bromoalkenyl phenyl tellurides give coupled products that retain the tellurium group.[4]

[1]Powell, N.A., Rychnovsky, S.D. *TL* **37**, 7901 (1996).
[2]Botta, M., Summa, V., Corelli, F., Di Pietro, G., Lombardi, P. *TA* **7**, 1263 (1996).
[3]Chowdhury, C., Kundu, N.G. *CC* 1067 (1996).
[4]Huang, X., Wang, Y.-P. *SC* **26**, 3087 (1996).

Palladium(II) chloride–tertiary phosphine.

N-Arylation. Primary amines are converted into secondary anilines with $(R_3P)_2PdCl_2$. The method can be used to *N*-arylate α-amino acids.[1] A second-generation catalyst consisting of (dppf)$PdCl_2$–NaOBu^t [dppf = 1,1′-bis(diphenylphosphino)-ferrocene][2] is a significant improvement over systems using $(o\text{-Tol})_3P$-ligated catalysts.

Modified Suzuki coupling.[3] Aryl bromides are converted into ethynylarenes by treatment with sodium acetylide in the presence of trimethyl borate and a catalytic amount of (dppf)$PdCl_2$.

Cross-couplings. Organosilyl chlorides are reactive coupling partners for aryl chlorides.[4] Arylation and alkenylation are readily achieved.

SiMeCl2 + Cl–C6H4–CN → $(Et_3P)_2PdCl_2$ / Bu_4NF / THF, 90°, 20 h → 91%

A three-component coupling that can be considered as tandem Heck and Stille reactions gives rise to *cis*-4,5-diaryl-1,3-dioxolane.[5] The reactants are aryl bromides, 1,3-dioxole, and phenyltributylstannane. (Other aryltributylstannanes should also be applicable.)

1,3-dioxole + Bu_3SnPh + ArBr → $[(o\text{-Tol})_3P]_2PdCl_2$ / THF, 80°, 24 h → 4-Ph-5-Ar-1,3-dioxolane, 41 - 74%

[1]Ma, D., Yao, J. *TA* **7**, 3075 (1996).
[2]Driver, M.S., Hartwig, J.F. *JACS* **118**, 7217 (1996).
[3]Fürstner, A., Nikolakis, K. *LA* 2107 (1996).
[4]Gouda, K.-I., Hagiwara, E., Hatanaka, Y., Hiyama, T. *JOC* **61**, 7232 (1996).
[5]Oda, H., Hamataka, K., Fugami, K., Kosugi, M., Migita, T. *SL* 1225 (1995).

Palladium(II) chloride–copper(I) chloride-oxygen. 18, 283

β-Oxy carbonyl compounds.[1] The regioselectivity of the Wacker–Tsuji oxidation of 1-alkenes is completely changed when the allylic and homoallylic positions are oxygenated. Some 2-alkenyl analogs also behave similarly, to give β-oxy carbonyl products.

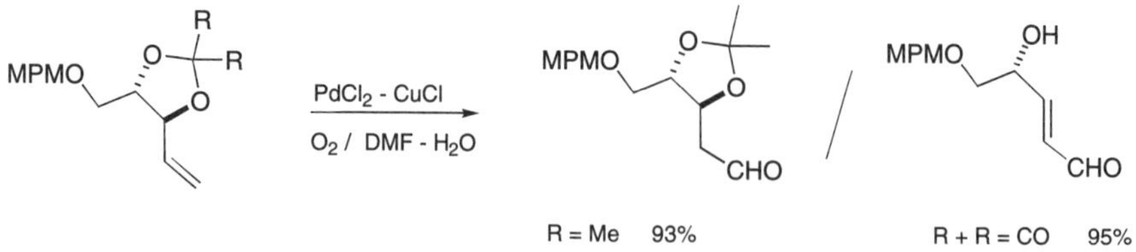

[1]Kang, D.-K., Jung, K.-Y., Chung, J.-U., Namkoong, E.-Y., Kim, T.-H. *JOC* **60**, 4678 (1995).

Palladium(II) hydroxide/carbon.

Hydrostannylation.[1] Alkenes react with tin hydrides with the help of this catalyst. Thus, allylic alcohols furnish γ-hydroxyalkylstannanes.

[1]Lautens, M., Kumanovic, S., Meyer, C. *ACIEE* **35**, 1329 (1996).

Palladium(II) iodide–thiourea.

Carbonylation.[1] Terminal alkynes give a mixture of 2-substituted dimethyl maleates and butenolides. Both oxidative and reductive pathways are followed.

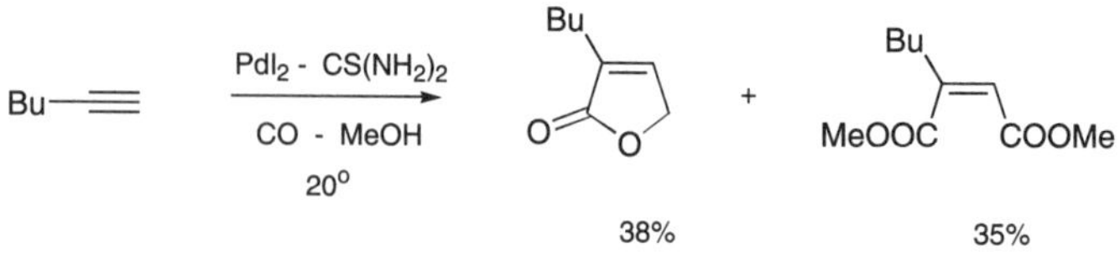

[1]Gabriele, B., Salerno, G., Costa, M., Chiusoli, G.P *JOMC* **503**, 21 (1995).

Paraformaldehyde. 18, 284

Hydroxymethylation. Hydroxymethylation of 1-alkynes and lactams, including β-lactams, is easily achieved. The alkynes are activated by treatment with EtMgBr,[1] and the lactam alkylation is aided by ultrasound.[2]

4-Dialkylamino-1-(trimethylsilyl)-1,2-butadienes.[3] Iminium ions generated by the condensation of secondary amines with $(HCHO)_n$ react readily with 1,1-bis(trimethylsilyl)-2-propyne, with expulsion of a trimethylsilyl group.

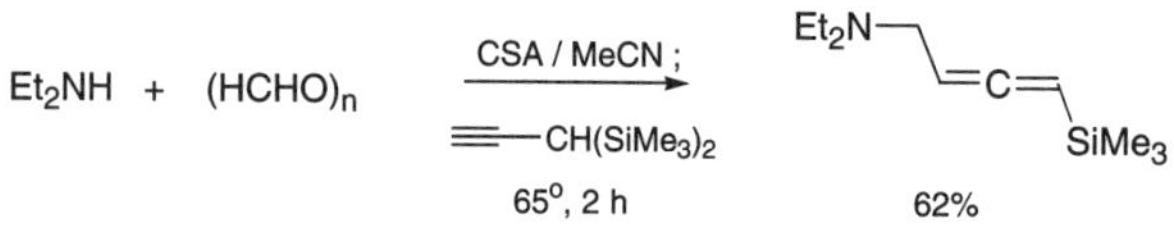

Azomethine ylides. These ylides can be prepared from *N*-alkylglycine[4] or *N*-alkyl(trimethylsilylmethyl)amines[5] by reaction with $(HCHO)_n$. The reactive species are trapped as 1,3-dipolar cycloadducts, namely, and 3,4-disubstituted pyrrolidines.

2-Functionalized acrylic acids.[6] In a one-step reaction, malonic acid, paraformaldehyde, and a nucleophile (e.g., RR′NH) give such products in good yields (5 examples, 78–92%).

COOH COOH + $(HCHO)_n$ + H–N(pyrrolidine) —EtOH, Δ→ COOH … N(pyrrolidine)

73%

2-Methylenecycloalkane-1-methanols.[7] Activation with Me_2AlCl makes formaldehyde (from paraformaldehyde) receptive to attack by 1-trimethylsilylmethylcycloalkenes in the presence of $SnCl_4$. The reaction is a silicon-directed ene reaction.

R R' SiMe$_3$ —$(HCHO)_n$ - Me_2AlCl; $SnCl_4$ / CH_2Cl_2 ; $NaHCO_3$ / H_2O→ R R' OH

90 - 95%

RCONH2 → RCN.[8] This dehydration of primary amides is performed by heating with formic acid and $(HCHO)_n$ in MeCN. Amides in which the α-position is fully substituted undergo the dehydration without any problem.

[1]Zwierzak, A., Tomassy, B. *SC* **26**, 3593 (1996).
[2]Jouglet, B., Oumoch, S., Rousseau, G. *SC* **25**, 3869 (1995).
[3]M'Baze Meva'a, L., Pornet, J. *SC* **26**, 3351 (1996).
[4]Nyerges, M., Balazs, L., Kadas, I., Bitter, I., Kovesdi, I., Toke, L. *T* **51**, 6783 (1995).
[5]Torii, S., Okumoto, H., Genba, A. *CL* 747 (1996).
[6]Krawczyk, H. *SC* **25**, 641 (1995).
[7]Monti, H., Feraud, M. *SC* **26**, 1721 (1996).
[8]Heck, M.-P., Wagner, A., Mioskowski, C. *JOC* **61**, 6486 (1996).

Pentaarylantimony.

Arylations.[1] Reaction of these reagents with carbonyl compounds and acid chlorides furnishes alcohols and aryl ketones, respectively. The former reaction requires a Lewis acid at a low temperature (–78° to room temperature).

[1]Fujiwara, M., Tanaka, M., Baba, A., Ando, H., Souma, Y. *JOMC* **508**, 49 (1996).

Pentamethylcyclopentadienylrhenium(V) trioxide.

Alkenes from 1,2-diols.[1] The catalytic elimination using Cp^*ReO_3 in the presence of Ph_3P in hot chlorobenzene is stereoselective. This method is potentially competitive with the standard Corey–Winter procedure.

PPh3
PhCl , 120°, 80 h
86%

[1]Cook, G.K., Andrews, M.A. *JACS* **118**, 9448 (1996).

Perfluoro-2,3-dialkyloxaziridines. 18, 285

Ketones from ethers.[1] Alkyl ethers—particularly methyl alkyl ethers—are oxidized to ketones in Freon-11 at room temperature by the oxaziridines. Thus, 2-adamantanone is obtained in 91% yield from 2-methoxyadamantane.

[1]Arnone, A., Bernardi, R., Cavicchioli, M., Resnati, G. *JOC* **60**, 2314 (1995).

Phase-transfer catalysts. 13, 239–240; **15**, 252–253; **18**, 286–289

Etherification. With the use of poly(ethylene glycol) [PEG] as a phase-transfer catalyst (PTC), diaryloxymethanes are readily prepared[1] from ArOK and CH_2Cl_2 in methanol at room temperature. The same catalyst is used in a synthesis of triaryl cyanurates[2] from cyanuric chloride.

A useful method for access to vinyl ethers involves the addition of alcohols to alkynes in the presence of 18-crown-6.[3] The *(Z)*-isomers are predominant.

The monobenzylation of symmetrical diols with BnBr, KOH and 18-crown-6 is highly efficient (7 examples, 82–91%).[4]

N-Alkylations. Smooth alkylation of pyrrole without solvent,[5] synthesis of diarylamines by *N*-arylation with activated aryl halides,[6] and the direct conversion of *o*-nitrotrifluoroacetanilides to *N*-alkylamines[7] are performed under PTC.

2-Bromoethyl methacrylate is converted into methyl *N*-(methacryloyloxy)ethyl carbamate on treatment with KOCN in methanol in the presence of Bu_4NBr.[8]

Synthesis of organosulfur compounds. When catalyzed by PEG-400, the reaction of alkyl and acyl halides with sulfur gives dialkyl and diacyl disulfides, respectively.[9,10] The opening of epoxide with sodium arenesulfinates using polysorbate-80 as a PTC constitutes a simple route to β-hydroxy sulfones.[11]

C-Alkylations. Allylic phenyl sulfones undergo alkylation very readily when the anions are generated in the presence of aqueous NaOH and tetrabutylammonium bromide: With dibromoalkanes, cyclic products are obtained.[12] 2,2-Dibromo-1-phenylcyclopropane and α-substituted phenylacetonitriles react to give cyclopropene derivatives.[13]

50% NaOH, CH_2Cl_2, $[BnNEt_3]^+ Cl^-$, 45°, 9 h — 53%

The PTC technique has been advantageously applied to alkylation[14] and Pd-catalyzed arylation[15] of alkynes. For the Heck reaction, the proper selection of Pd, base, and PTC allows a reaction in water, aqueous/organic solvent mixture, or strictly anhydrous conditions.[16]

In the presence of 18-crown-6, concurrent cyclization and arylation is favored over the Heck reaction of a terminal double bond of alkenyl cyanoacetic esters.[17]

KH / 18-c-6, Pd(dppe) - PhBr, 60°, 19 h — 47%

Carbonylation and hydroformylation. With PEG as a PTC, $Co_2(CO)_8$ mediates the carbonylation of benzylic halides.[18] A more complex system is used in the transformation of propargylic alcohols into 2-alkylidenesuccinic acids (9 examples, 84–97%).[19] Very similar results are obtained from the carbonylation of alkynyl ketones.[20]

$Ni(CN)_2$ - $LaCl_3 \cdot 7H_2O$, CO / PEG-400, NaOH / PhMe, 95°, 4 - 8 h — (86 : 14) 95%

A breakthrough in biphasic, $Rh(acac)(CO)_2$-catalyzed hydroformylation is the use of per(2,6-di-*O*-methyl)-β-cyclodextrin as an inverse PTC.[21] Highly selective reactions of terminal double bonds are observed.

Additions and condensations. Nitroalkanes form adducts with enones in aqueous media in the presence of cetyltrimethylammonium chloride.[22]

Wittig reactions conducted in a solid (KOH)–liquid (CH_2Cl_2) two-phase system[23] containing 18-crown-6 give rise to *(E)*-stilbenes as major products when one of the three phenyl substituents of the phosphonium salt precursors is replaced with a chlorine atom. Both the lower steric demand of the Cl and a stronger P–Cl bond tend to shift the transition state toward a nearly planar four-centered structure.

Cycloadditions. 2-Vinylaziridines can be prepared directly from allyldimethylsulfonium bromide and *N*-sulfonylimines.[24] The sulfonium ylides are generated in situ.

A two-site PTC (**1**) prepared from acetophenone, formaldehyde, HCl, and triethylamine in three steps is effective in promoting dichlorocyclopropanation.[25] It is of interest to note that 1,1,2-tribromocyclopropane undergoes decomposition to give the diethyl acetal of propargylaldehyde under PTC in the presence of an alcohol.[26]

Ph $\overset{+}{N}Et_3$ $Et_3\overset{+}{N}$ 2 Cl^-

(**1**)

Oxidations. Alcohols are oxidized in dichloroethane with catalytic amounts of a dichromate salt using sodium percarbonate as a recycling agent and PTC.[27,28] Epoxidations of enones by sodium perborate[29] or NaOCl[30] under PTC conditions give high yields. However, with NaOCl and hexaethylguanidinium chloride, cyclohexenones give 6,6-dichloro-2,3-epoxycyclohexanones.

The regeneration of carbonyl compounds from semicarbazones by $Me_3SiCl–NaNO_x$[31] also benefits from a PTC.

[1]Salunkhe, M.M., Kavitake, B.P., Patil, S.V., Wadgaonkar, P.P. *JCR(S)* 503 (1995).
[2]Kavitake, B.P., Patil, S.V., Salunkhe, M.M., Wadgaonkar, P.P. *BSCB* **104**, 675 (1995).
[3]Bellucci, G., Chiappe, C., Lo Moro, G. *SL* 880 (1996).
[4]Bessodes, M., Boukarim, C. *SL* 1119 (1996).
[5]Diez-Barra, E., de la Hoz, A., Loupy, A., Sanchez-Migallon, A. *JHC* **31**, 1715 (1994).
[6]Durantini, E.N., Chiacchiera, S.M., Silber, J.J. *SC* **26**, 3849 (1996).
[7]Brown, S.A., Rizzo, C.J. *SC* **26**, 4065 (1996).
[8]Dubosclard-Gottardi, C., Caubere, P., Fort, Y. *T* **51**, 2561 (1995).
[9]Wang, J.-X., Cui, W., Hu, Y. *SC* **25**, 3573 (1995).
[10]Wang, J.-X., Wang, C.-H., Cui, W., Hu, Y., Zhao, K. *SC* **25**, 889 (1995).
[11]Maiti, A.K., Bhattacharyya, P. *IJC(B)* **35B**, 67 (1996).
[12]Jonczyk, A., Radwan-Pytlewski, T. *G* **126**, 111 (1996).
[13]Arct, J., Fedorynski, M., Minksztym, K., Jonczyk, A. *S* 1073 (1996).

[14]Dehmlow, E.V., Fastabend, U. *G* **126**, 53 (1996).
[15]Nguefack, J.-F., Bolitt, V., Sinou, D. *TL* **37**, 5527 (1996).
[16]Jeffery, T. *T* **52**, 10113 (1996).
[17]Bouyssi, D., Coudanne, I., Uriot, H., Gore, J., Balme, G. *TL* **36**, 8019 (1995).
[18]Zucchi, C., Palyi, G., Galamb, V., Sampar-Szerencses, E., Marko, L., Li, P., Alper, H. *OM* **15**, 3222 (1996).
[19]Zhou, Z., Alper, H. *OM* **15**, 3282 (1996).
[20]Arzoumanian, H., Jean, M., Nuel, D., Cabrera, A., Gutierrez, J.L.G., Rosas, N. *OM* **14**, 5438 (1995).
[21]Monflier, E., Tilloy, S., Fremy, G., Castanet, Y., Mortreux, A. *TL* **36**, 9481 (1995).
[22]Ballini, R., Bosica, G. *TL* **37**, 8027 (1996).
[23]Bellucci, G., Chiappe, C., Lomoro, G. *TL* **37**, 4225 (1996).
[24]Li, A.-H., Dai, L.-X., Hou, X.-L., Chen, M.-B. *JOC* **61**, 4641 (1996).
[25]Balakrishnan, T., Jayachandran, J.P. *SC* **25**, 3821 (1995).
[26]Sydnes, L.K., Bakstad, E. *ACS* **50**, 446 (1996).
[27]Mohand, S.A., Levina, A., Muzart, J. *SC* **25**, 2051 (1995).
[28]Mohand, S.A., Muzart, J. *SC* **25**, 2373 (1995).
[29]Straub, T.S. *TL* **36**, 663 (1995).
[30]Schlama, T., Alcaraz, L., Mioskowski, C. *SL* 571 (1996).
[31]Khan, R.H., Mathur, R.K., Ghosh, A.C. *JCR(S)* 506 (1995).

Phenyl(cyano)iodine(III) tosylate. 18, 289

Bicyclization.[1] Alkynylstannanes give alkynyliodonium species that are receptive to attack by a tosylamide anion, forming iodonium ylides. A vinylidenecarbene generated by the decomposition of such an iodonium ylide is liable to undergo further reactions, such as insertion into a remote C–H bond, resulting in a cycloalkene.

TsNH / $CH(OEt)_2$ / $SnBu_3$ — PhI(CN)OTf ; *t*-BuOK ; H_3O^+ → N-Ts bicyclic enone, 39%

[1]Schildknegt, K., Bohnstedt, A.C., Feldman, K.S., Sambandam, A. *JACS* **117**, 7544 (1995).

Phenyliodine(III) bis(trifluoroacetate). 13, 241–242; **14**, 257; **15**, 257–258; **16**, 274–275; **18**, 289–290

Benzylic oxidations.[1] *o*-Alkylbenzoic acids form lactones in a photoinduced reaction with $PhI(OCOCF_3)_2$ and iodine.[1] Trimethylsilylmethylpyridines and -quinolines undergo desilylative oxidation to afford the alcohols.[2]

Oxidation of phenols and derivatives. Facile dearomatization[3,4] and nonphenolic oxidative coupling[5] lead to some novel structures.

PhI(OCOCF$_3$)$_2$, CF$_3$CH$_2$OH, rt, 0.5 h — 74%

PhI(OCOCF$_3$)$_2$, Me$_3$SiOTf, MeOH - CF$_3$CH$_2$OH — 94%

Nucleophilic sulfenylation.[6] Under oxidizing conditions, phenol ethers are susceptible to attack by thiols. Trimethylsilyl thiocyanate can also be used to introduce a thiocyanate to the aromatic ring.

[1]Togo, H., Muraki, T., Yokoyama, M. *TL* **36**, 7089 (1995).
[2]Andrews, I.P., Lewis, N.J., McKillop, A., Wells, A.S. *H* **43**, 1151 (1996).
[3]Kita, Y., Takada, T., Ibaraki, M., Gyoten, M., Mihara, S., Fujita, S., Tohma, H. *JOC* **61**, 223 (1996).
[4]Kita, Y., Egi, M., Okajima, A., Ohtsubo, M., Takada, T., Tohma, H. *CC* 1491 (1996).
[5]Kita, Y., Gyoten, M., Ohtsubo, M., Tohma, H., Takada, T. *CC* 1481 (1996).
[6]Kita, Y., Takada, T., Mihara, S., Whelan, B.A., Tohma, H. *JOC* **60**, 7144 (1995).

Phenyliodine(III) diacetate. 13, 242–243; **14**, 258–259; **15**, 258; **16**, 275–276; **17**, 280–281; **18**, 290–291

Oxidative heterocyclizations. 2-Hydroxychalcones are converted into flavones.[1] 2-(3-Hydroxypropyl) sugars undergo photoinduced oxidation to give spiroacetals[2] with PhI(OAc)$_2$–I$_2$.

An interesting application of this oxidizing agent is in the formation of spiroisoxazolines from *o*-hydroxyarylacetoxime derivatives,[3] as well as the more well-known transformation of 2-(1-hydroxyalkyl)furans to the pyranones.[4] The cycloamination[5] of an aromatic compound bearing a sulfonamide side chain provides access to *N*-heterocycles.

PhI(OAc)$_2$, MeCN, 0°, 1 h — 76%

aerophobins

PhI(OAc)$_2$ - I$_2$

hv / ClCH$_2$CH$_2$Cl

65°, 2 h

NHTf

N Tf

80%

Dibenzocyclooctene lignans.[6] The skeleton of these natural products is rapidly formed in high yields by the direct oxidation of phenolic dibenzylbutyrolactones. The oxidation process is biomimetic, although the reagent is not.

MeO MeO H H O O OH

PhI(OAc)$_2$

MeOH - H$_2$O

15 min

MeO MeO H O O H HO + MeO MeO H O O H HO

(88 : 12)

87%

1,2-Dithiocyanatoalkanes.[7] The formal addition of $(SCN)_2$ to an alkene is achieved on in situ oxidation of KSCN by PhI(OAc)$_2$ in the presence of magnesium perchlorate or TEMPO. The process involves radicals.

+ KSCN

PhI(OAc)$_2$

Mg(ClO$_4$)$_2$

SCN SCN

60%

Oxidative displacement of propargylic silanes.[8] Such silanes are converted into ethers, esters, and amides on reaction with the proper nucleophiles. The reaction is mediated by *m*-nitrophenyliodine(III) acetate. Allenyliodane intermediates are involved.

R SiMe$_3$ R'

NO$_2$ I(OAc)$_2$

MeOH / BF$_3$• OEt$_2$

rt

R OMe R'

63 - 89%

[1]Litkei, G., Gulacsi, K., Antus, S., Blasko, G. *LA* 1711 (1995).
[2]Martin, A., Salazar, J.A., Suarez, E. *JOC* **61**, 3999 (1996).
[3]Murakata, M., Yamada, K., Hoshino, O. *T* **52**, 14713 (1996).
[4]De Mico, A., Margarita, R., Piancatelli, G. *TL* **36**, 3553 (1995).
[5]Togo, H., Hoshina, Y., Yokoyama, M. *TL* **37**, 6129 (1996).
[6]Pelter, A., Satchwell, P., Ward, R.S., Blake, K. *JCS(P1)* 2201 (1995).
[7]De Mico, A., Margarita, R., Mariani, A., Piancatelli, G. *TL* **37**, 1889 (1996).
[8]Kida, M., Sueda, T., Goto, S., Okuyama, T., Ochiai, M. *CC* 1933 (1996).

Phenyliodine(III) diazide.

Azide transfer to 3-deoxyglycals.[1] This explosive-prone reagent is prepared in situ from iodosobenzene and Me_3SiN_3, and the oxidative azidation in oxygen-free CH_2Cl_2 gives 3-azidoglycals. When C-3 is substituted, anomeric azides are formed.

BnO PhI(N3)2 → BnO N3 51%

AcO AcO PhI(N3)2 → AcO AcO N3 + AcO AcO N3 (1 : 1) 37%

[1]Kirschning, A., Domann, S., Dräger, G., Rose, L. *SL* 767 (1995).

Phenyl(propynyl)iodonium triflate.

Dihydropyrroles and indoles.[1] The salt reacts with tosylamide anions to provide five-membered heterocycles via carbene intermediates. When a nucleofugal group (methoxy is sufficient) is present at the bond insertion point, automatic elimination gives a pyrrole as the product. Insertion into an aryl C–H bond leads to an indole.

NHTs BuLi ; —≡—$\overset{+}{I}$Ph TfO⁻ → N Ts 66%

[1]Feldman, K.S., Bruendl, M.M., Schildknegt, K. *JOC* **60**, 7722 (1995).

Phenylselenium pentafluoride.

vic-Fluorination.[1] $PhSeF_5$ surrenders two fluorine atoms to C=C double bonds at room temperature. From *(E)*-stilbene, a mixture of *erythro-* and *threo*-difluoride is obtained in a 2:1 ratio.

[1]Lermontov, S.A., Zavorin, S.I., Bakhtin, I.C., Zefirov, N.S., Stang, P.J. *PSS* **102**, 283 (1995).

Phenylselenium trichloride.

Vinylic chlorination.[1] The reagent $PhSeCl_3$ is prepared from benzeneseleninic acid and thionyl chloride under ultrasonic conditions at room temperature. It can be used as a chlorinating agent for alkenes.

$PhSeCl_3$ / CH_2Cl_2 ; $NaHCO_3$ / H_2O

50%

[1]Stuhr-Hansen, N., Henriksen, L., Kodra, J.T. *SC* **26**, 3345 (1996).

Phenylselenoethynyl *p*-tolyl sulfone.

Ketene equivalent.[1] The sulfone is a reactive dienophile. The Diels–Alder adducts can be converted into cyclic ketones.

Ts, SePh; PhMe, 60°, 6 h; Ts, SePh

90%

[1]Back, T.G., Wehrli, D. *SL* 1123 (1995).

Phthalimidesulfenyl chloride.

α,α′-Dioxothiones.[1] Sulfenylation of β-diketones with PhthN–SCl followed by treatment with a mild base (e.g., pyridine) generates the reactive species, which behave as heterodienes. 1,4-Oxathiins are readily obtained.

py, $CHCl_3$, rt; OEt; OEt

78%

Aromatic phthalimidosulfenylation.[2] The reaction usually takes place at room temperature. No catalyst is required for activated arenes. From the products, a synthesis of *o*-thioquinones is developed.

PhtN-SCl
$CHCl_3$, rt;
py, 60°

[1]Capozzi, G., Franck, R.W., Mattioli, M., Menichetti, S., Nativi, C., Valle, G. *JOC* **60**, 6416 (1995).
[2]Capozzi, G., Menichetti, S., Nativi, C., Simonti, M.C. *TL* **35**, 9451 (1994).

Piperidine.

Fmoc group cleavage.[1] For deprotection of the amine derivatives in the solid-phase synthesis of *O*-linked glycopeptides, piperidine is preferable to morpholine.

[1]Kihlberg, J., Vuljanic, T. *TL* **34**, 6135 (1993).

Platinum(II) chloride.

1-Vinylcycloalkenes. The cyclization of 1,6- and 1,7-enynes is catalyzed by $PtCl_2$. The terminal carbon atom of the product is transposed.[1]

Me_3SiO, E, E → $PtCl_2$, PhMe, 80° → Me_3SiO, E, E — 93%

E, E → $PtCl_2$, PhMe, 80° → E, E — 97%

[1]Chatani, N., Furukawa, N., Sakurai, H., Murai, S. *OM* **15**, 901 (1996).

Platinum divinyltetramethyldisiloxane.

Reductive O-silylation.[1] Conjugated ketones give silyl enol ethers of the corresponding saturated ketones on reaction with a hydrosilane using the Karstedt catalyst (**1**). The method is particularly valuable for the acquisition of those derivatives with a bulky silyl group. Most reactions can be carried out at room temperature, although enones with substituents at the β or γ position require elevated temperatures (70°) to complete the reaction. Both a higher temperature and a longer reaction time are needed in the preparation of triphenylsiloxyalkenes. The relative reactivity of hydrosilanes follows the order of $Et_3SiH > t\text{-}BuMe_2SiH >> PhMe_2SiH, (EtO)_3SiH, i\text{-}Pr_3SiH >> Ph_3SiH$.

i-Pr3SiH; Pt[divinyltetramethyldisiloxane]2 (**1**); rt, 4 h; i-Pr3OSi; 93%

[1]Johnson, C.R., Raheja, R.K. *JOC* **59**, 2287 (1994).

Poly-η-(pyrazine)zinc borohydride.

Reductions.[1] This reagent is prepared from the unstable $Zn(BH)_4$ by coordination with pyrazine in ether. It is stable and efficiently reduces many types of organic compounds.

[1]Tamami, B., Lakouraj, M.M. *SC* **25**, 3089 (1995).

Potassium *t*-butoxide. 13, 252–254; 15, 271–272; 17, 289–290; 18, 296–297

Eliminations. A convenient one-pot synthesis of 1-alkynylphosphonates[1] involves enolphosphorylation of β-keto phosphonates and elimination with *t*-BuOK.

(EtO)2P; NaH / THF ; (EtO)2POCl; OPO(OEt)2; t-BuOK; -78°; 90% (two steps)

Selective saponification.[2] An ester group can be selectively hydrolyzed with *t*-BuOK in wet THF at 0° without affecting a malonate unit having at least one active methine hydrogen.

EtOOC / *t*BuOOC / COOEt / COOEt —*t*-BuOK / THF - H_2O ; 0.5 N HCl→ EtOOC / *t*BuOOC / COOH / COOH 71%

Alkynylation.[3] The *t*-BuOK–DMSO system is useful for inducing the condensation of terminal alkynes with ketones to afford propargylic alcohols.

De-O-silylation.[4] Silyl dienol ethers are cleaved, and the dienolate anions can be trapped as dienol phosphates and as Diels–Alder adducts with aldehydes.

$OSiMe_3$ + *t*-BuOK —-78° ; $ClPO(OEt)_2$→ $OPO(OEt)_2$
—-78° ; PhCHO→ OH, O, Ph 86%

Dehydrogenation of stilbenes.[5] With an excess of *t*-BuOK, stilbenes are converted into tolanes in the presence of air. A larger excess of the base causes hydroxylation at an *o*-position of the nitro group. It is essential to use DMSO or DMF as a solvent.

MeO– –NO_2 —*t*-BuOK, DMF (air), rt→ MeO– –NO_2 (X) X = H, OH

Rearrangement of O-propargyl ketoximes.[6] Treatment with *t*-BuOK converts these oxime ethers into *N*-(1-alkenyl)acrylamides.

Ph, N, O —*t*-BuOK, THF (DMSO)→ Ph, NH, O 41%

[1]Hong, J.E., Lee, C.W., Kwon, Y., Oh, D.Y. *SC* **26**, 1563 (1996).
[2]Wilk, B.K. *SC* **26**, 3859 (1996).
[3]Babler, J.H., Liptak, V.P., Phan, N. *JOC* **61**, 416 (1996).
[4]Duhamel, P., Cahard, D., Poirier, J.-M. *JCS(P1)* 2509 (1993).
[5]Akiyama, S., Tajima, K., Nakatsuji, S., Nakashima, K., Abiru, K., Watanabe, M. *BCSJ* **68**, 2043 (1995).
[6]Trofimov, B.A., Tarasuua, O.A., Sigalou, M.U., Mikhaleva, A.I. *TL* **36**, 9181 (1995).

Potassium cyanide.

Retro-benzoin condensation.[1] KCN promotes C–C bond cleavage in benzoins, and therefore, ketones are obtained by using this reaction in conjunction with prior alkylation.

KCN / DMF, 80°, 1 h; 98%

ω-Cyanoalkyl methacrylates.[2] By using KCN for displacement of the haloalkyl esters, a phase transfer catalyst, methyltriphenylphosphonium, is added to advantage.

[1]Miyashita, A., Suzuki, Y., Okumura, Y., Higashino, T. *CPB* **44**, 252 (1996).
[2]Fort, Y., Dubosclard-Gottardi, C. *SC* **26**, 2811 (1996).

Potassium fluoride. 13, 256–257; **15**, 272; **18**, 297–298

Highly active KF is obtained by slow evaporation of a methanolic solution, followed by drying at 100°.[1]

1,3-Diols from β-hydroxyacylsilanes.[2] Desilylation of the acylsilanes and decomposition of phenyldimethylsilyl fluoride to furnish a nucleophilic phenylating reagent unveil a pair of new reactants that combine to give 1,3-diols.

OH O, $SiMe_2Ph$, C_6H_{13}; KF / DMSO, rt; OH OH, Ph, C_6H_{13}; 81%

As base.[3] KF on alumina effects the formation of trichloroacetimino ethers, and those derived from allylic alcohols readily undergo rearrangement to afford *N*-allyl trichloroacetamides. Xanthates are similarly prepared.[4]

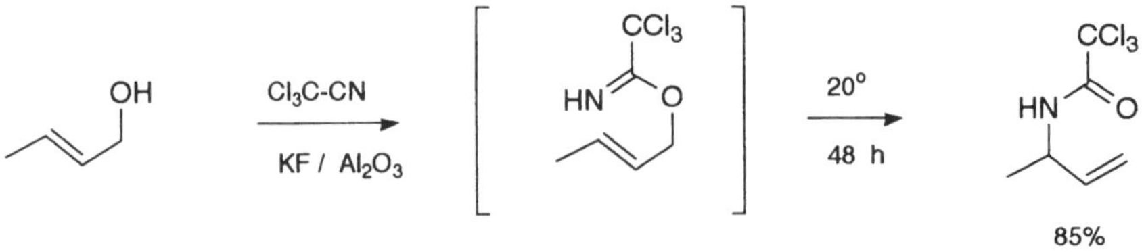

Amides from nitriles.[5] This hydration is catalyzed by KF on natural phosphate and sodium phosphate dodecahydrate. It is done in refluxing butanol (3 examples, 90–98%).

[1]Smyth, T.P., Carey, A., Hodnett, B.K. *T* **51**, 6363.
[2]Morihata, K., Horiuchi, Y., Taniguchi, M., Oshima, K., Utimoto, K. *TL* **36**, 5555 (1995).
[3]Villemin, D., Hachemi, M. *SC* **26**, 1329 (1996).
[4]Villemin, D., Hachemi, M. *SC* **26**, 2449 (1996).
[5]Sebti, S., Rhilil, A., Saber, A., Hanafi, N. *TL* **37**, 6555 (1996).

Potassium hexamethyldisilazide. 13, 257; **16**, 282–283; **18**, 298–299

Cyclization of alkyl o-(1-alkynyl)aryl ketones. The formation of a 3-alkyl-1-naphthol on treatment of the ketone (6 examples, 70–95%) with KHMD, initially at a low temperature, involves a rather unusual intramolecular addition of an enolate ion to an alkyne without any activator directly attached.

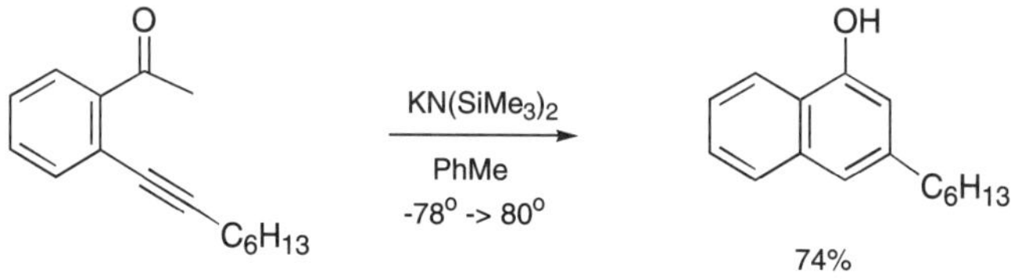

[1]Makra, F., Rohloff, J.C., Muehldorf, A.V., Link, J.O. *TL* **36**, 6815 (1995).

Potassium hydride. 13, 257–258; **14**, 265; **17**, 290; **18**, 299–300

N-Alkylations.[1] Potassium hydride and Et_3N make up a good system for *N*-alkylation of secondary amines. The presence of Et_3N apparently prevents quaternization of the products.

1,5-Hydride shift. An alkoxide-accelerated hydride shift establishes the [in,out] bridgehead configurations that are characteristic of ingenol. Thus, heating a dicarbocyclic dienol with KH/18-crown-6 in dioxane gives the desired enone in 64% yield.[2]

KH / 18-c-6

dioxane Δ

64%

"Instant ylides."[3] Equimolar mixtures of KH and finely powdered phosphonium salts are stable (at 0°) and ready to use Wittig reagent precursors. Ylides are formed when *t*-BuOMe is added to the mixtures.

[1]Mohri, K., Suzuki, K., Usui, M., Isobe, K., Tsuda, Y. *CPB* **43**, 159 (1995).
[2]Rigby, J.H., de Sainte Claire, V., Cuisiat, S.V., Heeg, M.J. *JOC* **61**, 7992 (1996).
[3]El-Khoury, M., Wang, Q., Schlosser, M. *TL* **37**, 9047 (1996).

Potassium monoperoxysulfate. 13, 259; **14**, 267; **15**, 274–275; **16**, 285; **18**, 300

Oxidation of hetero functionalities. Oxone® converts aromatic amines to nitroarenes,[1] selenides to selenones,[2] dithioacetals to carbonyl compounds,[3] and boronic acids or esters to alcohols.[4]

Epoxidation.[5] Oxone is used to epoxidize alkenes under phase-transfer conditions (9 examples, 83–96%). *N*-Methyl-*N*-dodecyl-4-oxopiperidinium triflate is a suitable catalyst.

Nascent halogens.[6] Halogens generated in situ from NaX (X = Cl, Br) by Oxone have been used to halogenate double bonds. Thus, enones are converted into α-halo enones.

[1]Webb, K.S., Seneviratne, V. *TL* **36**, 2377 (1995).
[2]Ceccherelli, P., Curini, M., Epifano, F., Marcotullio, M.C., Rosati, O. *JOC* **60**, 8412 (1995).
[3]Ceccherelli, P., Curini, M., Marcotullio, M.C., Epifano, F., Rosati, O. *SL* 767 (1996).
[4]Webb, K.S., Levy, D. *TL* **36**, 5117 (1995).
[5]Denmark, S.E., Forbes, D.C., Hays, D.S., DePue, J.S., Wilde, R.G. *JOC* **60**, 1391 (1995).
[6]Dieter, R.K., Nice, L.E., Velu, S.E. *TL* **37**, 2377 (1996).

Potassium permanganate. 13, 258–259; **14**, 267; **15**, 273–274; **18**, 301

Epoxidation of steroids. In such reactions, either $CuSO_4$1 or $Ag_2SO_4$2 is also present.

α-Keto esters.[3] Good yields of these products are obtained from oxidation of ethoxyethynyl alcohols by $KMnO_4$ oxidation.

OH

OEt

$KMnO_4$ - $MgSO_4$

$NaHCO_3$ / Me_2CO

OH

COOEt

O

98%

Silanols.[4] Hindered hydrotriorganosilanes, such as triphenylsilane, are susceptible to $KMnO_4$ oxidation in THF to give silanols (4 examples, 72–92%).

Methyl ketones.[5] $KMnO_4$ mediates the free-radical addition [also in the presence of $Mn(OAc)_2$] of acetone to alkenes in acetic acid, forming methyl ketones in moderate yields.

[1]Parish, E.J., Li, H., Li, S. *SC* **25**, 927 (1995).
[2]Hanson, J.R., Nagaratnam, S., Stevens, J. *JCR(S)* 102 (1996).
[3]Tatlock, J.H. *JOC* **60**, 6221 (1995).
[4]Lickiss, P.D., Lucas, R. *JOMC* **521**, 229 (1996).
[5]Linker, U., Kersten, B., Linker, T. *T* **51**, 9917 (1995).

Potassium permanganate–sodium periodate.

Oxidative cleavage of double bonds.[1] The oxidation in water is catalyzed by sand. Sodium perchlorate may be used instead of $NaIO_4$.

[1]Huang, B., Gupton, J.T., Hansen, K.C., Idoux, J.P. *SC* **26**, 165 (1996).

Potassium trifluoroacetate.

Aryl trifluoromethyl sulfides.[1] Heating mixtures of CF_3COOK and diaryl disulfides constitutes a simple preparation of the sulfides.

[1]Quiclet-Sire, B., Saicic, R.N., Zard, S.Z. *TL* **37**, 9057 (1996).

Pyridinium fluorochromate.

Allylic oxidation.[1] Steroidal 5-en-7-ones are produced in good yields (6 examples, 86–88%) from Δ^5-steroids on heating with pyridinium fluorochromate with molecular sieves in refluxing benzene.

[1]Parish, E.J., Sun, H., Kizito, S.A. *JCR(S)* 544 (1996).

2-Pyridyl cyanate.

Nitriles.[1] A convenient synthesis of nitriles is based on the Grignard reaction with 2-PyOCN. The latter compound is available from the reaction of 2-hydroxypyridine with cyanogen bromide (with pyridine as a base).

[1]Koo, J.S., Lee, J.I. *SC* **26**, 3709 (1996).

Q

Quinolinium bromochromate.

Bromination and oxidation.[1] This reagent has a dual capability, acting as a brominating agent of activated aromatic compounds in HOAc and as an oxidant.

[1]Ozgun, B., Degirmenbasi, N. *SC* **26**, 3601 (1996).

R

Rhenium(VII) oxide. 17, 296–297; **18**, 305

Tetrahydrofuran rings. δ,ε-Unsaturated alcohols are oxidized by Re_2O_7-2,6-lutidine to give 2-(1-hydroxyalkyl)tetrahydrofurans. This method, in conjunction with the asymmetric dihydroxylation, constitutes a versatile synthetic approach to bioactive natural products from plants of the Annonaceae family.[1]

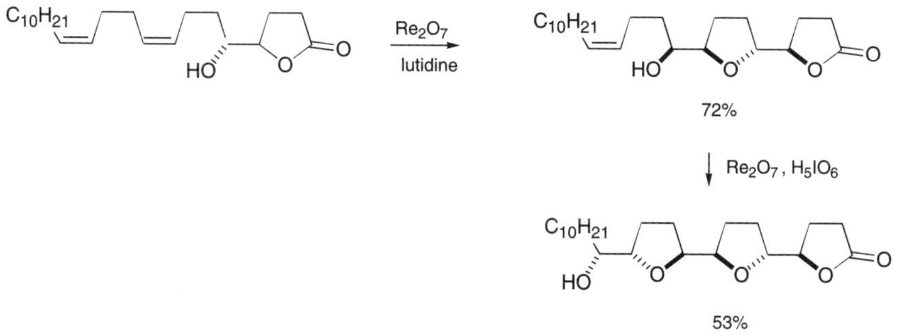

A more general and effective modification[2] of the reagent is the system $(RCOO)ReO_3/(RCO)_2O$, particularly the (dichloroacetyl)perrhenate with excess dichloroacetic anhydride, which is suitable for *syn*-oxidative cyclization of acid-sensitive substrates (in the same reaction as described above).

[1]Sinha, S.C., Sinha-Bagchi, A., Keinan, E. *JACS* **117**, 1447 (1995).
[2]McDonald, F.E., Towne, T.B. *JOC* **60**, 5750 (1995).

Rhodium carbonyl clusters. 13, 288; **15**, 334; **18**, 305–306

Reductions. Polymer-bound $Rh_6(CO)_{16}$ catalyzes the reduction of conjugated carbonyl compounds leading to allylic alcohols.[1] On the other hand, the bimetallic catalyst $Rh_6(CO)_{16}$-$Re_2(CO)_{10}$ converts amides to amines.[2]

Hydroformylation and silaformylation. High α-regioselectivity for hydroformylation of vinylpyrroles is mediated by $Rh_4(CO)_{12}$. Thus, this method is very useful for the preparation of compounds with a branched chain.[3]

The regio- and stereoselective intramolecular silaformylation of alkynes is the result of an *exo-dig* patterned pathway.[4] The intermolecular reaction is still *syn*-selective, but the formyl group is placed in the more hindered position.

Cyclocarbonylation of functionalized o-alkynylarenes. Oxindoles and tricyclic lactones are obtained as major products from the $Rh_6(CO)_{16}$-catalyzed reactions of 2-alkynylanilines[5] and 2-alkynylbenzaldehydes,[6] respectively.

[1]Kaneda, K., Mizugaki, T. *OM* **15**, 3247 (1996).
[2]Hirosawa, C., Wakasa, N., Fuchikami, T. *TL* **37**, 6749 (1996).
[3]Settambolo, R., Caiazzo, A., Lazzaroni, R. *JOMC* **506**, 337 (1996).
[4]Monteil, F., Matsuda, I., Alper, H. *JACS* **117**, 4419 (1995).
[5]Hirao, K., Mori, N., Joh, T., Takahashi, S. *TL* **36**, 6243 (1995).
[6]Sugioka, T., Zhang, S.-W., Morii, N., Joh, T., Takahashi, S. *CL* 249 (1996).

Rhodium carboxylates. 13, 266; **15**, 278–286; **16**, 289–292; **17**, 298–302; **18**, 306–307

α-Heterosubstituted esters. Rhodium carbenoids derived from α-diazo esters undergo X–H bond insertion. These processes are the basis for the synthesis of α-amino esters.[1,2] The insertion into Si–H bonds can be repeated, as shown in a preparation of 3-silaglutarate esters.[3]

$PhSi(Me)H_2$ + $N_2CHCOOEt$ $\xrightarrow[\text{CH}_2\text{Cl}_2,\ \text{rt}]{Rh_2(OAc)_4}$ $PhSi(Me)(CH_2COOEt)_2$ 64%

Formation of five-membered rings. Intramolecular insertion of Rh carbenoids into C–H bonds preferentially results in cyclopentanes. Chiral ligands on the Rh complexes have some effects on asymmetric induction,[4] but presently the enantioselectivities are moderate at best. The use of menthyl esters and chiral catalysts has not fared any better.[5]

$(PhCH_2CH(NPht)COO)_4Rh_2$; CH_2Cl_2, 0°

96% (46% ee, determined after removal of the ester group)

Note that the C–H bond to be inserted by an acylcarbenoid must be activated, e.g., being benzylic, allylic, or α to a heteroatom. A general route to cyclopentanones probably requires the decomposition of α′,β′-unsaturated α-diazocarbonyl compounds, followed by reduction of the resulting 2-cyclopentenones.[6]

$Rh_2(OAc)_4$

62%

3(2H)-Furanones are formed from alkoxyalkyl diazomethyl ketones. The insertion proceeds with retention of configuration at the carbinyl center of α′-alkoxymethyl α-diazomethyl ketones.[7] A surprising stereoselectivity is found during the formation of 2-deoxyxylolactone from a symmetrical diazoacetic ester.[8]

(93 : 7)

97% ee 50% ee

γ-Benzyloxy-α-diazoesters yield 2-phenyltetrahydrofuran-3-carboxylic esters.[9] Interestingly, although there is only marginal diastereoselectivity (*trans:cis* ca. 3:1) about the C-2/C-3 centers, the phenyl group at C-2 of the products is *cis* to the C-5 substituent (the main chain of the substrates). On the other hand, diazomethyl tetrahydrofuran-2-ylmethyl ketone gives an oxacyclooctanone via a bicyclic oxonium ylide intermediate.[10] Thus, mesocyclic ethers are quite readily available in this manner.

MeOOC OPh N_2 O Ph $Rh_2(OCOC_7H_{15})_4$ CH_2Cl_2, rt MeOOC OPh O Ph + MeOOC OPh O Ph (3 : 1) 92%

O O N_2 $Rh_2(OAc)_4$ HOAc OAc O O > 90%

(Z)-α,β-Unsaturated carbonyl compounds.[11] At low temperatures and in the presence of $Rh_2(OCOCF_3)_4$, α-diazocarbonyl compounds decompose to afford conjugated carbonyl compounds that have a *(Z)*-configuration (6 examples, 80–94%).

Bicyclo[n.1.0]alkanes. Compounds with either a small[12] or a large[13] ring fused to a cyclopropane unit are accessible by the Rh-catalyzed decomposition of unsaturated diazoesters. The double bond may be present in the carboxyl moiety or in the alcohol constituent.

O O N_2 $Rh_2(OAc)_4$ CH_2Cl_2 O O 65%

Epoxides and aziridines. The Corey–Franzen procedure is not suitable for the conversion of base-sensitive carbonyl compounds to epoxides. The transformation is now achieved by slowly adding a diazoalkane to the substrate and Me_2S in the presence of $Rh_2(OAc)_4$ at room temperature.[14] Aziridination of alkenes can use the same catalyst to decompose PhI=NNs.[15]

Cycloadditions. [4+1]-Cycloadducts are formed from 2-siloxy-1,3-dienes and the Rh-carbenoid generated from dimethyl diazomalonate.[16] Novel cyclo- pentenedicarboxylic esters are obtained from β-(diazocarbonyl)amides and dimethyl acetylenedicarboxylate.[17]

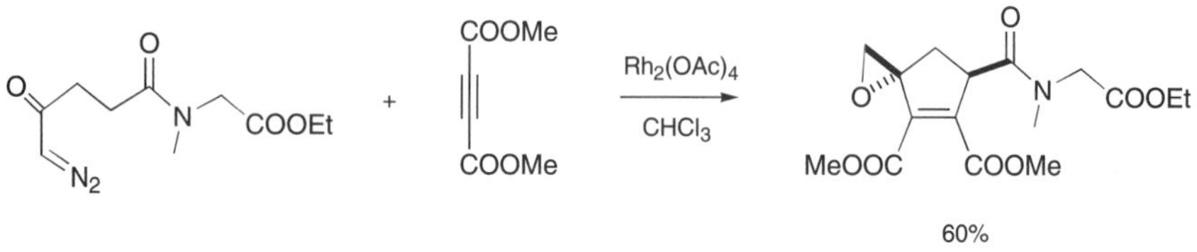

Acyl nitrile ylides are generated from the Rh-catalyzed decomposition of diazo ketones in the presence of nitriles.[18] However, the synthetic use of this reaction appears limited.

Chiral carboxamidate complexes.[19] After exchanging the ligands of dirhodium tetraacetate to chiral pyrrolidinones (as well as their heteroatom analogs) bearing a methyl ester at C-5 new carboxamidate complexes are formed. These are catalysts of choice for enantioselective intramolecular metal carbene transformations. One such complex is particularly effective for the formation of β-benzyl-γ-butyrolactones from hydrocinnamyl diazoacetates. The lactones are useful for the synthesis of some lignans.

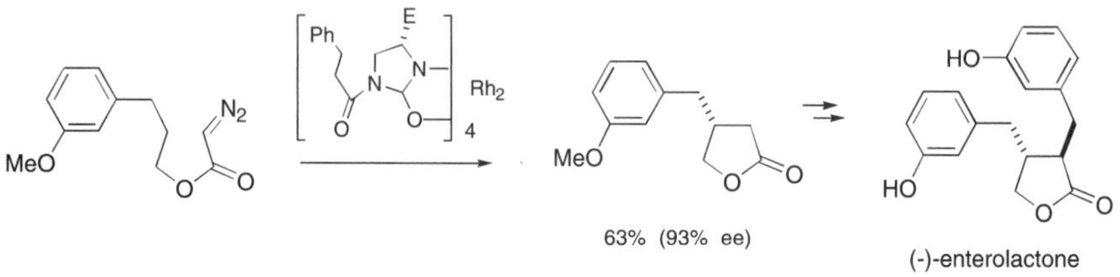

63% (93% ee)

(-)-enterolactone

[1]Aller, E., Buck, R.T., Drysdale, M.J., Ferris, L., Haigh, D., Moody, C.J., Pearson, N.D., Sanghera, J.B. *JCS(P1)* 2879 (1996).
[2]Bagley, M.C., Buck, R.T., Hind, S.L., Moody, C.J., Slawin, A.M.Z. *SL* 825 (1996).
[3]Barnier, J.-P., Blanco, L. *JOMC* **514**, 67 (1996).
[4]Hashimoto, S.-I., Watanabe, N., Sato, T., Shiro, M., Ikegami, S. *TL* **34**, 5109 (1993).
[5]Hashimoto, S.-I., Watanabe, N., Kawano, K., Ikegami, S. *SC* **24**, 3277 (1994).
[6]Mateos, A.F., Coca, G.P., Alonso, J.J.P., Gonzalez, R.R., Hernandez, C.T. *SL* 1134 (1996).
[7]Lee, E., Choi, I., Song, S.Y. *CC* 321 (1995).
[8]Doyle, M.P., Dyatkin, A.B., Tedrow, J.S. *TL* **35**, 3853 (1994).
[9]Taber, D.F., Song, Y. *JOC* **61**, 6706 (1996).
[10]Oku, A., Ohki, S., Yoshida, T., Kimura, K. *CC* 1077 (1996).
[11]Taber, D.F., Herr, R.J., Pack, S.K., Geremia, J.M. *JOC* **61**, 2908 (1996).

[12]Shi, G.-Q., Cai, W.-L. *JCS(P1)* 2337 (1996).
[13]Doyle, M.P., Protopopova, M.N., Poulter, C.D., Rogers, D.H. *JACS* **117**, 7281 (1995).
[14]Aggarwal, V.K., Abdel-Rahman, H., Jones, R.V.H., Standen, M.C.H. *TL* **36**, 1731 (1995).
[15]Müller, P., Baud, C., Jacquier, Y. *T* **52**, 1543 (1996).
[16]Schnaubelt, J., Marks, E., Reissig, H.-U. *CB* **129**, 73 (1996).
[17]Padwa, A., Price, A.T., Zhi, L. *JOC* **61**, 2283 (1996).
[18]Fukushima, K., Ibata, T. *BCSJ* **68**, 3469 (1995).
[19]Doyle, M.P., Protopopova, M.N., Zhou, Q.-L., Bode, J.W., Simonsen, S.H., Lynch, V. *JOC* **60**, 6654 (1995).

Ruthenium carbene complexes. 18, 308

Metathetic ring closure. The powerful technique of ring-closing metathesis (RCM) using ruthenium carbene complexes such as **1** and **2** is shown in the preparation of a highly strained tricyclic compound[1] that cannot be realized using $MeReO_3$.

PCy3, Cl, Cl, Ru, Ph, Ph, PCy3

(1)

PCy3, Cl, Cl, Ru, Ph, PCy3

(2)

(1)

PhH , rt

99%

Ring strain apparently determines the competitive pathway of RCM or metathetic polymerization to be adopted in the case of 1,2-diallyloxycyclohexane.[2] Ring formation from the *trans*-isomer is favored (60% yield), while the *cis*-isomer gives only 20% yield of the more strained bicyclic product.

The bulky trityloxy group of *N*-(1-trityloxymethyl)alkenyl acrylamides favors ring closure to give conjugated lactams.[3]

(1)

OTr

n = 0, 1

76 - 95%

From succinimide and glutarimide, it is quite simple to create indolizidine and quinolizidine units, respectively, by the RCM method.[4] Accordingly, many alkaloids can be synthesized, e.g., (+)-castanospermine.[5]

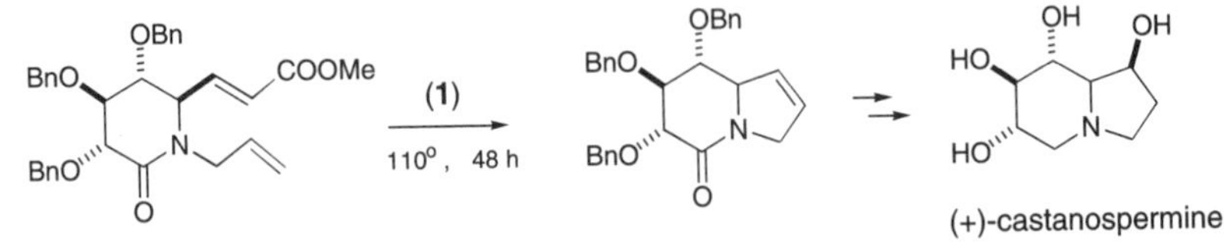

OBn BnO BnO COOMe N O (1) 110°, 48 h OBn BnO BnO N O OH OH HO HO N

(+)-castanospermine

One of the most important attributes of RCM is its suitability for forming macrocycles. RCM constitutes the key step of the synthesis of lasiodiplodin,[6] epothilone-A,[7] and the macrolactam segment of manzamine-A.[8] Except in the last case, the RCM reaction is usually very efficient.

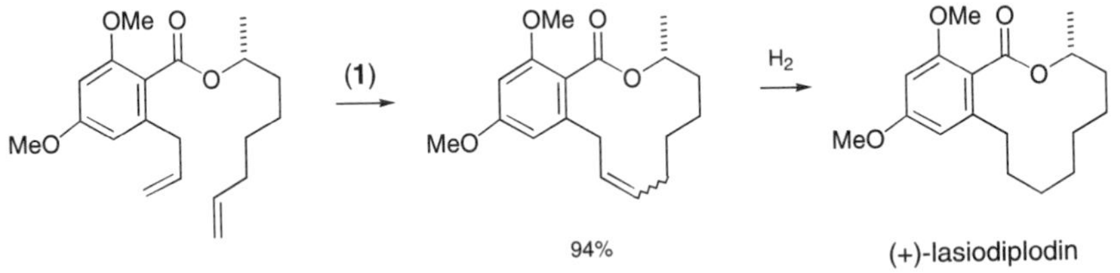

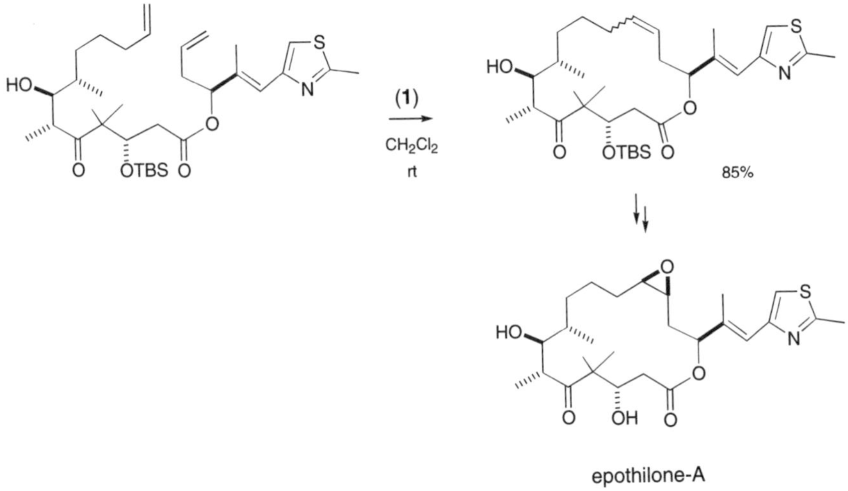

OMe O MeO O (1) OMe O MeO O 94% H_2 OMe O MeO O

(+)-lasiodiplodin

HO O OTBS O O S N (1) CH_2Cl_2 rt HO O OTBS O O S N 85%

HO O OH O O O S N

epothilone-A

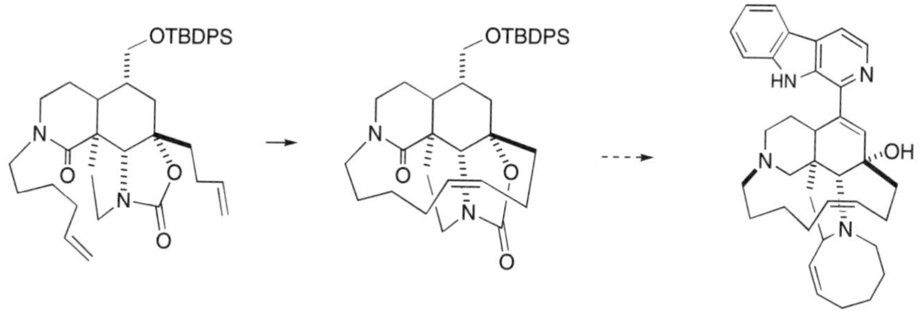

OTBDPS OTBDPS

manzamine-A

We still do not know all the reaction parameters of RCM. For example, the site of ring closure leading to 14-membered lactone precursors of 12-methyl-13-tridecanolide seems to be the determining factor for the effectiveness.[9]

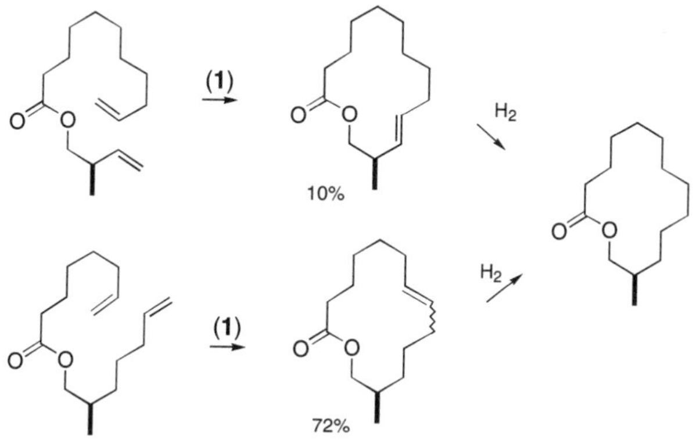

(1) 10% H_2

(1) 72% H_2

A new route to crown ethers is by ring-closing metathesis.[10] Good yields are obtained with a 1% catalyst at a relatively high concentration of 0.1 M.

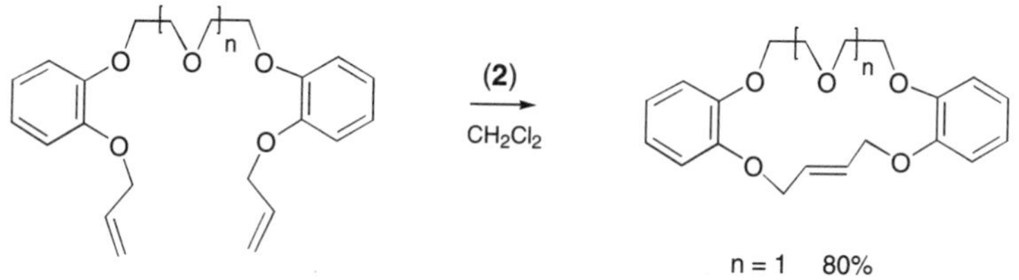

(2) CH_2Cl_2

n = 1 80%
n = 2 66%
n = 4 72%

Bridged calix[4]arenes in which two phenolic hydroxyl groups are derivatized as 4-pentenyl ethers undergo RCM in the presence of a carbene complex **2**.[11] Another impressive result concerns the construction of a cylindrical tricyclic peptide.[12] This compound contains a 38-membered ring and is formed in 65% yield on exposure to **1**. The two-step reaction involves RCM in the final stage.

An RCM of an enyne giving a diene that is eminently suitable for elaboration into (–)-stemoamide has been reported.[13]

MeOOC (2) MeOOC 87% (-)-stemoamide

A remarkable observation is that [Ru] and [Mo] catalysts can show divergent diastereoselectivity.[14]

CF_3CO–N OTBS 88% [Ru] CF_3CO–N OTBS [Mo] CF_3CO–N OTBS 97%

Tandem ring opening transforms a cycloalkene with unsaturated side chains into molecules containing multiple cycloalkenes linked by atoms inside the original cyclic system.[15]

(2) PhH, 60° 70%

Ring-opening cross-metathesis. The metathesis between a cyclobutene and a terminal alkene leads only to a cross product. This observation is helpful for the design of short syntheses of multifidene and viridiene.[16]

(1) ← (1) →

multifidene viridiene

Ring-opening metathetic polymerization. This process is being actively pursued in the synthesis of novel polymers. To illustrate the point, note that in an aqueous-organic two-phase system, unprotected sugar-substituted norbornenes are polymerized virtually quantitatively.[17]

(1)
H_2O - CH_2Cl_2
(6 : 1)

99%

[1]Schneider, M.F., Junga, H., Blechert, S. *T* **51**, 13003 (1995).
[2]Miller, S.J., Kim, S.-H., Chen, Z.-R., Grubbs, R.H. *JACS* **117**, 2108 (1995).
[3]Huwe, C.M., Kiehl, O.C., Blechert, S. *SL* 65 (1996).
[4]Huwe, C.M., Blechert, S. *TL* **36**, 1621 (1995).
[5]Huwe, C.M., Blechert, S. *S* 61 (1997).
[6]Fürstner, A., Kindler, N. *TL* **37**, 7005 (1996).
[7]Yang, Z., He, Y., Vourloumis, D., Vallberg, H., Nicolaou, K.C. *ACIEE* **36**, 166 (1997).
[8]Borer, B.C., Deerenberg, S., Bieräugel, H., Pandit, U.K. *TL* **35**, 3191 (1994).
[9]Fürstner, A., Langemann, K. *JOC* **61**, 3942 (1996).
[10]König, B., Horn, C. *SL* 1013 (1996).
[11]McKervey, M.A., Pitarch, M. *CC* 1689 (1996).
[12]Clark, T.D., Ghadiri, M.R. *JACS* **117**, 12364 (1995).
[13]Kinoshita, A., Mori, M. *JOC* **61**, 8356 (1996).
[14]Huwe, C.M., Velder, J., Blechert, S. *ACIEE* **35**, 2376 (1996).
[15]Zuercher, W.J., Hashimoto, M., Grubbs, R.H. *JACS* **118**, 6634 (1996).
[16]Randall, M.L., Tallarico, J.A., Snapper, M.L. *JACS* **117**, 9610 (1995).
[17]Fraser, C., Grubbs, R.H. *Macromolecules* **28**, 7248 (1995).

Ruthenium(III) chloride. 13, 268; **14**, 271–272

Hydrogenation.[1] Substituted aromatic compounds are readily reduced completely under relatively mild conditions in the presence of $RuCl_3$ and Oc_3N.

[1]Fache, F., Lehuede, S., Lemaire, M. *TL* **36**, 885 (1995).

Ruthenium(III) chloride–peracetic acid. 18, 310

Oxidation of alcohols.[1] Oxidation of alcohols in EtOAc by peracetic acid at room temperature is catalyzed by $RuCl_3$ and various other Ru compounds. This oxidation can be used to prepare α-keto esters without incurring C–C bond cleavage. Primary alcohols are oxidized to carboxylic acids. The use of *t*-BuOOH is much less satisfactory in terms of conversion, and H_2O_2 is quite ineffective. Other transition metal salts, such as $CuCl_2$, $Cu(OAc)_2$, $PdCl_2$, and $NiCl_2$, are not suitable.

[1]Murahashi, S.-I., Naota, T., Oda, Y., Hirai, N. *SL* 733 (1995).

Ruthenium(III) chloride–sodium periodate. 18, 310

Cleavage of double bonds. The reagent combination forms RuO_4 in situ, which is capable of cleaving double bonds. A convenient synthesis of 1-chlorocyclopropanecarboxylic acids is developed.[1]

[1]Coudret, J.L., Ernst, K., de Meijere, A., Waegell, B. *S* 920 (1994).

S

Samarium. 14, 275; **17**, 305–307; **18**, 311

Reduction.[1] Heteroaromatic rings of pyridine, quinoline, and isoquinoline derivatives are reduced very rapidly by Sm and hydrochloric acid at room temperature. Thus, 2-phenylpyridine gives 2-phenylpiperidine in 96% yield.

The combination of Sm with titanocene chloride in *t*-butanol constitutes a very effective reducing system for organic azides.[2]

Dehalogenation.[3] *vic*-Dihalides are converted into alkenes with Sm in methanol at room temperature (15 examples, 92–100%).

Homoallylic amines.[4] Samarium powder forms allylsamarium halides, which add to imines under mild conditions (7 examples, 52–80%).

Cyclopropanation.[5] The Sm–Me_3SiCl combination promotes Simmons–Smith reactions on allylic alcohols, including allenic alcohols. Thus, 2-methylenecyclopropyl carbinols are readily obtained. The diastereoselectivity could be affected by using different promoters.

Sm - CH_2I_2, THF, -78° -> rt

+ TBS-Cl	60%	44 : 1
+ $(i\text{-PrO})_4Ti$	59%	>50 : 1

Alkylidenation.[6] With the use of Sm mixed with SmI_2 and a catalytic amount of $CrCl_3$, *gem*-dibromoalkanes and ketones form olefins at room temperature (14 examples, 38–71%).

Silylation of 1-alkynes.[7] The in situ generation of active zinc from $ZnCl_2$ to effect the formation of trialkylsilylalkynes from R_3SiCl and 1-alkynes is achieved using samarium in MeCN (sealed tube, 100°). Samarium(II) iodide is not useful.

[1]Kamochi, Y., Kudo, T. *CPB* **43**, 1422 (1995).
[2]Huang, Y., Zhang, Y., Wang, Y. *SC* **26**, 2911 (1996).
[3]Yamada, R., Negoro, N., Yanada, K., Fujita, T. *TL* **37**, 9313 (1996).
[4]Wang, J., Zhang, Y., Bao, W. *SC* **26**, 2473 (1996).
[5]Lautens, M., Ren, Y. *JOC* **61**, 2210 (1996).
[6]Matsubara, S., Horiuchi, M., Takai, K., Utimoto, K. *CL* 259 (1995).
[7]Sugita, H., Hatanaka, Y., Hiyama, T. *SL* 637 (1996).

Samarium–iodine.

Deprotection.[1] Samarium and iodine in methanol is effective for the deacylation and removal of ester groups.

Carbonyl ylides.[2] The parent carbonyl ylide is generated from bischloromethyl ether and trapped by dipolarophiles to give tetrahydrofurans.

[1]Yanada, R., Negoro, N., Bessho, K., Yanada, K. *SL* 1261 (1995).
[2]Hojo, M., Aihara, H., Ito, H., Hosomi, A. *TL* **37**, 9241 (1996).

Samarium–titanium(IV) chloride.

Desulfonylation.[1] The reagent combination in THF at room temperature removes the sulfonyl group from β-keto sulfones.

Reduction of sulfonyl chlorides and sulfoxides. Arenesulfonyl chlorides (and sodium arenesulfonates) are reduced to diaryl disulfides,[2] whereas sulfoxides undergo deoxygenation.[3] Titanocene dichloride can replace $TiCl_4$ in the latter reaction.[4]

[1]Wang, J., Zhang, Y. *SC* **26**, 1931 (1996).
[2]Wang, J., Zhang, Y. *SC* **26**, 135 (1996).
[3]Wang, J., Zhang, Y. *SC* **25**, 3545 (1995).
[4]Zhang, Y., Yu, Y., Bao, W. *SC* **25**, 1825 (1995).

Samarium(III) chloride. 14, 275–276; **15**, 282; **18**, 312

Aldol reaction.[1] $SmCl_3$ with Et_3GeNa in THF–HMPA has been used to promote the enolization of ketones and subsequent condensation with aldehydes (10 examples, 93–99%). Kinetic enolates are formed, and in the cases of acyclic ketones and amides, the *(Z)*-enolates that are formed react with aldehydes to provide *syn* adducts selectively.

[1]Yokoyama, Y., Mochida, K. *SL* 445 (1996).

Samarium(II) iodide. 13, 270–272; **14**, 276–281; **15**, 282–284; **16**, 294–300; **17**, 307–311; **18**, 312–316

Barbier reactions.[1] In nitrile solvents, the condensation of alkyl halides with carbonyl compounds gives alcohols. The reaction between esters and iodoalkanes proceeds only in the presence of NiI_2. With an iron(III) salt as a catalyst, cyclic *N*-iodoalkyl imides form polycyclic enamides via carbinol lactams.[2]

SmI_2, $Fe(dbm)_3$ / THF / 0°, 3 h → 95%

Condensation of α-haloketones. α-Haloketones react with acid chlorides and anhydrides to afford β-diketones when the reaction is promoted by SmI_2 (14 examples, 71–90%).[3] Similarly, aldol condensation occurs with aliphatic α-diketones,[4] while phenacyl bromide condenses with aldehydes if Et_2AlCl is present.[5]

Reformatsky reactions. A practical synthesis of β-hydroxy-α,α-difluoroalkanoic esters[6] involves halodifluoroacetic esters and induction by SmI_2. α-Bromoalkanoic esters undergo self-condensation,[7] and if they are quenched by carbonyl compounds, δ-hydroxy-β-keto esters are formed as products.[8]

Br COOEt
SmI_2 / THF, -50°; cyclohexanone
OH O COOEt
98%

C-glycosides. Glycosyl sulfones[9,10] and phosphates[11] are suitable substrates for condensation with carbonyl compounds to afford *C*-glycosides. Glycosylsamarium(III) species are involved.

2,3-Pentadiene-1,5-diols.[12] Alkynylepoxides undergo reductive ring opening to generate allenylsamarium iodides, which, on reaction with carbonyl compounds, deliver the unusual products.

O
SmI_2 / THF, -5°; cyclohexanone
=C= OH
OH
83%

Reduction of N–X bonds. Both aliphatic and aromatic organoazides[13–15] are readily reduced by SmI_2 at room temperature. Removal of the oxygen functionalities from derivatives of hydroxylamine and hydroxamic acids is also achieved.[16,17]

Pyridine-2-sulfonamides are cleaved by SmI_2 to furnish amines.[18] Therefore, such amine derivatives are synthetically useful.

Cleavage of three-membered heterocycles. 2-Acylaziridines afford β-amino ketone or ester derivatives regioselectively.[19] The ring opening of epoxides by nucleophiles is catalyzed by SmI_2.[20]

Pyrrolidines.[21] Homolysis of benzotriazol-1-ylmethylamines leads to α-amino radicals. This SmI_2-mediated reaction is followed by cyclization when the amino nitrogen atom is located at the δ-carbon of an α,β-unsaturated ester.

BnNH⌒⌒⌒COOMe —[HCHO - benzotriazole ; SmI_2 / THF, -10° -> rt]→ BnN-pyrrolidine-CH₂COOMe

70%

Cleavage of S–C bonds. Benzyloxymethyl 2-pyridyl sulfone can be used to benzyloxymethylate carbonyl compounds to afford RR′C(OH)CH_2OBn (8 examples, 75–91%).[22] Aryl thiocyanates suffer loss of the CN group. The thiolate anions can be alkylated in situ[23] or coupled with aryl iodides in a Pd-catalyzed process for the preparation of unsymmetrical diaryl sulfides.[24]

Reductive couplings. α-Keto amides furnish tartaric amides,[25] and the dimethylacetals of aromatic aldehydes under anhydrous conditions give α,α′-dimethoxybibenzyls.[26] In the presence of water, demethoxylation occurs. When the acetals contain electron-withdrawing groups (CN, COOR) in the aromatic nuclei, the reaction products are ArCH_2OMe; acetals of aliphatic aldehydes behave similarly.

Acid chlorides give α-ketols as coupling products extremely rapidly on treatment with SmI_2 in the presence of *t*-BuNC and NiI_2.[27]

2-Arylcyclopropane-1,1-dicarboxylic esters undergo ring-opening coupling to give 3,4-diaryl-1,1,6,6-hexanetetracarboxylic esters[28] in moderate to good yields on heating with SmI_2 in THF.

γ-Hydroxy amides. Acrylamides undergo reduction by SmI_2, and in the presence of carbonyl compounds condensation follows. This method can be used for preparing δ-amino alcohols when it is applied to *N,N*-dibenzylacrylamide.[29]

Bn_2NC(O)CH=CH₂ —[SmI_2 / THF; cyclohexanone; Et_2AlCl / *t*-BuOH]→ Bn_2NC(O)CH₂CH₂-C(OH)(cyclohexyl) (75%) ⇀ H_2NCH₂CH₂CH₂-C(HO)(cyclohexyl)

75%

Ene reactions.[30] SmI_2 is a remarkable ene reaction catalyst that promotes the cyclization of some unsaturated aldehydes and of alkylidenemalonates at room temperature.

1-Aza-1,3-dienes.[31] Imines react in the presence of aldehydes and SmI_2. The reaction is essentially an aldol condensation in which the imines behave as the enamine tautomers.

imine + CHO —[SmI_2 / THF]→ 1-aza-1,3-diene

87%

Expansion of cyclobutanones. With CH_2I_2 and SmI_2, cyclobutanones undergo expansion to furnish cyclopentanones.[32]

Tetrahydrofurans.[33] α-Iodoalkyl trialkylsilyl ethers undergo a very unusual reaction with SmI_2 in an ethylene atmosphere.

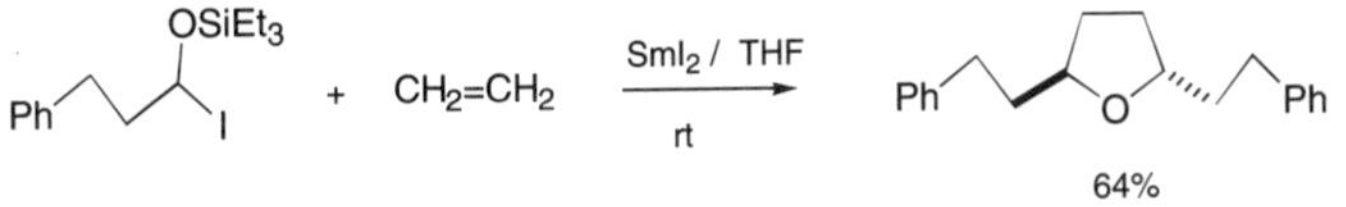

Reductive alkylation. In a synthesis of *meso*-chimonanthine and *meso*-calycanthine, two acetaldehyde chains in the masked form are introduced as a *(Z)*-2-buten-1,4-diyl unit by reductive alkylation of *N,N′*-dibenzylisoindigo with *(Z)*-1,4-dichloro-2-butene.[34] The reaction is mediated by SmI_2–LiCl.

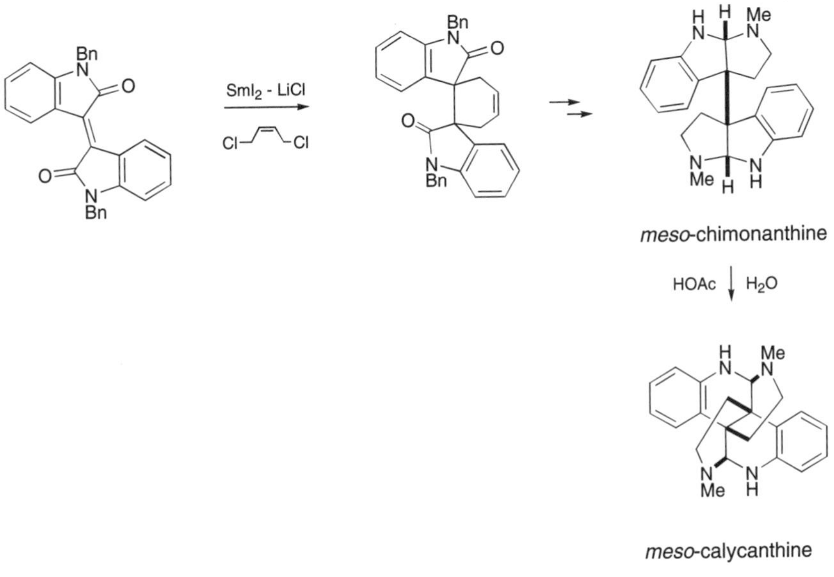

Reductive coupling of β-hydroxy aldehydes and ketones with acrylonitrile leads to *anti*-4,6-dihydroxyalkanenitriles only.[35] Chelation control in the formation of ketyl radicals is responsible for the results.

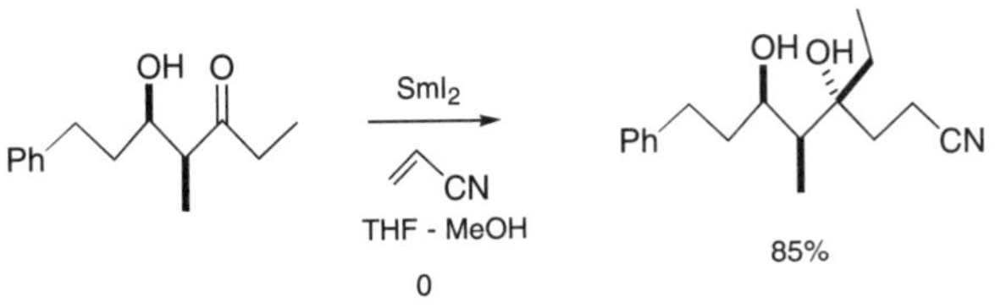

Cyclic 1,2-amino alcohols.[36] Intramolecular reductive coupling of carbonyl-tethered oxime ethers with SmI_2 forms the cyclic products even in the absence of HMPA. Only one diastereomer is produced. After water has been added (20–25 equiv), the N–O bond is cleaved in a subsequent reduction with the excess reagent at room temperature. A previous report indicates the failure of the intermolecular version.

[1]Hamann, B., Namy, J.-L., Kagan, H.B. *T* **52**, 14225 (1996); Machrouhi, F., Hamann, B., Namy, J.-L., Kagan, H.B. *SL* 633 (1996).
[2]Ha, D.-C., Yun, C.-S., Yu, E. *TL* **37**, 2577 (1996).
[3]Ying, T., Bao, W., Zhang, Y., Xu, W. *TL* **37**, 3885 (1996).
[4]Arime, T., Takahashi, H., Kobayashi, S., Yamaguchi, S., Mori, N. *SC* **25**, 389 (1995).
[5]Aoyagi, Y., Yoshimura, M., Tsuda, M., Tsuchibuchi, T., Kawamata, S., Tateno, H., Asano, K., Nakmura, H., Obokata, M., Ohta, A., Kodama, Y. *JCS(P1)* 689 (1995).
[6]Yoshida, M., Suzuki, D., Iyoda, M. *SC* **26**, 2523 (1996).
[7]Park, H.S., Lee, I.S., Kim, Y.H. *TL* **36**, 1673 (1995).
[8]Utimoto, K., Matsui, T., Takai, T., Matsubara, S. *CL* 197 (1995).
[9]Mazeas, D., Skrydstrup, T., Beau, J.-M. *ACIEE* **34**, 909 (1995).
[10]Urban, D., Skrydstrup, T., Riche, C., Chiaroni, A., Beau, J.-M. *CC* 1883 (1996).
[11]Hung, S.-C., Wong, C.-H. *ACIEE* **35**, 2671 (1996).
[12]Aurrecoechea, J.M., Solay, M. *TL* **36**, 2501 (1995).
[13]Goulaouic-Dubois, C., Hesse, M. *TL* **36**, 7427 (1995).
[14]Benati, L., Montevecchi, P.C., Nanni, D., Spagnolo, P., Volta, M. *TL* **36**, 7313 (1995).
[15]Huang, Y., Zhang, Y., Wang, Y. *SC* **26**, 2911 (1996).
[16]Keck, G.E., McHardy, S.F., Wager, T.T. *TL* **36**, 7419 (1995).
[17]Chiara, J.L., Destabel, C., Gallego, P., Marco-Contelles, J. *JOC* **61**, 359 (1996).
[18]Goulaouic-Dubois, C., Guggisberg, A., Hesse, M. *JOC* **60**, 5969 (1995).
[19]Molander, G.A., Stengel, P.J. *JOC* **60**, 6660 (1995).
[20]Van de Weghe, P., Collin, J. *TL* **36**, 1649 (1995).
[21]Aurrecoechea, J.M., Fernandez-Acebes, A. *SL* 39 (1996).
[22]Skrydstrup, T., Jespersen, T., Beau, J.-M., Bols, M. *CC* 515 (1996).
[23]Toste, F.D., LaRonde, F., Still, I.W.J. *TL* **36**, 2949 (1995).
[24]Still, I.W.J., Toste, F.D. *JOC* **61**, 7677 (1996).
[25]Yamashita, M., Okuyama, K., Kawasaki, I., Ohta, S. *TL* **37**, 7755 (1996).
[26]Studer, A., Curran, D.P. *SL* 255 (1996).
[27]Hamann, B., Namy, J.L., Kagan, H.B. *T* **52**, 14225 (1996).
[28]Yamashita, M., Okuyama, K., Ohhara, T., Kawasaki, I., Ohta, S. *SL* 547 (1996).
[29]Aoyagi, Y., Maeda, M., Moro, A., Kubota, K., Fujii, Y., Fukaya, H., Ohta, A. *CPB* **44**, 1812 (1996).
[30]Sarkar, T.K., Nandy, S.K. *TL* **37**, 5195 (1996).
[31]Shiraishi, H., Kawasaki, Y., Sakaguchi, S., Nishiyama, Y., Ishii, Y. *TL* **37**, 7291 (1996).
[32]Fukuzawa, S., Tsuchimoto, T. *TL* **36**, 5937 (1995).
[33]Hojo, M., Aihara, H., Hosomi, A. *JACS* **118**, 3533 (1996).
[34]Link, J.T., Overman, L.E. *JACS* **118**, 8166 (1996).
[35]Kawatsura, M., Hosaka, K., Matsuda, F., Shirahama, H. *SL* 729 (1995).
[36]Chiara, J.L., Marco-Contelles, J., Khiar, N., Gallego, P., Destabel, C., Bernabe, M. *JOC* **60**, 6010 (1995).

Samarium(II) iodide–hexamethylphosphoric triamide.

Barbier reaction.[1] The reaction between aryl halides and ketones using SmI_2–HMPA is reported.[1] Iodoalkynes afford propargylic alcohols in an analogous reaction.[2]

A tandem process transforming lactones bearing haloalkyl chains into substituted diquinanes[3] is both remarkable and efficient.

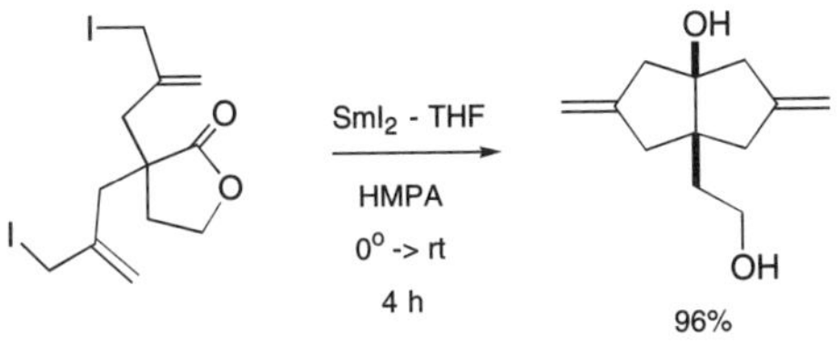

SmI2 - THF
HMPA
0° -> rt
4 h

96%

The product from an intramolecular reaction of a δ-bromo ester can form a second cyclopentane unit when an unsaturated carbon is provided to couple with an emerging ketyl group.[4] (Compare the group transfer reaction of 6-bromoalkynes to furnish bromoalkylidenecyclopentanes.)[5]

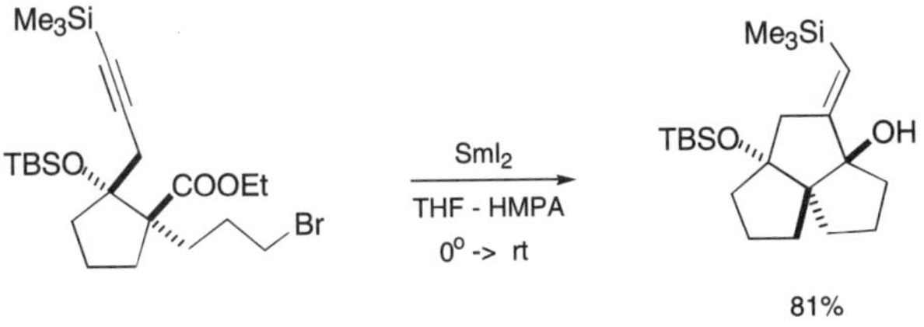

SmI2
THF - HMPA
0° -> rt

81%

Reductive addition of carbonyl to double bond. Some carbohydrate derivatives have been transformed into cyclopentanols[6] via fragmentation with SmI_2, which provides the susceptible 5-pentenals.

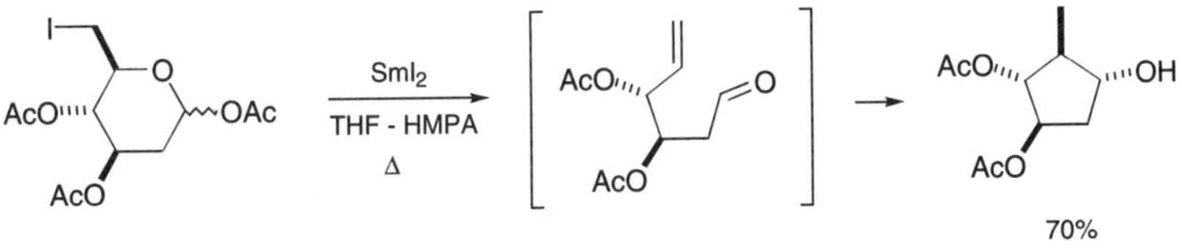

SmI2
THF - HMPA
Δ

70%

When the double bond is conjugated to an epoxide, an S_N2'-like reaction occurs.[7] The intermolecular version is also valuable for the synthesis of various 2-alkene-1,5-diols.[8]

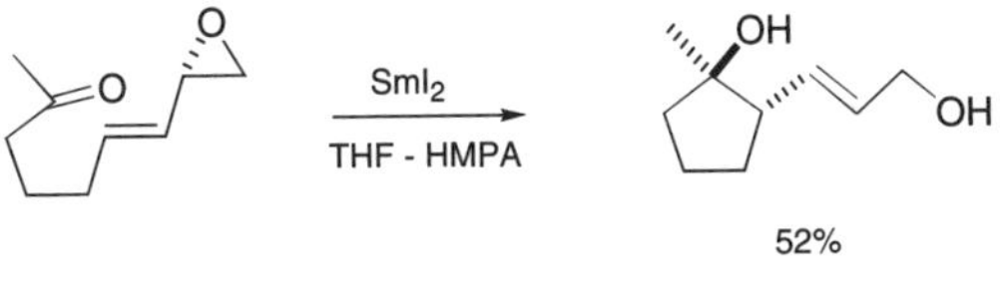

SmI2
THF - HMPA

52%

Reduction of α-oxy carbonyl compounds. The reductive cleavage of the C–O bond[9] may lead to more extensive structural changes, such as those that occur in the case of a carbohydrate aldehyde. The dialdehyde is labile, forming a cyclopentane derivative as a ring-contraction product.[10]

Coupling of carbonyl compounds. Cross-coupling of thiophenecarbaldehydes with other aromatic aldehydes[11] follows a course other than pinacol formation.

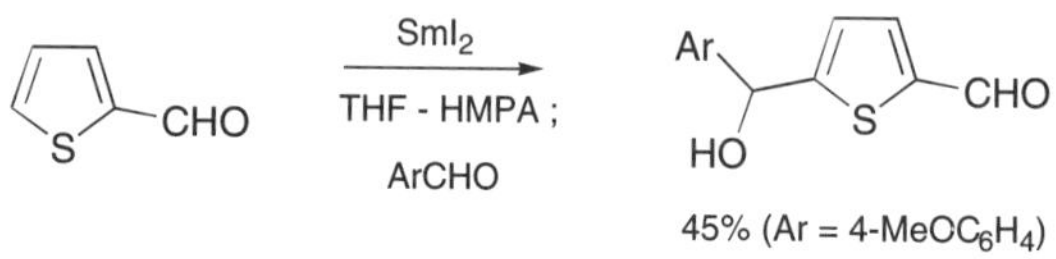

SmI_2; THF - HMPA ; ArCHO

45% (Ar = 4-$MeOC_6H_4$)

Efficient pinacol coupling can be effected on tricarbonylchromium-ligated benzaldehydes to give mainly the *threo*-diols.[12] For more common substrates, the coupling can be performed at room temperature with 10 mol% of SmI_2 if Mg and Me_3SiCl are present.[13]

Transpositional alkylations.[14,15] *N*-(*o*-iodobenzyl) heterocycles generate radicals that are translocated through hydrogen abstraction. The resulting α-amino radicals then react with carbonyl compounds. This process has previously been developed with Bu_3SnH as initiator.

Debromination and decyanation. A selective synthesis of 6α-bromopenicillinates is accomplished by partial reduction of the 6,6-dibromo compounds (7 examples, 88–100%).[16] α-Bromoacetic esters undergo reductive coupling to give succinic esters.[17]

The cyano groups of malononitriles and cyanoacetic esters are easily removed[18] using SmI_2–HMPA in THF at 0°.

Cleavage of cyclopropanecarboxylic esters.[19] The cyclopropane ring of both the mono esters and the *gem*-diesters are cleaved to provide the butyrates and ethylmalonates, respectively.

Desulfonylation. It is possible to complete the Julia–Lythgoe olefination by using the SmI_2–HMPA/THF system to eliminate the β-hydroxy sulfone unit.[20,21] An alternative method consists of dehydration (via the acetates) and reduction of the alkenyl sulfones. (Previously, this step was performed by using Na–Hg).[22]

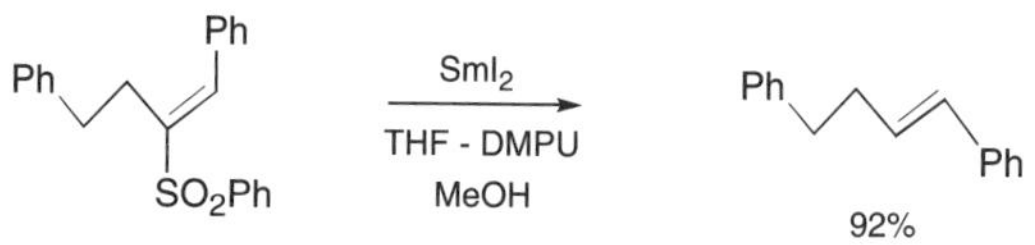

SmI_2; THF - DMPU; MeOH

92%

Reductive cleavage of diaryl dichalcogenides. Generation of the nucleophilic aryltelluride[23] and arylselenide ions[24] is accomplished at room temperature.

N-Alkoxy-2,2,6,6-tetramethylpiperidines.[25] The *N*-piperidinoxyl reacts with alkylsamarium reagents formed from alkyl halides and SmI_2–HMPA.

[1]Kunishima, M., Hioki, K., Kono, K., Sakuma, T., Tani, S. *CPB* **42**, 2190 (1994).
[2]Kunishima, M., Tanaka, S., Kono, K., Hioki, K., Tani, S. *TL* **36**, 3707 (1995).
[3]Molander, G.A., Harris, C.R. *JACS* **117**, 3705 (1995).
[4]Molander, G.A., Harris, C.R. *JACS* **118**, 4059 (1996).
[5]Zhou, Z., Larouche, D., Bennett, S.M. *T* **51**, 11623 (1995).
[6]Grove, J.J.C., Holzapfel, C.W., Williams, D.B.G. *TL* **37**, 5817 (1996).
[7]Molander, G.A., Shakya, S.R. *JOC* **61**, 5885 (1996).
[8]Aurrecoechea, J.M., Iztueta, E. *TL* **36**, 7129 (1995).
[9]Enholm, E.J., Schreier, J.A. *JOC* **60**, 1110 (1995).
[10]Chenede, A., Pothier, P., Sollogoub, M., Fairbanks, A.J., Sinay, P. *CC* 1373 (1995).
[11]Yang, S.-M., Fang, J.-M. *JCS(P1)* 2669 (1995).
[12]Taniguchi, N., Kaneta, N., Uemura, M. *JOC* **61**, 6088 (1996).
[13]Nomura, R., Matsuno, T., Endo, T. *JACS* **118**, 11666 (1996).
[14]Booth, S.E., Benneche, T., Undheim, K. *T* **51**, 3665 (1995).
[15]Murakami, M., Hayashi, M., Ito, Y. *AOMC* **9**, 385 (1995).
[16]Kang, H.-Y., Pae, A.N., Cho, Y.S., Choi, K.I., Koh, H.Y., Chung, B.Y. *H* **43**, 2337 (1996).
[17]Balaux, E., Ruel, R. *TL* **37**, 801 (1996).
[18]Kang, H.-Y., Hong, W.S., Cho, Y.S., Koh, H.Y. *TL* **36**, 7661 (1995).
[19]Yamashita, M., Okuyama, K., Ohhara, T., Kawasaki, I., Sakai, K., Nakata, S., Kawabe, T., Kusumoto, M., Ohta, S. *CPB* **43**, 2075 (1995).
[20]Ihara, M., Suzuki, S., Taniguchi, T., Tokunaga, Y., Fukumoto, K. *T* **51**, 9873 (1995).
[21]Marko, I.E., Murphy, F., Dolan, S. *TL* **37**, 2089 (1996).
[22]Keck, G.E., Savin, K.A., Weglarz, M.A. *JOC* **60**, 3194 (1995).
[23]Bao, W., Zhang, Y. *SC* **25**, 1913 (1995).
[24]Wang, H.-P., Zhang, Y.-M., Ruan, M.-D., Shi, S.-M. *YH* **16**, 38 (1996).
[25]Nagashima, T., Curran, D.P. *SL* 330 (1996).

Samarium(III) iodide. 18, 317

Enone synthesis. A convenient route to α,α'-dibenzylidenecycloalkanones is by an SmI_3-catalyzed condensation of siloxycycloalkenes with aromatic aldehydes.[1] β-Keto esters and β-diketones afford the α-benzylidene compounds.[2]

β-Diketones.[3] Acetylacetone can be converted into other β-diketones, $MeCOCH_2COR$ (12 examples, 73–83%), by reaction with RCOCl with SmI_3 as promoter.

α-Selenoketones.[4] These compounds are prepared from α-bromoketones and benzeneselenenyl bromide in MeCN using SmI_3 or SmI_2.

[1]Bao, W., Zhang, Y., Ying, T. *SC* **26**, 503 (1996).
[2]Bao, W., Zhang, Y., Wang, J. *SC* **26**, 3025 (1996).
[3]Hao, W., Zhang, Y., Ying, T., Lu, P. *SC* **26**, 2421 (1996).
[4]Ying, T., Bao, W., Zhang, Y. *SC* **26**, 1517 (1996).

Samarium(II) iodide–tetrakis(triphenylphosphine)palladium.

Deoxygenation of propargylic esters. Propargyl esters (phosphates[1] and acetates[2]) undergo reduction in isopropanol or *t*-butanol. Depending on the substrate used, either allenes or alkynes are the major products.

o-Bis(1-acetoxy-2-propynyl)benzene gives 2,3-naphthoquinodimethane, which can be trapped by dienophiles.[3]

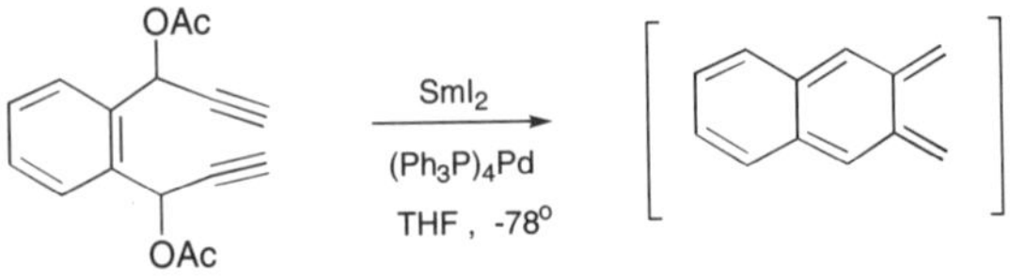

[1]Mikami, K., Yoshida, A., Matsumoto, S., Feng, F., Matsumoto, Y., Sugino, A., Hanamoto, T., Inanaga, J. *TL* **36**, 907 (1995).
[2]Marco-Contelles, J., Destabel, C., Chiara, J.L. *TA* **7**, 105 (1996).
[3]Sugimoto, Y., Hanamoto, T., Inanaga, J. *AOMC* **9**, 369 (1995).

Samarium(II) triflate. 18, 317

Alkylations.[1] The low-valent Sm triflate, prepared by treatment of $Sm(OTf)_3$ with *s*-BuLi, reacts with primary, secondary, and allylic alkyl halides to form $RSm(OTf)_2$. These species behave similarly to Grignard reagents.

[1]Fukuzawa, S., Mutoh, K., Tsuchimoto, T., Hiyama, T. *JOC* **61**, 5400 (1996).

Scandium(III) triflate. 18, 317–318

Acetalization.[1] As a Lewis acid, not only does $Sc(OTf)_3$ promote acetalization, but chiral acetals undergo stereoselective cleavage in its presence.

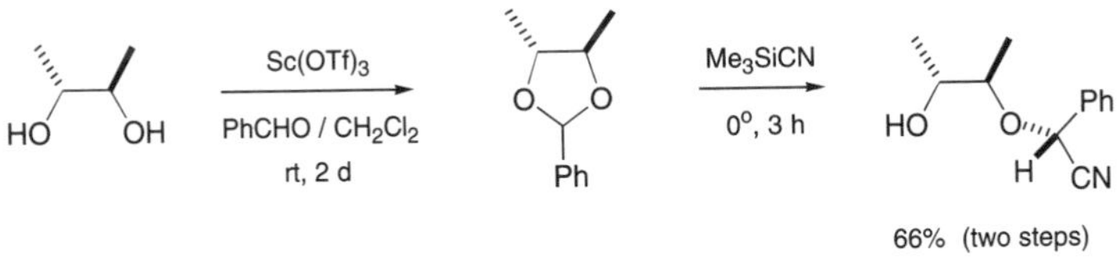

O-Acylation.[2] Alcohols, including tertiary alcohols, are acylated readily with $Sc(OTf)_3$ as a catalyst. The catalyst is superior to DMAP and Bu_3P.

Friedel–Crafts reactions. Alkylation of arenes with benzylic and allylic alcohols is achieved with $Sc(OTf)_3$ as catalyst.[3] An interesting variant is the reductive benzylation using arenecarbaldehydes and 1,3-propanediol.[4] The diol serves as a reductant.

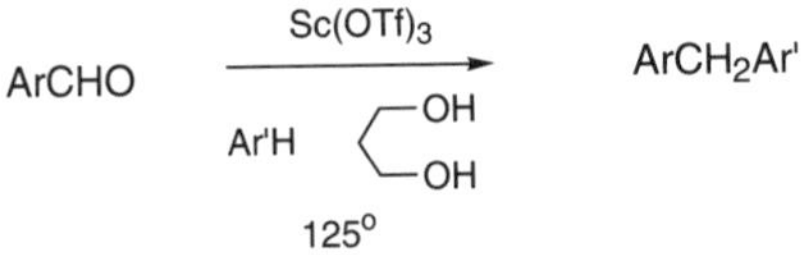

A reusable catalyst system for Friedel–Crafts acylation is $Sc(OTf)_3$–$LiClO_4$.[5] *o*-Acylation of substituted phenols and 1-naphthols catalyzed by $Sc(OTf)_3$ in a toluene–acetonitrile mixture is readily accomplished.[6] The same results are obtained by Fries rearrangement of acyloxynaphthalenes.[7]

Decarbonylation.[8] Aromatic aldehydes undergo decarbonylation on refluxing with $Sc(OTf)_3$ in methanol (forming HCOOMe as a by-product). The presence of electron-releasing substituents facilitates the reaction.

Allylation. The catalyzed addition of an allyl group to aldehydes by using allyltrimethylsilane,[9] and a similar reaction on imines with allyltributylstannane,[10] proceed at room temperature or below. Homoallylic alcohols and amines are formed, respectively.

Aldol reactions. The reaction of polymer-supported ketene *O*-silyl *S*-benzyl acetals with aldehydes[11] and imines[12] gives reasonable yields of the condensation products, which release 1,3-diols and amino alcohols, respectively, from the resin by treatment with $LiBH_4$.

γ-Hydroxy ketones. The opening of epoxides with lithium enolates in the presence of $Sc(OTf)_3$ is effective (5 examples, 78–95%).[13]

LiN(SiMe3)2 / PhMe ; Sc(OTf)3

95%

α,β-Unsaturated thioimidates.[14] The synthesis from *N*-tosylimines and alkynyl sulfides probably proceeds via the [2+2]-cycloadducts, which suffer ring opening immediately afterwards.

Sc(OTf)3, MeCN, rt

95%

Polymeric catalysts. A very useful modification of the catalyst is obtained by ligand exchange with a (polyallyl)triflamide, forming polymer-bound scandium triflamide bistriflate.[15] The catalyst has been used in the combinatorial synthesis of a tetrahydroquinoline library from anilines, aldehydes, and alkenes. A related catalyst prepared from $ScCl_3 \cdot 6H_2O$ and Nafion is effective in several useful synthetic reactions, including allylation, Diels–Alder reaction, and Friedel–Crafts acylation.[16]

[1]Fukuzawa, S., Tsuchimoto, T., Hotaka, T., Hiyama, T. *SL* 1077 (1995).
[2]Ishihara, K., Kubota, M., Kurihara, H., Yamamoto, H. *JOC* **61**, 4560 (1996).
[3]Tsuchimoto, T., Tobita, K., Hiyama, T., Fukuzawa, S. *SL* 557 (1996); El Gihani, M.T., Heaney, H., Shuhaibar, K.F. *SL* 871 (1996).

[4]Tsuchimoto, T., Hiyama, T., Fukuzawa, S. *CC* 2345 (1996).
[5]Kawada, A., Mitamura, S., Kobayashi, S. *CC* 183 (1996).
[6]Kobayashi, S., Moriwaki, M., Hachiya, I. *SL* 1153 (1995).
[7]Kobayashi, S., Moriwaki, M., Hachiya, I. *CC* 1527 (1995).
[8]Castellani, C.B., Carugo, O., Giusti, M., Leopizzi, C., Perotti, A., Invernizzi, A.G., Vidari, G. *T* **52**, 11045 (1996).
[9]Aggarwal, V.K., Vennall, G.P. *TL* **37**, 3745 (1996).
[10]Bellucci, C., Cozzi, P.G., Umani-Ronchi, A. *TL* **36**, 7289 (1995).
[11]Kobayashi, S., Hachiya, I., Yasuda, M. *TL* **37**, 5569 (1996).
[12]Kobayashi, S., Hachiya, I., Suzuki, S., Moriwaki, M. *TL* **37**, 2809 (1996).
[13]Crotti, P., Di Bussolo, V., Favero, L., Pineschi, M., Pasero, M. *JOC* **61**, 9548 (1996).
[14]Ishitani, H., Nagayama, S., Kobayashi, S. *JOC* **61**, 1902 (1996).
[15]Kobayashi, S., Nagayama, S. *JACS* **118**, 8977 (1996).
[16]Kobayashi, S., Nagayama, S. *JOC* **61**, 2256 (1996).

Scandium(III) triflimide.

Acylation.[1] Alcohols undergo acylation with either anhydrides or free carboxylic acids in the presence of $Sc(NTf_2)_3$, which is prepared from $Sc(OAc)_3$ and Tf_2NH.

Acetalization.[2] The catalyzed reaction between a carbonyl compound and an alcohol (diol) uses trimethyl orthoformate as a dehydrating agent. Dioxolanone derivatives are also readily formed by replacing the alcohol with hydroxy acids, such as lactic acid.

[1]Ishihara, K., Kubota, M., Yamamoto, H. *SL* 265 (1996).
[2]Ishihara, K., Karumi, Y., Kubota, M., Yamamoto, H. *SL* 839 (1996).

Selenium dioxide. **13**, 272–273; **17**, 312–313; **18**, 318–319

Dihydrooxazinones.[1] 2-Alkyloxazolines undergo oxidative rearrangement with SeO_2 in dioxane to give 5,6-dihydro-2*H*-1,4-oxazin-2-ones, most likely proceeding from the intermediate 2-acyloxazoline products.

SeO_2, dioxane Δ

75%

Enollactones.[2] 2-Benzylidenecycloalkanones are subject to Baeyer–Villiger oxidation on reaction with SeO_2–H_2O_2 at room temperature. Selenium dioxide is present in a catalytic amount.

3-Hydroxy-1-alkenyl carbamates.[3] Transpositional oxygenation of allylic carbamates with SeO_2 cannot be achieved directly. However, after stannylation, the oxidation proceeds well (using SeO_2/*t*-BuOOH or SeO_2 alone).

iPr$_2$N Bu$_3$Sn– SeO$_2$ / CH$_2$Cl$_2$ → HO– iPr$_2$N 96%

[1]Shafer, C.M., Molinski, T.F. *JOC* **61**, 2044 (1996).
[2]Guzman, J.A., Mendoza, V., Garcia, E., Garibay, C.F., Olivares, L.Z., Maldonado, L.A. *SC* **25**, 2121 (1995).
[3]Madec, D., Ferezou, J.-P. *SL* 867 (1996).

Silica gel. 15, 282; **18**, 319

Ether cleavage. Silica gel on which oxalic acid is deposited hydrolyzes enol ethers. This reagent is useful for converting 3-methoxy-2,5(10)-diene steroids (Birch reduction products of A-aromatic steroids) into the corresponding 5(10)-en-3-ones.[1] For the cleavage of other ethers in methanol, the silica–alumina gel (prepared by the sol–gel method) is very valuable.[2] The ease of deprotection follows the order TMS > 1-methyl-1-methoxyethyl >> 1-ethoxyethyl > THP >> methoxymethyl.

OMe SiO$_2$ - Al$_2$O$_3$ / MeOH , rt → OH 97%

Opening of epoxides. Epoxides alkylate *N*-heterocycles[3] such as indoles, pyrroles, imidazoles, and pyrazoles. For example, styrene oxide and indole give 2-(3-indolyl)-2-phenylethanol in 88% yield (room temperature, 7 days).

β-Halohydrins are formed readily from epoxides on contact with silica-gel-supported LiX.[4]

Removal of N-Boc groups. Selective deprotection of *t*-butoxycarbonyl groups[5] attached to nitrogen atoms in conjugation with an aromatic or a carbonyl group is performed using silica gel.

COOBu^t NHBoc N Boc SiO$_2$ (vac.) 50°, 7 h → COOBu^t NHBoc N H 80%

Isomerization of epoxides.[6] Stirring epoxides with silica gel in acetone or ethyl acetate at room temperature completes the isomerization to aldehydes (7 examples, 62–92%).

Wittig reactions.[7] Stabilized Wittig reagents react with aldehydes (not ketones) in the presence of silica gel at room temperature (8 examples, 78–90%).

[1]Liu, L.-G., Zhang, T., Li, Z.-S. *SC* **26**, 2999 (1996).
[2]Matsumoto, Y., Mita, K., Hashimoto, K., Iio, H., Tokoroyama, T. *T* **52**,9387 (1996).
[3]Kotsuki, H., Hayashida, K., Shimanouchi, T., Nishizawa, H. *JOC* **61**, 984 (1996).
[4]Kotsuki, H., Shimanouchi, T. *TL* **37**, 1845 (1996).
[5]Apelquist, T., Wensbo, D. *TL* **37**, 1471 (1996).
[6]Lemini, C., Ordonez, M., Perez-Flores, J., Cruz-Almanza, R. *SC* **25**, 2695 (1995).
[7]Patil, V.J., Mavers, U. *TL* **37**, 1281 (1996).

Silver bromate. 18, 319–320

Oxidation.[1] In MeCN and in combination with a Lewis acid (e.g., $AlCl_3$), sodium bromate oxidizes aromatic aldehydes to carboxylic acids.

[1]Firouzabadi, H., Mohammadpoor-Baltork, I. *BCSJ* **68**, *24,* 2319 (1995).

Silver carboxylates.

Nitrile oxides.[1] Silver acetate is used to generate nitrile oxides from hydroximoyl chlorides at room temperature. The reactive species are trapped as 1,3-dipolar cyloadducts (7 examples, 76–98%).

β-Amino acid derivatives.[2] Diazo ketones derived from α-amino acids undergo Wolff rearrangement in the presence of silver benzoate–Et_3N, and the ketenes are trapped as *N*-protected β-amino esters.

BocNH ... N_2 + HO ... OH —(AgOBz - Et_3N; THF, -5°)→ BocNH ... O ... OH

53%

[1]Tokunaga, Y., Ihara, M., Fukumoto, K. *H* **43**, 1771 (1996).
[2]Guibourdenche, C., Podlech, J., Seebach, D. *LA* 1121 (1996).

Silver cyanate.

Ureas.[1] Acylisocyanates are formed when silver cyanate reacts with acid chlorides. With proper substituents, such products are precursors of uracils.

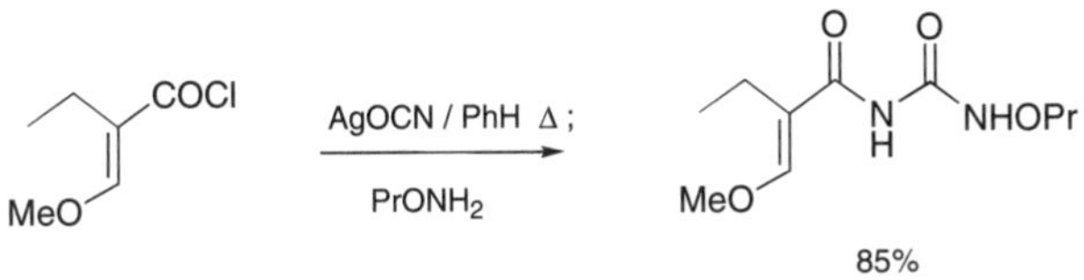

[1]Kim, D.-K., Kim, G., Lim, J., Kim, K.H. *JHC* **32**, 1625 (1995).

Silver iodide.

γ-Alkylidene γ-lactones.[1] 4-Alkynoic acids, including the α,β-unsaturated congeners, are cyclized by the catalysis of AgI in DMF. Similarly, 3-alkylidenephthalides are synthesized from 2-alkynylbenzoic acids.

Epoxy enynes. The coupling of enol triflates with epoxyalkynes is catalyzed by AgI–$Pd(PPh_3)_4$ (5 examples, 52–90%).

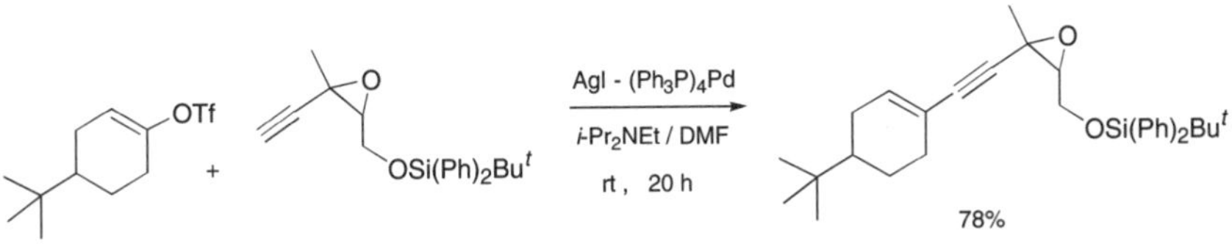

[1]Ogawa, Y., Maruno, M., Wakamatsu, T. *H* **41**, 2587 (1995).
[2]Bertus, P., Pale, P. *TL* **37**, 2019 (1996).

Silver nitrate. 18, 320

Furans.[1] Isomerization of allenones, alkynyl allylic alcohols, and allenylcarbinols to give furan derivatives is catalyzed by $AgNO_3$/silica gel (10 examples, 73–99%).

Free-radical reactions. Silver nitrate, sodium persulfate, and iron(III) nitrate constitute an oxidizing system that degrades carboxylic acids to radicals.[2] Adding these reactive intermediates to radical acceptors such as methyl vinyl ketone, acrylic esters, and acrylonitrile initiates synthetically useful processes. Monoamides of oxalic acid undergo oxidative degradation by $(NH_4)_2S_2O_8$ in the presence of $AgNO_3$-$Cu(OAc)_2$ to afford isocyanates in a biphasic system (11 examples, 45–87%).[3]

[1]Marshall, J.A., Sehon, C.A. *JOC* **60**, 5966 (1995).
[2]Araneo, S., Fontana, F., Minisci, F., Recupero, F., Serri, A. *TL* **36**, 4307 (1995).
[3]Minisci, F., Fontana, F., Coppa, F., Yan, Y.M. *JOC* **60**, 5430 (1995).

Silver(I) triflate. 13, 274–275; **14**, 282–283; **16**, 302; **17**, 314; **18**, 322–323

Addition to conjugated alkynoate esters.[1] The addition of alcohols to the esters to give enol derivatives is catalyzed by AgOTf. While this reaction is suitable for the preparation of 3-alkoxyacrylic esters and 2-alkoxyfumaric esters, unsymmetrical alkynoic esters afford mixtures of products.

[1]Kataoka, Y., Matsumoto, O., Tani, K. *CL* 727 (1996).

Sodium amalgam. 18, 324

Desulfonylative dimerization.[1] α-Alkoxymethyl sulfones undergo C–S bond cleavage on treatment with sodium amalgam. This reaction can be used to prepare polyethers.

MeO~~O~~SO_2Ph —(Na-Hg, PhH Δ)→ MeO~~O~~O~~OMe 79%

[1]Julia, M., Uguen, D., Zhang, D. *AJC* **48**, 279 (1995).

Sodium–ammonia. 16, 303–304; **18**, 324

Birch reductions. Derivatives of pyrrole[1] and furan[2] are reduced to the dihydro level, opening up many synthetic possibilities.

N-Boc pyrrole-2-carbonyl pyrrolidine —(Na / NH_3 ; MeI)→ 2-methyl-2,5-dihydro N-Boc product 85%

3-COOH-2-$SiMe_3$ furan —(Na / NH_3, i-PrOH / NH_4Cl ; CH_2N_2)→ 3-COOMe-2-$SiMe_3$ dihydrofuran

Aryldemethylation of $ArSnMe_3$. The replacement of one or all of the methyl groups attached to tin can be achieved via cleavage of the Sn–Me bond followed by photoinduced $S_{RN}1$ reaction with chloroarenes.[3]

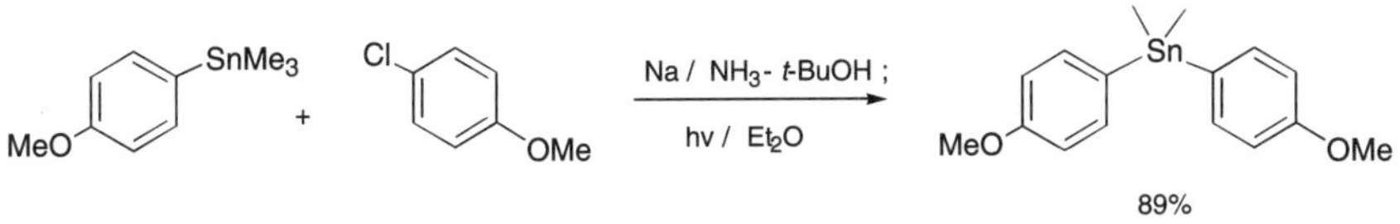

[1]Donohoe, T.J., Guyo, P.M. *JOC* **61**, 7664 (1996).
[2]Beddoes, R.L., Lewis, M.L., Gilbert, P., Quayle, P., Thompson, S.P., Wang, S., Mills, K. *TL* **37**, 9119 (1996).
[3]Yammal, C.C., Podesta, J.C., Rossi, R.A. *JOMC* **509**, 1 (1996).

Sodium azide. 18, 325–326

β-Azido alcohols.[1] 1,2-Diol thionocarbonates behave in a manner similar to that of their corresponding cyclic sulfates, and their reaction with NaN_3 in DMF leads to β-azido alcohols.

Amino acids.[2] Hydrazoic acid generated in situ from NaN_3 and HOAc adds to γ-keto-α,β-unsaturated carbonyls.

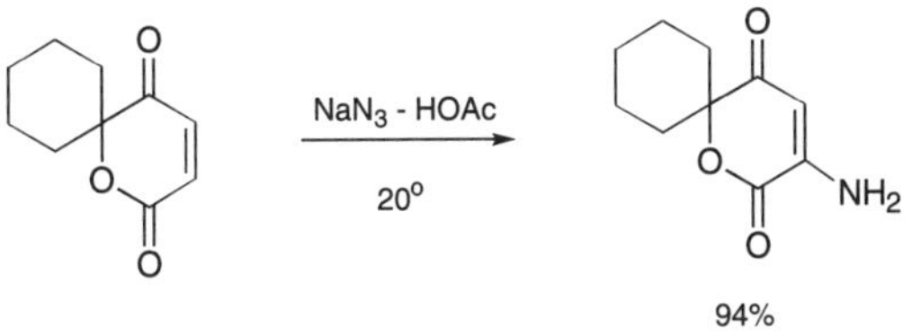

Tetrazoles.[3] Ketones are converted into substituted tetrazoles, as depicted in the following equation.

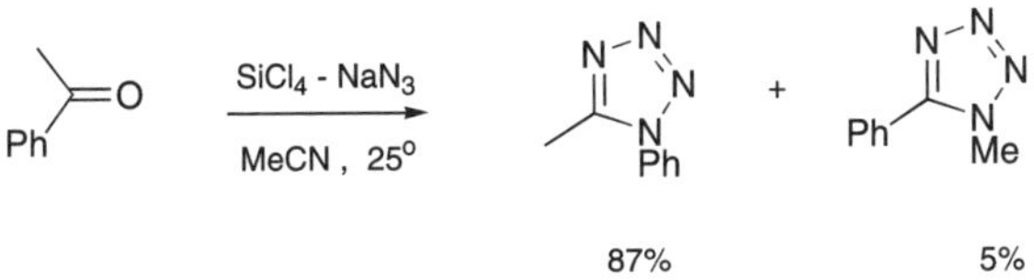

[1]Ko, S.Y. *JOC* **60**, 6250 (1995).
[2]Couladouros, E.A., Apostolopoulos, C.D. *SL* 341 (1996).
[3]El-Ahl, A.-A.S., Elmorsy, S.S., Soliman, H., Amer, F.A. *TL* **36**, 7337 (1995).

Sodium borohydride. 13, 278–279; **15**, 290; **16**, 304; **18**, 326–327

syn-1,3-Diols.[1] The reduction of β-hydroxy ketones with $NaBH_4$ is highly stereoselective by precoordination of the substrates with terphenylboronic acid.

Reduction of hindered ketones.[2] The addition of Amberlyst-15 (in the H^+ form) resin to the reducing system in THF permits the reduction of hindered ketones. Other functionalities, such as acetals or silyl ethers, remain unaffected.

Regioselective reductions.[3] The deoxygenation of *o*-hydroxyphenones can be accomplished. Thus, mixed carbonates of 4-acylresorcinols are converted into monoprotected alkylresorcinols.

OCOOMe, OCOOMe, acetyl-substituted benzene → ($NaBH_4$, THF) → OCOOMe, OH, ethyl-substituted benzene, 76%

RCOOH → RCH₂OH. On conversion into the *N*-acylimidazole, a carboxylic acid becomes reducible to the primary alcohol by $NaBH_4$ in the presence of water.[4]

Reduction of C=N bonds. A convenient preparation of camphorsultam involves the treatment of 10-camphorsulfonyl chloride with ammonia, followed by reduction with $NaBH_4$.[5] 3-*exo*-aminoisoborneol is available from 3-oximinocamphor by a two-step reduction process,[6] first with $NaBH_4$ alone and then with $NaBH_4–NiCl_2·6H_2O$.

($NaBH_4$, MeOH - H_2O) → 66%

Reductive alkylation of amines. Selective benzylation of terminal amino groups of a polyamine[7] is achieved by $NaBH_4$ reduction of the derived Schiff bases.

Secondary amines are methylated on treatment with paraformaldehyde, $ZnCl_2$, and $NaBH_4$ in dichloromethane (19 examples, 60–96%).[8] Note the unusual solvent used.

N-Aroylpyrroles are deoxygenated by $NaBH_4–BF_3·OEt_2$.[9] Apparently, this is a borane reduction. More interesting is the transformation of aroyl azides into *N,N*-bis(2,2,2-trifluoroethyl)anilines (with $NaBH_4–CF_3COOH$),[10] which proceeds by a Curtius rearrangement followed by trifluoroacetylation and reduction.

Conjugate reductions.[11] Borohydride exchange resin with catalytic amounts of $CuSO_4$ constitutes a valuable reagent for the saturation of the conjugated double bond of α,β-unsaturated esters, amides, and nitriles in methanol at room temperature. The similar reduction of enones has been reported previously.

Other reactions. The deoxygenation of bis(tributyltin)oxide to afford hexabutylditin (in EtOH, 2h, 83% yield)[12] and the conversion of *N*-alkenylpyridinium salts to ketones[13] are accomplished using $NaBH_4$ as reagent. Anilines are obtained by reduction of nitroarenes[14] with $NaBH_4$–$(NH_4)_2SO_4$.

[1]Yamashita, H., Narasaka, K. *CL* 539 (1996).
[2]Caycho, J.R., Tellado, F.G., de Armas, P., Tellado, J.J.M. *TL* **38**, 277 (1997).
[3]Mitchell, D., Doecke, C.W., Hay, L.A., Koenig, T.M., Wirth, D.D. *TL* **36**, 5335 (1995).
[4]Sharma, R., Voynov, G.H., Ovaska, T.V., Marquez, V.E. *SL* 839 (1995).
[5]Capet, M., David, F., Bertin, L., Hardy, J.C. *SC* **25**, 3323 (1995).
[6]Przeslawski, R.M., Newman, S., Thornton, E.R., Joullie, M.M. *SC* **25**, 2975 (1995).
[7]Sclafani, J.A., Maranto, M.T., Sisk, T.M., Van Arman, S.A. *JOC* **61**, 3221 (1996).
[8]Bhattacharyya, S. *SC* **25**, 2061 (1995).
[9]D'Silva, C., Iqbal, R. *S* 457 (1996).
[10]Krein, D.M., Sullivan, P.J., Turnbull, K. *TL* **37**, 7213 (1996).
[11]Sim, T.B., Yoon, N.M. *SL* 726 (1995).
[12]McAlonan, H., Stevenson, P.J. *OM* **14**, 4021 (1995).
[13]Al-Abed, Y., Naz, N., Khan, K.M., Voelter, W. *ACIEE* **35**, 523 (1996).
[14]Gohain, S., Prajapati, D., Sandhu, J.S. *CL* 725 (1995).

Sodium borohydride–antimony(III) halide.

Hydrodebromination.[1] Removal of the halogen atom from an α-halo carbonyl at low temperatures is accomplished with $NaBH_4$–$SbBr_3$.

$ArNO_2 \rightarrow ArNH_2$.[2] This reduction can be performed with either $NaBH_4$–$SbCl_3$ or $NaBH_4$–$BiCl_3$ at room temperature.

[1]Sayama, S., Inamura, Y. *CL* 633 (1996).
[2]Ren, P.-D., Pan, S.-F., Dong, T.-W., Wu, S.-H. *SC* **25**, 3799 (1995).

Sodium borohydride–bismuth(III) chloride. 18, 327

Reduction of double bonds. This reducing system is capable of saturating double bonds in such compounds as styrene[1]and α,β-unsaturated esters.[2]

[1]Ren, P.-D., Pan, S.-F., Dong, T.-W., Wu, S.-H. *SC* **26**, 763 (1996).
[2]Ren, P.-D., Pan, S.-F., Dong, T.-W., Wu, S.-H. *SC* **25**, 3395 (1995).

Sodium borohydride–iodine. 17, 316; 18, 328

Reduction of C=N and N=N bonds. The reduction of *O*-acyl oximes to primary amines[1] and of both azoarenes and azoxyarenes to *N,N*-diarylhydrazines[2] is readily achieved.

[1]Barbry, D., Champagne, P. *SC* **25**, 3503 (1995).
[2]Karmakar, D., Prajapati, D., Sandhu, J.S. *JCR(S)* 464 (1996).

Sodium cyanoborohydride. 14, 287–288; **16**, 305–306; **18**, 331

Nitroalkanes from nitroalkenes.[1] The reduction of the conjugated double bond with $NaBH_3CN$ is catalyzed by a zeolite (H-ZSM-5) in MeOH at room temperature (10 examples, 65–79%).

Reductive alkylation. Alkylation of an amine simply involves treatment with an aldehyde in the presence of trimethyl orthoformate as a dehydrant and in situ reduction.[2] An analogous method applying to α-amino esters with glyoxal results in optically active *N,N′*-disubstituted piperazines,[3] and a convergent reductive amination of a ketodialdehyde to form an indolizidine skeleton in 53% yield makes possible a concise synthesis of castanospermine from α-D-glucopyranoside.[4]

BnO, CHO, =O, BnO, CHO, BnO, OBn → ($HCOONH_4$; $NaBH_3CN$, MeOH) → BnO, N, BnO, BnO, OBn → (H_2; Pd/C) → HO, N, HO, HO, OH

castanospermine

Desulfonylation. The facile detachment of the arenesulfonyl moiety from allylic benzothiazol-2-yl sulfones with the use of $NaBH_3CN$ is crucial to the preparation of alkenes by chain homologation based on alkylation of benzothiazol-2-yl methallyl sulfone.[5] Transposition of the double bond occurs during the desulfonylation.

N, S, SO_2, Ph → ($NaBH_3CN$; aq HCl / THF) → Ph 84%

Bz_2O_2 ↓ *t*-BuOH Δ

N, S, SO_2, $CHCH_2Ph$ 78% → ($NaBH_3CN$; aq HCl / THF) → Ph 63%

Removal of organotin residues. Organotin reagents often cause operational difficulties in that clean removal of their residue is problematical. Brief treatment of the crude reaction mixture with $NaBH_3CN$ (reduction to R_3SnH) facilitates purification.[6]

[1]Gupta, A., Haque, A., Vankar, Y.D *CC* 1653 (1996).

[2]Szardenings, A.K., Burkoth, T.S., Look, G.C., Campbell, D.A. *JOC* **61**, 6720 (1996).
[3]Watanabe, M., Kojima, Y., Kawabe, K., Hatamoto, E., Tsuru, E., Miyake, H., Yamashita, T. *S* 452 (1996).
[4]Zhao, H., Mootoo, D.R. *JOC* **61**, 6762 (1996).
[5]Phillips, E.D., Warren, E.S., Whitham, G.H. *T* **53**, 307 (1997).
[6]Crich, D., Sun, S. *JOC* **61**, 7200 (1996).

Sodium cyanoborohydride–boron trifluoride etherate. 18, 331

De-O-protection. Both tetrahydropyranyl ethers[1] and *p*-methoxybenzyl ethers[2] are said to be cleaved with this reagent. However, the method does not seem to have any advantage over existing procedures.

Reductive dealkoxylation of acetals.[3] The major products from mixed acetals are ethers that retain the larger alkyl groups.

$NaBH_3CN$ / $BF_3 \cdot OEt_2$

96%

1,2,3,4-Tetrahydroquinolines.[4] Partial saturation of quinolines is observed.

[1]Srikrishna, A., Sattigeri, J.A., Viswajanani, R., Yelamaggad, C.V. *JOC* **60**, 2260 (1995).
[2]Srikrishna, A., Viswajanani, R., Sattigeri, J.A., Vijaykumar, D. *JOC* **60**, 5961 (1995).
[3]Srikrishna, A., Viswajanani, R. *T* **51**, 3339 (1995).
[4]Srikrishna, A., Reddy, T.J., Viswajanani, R. *T* **52**, 1631 (1996).

Sodium cyanoborohydride–mercury(II) trifluoroacetate.

Decyanation.[1] α-Amino nitriles are converted into amines with this reagent combination in the presence of DABCO in methanol.

[1]Sassaman, M.B. *T* **52**, 10835 (1996).

Sodium cyanoborohydride–tributyltin chloride.

Desulfonylation.[1] The removal of the sulfonyl moiety from β-keto sulfones with this couple in the presence of AIBN in refluxing *t*-butanol relies on a free-radical process, apparently involving a tin hydride reagent.

[1]Giovannini, R., Petrini, M. *SL* 973 (1995).

Sodium cyanoborohydride–zinc iodide.

Deoxygenation of 1-indanols.[1] This is just an alternative method involving an acid-stable reagent source to deliver hydride to an incipient carbocation.

[1]Alesso, E.N., Bianchi, D.E., Finkielsztein, L.M., Lantano, B., Moltrasio, G.Y., Aguirre, J.M. *TL* **36**, 3299 (1995).

Sodium dialkynyl(diethyl)aluminate.

Propargylic alcohols.[1] These reagents are a new class of alkynylating agents obtained from 1-alkynes and $Na(Et_2AlH_2)$. They are superior to alkynyllithiums and Grignard reagents.

[1]Ahn, J.H., Joung, M.J., Yoon, N.M. *JOC* **60**, 6173 (1995).

Sodium dithionite. 13, 281; **18**, 331

Debromination.[1] *vic*-Dibromides give *(E)*-alkenes by treatment with $Na_2S_2O_4$–$NaHCO_3$ in DMF at room temperature (17 examples, 76–98%).

Reduction of nitrogenous compounds. By using octylviologen as an electron transfer agent in MeCN, nitroalkanes are reduced to hydroxylamines[2] and azoarenes to hydrazoarenes.[3]

[1]Khurana, J.M., Sehgal, A. *SC* **26**, 3791 (1996).
[2]Park, K.K., Oh, C.H., Sim, W.-J. *JOC* **60**, 6202 (1995).
[3]Park, K.K., Han, S.Y. *TL* **37**, 6721 (1996).

Sodium hydride. 14, 288; **16**, 307–308; **18**, 333

Michael additions. *O*-Aroylmandelonitriles add to acceptors such as acrylonitrile in such a way that a transfer of the aroyl group also occurs in situ, resulting in the formation of 1,4-diketones.[1] Yields of this reaction range from moderate to good (13 examples, 30–84%).

Ph(CN)CH–OBz + CH_2=CH–CN —(NaH / DMSO, 20°)→ PhCO–CH_2–CH(CN)–COPh

62%

An intramolecular addition of an enolate ion to a conjugated alkyne gives a furan derivative.[2]

R–CO–CH_2CH_2–C≡C–COOMe —(NaH, THF - DMF)→ 2-R-5-(CH_2COOMe)furan

57 - 91%

Acetals and thioacetals.[3]

Complex reagent for dehalogenation.[4] The aggregate prepared from NaH, *i*-PrONa, $Ni(OAc)_2$, and Ti(0) reduces haloarenes quantitatively.

[1]Miyashita, A., Matsuoka, Y., Numata, A., Higashino, T. *CPB* **44**, 448 (1996).
[2]Vieser, R., Eberbach, W. *TL* **36**, 4405 (1995).
[3]Cossu, S., De Lucchi, O., Fabris, F., Ballini, R., Bosica, G. *S* 1481 (1996).
[4]Li, H., Liao, S., Xu, Y. *CL* 1059 (1996).

Sodium hydrogen telluride. 13, 282; **18**, 333–334

Reduction.[1] Aromatic carbonyl compounds carrying electron-withdrawing groups on the ring are selectively reduced.

Elimination.[2] *vic*-Disulfonates react with NaTeH in DMF to give alkenes at room temperature (11 examples, 85–96%). *threo*-Isomers yield *(E)*-alkenes, and *erythro*-isomers *(Z)*-alkenes.

Dealkylation.[3] Quaternary ammonium ions containing benzyl, allyl and methyl groups are decomposed by NaTeH. Accordingly, such salts should not be used in phase-transfer reactions involving NaTeH.

[1]Yamashita, M., Tanaka, Y. *AOMC* **10**, 791 (1996).
[2]Bargues, V., Blay, G., Fernandez, I., Pedro, J.R. *SL* 655 (1996).
[3]Li, W., Zhou, X.-J. *SC* **25**, 3635 (1995).

Sodium hypochlorite–chiral nitroxyl.

Enantioselective oxidation.[1] The presence of (**1**) makes oxidation of a prostereogenic secondary alcohol with NaOCl enantioselective. The unoxidized alcohol that remains is optically active.

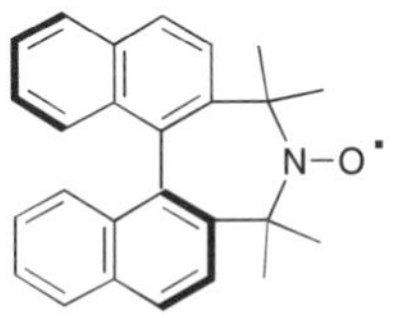

N–O•

(**1**)

[1]Rychnovsky, S.D., McLernon, T.L., Rajapakse, H. *JOC* **61**, 1194 (1996).

Sodium nitrite.

Binaphthyls.[1] Naphthalene derivatives undergo oxidative dimerization on treatment with $NaNO_2$ and triflic acid in MeCN. Radical cation intermediates are involved.

Alkynes from isoxazolin-5-ones.[2] The oxidative degradation is achieved with $NaNO_2$ and $FeSO_4$ in aqueous acetic acid.

N–O OHC S + NaBH4 EtOH N–O O S 79% NaNO2 FeSO4 HOAc - H2O S 62%

[1]Tanaka, M., Nakashima, H., Fujiwara, M., Ando, H., Souma, Y. *JOC* **61**, 788 (1996).
[2]Boivin, J., Huppe, S., Zard, S.Z. *TL* **37**, 8735 (1996).

Sodium perborate. 14, 290–291; **16**, 310; **18**, 337–338

Deoximation.[1] Oxidation with $NaBO_3$ in HOAc provides another method for the cleavage of oximes.

$Ar_3Bi \rightarrow Ar_3Bi(OAc)_2$.[2] Triarylbismuth diacetates are readily obtained using sodium perborate to react with triarylbismuth in HOAc.

Oxidative cleavage of cyclic acetals.[3] Good yields of the acetoxyalkyl alkanoates are achieved when cyclic acetals are treated with a combination of $NaBO_3$ and Ac_2O.

O O O NaBO3• 4H2O, Ac2O Na2CO3, PhH 55°, 10 h O O O OAc 72%

[1]Bandgar, B.P., Shaikh, S.I., Iyer, S. *SC* **26**, 1163 (1996).
[2]Combes, S., Finet, J.-P. *SC* **26**, 4569 (1996).
[3]Bhat, S., Ramesha, A.R., Chandrasekaran, S. *SL* 329 (1995).

Sodium percarbonate.

Oxidation of alcohols.[1] The oxidation is carried out with $Na_2CO_3 \cdot 1.5H_2O_2$ in the presence of molybdenyl acetylacetonate and a phase-transfer catalyst.

Cleavage of C–C bonds. α-Substituted carbonyl compounds such as α-ketols[2] and α-halo ketones[3] suffer bond scission to give carboxylic acids when they are subjected to percarbonate oxidation and simultaneously irradiated with ultrasound.

For Baeyer–Villiger oxidation of ketones, acetic anhydride is also added.[4]

[1]Maignien, S., Ait-Mohand, S., Muzart, J. *SL* 439 (1996).
[2]Yang, D.T.C., Cao, Y.H., Evans, T.T., Kabalka, G.W. *SC* **26**, 4275 (1996).

[3]Yang, D.T.C., Cao, Y.H., Kabalka, G.W. *SC* **25**, 3695 (1995).
[4]Zhang, Y., Hu, H., Fang, Y., Ai, H., Tao, F.-G. *YH* **16**, 64 (1996).

Sodium periodate. 15, 294; **18**, 338–339

Sulfide oxidation. The well-known oxidation of sulfides to give sulfoxides has been extended to compounds of type (**1**), which, on further treatment with H_2O_2 and KF in weakly basic conditions, are degraded to give secondary alcohols.[1] Thus, the cyclopropylsilyl residue can be considered to be a masked hydroxyl group.

$NaIO_4$ / MeOH - H_2O ; PhH Δ ; H_2O_2 - KF

(**1**) → 86%

[1]Angelaud, R., Landais, Y., Maignan, C. *TL* **36**, 3861 (1995).

Sodium tetracarbonylhydridoferrate.

3-Cyclobutene-1,2-diones.[1] The reaction of $NaFeH(CO)_4$ with MeI liberates methane. The simultaneously generated $Fe(CO)_4$ species is trapped by various alkynes to provide the cyclobutenediones (7 examples, 27–42%) after oxidative demetallation.

R—≡—R' → $NaFeH(CO)_4$ - MeI ; $CuCl_2$; H_2O

[1]Periasamy, M., Radhakrishnan, U., Brunet, J.-J., Chauvin, R., El Zaizi, A.W. *CC* 1499 (1996).

Sodium triacetoxyborohydride. 13, 283; **16**, 309–310; **18**, 340

Reduction.[1] The differentiation of two aldehyde groups is important while (+)-digitoxigenin is being synthesized. During reduction with $NaBH(OAc)_3$ in benzene, the neopentyl aldehyde can be retained.

$NaBH(OAc)_3$ / PhH

67%

[1]Stork, G., West, F., Lee, H.Y., Isaacs, R.C.A., Manabe, S. *JACS* **118**, 10660 (1996).

Sodium triethylgermanate.

Trifluoromethylation.[1] Aldehydes undergo trifluoromethylation by reaction with trifluoromethylsodium, which is generated from phenyl trifluoromethyl sulfide on reaction with Et_3GeNa. The yields of trifluoromethylcarbinols are excellent (9 examples, 91–96%). A temperature range between –50° and –60° must be maintained, because no significant formation of $NaCF_3$ at lower temperatures takes place, and at higher temperatures the reagent decomposes. Note that Me_3SiNa and Et_3SnNa cannot be used to generate trifluoromethylsodium and that $PhOCF_3$ is not a surrogate for $PhSCF_3$.

$PhSCF_3$ —(Et_3GeNa / THF - HMPA / -60°)→ [$NaCF_3$] —(PhCHO / -60°)→ PhCH(OH)CF_3 92%

Aldol reactions.[2] The complex of $NaGeEt_3$ with $SmCl_3$ promotes a *syn*-selective aldol reaction of ketones and amides with aldehydes in the presence of HMPA in THF. The enolization step at 0° is followed by treatment of the aldehydes at –78°.

[1]Yokoyama, Y., Mochida, K. *SL* 1191 (1996).
[2]Yokoyama, Y., Mochida, K. *SL* 445 (1996).

Sulfuric acid. 18, 342

Detosylation.[1] A rapid removal of *N*-tosyl groups is accomplished by heating the amides with sulfuric acid at high temperatures. The protonated amines apparently survive such harsh conditions.

MeN(Ts)CH$_2$CH$_2$N(Ts)CH$_2$CH$_2$N(Ts)Me —(H_2SO_4, 180°; base)→ MeNHCH$_2$CH$_2$NHCH$_2$CH$_2$NHMe 90%

Dibenzofurans.[2] Aqueous sulfuric acid is used as a medium for the free-radical cyclization of *o*-aryloxyarenediazonium salts, with hydroquinone or metal salts ($SnCl_2$, $FeSO_4$, $CuSO_4$, NaI) as a promoter.

$\overset{+}{N_2}$ BF_4^- — H_2SO_4, hydroquinone, H_2O, Δ → 70%

[1]Lazar, I.. *SC* **25**, 3181 (1995).
[2]Wassmundt, F.W., Pedemonte, R.P. *JOC* **60**, 4991 (1995).

T

Tantalum(V) chloride–zinc. 16, 312; **17**, 321; **18**, 343

Reductive coupling of alkynes with allylic alcohols. Alkynes form $TaCl_3$ complexes, which react with allylic alcohols to form 1,4-dienes.[1]

Cyclotrimerization of alkynes.[2] An alkyne is activated by $TaCl_5$–Zn, and subsequent treatment with a diyne gives rise to condensed bicyclic products containing one aromatic ring. Adding an (ω-1)-alkynenitrile in the second stage results in the formation of a pyridine derivative.

R—≡—R → $TaCl_5$ - Zn / DME - PhH ; py, THF ; ≡—$(CH_2)_n$—≡ ; NaOH / H_2O

n = 2, 3, 4 56 - 82%

[1]Takai, K., Yamada, M., Odaka, H., Utimoto, K., Fujii, T., Furukawa, I. *CL* 315 (1995).
[2]Takai, K., Yamada, M., Utimoto, K. *CL* 851 (1995).

Tantalum(V) fluoride.

Chlorination of perfluoroalkylethenes.[1] TaF_5 catalyzes chlorination with chlorosulfonic acid. NbF_5 is also effective. Since the allylic chlorides can be easily transformed into other compounds, this reaction serves to mediate the functionalization of the alkenes.

C_nF_{2n+1}—CF_2—CH=CH$_2$ → TaF_5 - $ClSO_3H$, 25 - 30°, 5- 16 h → C_nF_{2n+1}(F)C=CH–CH$_2$Cl

~ 75%

[1]Petrov, V.A. *JOC* **60**, 3423 (1995).

Tellurium(IV) chloride. 16, 316; **18**, 343

Knoevenagel reaction.[1] The catalytic activity of $TeCl_4$ in the condensation is demonstrated (17 examples, 80–95%).

Dehydrosulfurization.[2] Under mild conditions ($TeCl_4$–Et_3N, room temperature), thioamides are converted into nitriles (7 examples, 92–98%).

Phosphorothioate esters.[3] Mixtures of thiols and trialkyl phosphites react by a redox pathway in the presence of $TeCl_4$ and lutidine (9 examples, 90–96%).

$(MeO)_3P$ + HS-$C_{12}H_{25}$ —[$TeCl_4$; lutidine / CH_2Cl_2; -42° -> rt; 1 h]→ $(MeO)_2P(O)$-S-$C_{12}H_{25}$ 96%

Reduction of sulfoxides.[4] This deoxygenation uses $TeCl_4$ and NaI in MeCN. The reaction proceeds at room temperature.

[1]Khan, R.H., Mathur, R.K., Ghosh, A.C. *SC* **26**, 683 (1996).
[2]Aso, Y., Omote, K., Takagi, S., Otsubo, T., Ogura, F˙ *JCR(S)* 152 (1995).
[3]Watanabe, Y., Inoue, S., Yamamoto, T., Ozaki, S. *S* 1243 (1995).
[4]Khan, R.H., Rastogi, R.C. *IJC(B)* **33B**, 293 (1994).

Tetrabutylammonium fluoride (TBAF). 13, 286–287; **14**, 293–294; **15**, 298,304; **17**, 324–326; **18**, 344–345

As base. For vicarious substitution of *m*-dinitrobenzene with carbon nucleophiles under photochemical conditions,[1] TBAF serves as the base. The nitroaldol condensation with α-amino aldehydes constitutes the key step of a stereoselective synthesis of 1,3-diamino-2-alkanols.[2]

Ph(CH₂)CH(NBn_2)CHO + O_2NCH₂CH₂COOMe —[$Bu_4NF \cdot H_2O$; 0°, 10 - 30 min]→ Ph-CH₂-CH(NBn_2)-CH(OH)-CH(NO_2)-CH₂-COOMe 66%

A procedure using TBAF hydrate in combination with a large excess of 1-octanethiol or phenylmethanethiol to deprotect 9-fluorenylmethoxycarbonylamines has been developed.[3] Subsequent oxidation of the thiol with bis(1-methyl-1*H*-tetrazol-5-yl) disulfide inactivates the TBAF and allows one-pot peptide bond formation.

gem-Difluorocyclopropanes.[4] Difluorocyclopropanes bearing electron-withdrawing substituents that are not available directly can be prepared by a halogen exchange reaction from the corresponding dichlorocyclopropanes, using TBAF in DMF at 0°, although the yields are moderate (5 examples, 40–46%).

Desilylation. The mild conditions associated with the cleavage of the C–Si bond by TBAF are favorable to many synthetic processes that involve carbanions. The method has been applied to the alkylation of aldehydes with trimethylsilyl-*o*-carborane[5] and to the allylation of sulfines.[6]

Bu_4NF / THF, PhCHO, 25°; 98%

Bu_4NF / DMF, allyl-$SiMe_3$, 4A mol. sieves; 30 - 76%

An intramolecular aldol/Claisen condensation leading to a perhydroindanolone is interesting in that the configuration of the ester group that flanks the departing silyl group is dependent on workup conditions.[7]

Bu_4NF / THF, -78°; NH_4Cl quench

quenching T = -78° 64% -
quenching T = 20° - 41%

A new route to *o*-allylanilines[8] consists of the silylmethylation of 3,4-dihydro-2,1-benzothiazin-2-ones at C-3 and fragmentation induced by TBAF.

BuLi ; Me_3SiCH_2I → Bu_4NF / THF ; KOH; 69%

Disilanes can now be considered as masked alcohols. The hydroxydesilylation is accomplished by treatment with TBAF and then basic hydrogen peroxide.[9]

MeO–C6H4–Si(Me)2–SiMe3 → (Bu_4NF / THF, 25°; H_2O_2 - $KHCO_3$ / MeOH, 40°, 2 h) → MeO–C6H4–OH 85%

Cleavage of enol ethers. The combination of TBAF and an acid such as $BF_3{\cdot}OEt_2$ is selective for the deprotection of enol ethers.[10] In the presence of an acid fluoride, such an enolate ion can be *O*-acylated with retention of configuration.[11]

Cl–(CH2)n–CH=CH–OSiMe3 → (Bu_4NF - $C_{15}H_{31}COF$, THF, 0°, 2 h) → Cl–(CH2)n–CH=CH–$OCOC_{15}H_{31}$ 84%

[1]Cervera, M., Marquet, J. *TL* **37**, 7591 (1996).
[2]Hanessian, S., Devasthale, P.V. *TL* **37**, 987 (1996).
[3]Ueki, M., Nishigaki, N., Aoki, H., Tsurusaki, T., Katoh, T. *CL* 721 (1993).
[4]Jonczyk, A., Kaczmarczyk, G. *TL* **37**, 4085 (1996).
[5]Cai, J., Nemoto, H., Nakamura, H., Singaram, B., Yamamoto, Y. *CL* 791 (1996).
[6]Capperucci, A., Degl'Innocenti, A., Leriverend, C., Metzner, P. *JOC* **61**, 7174 (1996).
[7]Schinzer, D., Blume, T., Jones, P.G. *ACIEE* **35**, 2500 (1996).
[8]Harmata, M., Jones, D.E. *TL* **36**, 4769 (1995).
[9]Suginome, M., Matsunaga, S.-I., Ito, Y. *SL* 941 (1995).
[10]Gevorgyan, V., Yamamoto, Y. *TL* **36**, 7765 (1995).
[11]Limat, D., Schlosser, M. *T* **51**, 5799 (1995).

Tetrabutylammonium hexanitrocerate.

γ-Keto esters.[1] Radicals derived from α-tributylstannylacetic esters are formed by oxidation with the cerate ion. They react with electron-rich alkenes, such as silyl enol ethers under the formation of γ-keto esters.

[1]Kohno, Y., Narasaka, K. *BCSJ* **68**, 322 (1995).

Tetrabutylammonium nitrate.

Alkyl nitrates.[1] The preparation of RCH_2ONO_2 by the nucleophilic substitution of alkyl toluenesulfonates by sodium nitrate is catalyzed by Bu_4NNO_3 (7 examples, 67–92%).

[1]Hwu, J.R., Vyas, K.A., Patel, H.V., Lin, C.-H., Yang, J.-C. *S* 471 (1994).

Tetrabutylammonium peroxydisulfate.

Oxidations. This reagent is able to cleave tosylhydrazones in good yields (18 examples, 90–98%)[1] in refluxing dichloroethane. Oxidation of sulfides to sulfoxides occurs at room temperature (11 examples, 98–99.5%).[2]

Tetrahydropyranyl ethers.[3] Alcohols in refluxing tetrahydropyran (*not* dihydropyran) form THP ethers in the presence of the ammonium persulfate (9 examples, 86–95%). Oxidation of the cyclic ether precipitates the reaction.

[1]Chen, F., Yang, J., Zhang, H., Guan, C., Wan, J. *SC* **25**, 3163 (1995).
[2]Chen, F., Wan, J., Guan, C., Zhang, H. *SC* **26**, 253 (1996).
[3]Choi, H.C., Cho, K.I., Kim, Y.H *SL* 207 (1995).

Tetrabutylammonium polyoxotungstates.

Carbonylation.[1] Homologous aldehydes are formed as the major products from alkanes under radical carbonylation conditions. Thus, magnetically stirred acetonitrile solutions of $Bu_4N^+(W_{10}O_{32})^{4-}$ or $Bu_4N^+(PW_{12}O_{40})^{3-}$ and alkanes are saturated with CO (1 atm) and are irradiated (550-W medium-pressure Hg lamp and a Pyrex filter) at room temperature for 16 hours to complete the reaction.

[1]Jaynes, B.S., Hill, C.L. *JACS* **117**, 4704 (1995).

Tetrabutylammonium (triphenyl)difluorosilicate (TBAT).

Alkyl fluorides.[1] The reagent $Bu_4N^+Ph_3SiF_2^-$ is prepared in two steps (95% yield) from Ph_3SiOH by treatment with aqueous HF in methanol and then with TBAF in a mixture of THF and dichloromethane. It is a good source of fluoride for nucleophilic substitution of alkyl sulfonates. Alkyl halides undergo elimination preferentially.

COOMe
OTf
Bu_4N^+ $Ph_3SiF_2^-$
MeCN
Δ, 48 h
COOMe
F
76%

Propargyl alcohols.[2] The condensation of alkynylsilanes and carbonyl compounds is mediated by TBAT.

[1]Pilcher, A.S., Ammon, H.L., DeShong, P. *JACS* **117**, 5166 (1995).
[2]Pilcher, A.S., DeShong, P. *JOC* **61**, 6901 (1996).

Tetrachlorophthalic anhydride.

Amine protection.[1] The derived imides are stable to many reagents used in oligosaccharide transformations. Therefore, amino sugars protected as such imides are useful synthetic intermediates. Ester groups and anomeric penten-4-yloxy residues survive the cleavage conditions involving ethylenediamine treatment. The by-product of cleavage is insoluble in commonly used solvents (CH_2Cl_2, DMSO, DMF, EtOAc, Me_2CO, MeOH) and therefore can be removed by filtration.

[1]Debenham, J.S., Fraser-Reid, B. *JOC* **61**, 432 (1996).

Tetrachlorosilane.

Trichlorosilyl enol ethers.[1] The reaction of tributylstannyl enol ethers with $SiCl_4$ results in trichlorosilyl analogs that are highly reactive as donors in aldol reactions without catalysts. Asymmetric synthesis in the presence of chiral phosphoramides is realized.

[1]Denmark, S.E., Winter, S.B.D., Su, X., Wong, K.-T. *JACS* **118**, 7404 (1996).

Tetracyanoethylene (TCNE).

Acetonides from epoxides.[1] TCNE catalyzes the insertion reaction of acetone into a C–O bond of an epoxide. On the other hand, it acts as a Lewis acid in MeCN, inducing rearrangement to give carbonyl products. The presence of an alkoxy group in a side chain of the substrates is required.

Alcoholysis of epoxides.[2] Epoxide opening with alcohols to afford monoethers of 1,2-diols is promoted by TCNE at room temperature.

[1]Masaki, Y., Miura, T., Ochiai, M. *CL* 17 (1993).
[2]Masaki, Y., Miura, T., Ochiai, M. *BCSJ* **69**, 195 (1996).

Tetraethylammonium formate.

Monobenzylated Meldrum's acids.[1] The salt is a catalyst for the condensation of Meldrum's acid with aromatic aldehydes and is a reducing agent for the benzylidene derivatives.

[1]Toth, G., Köver, K.E. *SC* **25**, 3067 (1995).

Tetrahydro-2*H*-1,3-oxazin-2-ones.

Aminopropylation.[1] Reaction of these compounds with arylamines and arenethiols without a solvent at about 180° results in ring opening and loss of carbon dioxide.

[1]Poindexter, G.S., Strauss, K.M. *SC* **23**, 1329 (1993).

Tetrakis(triphenylphosphine)palladium(0). 13, 289–294; **14**, 295–299; **15**, 300–304; **16**, 317–323; **17**, 327–331; **18**, 347–349

Deallylation.[1] Amines are denuded of one or more allyl group(s) by the catalytic transfer to barbituric acid, which is a known allyl group scavenger.

Stille coupling. A synthesis of *(Z)*-cinnamyl alcohols can exploit this coupling using stannoxane derivatives.[2] The coupling is favored by nitrobenzene as solvent, dramatic improvement of yields observed in some cases.

SnBu2 + I– ; $(Ph_3P)_4Pd$ - LiCl, DMF, 100° ; –OH

78% (*Z* : *E* 97 : 3)

The Stille coupling between aryl halides and bis(tributylstannyl)ethyne furnishes symmetrical tolanes.[3] Dienyl triflones are available in three steps from 1-alkynes,[4] with the reaction sequence terminated by a Stille coupling.

C_6H_{13}, I, SO_2CF_3 ; $(Ph_3P)_4Pd$, $SnBu_3$, PhMe Δ ; C_6H_{13}, SO_2CF_3

93%

The cross-coupling of aryl triflates and halides in the presence of hexamethylditin is a tandem process.[5] From pyridyl triflates, the formation of pyridylstannanes is implied.

OTf, N + Br– ; $(Ph_3P)_4Pd$ - LiCl, Me_3Sn-$SnMe_3$, dioxane ; N

55%

Suzuki coupling. Unsymmetrical biaryls are prepared by Suzuki coupling[6] at ambient temperature in *N,N*-dimethylacetamide in the presence of TlOH. Arylcyclopropanes are most amenable to assembly by a Suzuki coupling,[7,8] due to the ready availability of cyclopropylboronic acids and esters.

$(Ph_3P)_4Pd$ - *t*-BuOK; DME, Δ, 48 h; 80%

Other cross-couplings. Arylation of alkynes in which the aryl group is originated from the triarylphosphine ligand has been discovered.[9]

$(Ph_3P)_4Pd$ - Ph_3P - CuI; THF, rt

+ piperidine 97%

+ diisopropylamine 95%

The coupling of arylzinc halides with alkyl *(E)*-2,3-dibromoalkenoates occurs at C-3.[10] 3-Arylindoles are obtained from 3-indolylzinc halides with haloarenes.[11] A synthesis of 1,4-disubstituted butadienes[12] starts from the hydrozirconation of ethynyl phenyl selenide. After Pd(0)-catalyzed coupling with alkenyl bromides, the seleno group is replaced via Grignard reaction in the presence of $(Ph_3P)_2NiCl_2$.

PhSe—≡ → $Cp_2Zr(H)Cl$; THF, 0°; 40 min → PhSe–CH=CH–$ZrCp_2Cl$ → $(Ph_3P)_4Pd$; Br–CH=CH–Ph; rt, 3 h → PhSe–CH=CH–CH=CH–Ph; 82% (one-pot reaction)

A stereoselective approach to 1,4-dienes[13] relies on the Pd-catalyzed coupling of tin-substituted allylic halides with alkenylaluminums and subsequent protodestannylation. The method is useful for synthesizing certain insect pheromones.

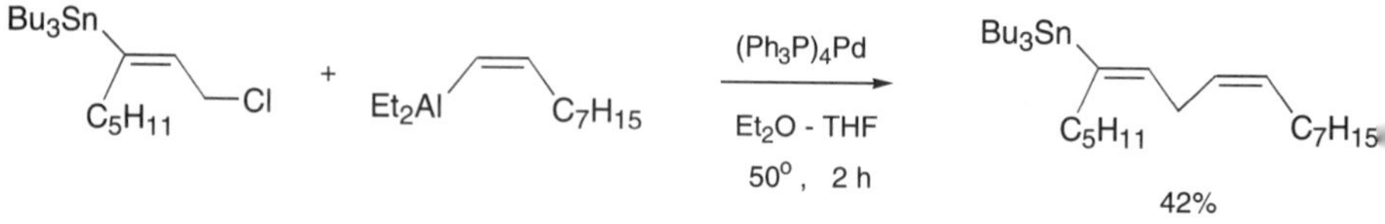

Coupling reactions of alkenyl halides with alkynes and allene are valuable for assembling enynes[14] and 2-sulfonylmethyl-1,3-dienes,[15] with sodium arenesulfinates present in the latter cases.

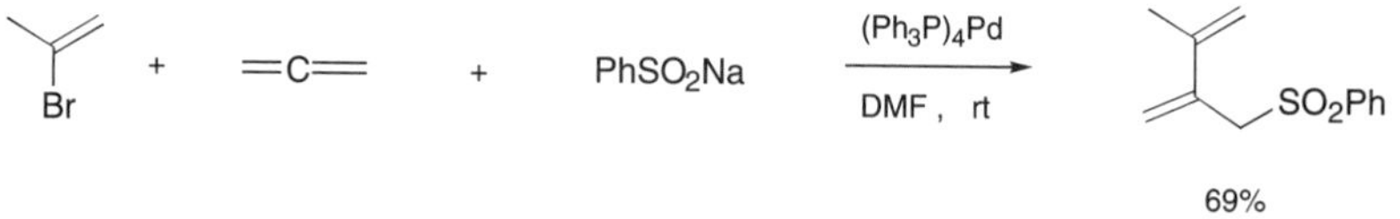

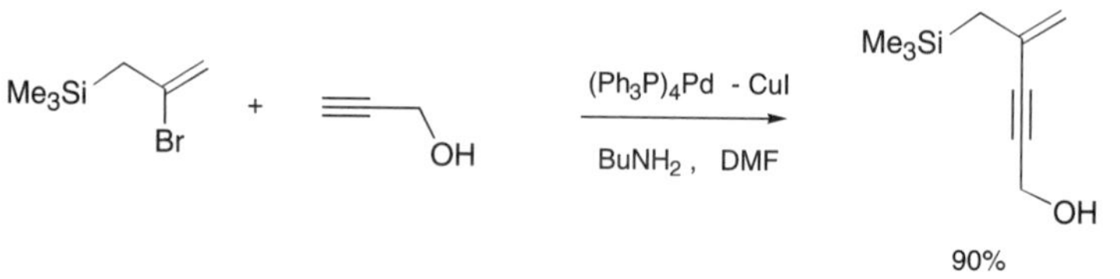

The chlorine atom of a 2-chloroquinoline is remarkably easy to displace in the Pd(0)-catalyzed reaction with 1-alkynes.[16]

Tetraarylphosphonium salts undergo the Heck reaction. Benzamides are produced when the phosphonium salts are treated with a secondary amine under CO.[17]

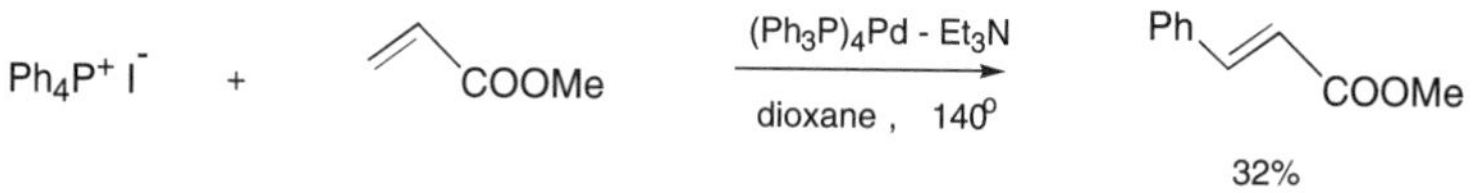

An intramolecular coupling featuring ring opening of a cyclopropane unit represents a novel way to functionalize indanes.[18]

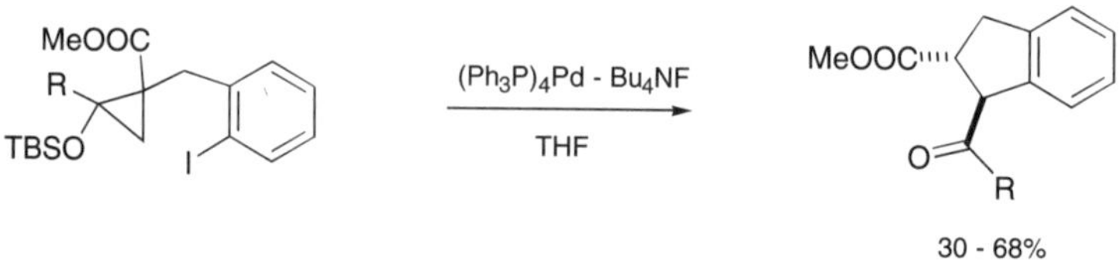

Carbonylations. Carbonylation reactions concomitant with cyclization of iodoalkenes leading to cyclopentenones[19] or *(Z)*-α-alkylidene-γ-butyrolactones[20] are of apparent interest to chemists in the field of synthesis. Other processes that unite three different components result in allyl ketones[21] and α-methylenephenones.[22]

C_6H_{13} $SiMe_3$ I — $(Ph_3P)_4Pd$ - Et_3N, CO (1 atm) / THF, 60° → C_6H_{13} $SiMe_3$ O 54%

— $(Ph_3P)_2PdCl_2$ - Et_3N, CO (40 atm) / MeOH, MeCN - PhH, 100° → C_6H_{13} $SiMe_3$ O COOMe 75%

C_5H_{11} I OH — $(Ph_3P)_4Pd$ - CO, PhMe, 50° → O C_5H_{11} O 67%

O Br + =C= HO — $(Ph_3P)_4Pd$ - CO, K_2CO_3 / DMF, 60°, 16 h → O O O 52%

OBz + CO + COOEt ZnI — $(Ph_3P)_4Pd$, rt, 18 h → O COOEt 62%

An annulation of a γ-butyrolactone ring to naphthols[23] recruits CO and an aldehyde to become the carbonyl group and the substituted α-carbon. Trifluoroacetic acid is a highly effective cocatalyst for this reaction, but HOAc, PhCOOH, and TsOH are not. The acid apparently promotes the formation of 2-(1-hydroxyalkyl)naphthols, which undergo carbonylation. That phenols are unreactive may be due to the difficulty in achieving this first step under the conditions employed.

Propargylic mesylates are converted to the transposed 2,3-alkadienoic esters and amides by $(Ph_3P)_4Pd$ and CO in the presence of alcohols and arylamines, respectively.[24] These allenoic acid derivatives are precursors of butenolides. In a synthesis of kallolide-B, this reaction is the critical step; $(Ph_3P)_4Pd$ is generated in situ from $(dba)_3Pd_2$ and Ph_3P in the presence of carbon monoxide.[25]

MsO
$(dba)_3Pd_2$ - PPh_3
CO , ROH / THF
H C O O Me_3Si
kallolide-B

Intramolecular reaction initiated by ionization of an allylic acetate and participated in by a distal allene linkage can have different ramifications when the molecule is also equipped with another double bond.[26] The separation between this latter double bond and the allene moiety determines the structure of the final product.

AcO
$(Ph_3P)_4Pd$ - CO
HOAc
80°
OAc
64%

OAc
$(Ph_3P)_4Pd$ - CO
HOAc
60°
COOH
28%

AcO
$(Ph_3P)_4Pd$ - CO
HOAc
75°
H O O
22%

Allylic substitutions. A notable exception to the general reactivity trend wherein carbonate is more reactive than acetate in the allylic substitution must be the consequence of a silicon effect.[27] Note that both $Pd(PPh_3)_4$ and $(dppe)_2Pd$ were employed.

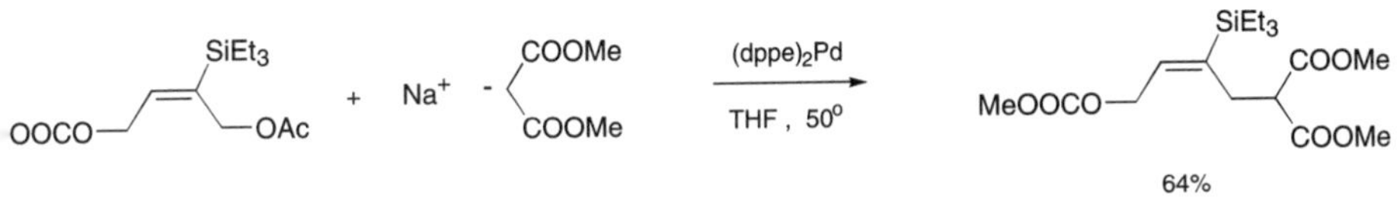

3-Alkoxyallyl acetates undergo substitution at the position α to the alkoxy group.[28] Allylic alcohols and amines become adequate allylating agents[29] under a moderate pressure of CO_2 and in the presence of $(Ph_3P)_4Pd$.

Umpolung substitutions in which allylic[30] or propargylic benzoates[31] are converted into nucleophiles are realized in the presence of diethylzinc.

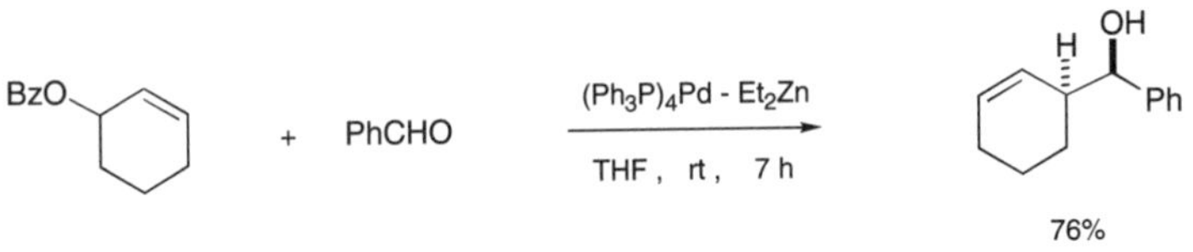

Rearrangements. Allylic sulfoximines undergo Pd-catalyzed rearrangement to give the tosylamides.[32] The reaction furnishes optically active amine derivatives from chiral sulfoximines.

The rearrangement of propargyl dienyl carbonates results in allenyl products.[33] The method is useful for synthesizing retinal.

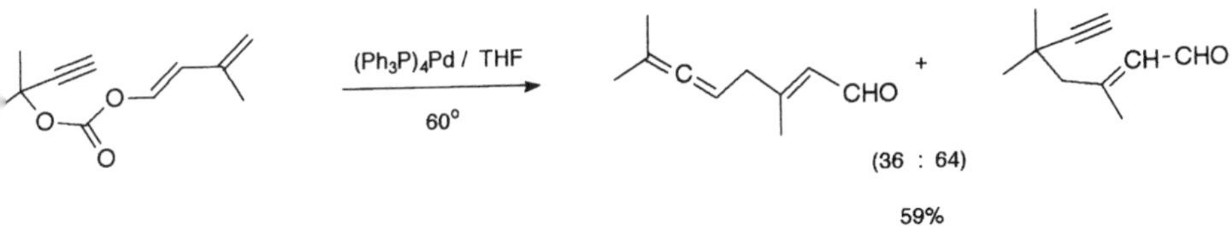

Cycloadditions. An intramolecular [2+3]-cycloaddition[34] involving a methylene-cyclopropane unit and a conjugated ester proceeds without stereochemical scrambling. A formal Diels–Alder reaction to rapidly build up the skeleton of rebeccamycin is reported.[35]

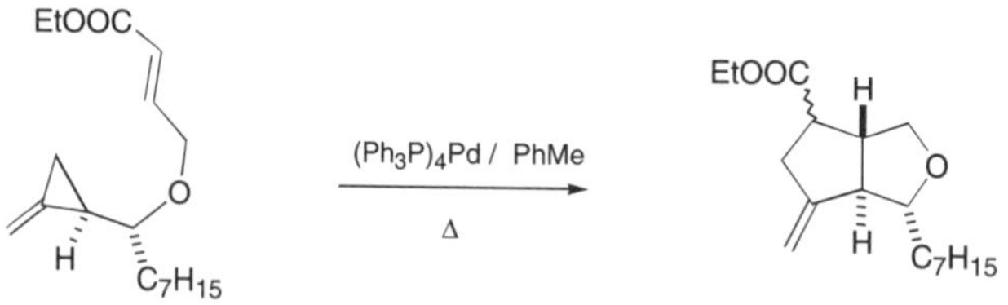

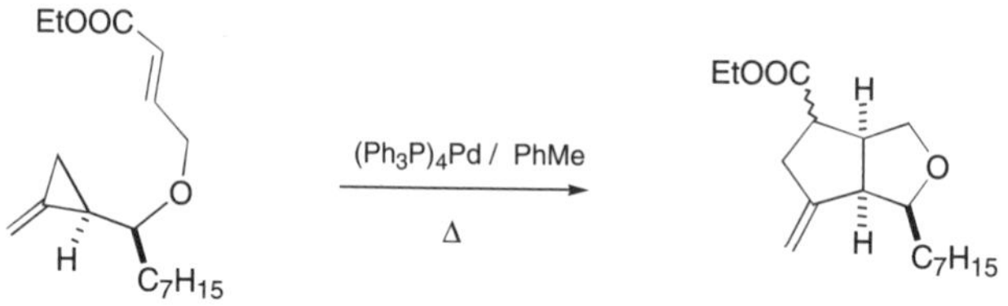

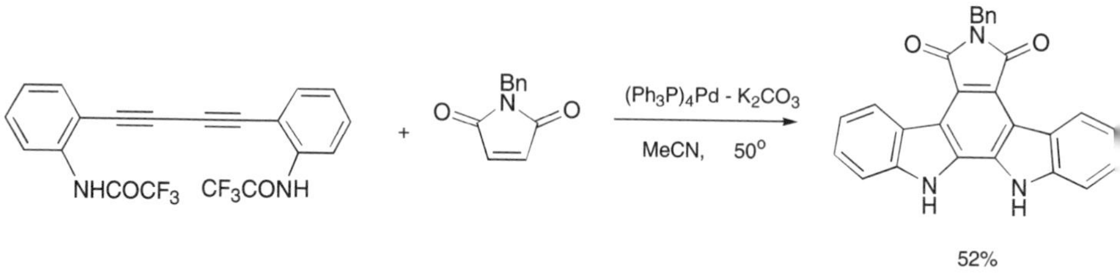

Conjugate enynes dimerize to give aromatic products.[36]

[1]Garro-Helion, F., Merzouk, A., Guibe, F. *JOC* **58**, 6109 (1993).
[2]Kraus, G.A., Watson, B.M. *TL* **37**, 5287 (1996).
[3]Cummins, C.H. *TL* **35**, 857 (1994).
[4]Xiang, J.S., Mahadevan, A., Fuchs, P.L. *JACS* **118**, 4284 (1996).
[5]Hitchcock, S.A., Mayhugh, D.R., Gregory, G.S. *TL* **36**, 9085 (1995).
[6]Anderson, J.C., Namli, H. *SL* 765 (1995).
[7]Wang, X.-Z., Deng, M.-Z. *JCS(P1)* 2663 (1996).
[8]Hildebrand, J.P., Marsden, S.P. *SL* 893 (1996).
[9]Buszek. K.R., Jeong, Y. *TL* **36**, 5677 (1995).
[10]Rossi, R., Bellina, F., Carpita, A., Mazzarella, F. *T* **52**, 4095 (1996).
[11]Sakamoto, T., Kondo, Y., Takazawa, N., Yamanaka, H. *JCS(P1)* 1927 (1996).
[12]Zhu, L.-S., Huang, Z.-Z., Huang, X. *T* **52**, 9819 (1996).
[13]Hutzinger, M.W., Oehlschlager, A.C. *JOC* **60**, 4595 (1995).
[14]Kabbara, J., Hoffmann, C., Schinzer, D. *S* 299 (1995).
[15]Vicart, N., Cazes, B., Gore, J. *T* **52**, 9101 (1996).
[16]Ciufolini, M.A., Mitchell, J.W., Roschangar, F. *TL* **37**, 8281 (1996).
[17]Sakamoto, M., Shimizu, I., Yamamoto, A. *CL* 1101 (1995).
[18]Khan, F.A., Czerwonka, R., Reissig, H.-U. *SL* 533 (1996).
[19]Negishi, E.-I., Ma, S., Amanfu, J., Coperet, C., Miller, J.A., Tour, J.M. *JACS* **118**, 5919 (1996).
[20]Luo, F.-T., Wang, M.-W., Liu, Y.-S. *H* **43**, 2725 (1996).
[21]Yasui, K., Fugami, K., Tanaka, S., Tamaru, Y. *JOC* **60**, 1365 (1995).
[22]Walkup, R.D., Guan, L., Kim, Y.S., Kim, S.W. *TL* **36**, 3805 (1995).
[23]Satoh, T., Tsuda, T., Kushino, Y., Miura, M., Nomura, M. *JOC* **61**, 6476 (1996).

[24]Marshall, J.A., Wolf, M.A. *JOC* **61**, 3238 (1996).
[25]Marshall, J.A., Wallace, E.M., Coan, P.S. *JOC* **60**, 796 (1995).
[26]Doi, T., Yanagisawa, A., Nakanishi, S., Yamamoto, K., Takahashi, T. *JOC* **61**, 2602 (1996).
[27]Thorimbert, S., Malacria, M. *TL* **37**, 8483 (1996).
[28]Vicart, N., Cazes, B., Gore, J. *TL* **36**, 535 (1995).
[29]Sakamoto, M., Shimizu, I., Yamamoto, A. *BCSJ* **69**, 1065 (1996).
[30]Tamaru, Y., Tanaka, A., Yasui, K., Goto, S., Tanaka, S. *ACIEE* **34**, 787 (1995).
[31]Tamaru, Y., Goto, S., Tanaka, A., Shimizu, M., Kimura, M. *ACIEE* **35**, 878 (1996).
[32]Pyne, S.G., Dong, Z. *JOC* **61**, 5517 (1996).
[33]Bienayme, H. *BSCF* **132**, 696 (1995).
[34]Lautens, M., Ren, Y. *JACS* **118**, 10668 (1996).
[35]Saulnier, M.G., Frennesson, D.B., Deshpande, M.S., Vyas, D.M. *TL* **36**, 7841 (1995).
[36]Saito, S., Salter, M.M., Gevorgyan, V., Tsuboya, N., Tando, K., Yamamoto, Y. *JACS* **118**, 3970 (1996).

2,2,3,3-Tetramethoxybutane and 9,9,10,10-tetramethoxy-9,10-dihydrophenanthrene.

Protection of 1,2-diols.[1,2] These reagents are prepared from the diketones and $HC(OMe)_3$ in methanol containing a little sulfuric acid. By an acid-catalyzed exchange reaction, 2,3-dimethoxy-1,4-dioxane derivatives are formed in the reaction with 1,2-diols. Diequatorial diols are selectively protected. Actually, the protection can be performed directly by the reaction of a 1,2-diol with an α-diketone, trimethyl orthoformate, in methanol in the presence of camphorsulfonic acid.[2]

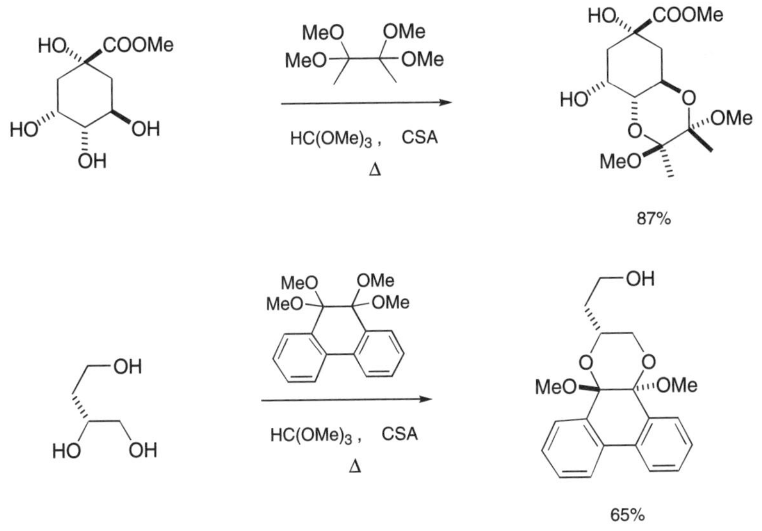

87%

65%

[1]Montchamp, J.-L., Tian, F., Hart, M.E., Frost, J.W. *JOC* **61**, 3897 (1996).

[2]Douglas, N.L., Ley, S.V., Osborn, H.M.I., Owen, D.R., Priepke, H.W.M., Warriner, S.L. *SL* 793 (1996).

Tetramethylammonium triacetoxyborohydride. 14, 299–300; **16**, 324; **17**, 331–332; **18**, 350

anti-Aminoalkyloxiranes.[1] Epoxy ketones undergo stereoselective reductive amination with primary amines in the presence of $Me_4NBH(OAc)_3$.

$Me_4NBH(OAc)_3$; $BnNH_2$, HOAc; mol. sieves; $ClCH_2CH_2Cl$; rt, 8 h

57% (*anti* : *syn* 94 : 6)

[1]Pegorier, L., Petit, Y., Larcheveque, M. *CC* 633 (1994).

***N,N,N′,N′*-Tetramethylethylenediammonium dichromate.**

Selective oxidation.[1] This salt oxidizes allylic and benzylic alcohols in refluxing dichloromethane.

[1]Chandrasekhar, S., Takhi, M., Mohapatra, S. *SC* **26**, 3947 (1996).

Tetramethylfluoroformamidinium hexafluorophosphate.

Peptide synthesis.[1] The reagent (**1**) is obtained from the corresponding chloro derivative by treatment with excess anhydrous KF in MeCN. It is nonhygroscopic and ideal for solid-phase coupling. Peptides of high purity are obtained in good yields.

$(Me_2N)_2C{=}O$ —($COCl_2$; NH_4PF_6)→ $(Me_2N)_2C^+Cl\ PF_6^-$ —(KF, MeCN)→ $(Me_2N)_2C^+F\ PF_6^-$ (**1**)

[1]Carpino, L.A., El-Faham, A. *JACS* **117**, 5401 (1995).

(2,2,6,6-Tetramethylpiperidino)magnesium halide.

As base.[1] These reagents are prepared in situ from an ordinary Grignard reagent such as EtMgX. They have been used to effect regioselective nuclear deprotonation of pyridinecarboxamides and pyridinyl carbamates. (For 3-substituted pyridines, metallation at C-4 is preferred.)

N-MgCl ; N-CHO / THF 0° CONEt2 CHO CONEt2 50%

Dithioacetals.[2] Sulfoxides bearing α-hydrogen atoms are converted into dithioacetals.

N-MgBr Et2O Δ 47 - 86%

[1]Schlecker, W., Huth, A., Ottow, E., Mulzer, J. *JOC* **60**, 8414 (1995).
[2]Kobayashi, K., Kawakita, M., Mannami, T., Morikawa, O., Konishi, H. *CL* 1551 (1994).

Tetramethyltin.

Michael reactions.[1] In the presence of Me_4Sn, the lithium enolate of cinchonine isobutyrate adds to enones with diastereoselectivity opposite to that arising from the iodozinc enolate, due to *re*-facial attack and *si*-facial attack on the enones, respectively. These results are explainable in terms of nonchelation and chelation transition states.

LDA / Me_4Sn / THF 43% (90% R)

$KN(SiMe_3)_2$ / ZnI_2 / DME 65% (92% S)

[1]Shimizu, M., Onogawa, Y., Fujisawa, T. *SL* 827 (1996).

Tetranitromethane. 18, 350

α-Nitro ketones.[1] Keeping a silyl enol ether and tetranitromethane in dichloromethane in the dark results in the formation of an α-nitro ketone. The photochemical nitration proceeds at –40°. Both processes involve radical cations as reactive intermediates.

[1]Rathore, R., Kochi, J.K. *JOC* **61**, 627 (1996).

Tetraphenylphosphonium hydrogendifluoride.

Aldol condensation.[1] $Ph_4P^+HF_2^-$ is a new catalyst for generating enolate ions from silyl enol ethers. This property allows its use in aldol reactions. $Ph_4P^+HF_2^-$ also cleaves silylalkynes.

[1]Bohsako, A., Asakura, C., Shioiri, T. *SL* 1033 (1995).

Thallium(III) nitrate. 16, 326; **18**, 351

Esters and amides.[1] The group exchange of hydrazides into esters and amides is mediated by TTN.

Ring contraction. The process is required to synthesize (+)-ferruginine based on a route starting from a chromium-promoted [6+2]photocycloaddition of azepines.[2] The proper bridged heterobicyclic system evolves from a TTN oxidation.

$Tl(ONO_2)_3$, MeOH → 85% → (+)-ferruginine

R* = (-)-8-phenylmenthyl

E = COOMe

[1]Kocevar, M., Mihorko, P., Polanc, S. *JOC* **60**, 1466 (1995).
[2]Rigby, J.H., Pigge, F.C. *JOC* **60**, 7392 (1995).

Thallium(III) trifluoroacetate.

2-(Trifluoromethyl)allylarenes.[1] Based on an oxidative pathway is a facile method for introducing a fluoromethylallyl group to arenes. It only requires the presence of $(CF_3COO)_3Tl$ and $CH_2{=}C(CF_3)CH_2SiMe_3$.

Thallation of phenol ethers.[2] Boron trifluoride etherate is added to a dichloroethane solution of the reagent and substrate in an improved procedure. The products are converted to the iodo compounds on treatment with aqueous KI.

Ring expansion and contraction. 1-Trimethylsiloxy-1-alkenylcycloalkanes undergo rearrangement on contact with $(CF_3COO)_3Tl$ to provide 2-methylenecycloalkanones (6 examples, 73–93%).[3]

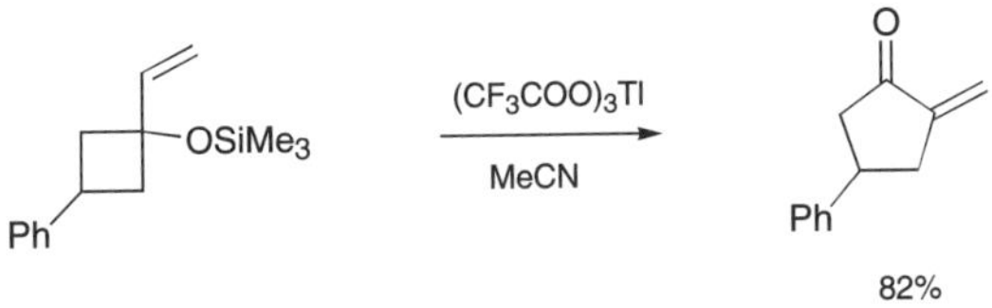

$(CF_3COO)_3Tl$, MeCN; 82%

[1]Fuchigami, T., Yamamoto, K. *CL* 937 (1996).
[2]dos Santos, M.L., de Magalhaes, G.C., Braz Filho, R. *JOMC* **526**, 15 (1996).
[3]Kim, S., Uh, K.H. *TL* **37**, 3865 (1996).

Thionyl chloride. 13, 297; **18**, 352

$RSO_3Na \rightarrow RCl$.[1] Alkanesulfonates are converted into alkyl chlorides by heating with $SOCl_2$ in DMF at 100° (5 examples, 46–81%).

1,3-Elimination.[2] 3,3-Bis(tributylstannyl)propanols lose $[Bu_3Sn/OH]$ on treatment with $SOCl_2$-pyridine in a stereoselective fashion to give cyclopropanes.

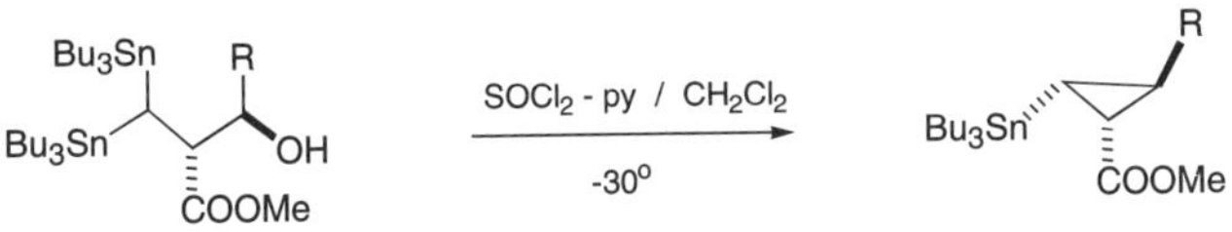

$SOCl_2$ - py / CH_2Cl_2, -30°

[1]Carlsen, P.H.J., Rist, O., Lund, T., Helland, I. *ACS* **49**, 701 (1995).
[2]Isono, N., Mori, M. *JOC* **61**, 7867 (1996).

***N,N′*-Thionyldiimidazole.**

Cyclic sulfites.[1] 1,2-Diols form cyclic sulfites very readily. Although there are few important synthetic uses for these compounds, their oxidation to cyclic sulfates provides excellent substrates for nucleophilic substitutions. Tetrahydrofuran ring closure in an intramolecular attack has enormous implication for the synthesis of a family of natural products, because proper design can precipitate a cascade process.

[1]Beauchamp, T.J., Powers, J.P., Rychnovsky, S.D. *JACS* **117**, 12873 (1995).

Thiophosgene.

Cyclic thionocarbonates.[1] 1,2-Diols form thionocarbonates on treatment with $CSCl_2$, DMAP, and pyridine. These products undergo ring opening with azide, thiolate, and acylthiolate ions. For those derived from 2,3-dihydroxy carboxylic esters, the ring opening occurs at C-2. Although the thionocarbonates are less reactive than the corresponding cyclic sulfates, the primary products decompose to the alcohols on aqueous workup, without the need for an extra hydrolysis step. On the other hand, the half-sulfate esters often require harsh conditions to release the hydroxyl group.

[1]Ko, S.Y. *JOC* **60**, 6250 (1995).

Thiourea.

(Z) → (E) Isomerization.[1] Conjugated aldehydes having a *(Z)*-configuration undergo isomerization in the presence of thiourea and K_2CO_3 in DMF at room temperature (3 examples, 72–90%).

Isothiocyanates. Both α-chloro aldoximes[2] and nitroalkanes[3] give isothiocyanates on reaction with thiourea. The former needs a base (e.g., Et_3N) to generate nitrile oxides in situ, and the latter requires both a base and a dehydrating agent, such as 4-chlorophenyl isocyanate.

[1]Phillips, O.A., Eby, P., Maiti, S.N. *SC* **25**, 87 (1995).
[2]Kim, J.N., Ryu, E.K. *TL* **34**, 8283 (1993).
[3]Kim, J.N., Song, J.H., Ryu, E.K. *SC* **24**, 1101 (1994).

Tin(II) bromide. 14, 303–304; 18, 352

Allylations. A two-phase reaction between allylic bromides and carbonyl compounds catalyzed by $SnBr_2$ is α-selective.[1] On the other hand, the reaction using SnI_2 and Bu_4NBr is γ-selective.[2]

Enantioselective acylation.[3] A nonenzymatic method for resolving secondary alcohols uses a chiral amine complex of $SnBr_2$.

OH / Ph + PhCOBr —($SnBr_2$ / CH_2Cl_2; chiral diamine (N-Me pyrrolidinylmethyl isoindoline), mol. sieves -78°)→ OCOPh / Ph, 44% (59% ee) + OH / Ph, 43% (61% ee)

[1]Masuyama, Y., Kishida, M., Kurusu, Y. *CC* 1405 (1995).
[2]Masuyama, Y., Kishida, M., Kurusu, Y. *TL* **37**, 7103 (1996).
[3]Oriyama, T., Hori, Y., Imai, K., Sasaki, R. *TL* **37**, 8543 (1996).

Tin(II) chloride. 13, 298–299; **15**, 309–310; **16**, 329; **18**, 353–354

Allylations. A direct preparation of allyltin reagents[1] from allyl halides is accomplished by treatment with $SnCl_2$. Allytributylstannane is activated by $SnCl_2$ in MeCN for reaction with carbonyl compounds and imines.[2] Allylic alcohols[3] and cyclic carbonates[4] can also be used as allylating agents for amines and aldehydes, respectively.

COOEt / Br —($SnCl_2$ - LiBr; CH_2Cl_2 - Et_2O, Δ)→ COOEt / $SnCl_2Br$ (74%) —(PhCHO; Et_2O)→ lactone, Ph (87%)

Dichloronitromethyl ketones.[5] These compounds can be prepared by the reaction of acid chlorides with a reagent derived from Cl_3CNO_2 and $SnCl_2$.

Cleavage of benzylidene acetals.[6] These derivatives are frequently encountered in sugar chemistry. A highly useful method for their cleavage involves treatment with $SnCl_2$ in dichloromethane at room temperature.

Hydrolysis of dimethylhydrazones.[7] The regeneration of carbonyl compounds is promoted by $Pd(OAc)_2$–$SnCl_2$.

2,3-Alkadienoic esters.[8] Ketenes are generated from unsaturated esters of 2,6-di-*t*-butyl-4-methylphenol on reaction with RLi–$SnCl_2$. An immediate Emmons– Wadsworth reaction provides the allenes.

RLi - $SnCl_2$, THF; $(MeO)_2P(O)CHCOOMe$, LDA; 52%

[1]Fouquet, E., Gabriel, A., Maillard, B., Pereyre, M. *BSCF* **132**, 590 (1995).
[2]Yasuda, M., Sugawa, Y., Yamamoto, A., Shibata, I., Baba, A. *TL* **37**, 5951 (1996).
[3]Masuyama, Y., Kagawa, M., Kurusu, Y. *CL* 1121 (1995).
[4]Kang, S.-K., Park, D.-C., Park, C.-H., Jang, S.-B. *SC* **25**, 1359 (1995).
[5]Demir, A.S., Tanyeli, C., Aksoy, H., Gulbeyaz, V., Mahasneh, A.S. *S* 1071 (1995).
[6]Xia, J., Hui, Y. *SC* **26**, 881 (1996).
[7]Mino, T., Hirota, T., Yamashita, M. *SL* 999 (1996).
[8]Tanaka, K., Otsubo, K., Fuji, K. *TL* **36**, 9513 (1995).

Tin(IV) chloride. 13, 300–301; **14**, 304–306; **15**, 311–313; **17**, 335–340; **18**, 354–356

Boc group removal.[1] Deprotection of Boc-amino acid residues can be performed with $SnCl_4$ in an organic solvent such as ethyl acetate at room temperature.

Desilylfunctionalizations. Aryltrichlorostannanes are formed[2] when activated silyl arenes (*o*- or *p*-methoxylated) are heated with $SnCl_4$. Catalyzed reactions of allylsilanes with tris(phenylchalcogeno)methanes give homologous chalcogenacetals.[3]

Heterocycloadditions. α-Alkoxy hydroperoxides undergo ionization on treatment with a Lewis acid (e.g., $SnCl_4$). Trapping such species with nucleophilic alkenes delivers 1,2-dioxolanes.[4] The use of $TiCl_4$ in this reaction leads to lower yields of the heterocyclic products or even only to the homoallylic ethers resulting from displacement of the peroxy group by an allyl group.

OMe, OOH; $SnCl_4$ / CH_2Cl_2, $SiMe_3$; $SiMe_3$, O–O; 42%

Aziridine-2-carboxylic esters are prepared[5] from hexahydro-1,3,5-triazines with alkyl diazoacetates in the presence of $SnCl_4$. *N*-Tosylmethylanilines apparently also undergo ionization to generate iminium species that are interceptable by alkenes. 1,2,3,4-Tetrahydroquinolines are formed.[6]

Two different Lewis acids, $SnCl_4$ and Me_2AlCl, exert opposite stereochemical results in a hetero-Diels–Alder reaction by conformational effects.[7]

EtO, AcO + O, OEt, O, N, O, O — CH_2Cl_2, -78° →

$SnCl_4$, 0.5 h	85%	1	:	15
Me_2AlCl, 48 h	82%	10	:	1

***o*-Alkenylation of phenols.**[8] 1-Alkynes are incorporated into the aromatic ring to form *o*-(alken-2-yl)phenols. The reaction is catalyzed by $SnCl_4$–Bu_4N.

Chlorination.[9] Tin(IV) chloride provides "Cl^{+}" when it is mixed with $Pb(OAc)_4$, and therefore, such a mixture is useful for chlorinating activated aromatic compounds.

Ene reactions. Intramolecular ene reaction of allenes with a carbonyl group or aldimine as the enophile is greatly facilitated by the presence of a silyl substituent on the allene moiety, reflecting a dipolar transition state.[10] Tin(IV) chloride lowers the reaction temperature to 0° to 25°.

$SiMe_2Ph$, H, C, O — $BnNH_2$; $SnCl_4$ / PhH → $SiMe_2Ph$, NHBn

70%

[1]Frank, R., Schutkowski, M. *CC* 2509 (1996).
[2]Lazarev, I.M., Dolgushin, G.V., Feshin, V.P., Voronkov, M.G. *MC* 150 (1996).
[3]Silveira, C.C., Fiorin, G.L., Braga, A.L. *TL* **37**, 6085 (1996).
[4]Dussault, P.H., Zope, U. *TL* **36**, 3655 (1995).
[5]Ha, H.-J., Kang, K.-H., Suh, J.-M., Ahn, Y.-G. *TL* **37**, 7069 (1996).
[6]Beifuss, U., Kunz, O., Ledderhose, S., Taraschewski, M., Tonko, C. *SL* 34 (1996).
[7]Tietze, L.F., Schneider, C., Grote, A. *CEJ* **2**, 139 (1996).
[8]Yamaguchi, M., Hayashi, A., Hirama, M. *JACS* **117**, 1151 (1995).
[9]Muathen, H.A. *T* **52**, 8863 (1996).
[10]Jin, J., Smith, D.T., Weinreb, S.M. *JOC* **60**, 5366 (1995).

Tin(II) hexamethyldisilazide. 18, 357

Allyl radical precursors.[1] The nucleophilic $Sn[N(SiMe_3)_2]$ reacts with various allylic halides, and the resulting allyltin(IV) species are a source of allyl radicals. Thus, in the presence of AIBN, coupling with alkyl halides occurs.

[1]Fouquet, E., Pereyre, M., Roulet, T. *CC* 2387 (1995).

Tin(II) triflate. 13, 301–302; **14**, 306–307; **15**, 313–314; **17**, 341–344; **18**, 357–358

Alkynylations.[1] α-Alkoxy amines undergo ionization in the presence of $Sn(OTf)_2$, allowing a reaction with stannylalkynes.

[1]Nagasaka, T., Nishida, S., Adachi, K., Kawahara, T., Sugihara, S., Hamaguchi, F. *H* ***36***, 2657 (1993).

Titanium(III) chloride. 13, 302; **16**, 330; **18**, 358–359

Pinacolization. In dichloromethane, aromatic aldehydes undergo reductive dimerization under the influence of $TiCl_3$. The reaction is highly stereoselective, giving the *syn*-1,2-diols predominantly (*syn:anti* ~200:1).[1] A cross-coupling is the major reaction when methyl phenylglyoxylate and pyridine are present,[2] and β-amino-α-hydroxy esters are obtained with the further introduction of an amine.[3]

ArCHO + $PhNH_2$ + PhC(O)COOMe —($TiCl_3$ - py / THF, rt)→ Ar–CH(NHPh)–C(OH)(Ph)–COOMe

53 - 67%

[1]Clerici, A., Clerici, L., Porta, O. *TL* **37**, 3035 (1996).
[2]Clerici, A., Clerici, L., Malpezzi, L., Porta, O. *T* **51**, 13385 (1995).
[3]Clerici, A., Clerici, L., Porta, O. *TL* **36**, 5955 (1995).

Titanium(III) chloride–lithium.

De-N-protection.[1] *N*-Allyl and *N*-benzyl derivatives undergo dealkylation on heating with $TiCl_3$–Li in THF under argon. The application of this method to the deoxygenation of aryl ketones[2] via reductive amination with aniline is too elaborate (taking three steps) to be of any synthetic significance.

[1]Talukdar, S., Banerji, A. *SC* **25**, 813 (1995).
[2]Talukdar, S., Banerji, A. *SC* **26**, 1051 (1996).

Titanium(III) chloride–sodium/alumina.

McMurry coupling.[1] Acylsilanes undergo coupling to give enedisilanes. It is interesting that the Brook rearrangement intervenes to afford β-silylalkenyl silyl ethers when the reagent is changed to Ti–Me_3SiCl.

COSiMe3 COSiMe3 → TiCl3, Na - Al2O3, DME → SiMe3 SiMe3 (49%) + OSiMe3 SiMe3 (44%)

[1]Fürstner, A., Seidel, G., Gabor, B., Kopiske, C., Krüger, C., Mynott, R. *T* **51**, 8875 (1995).

Titanium(III) chloride–zinc.

McMurry coupling. Pyrroles[1] and indoles[2] are prepared from keto amides. The reaction is suitable for the synthesis of lukianol-A and lamellarin O dimethyl ether. Titanium needs to be present only in substoichiometric quantities, and trimethylsilyl chloride is a useful activator.

Ph, O, NHCOCF3 → TiCl3 - Zn - Me3SiCl, MeCN Δ → Ph, CF3, N H (88%)

[1]Fürstner, A., Weintritt, H., Hupperts, A. *JOC* **60**, 6637 (1995).
[2]Fürstner, A., Hupperts, A. *JACS* **117**, 4468 (1995).

Titanium(IV) chloride. 13, 304–309; **14**, 309–311; **15**, 317–320; **16**, 332–337; **17**, 344–347; **18**, 359–361

Allylations. Double stereodifferentiation in the crotylation of aldehydes[1] permits the construction of polypropionate chain segments of natural products. The diastereoselection and absolute stereochemistry are determined by the local chiralities of the two components.

BnO CHO + COOMe Me2SiPh → TiCl4 / CH2Cl2, -78° → BnO OH COOMe (64%)

The temperature dependence of chemoselectivity[2] for lactone opening or displacement of a γ-alkoxy group has been discovered.

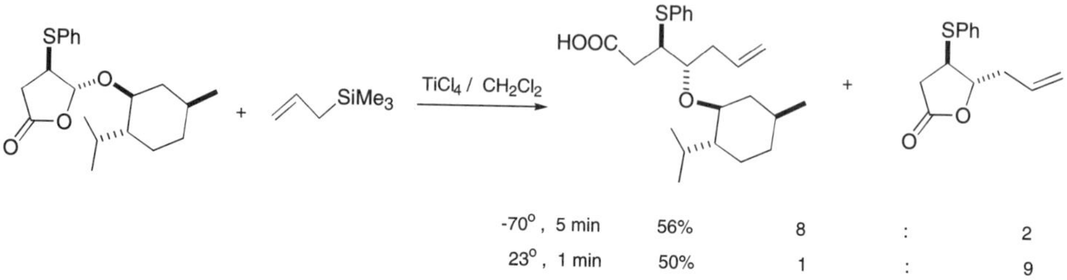

-70°, 5 min	56%	8	:	2
23°, 1 min	50%	1	:	9

Intramolecular electrophilic reaction of imines using a 2-propylidene-1,3-bis(silane) unit as nucleophile can lead to bridged heterocyclic frameworks.[3]

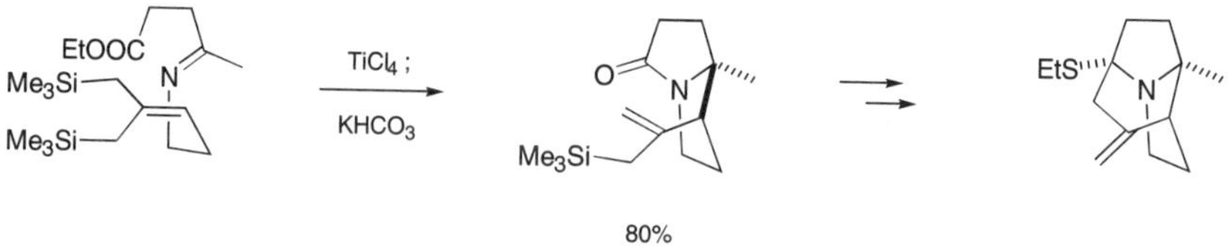

80%

Cycloadditions involving allylsilanes. $TiCl_4$ promotes [2+2]-cycloaddition of allylsilanes to α-keto esters, leading to oxetanes.[4] A [4+2]-cycloaddition process that gives 4-silylmethyl-1,2,3,4-tetrahydroquinolines is in contrast to the formation of *N*-(3-butenyl)anilines using allylstannanes instead of allylsilanes.[5] (Note that the cycloaddition need not be restricted to allylsilanes and that *N*-(benzenesulfenylmethyl)anilines undergo the same cycloaddition with alkenes in the presence of $TiCl_4$–PPh_3.)[6]

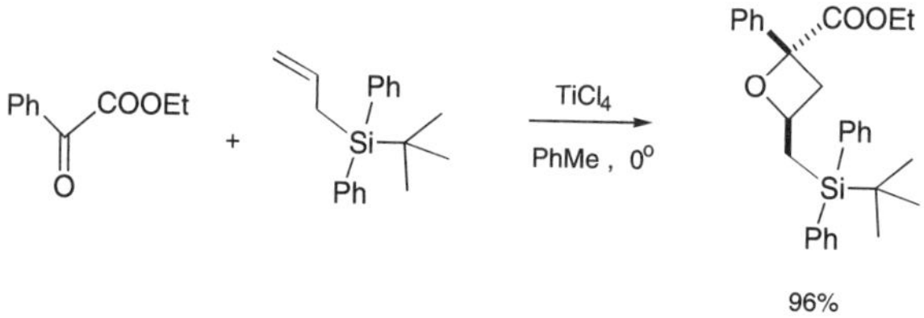

96%

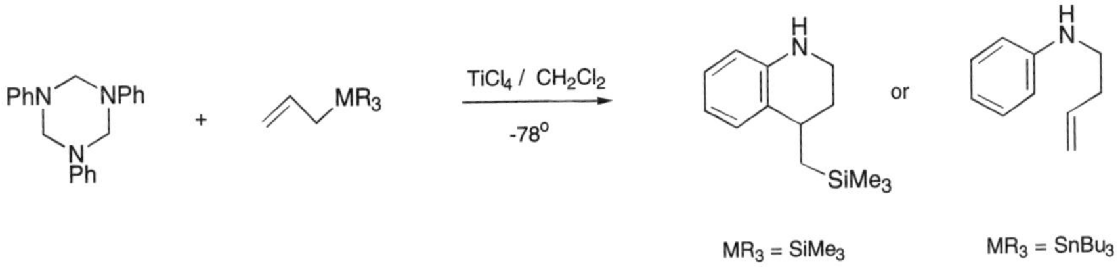

$MR_3 = SiMe_3$ $MR_3 = SnBu_3$

Michael and aldol reactions. The Mukaiyama–Michael reaction of ketene silyl acetals is diastereoselective.[7] Either the *syn*- or the *anti*-isomers are obtained at will.

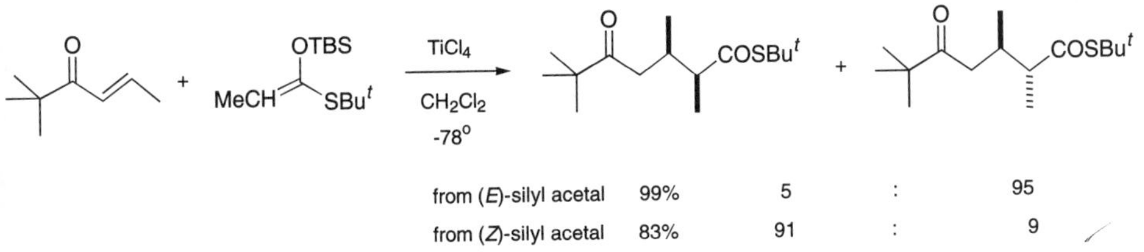

from (*E*)-silyl acetal	99%	5	:	95
from (*Z*)-silyl acetal	83%	91	:	9

As an alternative method to the foregoing allylation for the assembly of polypropionate chains, the aldol reaction[8] is useful. An analogous reaction of imines provides β-amino esters in which substituents at the 2,3-positions are *anti*.[9]

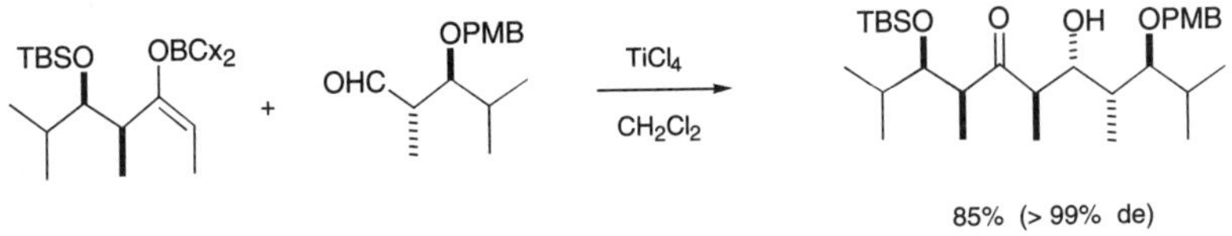

85% (> 99% de)

Carbocyclization. α-Phenylseleno ketones bearing an unsaturated side chain undergo cyclization accompanied by seleno group transfer.[10]

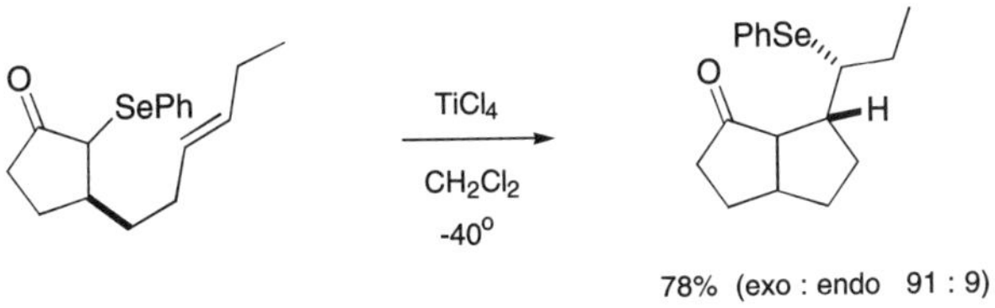

78% (exo : endo 91 : 9)

Anilinomethyl azides.[11] *N*-(Methoxymethyl)anilines readily undergo functional group exchange. With $TiCl_4$ and Me_3SiN_3 in CH_2Cl_2 at –78° the methoxy group is replaced by azide.

Nitrogenation.[12] A remarkable reaction that uses N_2 to incorporate nitrogen into organic molecules is achieved by the catalytic system composed of $TiCl_4$ (1.25 equiv), Li (12.5 equiv), and Me_3SiCl (12.5 equiv).

$TiCl_4$ - Li - Me_3SiCl; THF, N_2; Δ, 24 h

Simmons–Smith reaction. The formation of $RCH{=}CHCH_2OZnCH_2I$ is not necessarily followed by cyclopropanation. However, the addition of catalytic amounts of $TiCl_4$ has a dramatic effect.[13] Other Lewis acids, such as Et_2AlCl, BBr_3, and $SiCl_4$, are effective as well. Asymmetric induction is realized by using TADDOLate–$Ti(OPr^i)_2$.

[1] Jain, N.F., Takenaka, N., Panek, J.S. *JACS* **118**, 12475 (1996).
[2] van Oeveren, A., Feringa, B.L. *JOC* **61**, 2920 (1996).
[3] Kercher, T., Livinghouse, T. *JACS* **118**, 4200 (1996).
[4] Akiyama, T., Kirino, M. *CL* 723 (1995).
[5] Ha, H.-J., Ahn, Y.-G., Chon, J.-K. *JCS(P1)* 2631 (1995).
[6] Beifuss, U., Ledderhose, S. *CC* 2137 (1995).
[7] Otera, J., Fujita, Y., Fukuzumi, S. *T* **52**, 9409 (1996).
[8] Evans, D.A., Dart, M.J., Duffy, J.L., Rieger, D.L. *JACS* **117**, 9073 (1995).
[9] Shimizu, M., Kume, K., Fujisawa, T. *CL* 545 (1996).
[10] Toru, T., Kawai, S., Ueno, Y. *SL* 539 (1996).
[11] Ha, H.-J., Ahn, Y.-G. *SC* **25**, 969 (1995).
[12] Mori, M., Kawaguchi, M., Hori, M., Hamaoka, S.-i. *H* **39**, 729 (1994); Hori, M., Mori, M. *JOC* **60**, 1480 (1995).
[13] Charette, A.B., Brochu, C. *JACS* **117**, 11367 (1995).

Titanium(IV) chloride–triethylamine.

Chiral aziridine-2-carboxylates.[1] The generation of enolates and their subsequent cyclization occur when β-benzyloxyamino carbonyl compounds are treated with $TiCl_4$–Et_3N. Since the chiral substrates are readily available, this reaction realizes a valuable synthesis of chiral aziridines.

NHOBn; Ph; $TiCl_4$; Et_3N - CH_2Cl_2; Ph

97%

Dieckmann cyclization.[2] The direction of cyclization of 3-heteroadipic esters promoted by $TiCl_4$–Et_3N is dependent on the heteroatom. Thus, the isomeric 5-thia-2-oxocyclopentanecarboxylic esters and 4-aza-2-oxocyclopentanecarboxylic esters are formed selectively.

COOEt, X, COOR — $TiCl_4$ - Et_3N, CH_2Cl_2, -10° → (S ring, O, COOMe) or (EtOOC, O, N Bn)

X = S, R = Me X = NR, R = Et

[1]Cardillo, G., Casolari, S., Gentilucci, L., Tomasini, C. *ACIEE* **35**, 1848 (1996).
[2]Deshmukh, M.N., Gangakhedkar, K.K., Kumar, U.S. *SC* **26**, 1657 (1996).

Titanium(IV) chloride–zinc. 13, 310–311; 18, 364

Reduction of sulfoxides.[1] The applicability of this system in reducing cyclobutyl aryl sulfoxides extends the use of the latter compounds. Thus, condensation with ketones followed by the reduction and Lewis acid-catalyzed rearrangement leads to 2,2-disubstituted cyclopentanones, including spiroketones.

(O, S, Tol, HO) — $TiCl_4$ - Zn, THF, 5 - 10°, 3 h → (S, Tol, HO) 92% — $SnCl_4$, CH_2Cl_2, 0 - 20°, 20 min → (O) 63%

Amides.[2] The reduction of nitro compounds in the presence of acid chlorides leads to amides directly.

[1]Fitjer, L., Schlotmann, W., Noltemeyer, M. *TL* **36**, 4985 (1995).
[2]Shi, D.-Q., Zhou, L.-H., Dai, G.-Y., Chen, W.-X. *YH* **15**, 272 (1995).

Titanium(IV) chloride tris(triflate).

Esterification.[1] The catalyst $TiCl(OTf)_3$ promotes the formation of ester from equimolar amounts of carboxylic acids and alcohols in the presence of octamethylcyclotetrasiloxane.

Friedel–Crafts acylation.[2] The catalyst combined with TfOH in MeCN is an effective promoter of arene acylation with anhydrides at room temperature (14 examples, 61–98%).

[1]Izumi, J., Shiina, I., Mukaiyama, T. *CL* 141 (1995).
[2]Izumi, J., Mukaiyama, T. *CL* 739 (1996).

Titanium(IV) chloride–titanium tetraisopropoxide.

Cyclopropanone hemiacetals.[1] A convenient preparation of these compounds involves the reductive coupling of 1-alkenes with ethylene carbonate. The catalyst is derived from a combination of RMgCl and $(i\text{-PrO})_3TiCl$.

c-C_5H_9MgCl, $(i\text{-PrO})_3TiCl$, THF, 0°; OTIPS; 47%

Fragmentation.[2] Cyclobutyl sulfides in which an epoxide or oxetane ring is attached to the adjacent carbon atom are liable to fragmentation. Both rings undergo cleavage, leading to allylic and homoallylic alcohols, respectively.

$(i\text{-PrO})_2TiCl_2$, CH_2Cl_2, -78°, $PhSSiMe_3$; PhS, SPh, OH; 77% (*E* : *Z* 97 : 3)

[1]Lee, J., Kim, Y.G., Bae, J.G., Cha, J.K. *JOC* **61**, 4878 (1996).
[2]Fujiwara, T., Tsuruta, Y., Takeda, T. *TL* **36**, 8435 (1995).

Titanium tetraisopropoxide. **13**, 311–313; **14**, 311–312; **15**, 322; **16**, 339; **17**, 347–348; **18**, 363–364

Reduction.[1] An indirect reduction of β-ketols through a Tishchenko reaction with benzaldehyde proceeds with excellent stereoselectivity. The actual reagent is the ate complex $Li[(i\text{-PrO})_4TiBu]$, which is formed in situ from BuLi and $(i\text{-PrO})_4Ti$.

OH O, Ph + PhCHO; $(i\text{-PrO})_4Ti$, BuLi; BzO OH, Ph + HO OBz, Ph; (5 : 95)

Baeyer–Villiger oxidation.[2] The Sharpless asymmetric epoxidation reagent can be used for the conversion of cyclobutanones to chiral γ-lactones.

(i-PrO)$_4$Ti, t-BuOOH - (+)-DIPT, -20°, 6 h; 34% ee

[1]Mahrwald, R., Costisella, B. *S* 1087 (1996).
[2]Lopp, M., Paju, A., Kanger, T., Pehk, T. *TL* **37**, 7583 (1996).

Titanium(IV) iodide–lithium aluminum hydride–fluorotrichloromethane.

Homologation of ketones.[1] A synthesis of α,α-disubstituted acetic acids by the chain extension of ketones involves condensation with $CFCl_3$ using a low-valent titanium reagent.

[1]Garcia, M., del Campo, C., Llama, E.F., Sinisterra, J.V *JCS(P1)* 1771 (1995).

Titanocene borohydride.

Reduction.[1] The reagent is prepared from Cp_2TiCl_2 and $NaBH_4$ in DME. The reagent prepared in situ works equally well, and the reduction profile is not changed, even as borane is cogenerated, owing to the superior reactivity of titanocene borohydride.

Cp_2TiBH_4, DME, rt, 5 min; 97% (trans : cis 97 : 3)

[1]Barden, M.C., Schwartz, J. *JOC* **60**, 5963 (1995).

Titanocene dicarbonyl.

Intramolecular Pauson–Khand reaction.[1] The commercially available $Cp_2Ti(CO)_2$ catalyzes cyclopentenone formation in a moderate pressure (18 psig) of CO. The synthesis of a dienone from an allenyl alkyne is interesting.

$Cp_2Ti(PMe_3)_2$, CO / PhMe, 90°; 86%

[1]Hicks, F.A., Kablaoui, N.M., Buchwald, S.L. *JACS* **118**, 9450 (1996).

Titanocene dichloride. 18, 364–365

Allyltitanocenes. Titanocene dichloride is dechlorinated by *i*-PrMgCl, and the resulting species forms π-allyltitanocenes with 1,3-dienes. Treatment with ClCOOR completes the reductive alkoxycarbonylation to afford β,γ-unsaturated esters.[1]

Allyltitanocenes are also derived from allyl sulfides, Cp_2TiCl_2, and BuLi,[2] and they react with aldehydes to form homoallylic alcohols.

Epoxidation.[3] A copolycondensate of Cp_2TiCl_2 and $Si(OEt)_4$ prepared by a modified sol-gel method is useful for catalyzing alkene epoxidation using *t*-BuOOH. However, the yields are not high.

Catalyzed reduction.[4] The reduction of azoarenes by $NaBH_4$ with Cp_2TiCl_2 as a catalyst (probably forming Cp_2TiBH_4) leads to arylamines.

[1]Szymoniak, J., Pagneux, S., Felix, D., Moise, C. *SL* 46 (1996).
[2]Takeda, T., Miura, I., Horikawa, Y., Fujiwara, T. *TL* **36**, 1495 (1995).
[3]Thorimbert, S., Klein, S., Maier, W.F. *T* **51**, 3787 (1995).
[4]Dosa, P., Kronish, I., McCallum, J., Schwartz, J., Barden, M.C. *JOC* **61**, 4886 (1996).

***N*-(*p*-Toluenesulfonyl)aziridines.**

Pyrrolidines.[1] Reaction of the dianions derived from β-ketoesters with *N*-tosylaziridine gives ε-tosylamino-β-ketoesters, which cyclize on treatment with an H+-form ion exchange resin.

β-Amino carbonyl compounds.[2] *N*-Tosylaziridines serve in the alkylation of 2-lithio-1,3-dithianes to give the β-tosylaminoethyl derivatives. The dithioacetal unit of such products is readily cleaved.

Ph, NTs + Li (dithiane, S, S) —THF, -78° -> 0°→ Ph, NHTs, S, S

81%

[1]Lygo, B. *SL* 764 (1993).
[2]Osborn, H.M.I., Sweeney, J.B., Howson, B. *SL* 675 (1993).

***p*-Toluenesulfonylimino iodobenzene. 17**, 348; **18**, 365

N-Tosylaziridines.[1] Aziridination of alkenes in the presence of a (diimine)copper complex is accelerated by ligands, and evidence strongly implicates a redox mechanism.

[1]Li, Z., Quan, R.W., Jacobsen, E.N. *JACS* **117**, 5889 (1995).

p-Toluenesulfonyl isocyanate.

N-Tosylbenzamides.[1] These compounds are prepared from arylstannanes and TsNCO under Friedel–Crafts reaction conditions.

Oxazolidin-2-ones.[2] 3-Alkene-1,2-diols undergo stereoselective heterocyclization on reaction with TsNCO in the presence of a Pd(0) catalyst.

$[(i\text{-PrO})_3P]_4Pd$, THF, Δ, 2 h; 80%

α′-Functionalized enones from α-allenic alcohols. The spontaneous [2,3]-sigmatropic rearrangement of the adducts obtained from the alcohols and TsNCO leads to nucleophilic dienol derivatives.[3] For example, iodination with NIS gives α′-iodo-α,β-enones.

CH_2Cl_2; NIS; 61%

[1]Arnswald, M., Neumann, W.P. *JOC* **58**, 7022 (1993).
[2]Xu, D., Sharpless, K.B. *TL* **34**, 951 (1993).
[3]Friesen, R.W., Blouin, M. *JOC* **61**, 7202 (1996).

Trialkylboranes.

α-Alkylhydroxylamines.[1] Nitrones react with R_3B to produce the hydroxylamines regioselectively.

Aldol reactions.[2] Boron enolates are formed from α-iodoacyl silanes and condense with aldehydes.

NaI; 90%; Et_3B, MeCHO, hexane, rt, 5 h; 79% (erythro : threo 94 : 6)

[1]Hollis, W.G., Smith, P.L., Hood, D.K., Cook, S.M. *JOC* **59**, 3485 (1994).
[2]Horiuchi, Y., Taniguchi, M., Oshima, K., Utimoto, K. *TL* **36**, 5353 (1995).

Trialkylgallanes.

Nitro group replacement.[1] β-Nitrostyrenes are converted to β-alkylstyrenes in a rather unusual reaction.

Ar–CH=CH–NO_2 —(R_3Ga, hexane, rt)→ Ar–CH=CH–R

R = Et, Bu, *c*-Hx 49 - 81%

[1]Han, Y., Huang, Y.-Z., Zhou, C.-M. *TL* **37**, 3347 (1996).

Trialkylphosphine–1,1′-(azodicarbonyl)dipiperidine.

Dehydrative alkylation.[1] Employing alcohols as alkylating agents for active methylene compounds (e.g., methyl benzenesulfonylacetate) can be achieved by the reagent Me_3P–ADDP. Diols and bromoalkanols furnish cycloalkanecarboxylates.

Ph(CH₂)₃OH + $PhSO_2CH_2COOMe$ —(Me_3P / THF - PhMe, ADDP)→ Ph(CH₂)₃CH(COOMe)SO_2Ph 91%

Unsymmetrical polyfluoroalkyl ethers.[2] A convenient synthesis is based on the activation of an alcohol with Bu_3P–ADDP for subsequent reaction with the fluorinated alcohol.

[1]Yu, J., Lai, J.-Y., Falck, J.R. *SL* 1127 (1995).
[2]Falck, J.R., Yu, J., Cho, H.-S. *TL* **35**, 5997 (1994).

Triarylbismuthane tosylimides.

Oxidation.[1] The reagents Ar_3Bi=NTs are formed quantitatively from Ar_3Bi and PhI=NTs in dichloromethane at room temperature. They are mild oxidizing agents for alcohols.

[1]Suzuki, H., Ikegami, T. *JCR(S)* 24 (1996).

Triarylbismuth–copper(II) acetate.

N-Arylation of amides.[1] This convenient arylation is also applicable to imides, ureas, sulfonamides, and anilines. Of particular interest is the arylation of β-lactams.

$$\text{4-acetoxyazetidin-2-one (NH)} + Ar_3Bi \xrightarrow[CH_2Cl_2]{Cu(OAc)_2 - Et_3N} \text{N-Ar, 4-OAc } \beta\text{-lactam} \quad 75 - 93\%$$

[1]Chan, D.M.T. *TL* **37**, 9013 (1996).

Tributylarsin–magnesium.

α,β-Unsaturated nitriles.[1] The condensation of aldehydes with chloroacetonitrile provides access to unsaturated nitriles (7 examples, 73–81%).

[1]Shen, Y., Yang, B. *SC* **26**, 4693 (1996).

Tributylgermanium hydride.

Double carbonylation.[1] As a radical mediator in the carbonylation of 4-alkenyl iodides, Bu_3GeH behaves differently from Bu_3SnH, because it is a slower hydrogen donor, giving the second acyl radical time to react with the cyclic ketone, which was formed in the first step. Instead of the keto aldehydes, fused γ-lactones become the major products.

$$\text{alkenyl iodide} \xrightarrow{Bu_3MH - CO} \text{keto aldehyde (CHO)} + \text{fused } \gamma\text{-lactone}$$

	keto aldehyde	γ-lactone
M = Ge	12%	40%
M = Sn	45%	-

[1]Tsunoi, S., Ryu, I., Yamasaki, S., Fukushima, H., Tanaka, M., Komatsu, M., Sonoda, N. *JACS* **118**, 10670 (1996).

Tributylstannylmethylamine.

2-Azaallyl anions. The condensation of aldehydes with $Bu_3SnCH_2NH_2$ gives Schiff bases, which readily undergo Sn/Li exchange. With a juxtaposed multiple bond, cyclization occurs. The assembly of a perhydroindole intermediate for the synthesis of the amaryllidaceae alkaloids (–)-amabiline and (–)-augustamine has been realized.

[1]Pearson, W.H., Lovering, F.E. *JACS* **117**, 12336 (1995).

Tributyltin cyanide.

O-Acylcyanohydrins. An excellent catalyst for the conversion of aldehydes (including pivalaldehyde and enals) into RCH(OCOOR′)CN by reaction with methyl cyanoformate at room temperature is Bu_3SnCN (16 examples, 78–98%).

[1]Scholl, M., Lim, C.-K., Fu, G.C. *JOC* **60**, 6229 (1995).

Tributyltin ethoxide.

(E)-1-Alkenylstannanes. A stereoselective synthesis starts from the hydrozirconation of 1-alkynes. Metal–metal exchange with retention of configuration is achieved on reaction with Bu_3SnOEt.

[1]Kim, S., Kim, K.H. *TL* **36**, 3725 (1995).

Tributyltin hydride. 13, 316–319; **14**, 312–318; **15**, 325–333; **16**, 343–350; **17**, 351–361; **18**, 368–371

Desulfonylation. A convenient and economical preparation of phenyl vinyl sulfone[1] involves the reduction of *(Z)*-1,2-bis(benzenesulfonyl)ethene with Bu_3SnH.

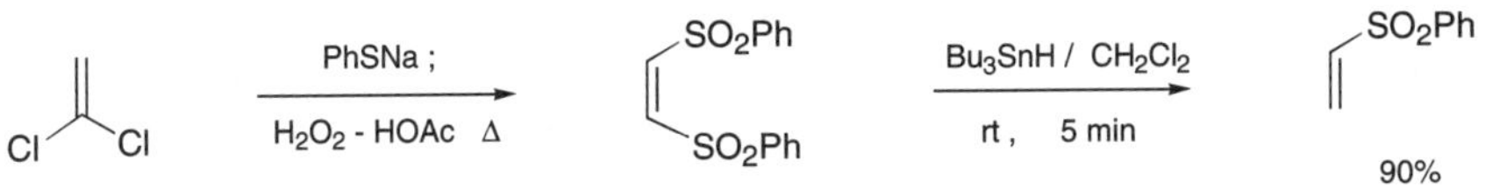

Halomethyltributylstannanes.[2] These valuable reagents (e.g., for Still–Wittig rearrangement) are prepared by a two-step procedure consisting of the hydroxymethylation of Bu_3SnH with paraformaldehyde and treatment with Ph_3P-CBr_4 or Ph_3P-NIS.

[1]Cossu, S., De Lucchi, O., Durr, R., Fabris, F. *SC* **26**, 211 (1996).
[2]Ahman, J., Somfai, P. *SC* **24**, 1117 (1994).

Tributyltin hydride–2,2′-azobisisobutyronitrile.

Desulfurization. A free-radical approach to the cleavage of *N*-sulfonyl amides[1] is applied by heating them with the tin hydride in toluene. However, the scope of this process is quite limited. Episulfides give the corresponding alkenes by a similar treatment.[2]

2-(Alkylamino)benzonitriles.[3] The reaction of α-(*o*-bromoarylamino)nitriles with Bu_3SnH results in debromination and cyano group migration.

Hydrodehalogenation. A method for preparing *exo*-2-methyl-1,3-dioxolanes of carbohydrates via the trichloromethyl derivatives has been described. Unlike acetalization with acetaldehyde, the bulky trichloromethyl group originating from chloral has a great tendency to be *exo*-oriented. Dechlorination with Bu_3SnH accomplishes the purpose.[4]

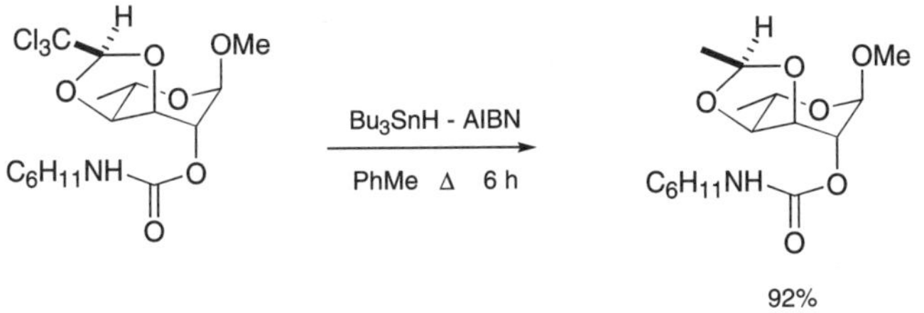

Ring contraction of lactones containing a bromine atom located at the carbon adjacent to the carbinyl position has been reported.[5] If such products are unwanted, the pathway can be suppressed by adding $(PhSe)_2$.

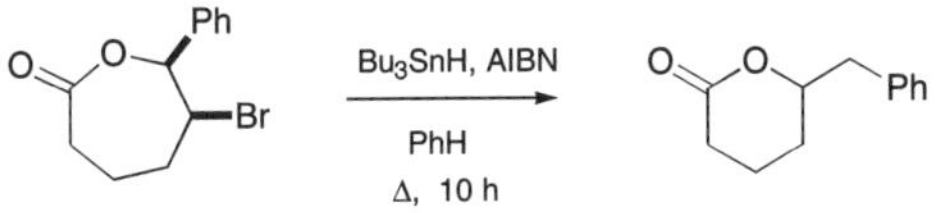

Radical cyclizations. When heated with tin hydride in benzene, thioesters[6] and selenoesters[7] bearing an alkenyl side chain give cyclic products. The side chain participates in the formation of a five- or six-membered ring with the acyl radical centers derived from such esters. Triethylborane can be used instead of AIBN as a radical initiator. *N*-(ω-Iodoalkenyl)acrylamides form 5/7-, 6/6-, or 6/8-fused bicyclic lactams of defined stereochemistry.[7a]

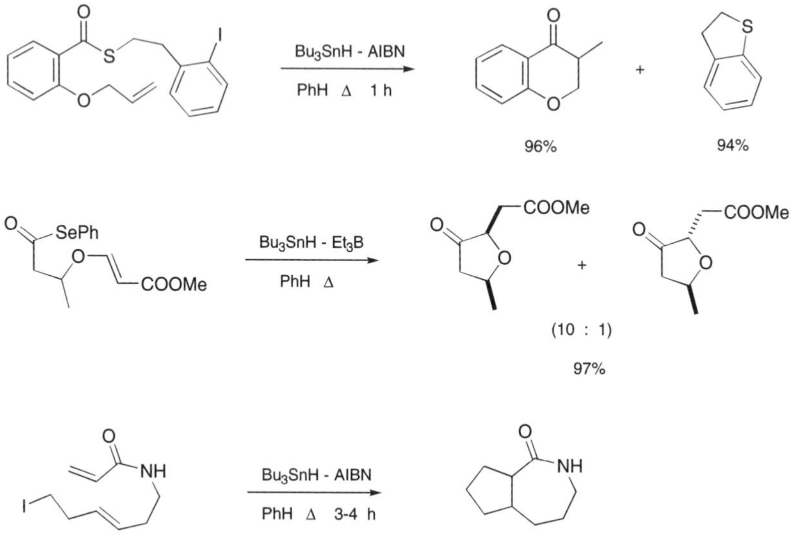

Bu_3SnH - AIBN, PhH Δ 1 h

96% + 94%

Bu_3SnH - Et_3B, PhH Δ

(10 : 1) 97%

Bu_3SnH - AIBN, PhH Δ 3-4 h

An interesting method of chain assembly is based on the formation of an oxasila heterobicycle and the subsequent oxidative cleavage of C–Si bonds with retention of configurations.[8]

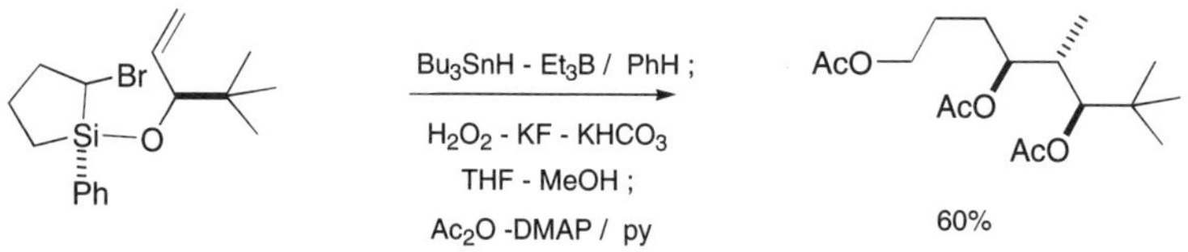

Bu_3SnH - Et_3B / PhH ; H_2O_2 - KF - $KHCO_3$ THF - MeOH ; Ac_2O -DMAP / py

60%

A γ-lactam radical can add intramolecularly to an acylsilane and trigger a radical version of the Brook rearrangement. Izidinones are readily prepared by this method.[9]

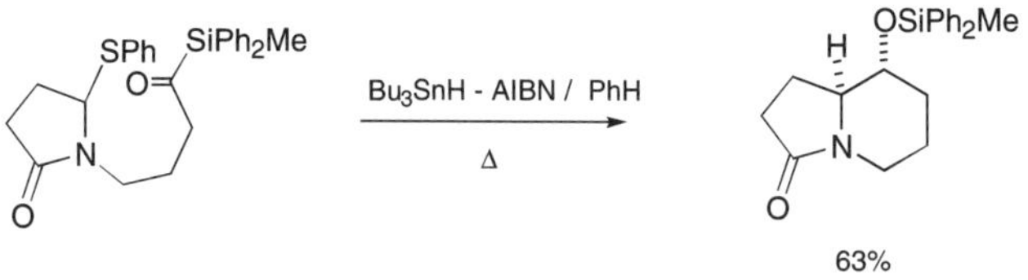

Bu_3SnH - AIBN / PhH, Δ

63%

A more conventional cyclization is involved in the synthesis of 2-benzenesulfonylmethylcyclopentanols from 6-bromo-3-hydroxy-1-alkenyl phenyl sulfones.[10] The major products from the *(E)*-sulfones are the *cis* cyclopentanols.

gem-Dialkoxy effects are manifested in radical reactions.[11]

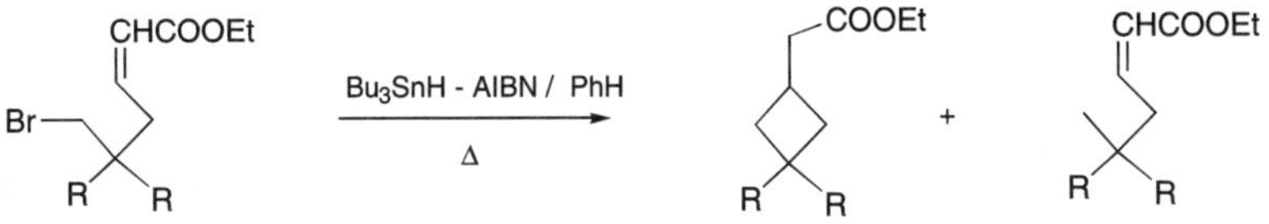

R = H	0	:	100
R = OEt	100	:	0

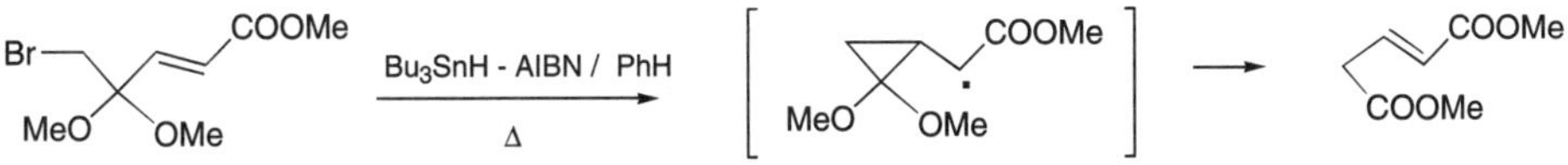

Cyclocarbonylation. Carbonylation followed by intramolecular addition of the acyl radical to a diphenylhydrazone[12] or an alkene (cf. tributylgermanium hydride)[13] furnishes a cyclopentanone product. The reaction of 4-alkenyl iodides is terminated by side-chain carbonylation.

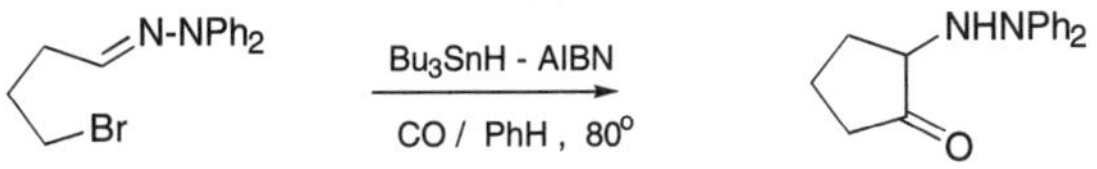

Small ring opening. The free-radical fragmentation of fused cyclobutanes allows the construction of fused medium rings[14] by a photocycloaddition approach.

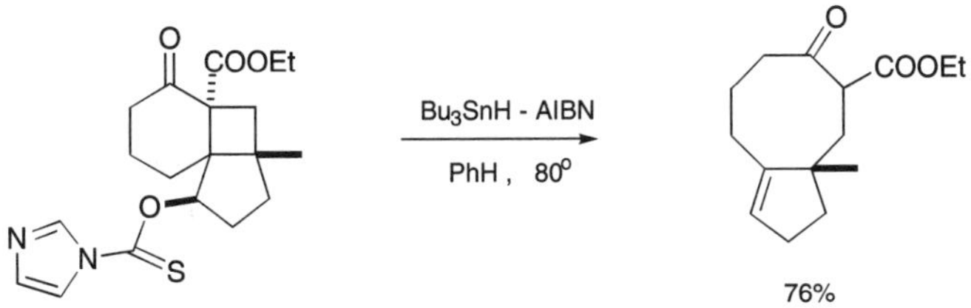

Vinylcyclopropanes undergo "homoconjugate" stannylformylation.[15] Ring opening occurs while the two functional groups are introduced at the termini.

Bu$_3$SnH - AIBN

CO / PhH, 80°

54% (*E* : *Z* 67 : 33)

Reductive alkylation of enones. Tin(IV) enolates are formed via the ketyls, which are derived from enones and Bu_3SnH. Reaction of the enolates with aldehydes furnishes aldols.[16] Quenching the enolates with alkyl halides in the presence of HMPA gives α-alkylated products.[17]

Bu$_3$SnH - AIBN / PhH ; PhCHO — 73% — TsOH, PhH — 92%

Nitriles. The oximation of nitroalkanes gives nitrolic acids, and further *O*-acylation and treatment with Bu_3SnH completes the transformation to nitriles.[18] Iminyl radicals are generated from *N*-benzotriazolylimines, and the fate of such radicals is either cyclization or fragmentation.[19]

Bu$_3$SnH - AIBN, PhH, 80° — 100%

Bu$_3$SnH - AIBN, PhH, 80° — 89%

Cyclic 1,2-diols.[20] Intramolecular pinacolization of 1,5- and 1,6-dialdehydes and keto aldehydes (but not diketones) can be effected with Bu_3SnH using AIBN thermally or photoinduced. The 1,2-cyclopentanediols thus generated are predominantly *cis*, whereas the *trans*-cyclohexanediols are slightly favored. Stannylenedioxy intermediates are implicated.

Hydrostannylation of conjugated esters.[21] The 8-phenylmenthyl esters of conjugated acids undergo stereoselective addition with Bu_3SnH in the presence of a Lewis acid (e.g., $BF_3 \cdot OEt_2$). The Lewis acid enforces the *s-trans* conformation in the transition state in which one face of the double bond is blocked by the phenyl ring. Accordingly, chiral β-stannyl esters are readily accessible. Their conversion to γ-hydroxy stannanes paves the way to optically active cyclopropanes. One report indicates that Et_3B is the initiator for the radical addition.

[1] Parsons, A.F., Pettifer, R.M. *TL* **37**, 1667 (1996).
[2] Izraelewicz, M.H., Nur, M., Spring, R.T., Turos, E. *JOC* **60**, 470 (1995).
[3] Cossy, J., Poitevin, C., Pardo, D.G., Peglion, J.L. *S* 1368 (1995).
[4] Rentsch, D., Miethchen, R. *CR* **293**, 139 (1996).
[5] Crich, D., Beckwith, A.L.J., Filzen, G.F., Longmore, R.W. *JACS* **118**, 7422 (1996).
[6] Crich, D., Yao, Q. *JOC* **61**, 3566 (1996).
[7] Evans, P.A., Roseman, J.D. *JOC* **61**, 2252 (1996).
[7a] Blake, A.J., Hollingworth, G.J., Pattenden, G. *SL* 643 (1996).
[8] Matsumoto, K., Miura, K., Oshima, K., Utimoto, K. *BCSJ* **68**, 625 (1995).
[9] Tsai, Y.-M., Nieh, H.-C., Pan, J.-S., Hsiao, D.-D. *CC* 2469 (1996).
[10] Adrio, J., Carretero, J.C., Arrayas, R.G. *SL* 640 (1996).
[11] Jung, M.E., Kiankarimi, M. *JOC* **60**, 7013 (1995).
[12] Brinza, I.M., Fallis, A.G. *JOC* **61**, 3580 (1996).
[13] Tsunoi, S., Ryu, I., Yamasaki, S., Fukushima, H., Tanaka, M., Komatsu, M., Sonoda, N. *JACS* **118**, 10670 (1996).
[14] Crimmins, M.T., Huang, S., Guise-Zawacki, L.E. *TL* **37**, 6519 (1996).
[15] Tsunoi, S., Ryu, I., Muraoka, H., Tanaka, M., Komatsu, M., Sonoda, N. *TL* **37**, 6729 (1996).
[16] Enholm, E.J., Whitley, P.E., Xie, Y. *JOC* **61**, 5384 (1996).
[17] Enholm, E.J., Whitley, P.E. *TL* **37**, 559 (1996).
[18] Chang, R.K., Kim, K. *TL* **37**, 7791 (1996).
[19] El Kaim, L., Meyer, C. *JOC* **61**, 1556 (1996).
[20] Hays, D.S., Fu, G.C. *JACS* **117**, 7283 (1995).
[21] Nishida, M., Nishida, A., Kawahara, N. *JOC* **61**, 3574 (1996).

Tributyltin hydride–organotin iodide.

Reductive opening of epoxides.[1] The selective opening of α,β-epoxy ketones leads to β-hydroxy ketones. A phosphine oxide is used as ligand.

Cl

Bu_3SnH - Bu_3SnI

$Ph_3P{=}O$

Cl

OH

59%

O

Ph

Bu_3SnH - Bu_3SnI

$Ph_3P{=}O$

OH O

Ph

58%

Reductive aldol reactions.[2] In situ aldol reaction of the tin enolates derived from the reduction of enones provides a method for the preparation of β-hydroxy ketones with the 2,3-*syn* isomers predominating.

[1]Kawakami, T., Tamizawa, D., Shibata, I., Baba, A. *TL* **36**, 9357 (1995).
[2]Kawakami, T., Miyatake, M., Shibata, I., Baba, A. *JOC* **61**, 376 (1996).

Tributyltin hydride–tetrakis(triphenylphosphine)palladium(0). 18, 372

(Z)-1-Bromo-1-alkenes.[1] The stereoselective debromination of the readily accessible 1,1-dibromoalkenes under mild conditions is rapid; it also allows the synthesis of enediynes and dienynes on subsequent coupling with alkynes, because the latter process is also catalyzed by Pd.

Bu_3SnH - $(Ph_3P)_4Pd$, PhH, rt, 1 h — 72%

(E)-Alkenyltributylstannanes.[2] Simultaneous debromination and hydrostannylation of 1-bromoalkynes occurs in high yields from a reaction at room temperature.

Bu_3SnH - $(Ph_3P)_4Pd$, THF, rt, 2 h — 80%

Ring scission of methylenecyclopropanes.[3] Hydrostannylation concomitant with ring opening gives homoallylic stannanes in a stereoselective manner. When the catalyst is changed to $Pd(OH)_2/C$, the products are distannanes due to further hydrostannylation of the vinyl group of the initial products.

Bu_3SnH / THF

+ $(Ph_3P)_4Pd$ — 68%

+ $Pd(OH)_2$ / C — 64%

[1]Uenishi, J., Kawahama, R., Shiga, Y., Yonemitsu, O., Tsuji, J. *TL* **37**, 6759 (1996); Uenishi, J., Kawahama, R., Yonemitsu, O., Tsuji, J. *JOC* **61**, 5716 (1996).
[2]Boden, C.D.J., Pattenden, G., Ye, T. *JCS(P1)* 2417 (1996).
[3]Lautens, M., Meyer, C., Lorenz, A. *JACS* **118**, 10676 (1996).

Tributyltin hydride–zirconium(IV) chloride.

(Z)-Alkenylstannanes.[1] The hydrostannylation of 1-alkynes in the presence of a Lewis acid is *anti*-selective.

tBuMe₂SiO (alkyne) → [Bu_3SnH -$ZrCl_4$ / PhMe, 0°; Et_3N, 0° -> rt] → tBuMe₂SiO (Z-alkenyl) $SnBu_3$ 87%

[1]Asao, N., Liu, J.-X., Sudoh, T., Yamamoto, Y. *CC* 2405 (1995).

Tributylphosphine–*N,N,N′,N′*-tetramethylazodicarboxamide.

Mitsunobu reaction.[1] This reagent system is useful for activating hindered secondary alcohols for inversion. 4-Methoxybenzoic acid is the nucleophile of choice.

[1]Tsunoda, T., Yamamiya, Y., Kawamura, Y., Ito, S. *TL* **36**, 2529 (1995).

Trichlorobis(triphenylphosphine)oxorhenium.

Sulfides/sulfoxide interchange. The reduction with triphenylphosphine requires the $(Ph_3P)_2RhOCl_3$ catalyst.[1] On the other hand, the same catalyst can be used to transfer the oxygen atom of diphenyl sulfoxide to other sulfides.[2]

[1] Arterburn, J.B., Perry, M.C. *TL* **37**, 7941 (1996).
[2]Arterburn, J.B., Nelson, S.L. *JOC* **61**, 2260 (1996).

Trichloroethylene.

Alkynyl sulfides.[1] A synthesis from alkyl- or arylthiols is accomplished through *S*-alkylation with trichloroethylene, complete dechlorination to generate the thioethynyllithium species, and *C*-alkylation.

SH → [KH / ClCH=CCl_2; BuLi; MeI / HMPA] → S (alkynyl sulfide) 98%

[1]Nebois, P., Kann, N., Greene, A.E. *JOC* **60**, 7690 (1995).

Trichlorosilane. 18, 373

Reduction of aldehydes and imines.[1] A mixture of $HSiCl_3$ and DMF forms a reducing system, converting aldehydes and imines to alcohols and amines, respectively, in dichloromethane.

Allylation. Allylic bromides are activated by converting them into allyltrichlorosilanes, which condense with aldehydes.[2] A convenient synthesis of α-methylene-γ-lactones is based on this process.

COOMe / Br —[Cl_3SiH; CuCl - iPr$_2$NEt, Et_2O Δ]→ COOMe / $SiCl_3$ —[RCHO; MeCN - DMF]→ OH COOMe / R — 63 - 84%

[1]Kobayashi, S., Yasuda, M., Hachiya, I. *CL* 407 (1996).
[2]Kobayashi, S., Yasuda, M., Nishio, K. *SL* 153 (1996).

Triethylsilane. 14, 322; **15**, 338; **16**, 356; **18**, 373–375

Defunctionalizations. With Pd(II) catalysts, Et_3SiH reduces organic halides[1] and aryl and enol triflates[2] to the hydrocarbons. Reductive cleavage of the C–O bond of a fused oxazolidine ring proceeds with retention of configuration.[3]

O, N, O, Ph —[Et_3SiH - $TiCl_4$]→ N, O, Ph, OH — 59% (80% de)

Aromatic ketones are deoxygenated[4] by reaction with $Et_3SiH–TiCl_4$.

Reduction of nitrogenous compounds. Tosylhydrazones are reduced to tosylhydrazines by $Et_3SiH–CF_3COOH$,[5] but the reduction of nitroarenes to arylamines is catalyzed by $(Ph_3P)_3RhCl$.[6]

Hydroboration and hydrosilylation. A mixture of Et_3SiH and BCl_3 is used for hydroboration of alkenes, leading to alkyldichloroboranes.[7] Conjugated dienes give *(Z)*-alkenyltriethylsilanes[8] on reaction with Et_3SiH photochemically in the presence of $Cr(CO)_6$.

[1]Boukherroub, R., Chatgilialoglu, C., Manuel, G. *OM* **15**, 1508 (1996).
[2]Kotsuki, H., Datta, P.K., Hayakawa, H., Suenaga, H. *S* 1348 (1995).
[3]Freville, S., Celerier, J.P., Thuy, V.M., Lhommet, G. *TA* **6**, 2651 (1995).

[4]Yato, M., Homma, K., Ishida, A. *H* **41**, 17 (1995).
[5]Wu, P.-L., Peng, S.-Y., Magrath, J. *S* 249 (1996).
[6]Brinkman, H.R., Miles, W.H., Hilborn, M.D., Smith, M.C. *SC* **26**, 973 (1996).
[7]Soundararajan, R., Matteson, D.S. *OM* **14**, 4157 (1995).
[8]Abdelqader, W., Chmielewski, D., Grevels, F.-W., Ozkar, S., Peynircioglu, N.B. *OM* **15**, 604 (1996).

Trifluoroacetic anhydride. 18, 376–377

Quinones.[1] *p*-Sulfinylphenols undergo Pummerer rearrangement on reaction with $(CF_3CO)_2O$. Hydrolysis of the products affords quinones.

HO—C₆Me₄—SOPh → $(CF_3CO)_2O$ / CH_2Cl_2; aq. $NaHCO_3$ - MeOH → tetramethyl-p-benzoquinone, 84%

Trifluoromethyl ketones.[2] The reaction of acid chlorides with $(CF_3CO)_2O$-pyridine gives $RCOCF_3$.

Vinylogous trifluoromethyl amides. The Potier–Polonovski rearrangement of tertiary amine *N*-oxides induced by $(CF_3CO)_2O$ is frequently followed by trifluoroacetylation.[3] This method is most suitable for the synthesis of (–)-altemicidin.[4]

$(CF_3CO)_2O$ - py; CH_2Cl_2 → 65%

[1]Akai, S., Takeda, Y., Iio, K., Yoshida, Y., Kita, Y. *CC* 1013 (1995).
[2]Boivin, J., El Kaim, L., Zard, S.Z. *T* **51**, 2573 (1995).
[3]Wenkert, E., Chauncy, B., Wentland, S.H. *SC* **3**, 73 (1973).
[4]Kende, A.S., Liu, K., Brands, K.M.J. *JACS* **117**, 10597 (1995).

Trifluoroacetonitrile.

4-Trifluoromethyl imidazolines.[1] CF_3CN is a 1,3-dipolarophile that forms 4-trifluoromethyl-Δ^3-imidazolines with azomethine ylides. Derived from simultaneous

N-acylation with Fmoc-protected amino acid fluorides, the adducts are amenable to peptide synthesis. The heterocycle is readily cleaved to generate 3,3,3-trifluoro-2-ketopropyl amides, the members of which are potent inhibitors of esterases and proteases.

Fmoc-NH ... F + N(=CHPh)CH₂SiMe₃ → [ylide intermediate] —CF_3CN→ product 70%

[1]Derstine, C.W., Smith, D.N., Katzenellenbogen, J.A. *JACS* **118**, 8485 (1996).

2-(Trifluoroacetylsulfenyl)pyridine–1-hydroxybenzotriazole.

Peptide synthesis.[1] The combined reagent is useful for acylating amino esters with N-protected amino acids with minimal racemization.

[1]Schmidt, U., Griesser, H. *CC* 1461 (1993).

(3,4,5-Trifluorobenzene)boronic acid.

Amides.[1] A carboxylic acid and an amine in refluxing toluene form the amide in the presence of this extremely active amidation catalyst (14 examples, 87–99%). Slightly higher temperatures are required for hindered acids, such as 1-adamantanecarboxylic acid, and for lactamization. Proline gives the tricyclic diketopiperazine (94% yield) in hot anisole.

[1]Ishihara, K., Ohara, S., Yamamoto, H. *JOC* **61**, 4196 (1996).

Trifluoromethanesulfonic acid (triflic acid). 14, 323–324; **15**, 339; **18**, 377

Diaryl sulfoxides.[1] A simple preparation is to store a mixture of an arene, sodium arenesulfonate, and triflic acid at room temperature.

N,N′-Alkylidene bisamides.[2] An improved condensation of an aldehyde with two molecules of an amide is promoted by triflic acid.

Schmidt reaction. Nascent carbocations generated from alcohols (using TfOH or $SnCl_4$) are trapped by the added alkyl azides.[3]

OH + RN_3 —TfOH ; $NaBH_4$→ NR

95% (R = n-Bu)

Lactonization.[4] Lactone formation from propargylic alcohols and homologs in which the other propargylic position is substituted with a tungsten residue is readily achieved by treatment with triflic acid (0.25 equiv) in CH_2Cl_2 at −40°. The π-allyl complexes are demetallated on further treatment with CF_3COOH.

CpW(CO)3Na, THF, 23° → (OC)3W(Cp) ... CF3SO3H, CH2Cl2, -40° → CpW(CO)2 ... 84%; CF3COOH, CHCl3 → 85%

9,10-Diarylphenanthrenes.[5] Tetraaryl-1,2-ethanediols are converted to phenanthrene derivatives on exposure to triflic acid. An exception is 1,2-diphenyl-1,2-di(3-pyridyl)ethanediol, which undergoes normal pinacol rearrangement.

TfOH; Ar = Ph, 4-ClC6H4, ... > 98%

[1]Yamamoto, K., Miyatake, K., Nishimura, Y., Tsuchida, E. *CC* 2099 (1996).
[2]Fernandez, A.H., Alvarez, R.M., Abajo, T.M. *S* 1299 (1996).
[3]Pearson, W.H., Fang, W.-K. *JOC* **60**, 4960 (1995).
[4]Chen, C.-C., Fan, J.-S., Lee, G.-H., Peng, S.-M., Wang, S.-L., Liu, R.-S. *JACS* **117**, 2933 (1995).
[5]Olah, G.A., Klumpp, D.A., Neyer, G., Wang, Q. *S* 321 (1996).

Trifluoromethanesulfonic anhydride (triflic anhydride). 13, 324–325; **14**, 324–326; **15**, 339–340; **16**, 357–358; **18**, 377–378

Cyclodehydration.[1] Hydroxyalkyl enol ethers undergo cyclization on treatment with Tf_2O. A remarkable solvent effect has been discovered.

Tf_2O, 25°

in CH_2Cl_2	13	:	87
in PhMe	97	:	3

Cyclopropanation.[2] Certain homoallylic alcohols cyclize in the presence of Tf_2O–collidine. The introduction of various functionalities at the α-position of the cyclopropane ring is possible by treating the cyclization mixture with nucleophiles (*C*- and *X*-). On the other hand, triethylamine effects a clean elimination to give high yields of alkenylcyclopropanes.

Tf_2O - collidine / CH_2Cl_2, -78° ; H_2O → 87%

Tf_2O - collidine / CH_2Cl_2, -78° ; Et_3N → 90%

Bischler–Napieralski cyclization.[3] Combined with DMAP, triflic anhydride is effective in inducing the cyclization of *N*-biphenylyl carbamates. The method is applicable to the total synthesis of several amaryllidaceae alkaloids.

N–COOMe

Tf_2O - DMAP

CH_2Cl_2 , 0° -> rt

92%

Phosphinimines.[4] A simple preparation of $R_3P{=}NH$ from a phosphine oxide involves the evaporation of a mixture of it with Tf_2O and NH_3. The process is repeated several times. Access to these unsubstituted imines by the Staudinger reaction is difficult, because the imines are hygroscopic and hydrolyzable.

Couplings.[5] Grignard reagents RMgX of various kinds, including alkyl-, aryl-, vinyl-, and allylmagnesium halides undergo homocoupling induced by Tf_2O in refluxing ether for a short period to give R–R (10 examples, 48–95%).

Activation of N,N-dimethylacrylamide. The *O*-triflyl salt of the amide is a bifunctional electrophile for electron-rich arenes. Thus, the salt is useful for the synthesis of 1-indanones.[6]

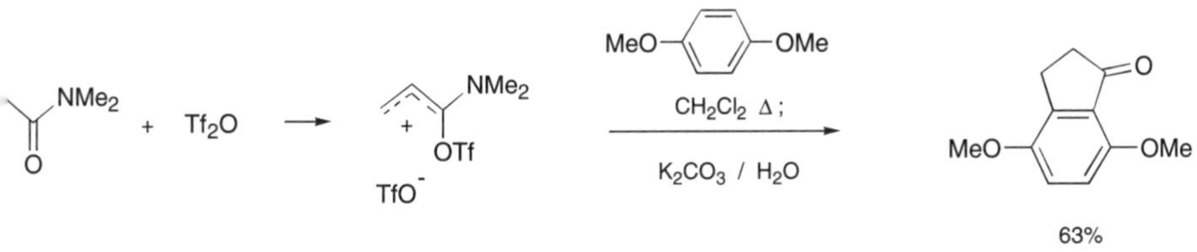

Nitration and nitratation.[7] Nitronium triflate is generated in situ from Bu_4NNO_3 and Tf_2O. Adamantane gives 1-nitro- or 1-nitratoadamantane on exposure to this reagent in nitromethane or dichloromethane, respectively. Thus, there is a marked solvent effect.

[1]Ishihara, K., Hanaki, N., Yamamoto, H. *CC* 1117 (1995).
[2]Nagasawa, T., Handa, Y., Onoguchi, Y., Suzuki, K. *BCSJ* **69**, 31 (1996).
[3]Banwell, M.G., Bissett, B.D., Busato, S., Cowden, C.J., Hockless, D.C.R., Holman, J.W., Read, R.W., Wu, A.W. *CC* 2551 (1995).
[4]Hendrickson, J.B., Sommer, T.J., Singer, M. *S* 1496 (1995).
[5]Nishiyama, T., Seshita, T., Shodai, H., Aoki, K., Kameyama, H., Komura, K. *CL* 549 (1996).
[6]Nenajdenko, V.G., Baraznenok, I.L., Balenkova, E.S. *T* **52**, 12993 (1996).
[7]Duddu, R., Damavarapu, R. *SC* **26**, 3495 (1996).

Trifluoromethanesulfonyl azide (triflic azide).

Organoazides from primary amines.[1] The diazo transfer is catalyzed by $CuSO_4$. Free hydroxyl groups do not disturb this reaction.

[1]Alper, P.B., Hung, S.-C., Wong, C.-H. *TL* **37**, 6029 (1996).

β-(Trifluoromethanesulfonyloxy)vinyliodonium trifluoromethanesulfonates.

Diaryliodonium triflates.[1] An efficient ligand exchange of the salts with aryllithiums leads to $(ArIAr')^+OTf^-$.

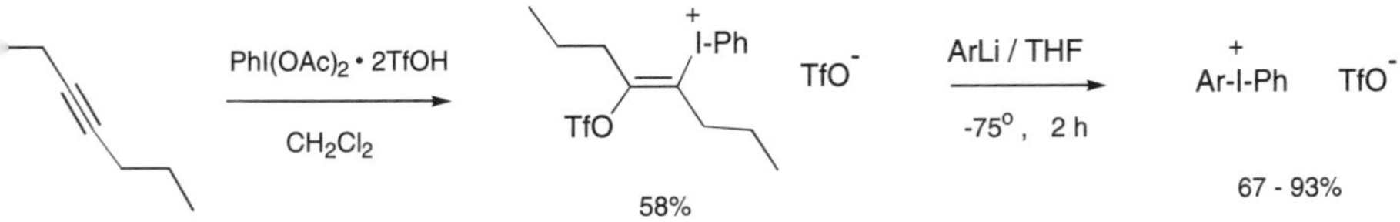

[1]Kitamura, T., Kotani, M., Fujiwara, Y. *TL* **37**, 3721 (1996).

***S*-(Trifluoromethyl)dibenzothiophenium salts.** **18**, 378

Trifluoromethylation. Reaction with enolate anions requires the presence of *B*-phenylcatecholborane.[1]

Related trifluoromethylating agents are analogues based on different chalcogen atoms (S, Se, Te).[2] The zwitterionic sulfonates **1** are also known.[3]

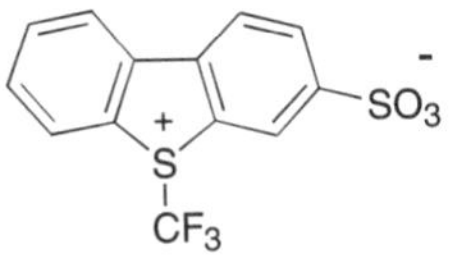

(**1**)

[1]Umemoto, T., Adachi, K. *JOC* **59**, 5692 (1994).
[2]Umemoto, T., Ishihara, S. *JACS* **115**, 2156 (1993).
[3]Umemoto, T., Ishihara, S., Adachi, K. *JFC* **74**, 77 (1995).

(Trifluoromethyl)trimethylsilane. **15**, 341; **18**, 378–379

Preparation.[1] A new preparative method involves the reaction of Me_3SiCl with CF_3Br and aluminum in NMP at room temperature. The yield is 62%.

Trifluoromethylation. Nascent trifluoromethyl anions are generated when CF_3SiMe_3 is in contact with a fluoride ion. Nucleophilic trifluoromethylation of oxazolidinones is for the synthesis of protected α-amino trifluoromethyl ketones.[2]

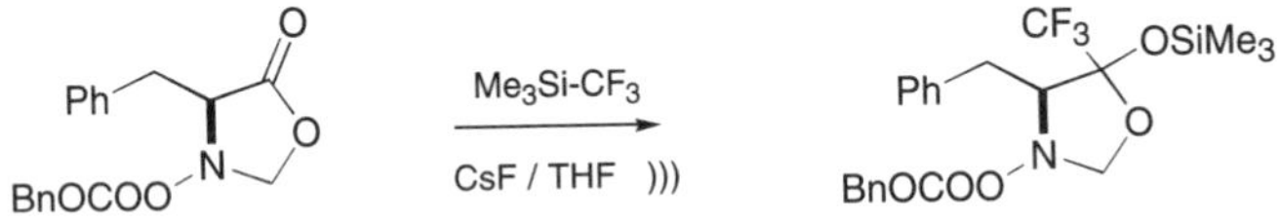

Me_3Si-CF_3 / CsF / THF))) → 95%

The formation of trifluoromethyl sulfides (and other chalcogenides) from thiocyanates[3] or disulfides,[4] CF_3SiMe_3, and Bu_4NF is possible. The combination of sulfenyl chlorides, CF_3SiMe_3, and TASF is also effective.[5]

(Trifluoromethyl)tributylstannane is readily prepared by the reaction of bis(tributyltin) oxide with CF_3SiMe_3, using Bu_4NF as an initiator.[6]

[1]Grobe, J., Hegge, J. *SL* 641 (1995).
[2]Walter, M.W., Adlington, R.M., Baldwin, J.E., Chuhan, J., Schofield, C.J. *TL* **36**, 7761 (1995).
[3]Billard, T., Large, S., Langlois, B.R. *TL* **38**, 65 (1997).
[4]Billard, T., Langlois, B.R. *TL* **37**, 6865 (1996).

[5]Movchum, V.N., Kolomeitsev, A.A., Yagupolskii, Y.L. *JFC* **70**, 255 (1995).
[6]Prakash, G.K.S., Yudin, A.K., Deffieux, D., Olah, G.A. *SL* 151 (1996).

Trifluoromethylzinc bromide.

Difluoromethyl ethers.[1] Primary and secondary alcohols are transformed into difluoromethyl ethers on reaction with CF_3ZnBr in $MeCN/CH_2Cl_2$ at room temperature.

[1]Miethchen, R., Hein, M., Naumann, D., Tyrra, W. *LA* 1717 (1995).

Triisobutylaluminum.

2-Hydroxytetrahydrofurans.[1] Triisobutylaluminum acts both as a Lewis acid to catalyze the isomerization of 4-methylene-1,3-dioxolanes to 2-oxotetrahydrofurans (via C–O bond cleavage and C–C bond formation) and as a reducing agent for the ketone group of the intermediate. The 4-methylene-1,3-dioxolanes are available by a Tebbe olefination.

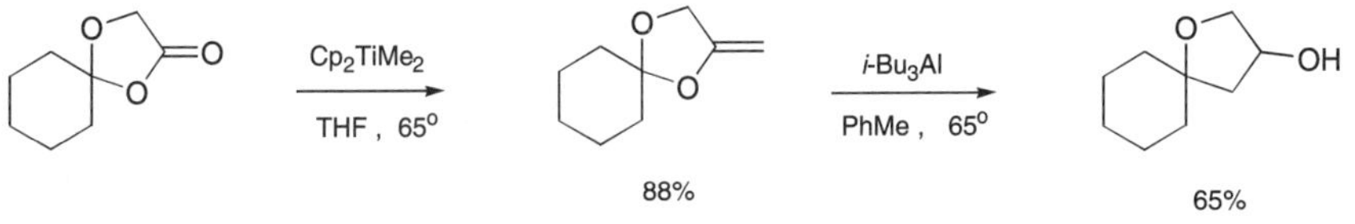

Thiazolines from esters.[2] 2-Mercaptoethylamine hydrochloride and an ester combine, with the concomitant elimination of the alcohol portion, by heating with i-Bu_3Al in toluene.

Sigmatropic rearrangements. The Claisen rearrangement of allyl vinyl ethers is promoted by i-Bu_3Al. Furthermore, the carbonyl group of the products also undergoes reduction to furnish 4-alkenols.[3] A stannyl group is lost in the products of distannyl- and silylstannyl-substituted allyl vinyl ethers when the rearrangement is conducted in the presence of i-Bu_3Al.[4]

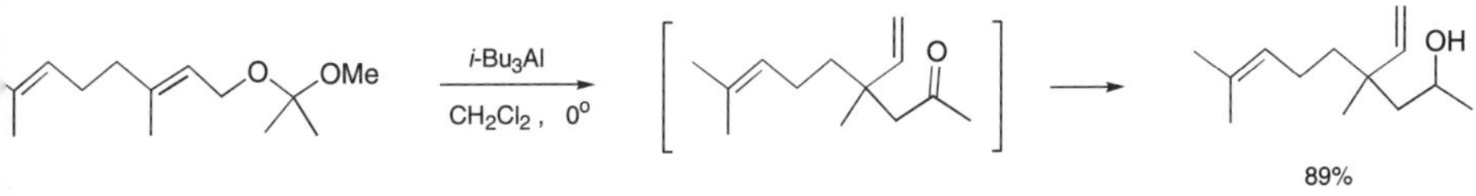

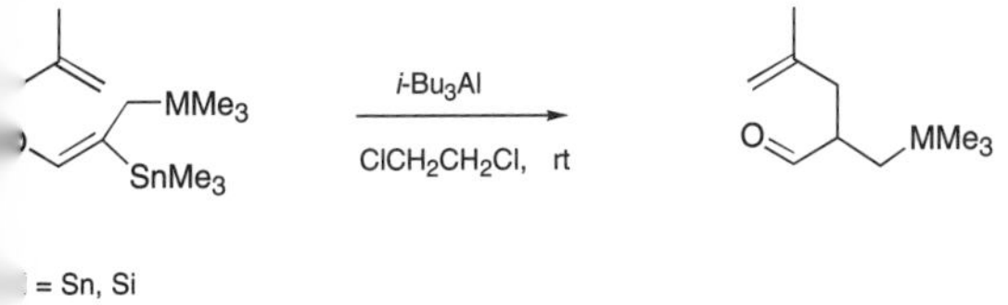

[1]Petasis, N.A., Lu, S.-P. *JACS* **117**, 6394 (1995).
[2]Busacca, C.A., Dong, Y., Spinelli, E.M. *TL* **37**, 2935 (1996).
[3]Rychnovsky, S.D., Lee, J.L. *JOC* **60**, 4318 (1995).
[4]Mitchell, T.N., Giesselmann, F. *SL* 475 (1996).

(2,4,6-Triisopropylphenyl)selenium bromide.

Tetrahydrofurans.[1] The reagent is prepared in situ from bis(2,4,6-triisopropylphenyl) diselenide and bromine in dichloromethane at –78°. It promotes a highly stereoselective cyclization of homoallylic alcohols.

[1]Lipshutz, B.H., Gross, T. *JOC* **60**, 3572 (1995).

Triisopropylsilyloxirane.

Triisopropylsilyl acyl silanes.[1] These silyl ketones with a bulky R_3Si group are difficult to prepare directly. A viable route involves reacting of organocuprates with the oxirane and subsequent Swern oxidation.

[1]Lipshutz, B.H., Lindsley, C., Susfalk, R., Gross, T. *TL* **35**, 8999 (1994).

Triisopropylsilyl tetrakis(trifloxy)borate.

Mukaiyama aldol reactions. The condensation catalyzed by *i*-$Pr_3SiB(OTf)_4$ exhibits Cram-type selectivity.[1] An analogous catalyst, $Me_3SiB(OTf)_3Cl$, is available from Me_3SiCl and $B(OTf)_3$.[2]

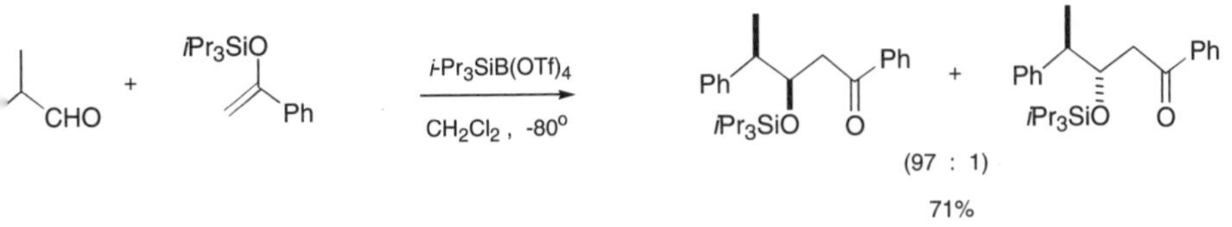

CHO + iPr3SiO / Ph → i-Pr3SiB(OTf)4 / CH2Cl2, -80° → Ph / iPr3SiO / Ph / O + Ph / iPr3SiO / Ph / O

(97 : 1)

71%

[1]Davis, A.P., Plunkett, S.J. *CC* 2173 (1995).
[2]Davis, A.P., Muir, J.E., Plunkett, S.J. *TL* **37**, 9401 (1996).

Trimethylaluminum.

Allylic substitutions.[1] The introduction of a methyl group by the transpositional displacement of an allylic system can use Me_3Al in the presence of a copper salt.

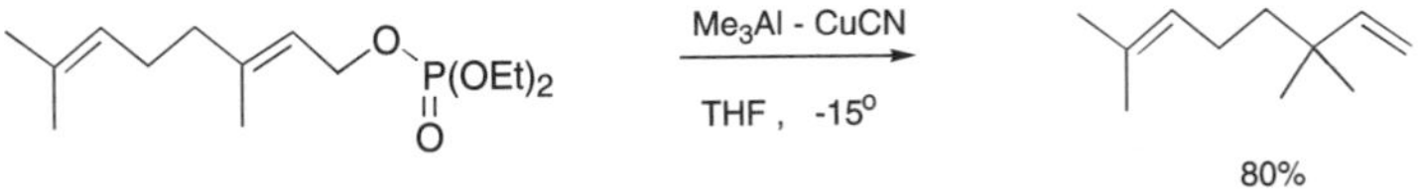

O-P(OEt)2 / O → Me3Al - CuCN / THF, -15° →

80%

Amino acid hydroxamates.[2] The group exchange from an ester to an *O*-benzylhydroxamate is promoted by Me_3Al. The product can be directly debenzylated by hydrogenolysis (8 examples, 65–99%).

Zwitterionic [3,3]-sigmatropic rearrangements. The zwitterionic aza-Claisen rearrangement is useful for ring expansion.[3] It is subject to 1,2-asymmetric induction.[4]

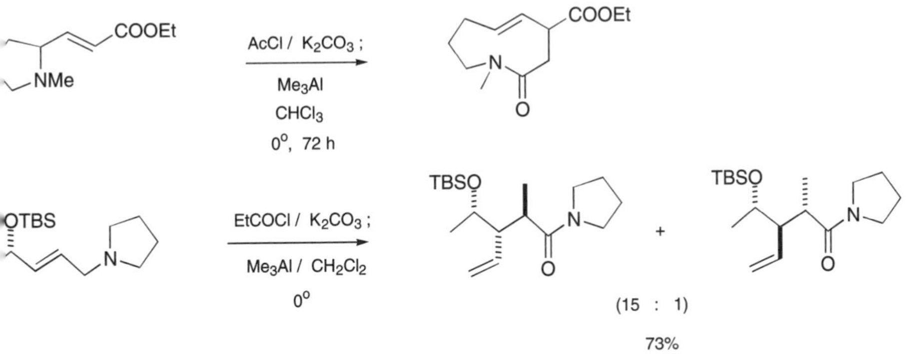

COOEt / NMe → AcCl / K2CO3 ; Me3Al / CHCl3 / 0°, 72 h → COOEt / N / O

OTBS / N → EtCOCl / K2CO3 ; Me3Al / CH2Cl2 / 0° → TBSO / N / O + TBSO / N / O

(15 : 1)

73%

β,γ-Alkynyl ketones.[5] The presence of a leaving group at the homopropargylic position of a propargyl alcohol makes possible a stereoselective 1,2-rearrangement after complexation with $Co_2(CO)_8$. In principle, ionization of the hydroxyl group is a favorable process, but it is not observed.

OH $Co_2(CO)_6$ — MsO, R, Bu — Me_3Al / CH_2Cl_2, -78°, 15 min → Bu, O, R, $(CO)_6Co_2$

R = $Ph(CH_2)_2$-

88% (>98% ee)

[1]Flemming, S., Kabbara, J., Nickisch, K., Westermann, J., Mohr, J. *SL* 183 (1995).
[2]Pirrung, M.C., Chau, J.H.-L. *JOC* **60**, 8084 (1995).
[3]Diederich, M., Nubbemeyer, U. *ACIEE* **34**, 1026 (1995).
[4]Nubbemeyer, U. *JOC* **60**, 3773 (1995).
[5]Nagasawa, T., Taya, K., Kitamura, M., Suzuki, K. *JACS* **118**, 8949 (1996).

Trimethylhydrazinium iodide.

Ar-aminations.[1] Nitroarenes undergo amination by vicarious nucleophilic substitution with $(Me_3NNH_2)^+I^-$ in DMSO upon addition of *t*-BuOK.

[1]Pagoria, P.F., Mitchell, A.R., Schmidt, R.D. *JOC* **61**, 2934 (1996).

2,8,9-Trimethyl-1-phospha-2,5,8,9-tetraazabicyclo[3.3.3]undecane.

Acylation.[1] This cage heterocycle (**1**) is an excellent catalyst for the acylation of hindered alcohols. Acetylation is accelerated in polar solvents (MeCN), but benzoylation is faster in solvents such as benzene.

P N N N N

(**1**)

Silylation.[2] For tertiary alcohols and hindered phenols, the conventional method of *t*-butyldimethylsilylation using TBS–Cl fails. A catalytic amount of **1** with 1.1 equiv. of Et_3N in MeCN is able to effect the silylation in high yields at room temperature.

[1]D'Sa, B.A., Verkade, J.G. *JOC* **61**, 2963 (1996).
[1]D'Sa, B.A., Verkade, J.G. *JACS* **118**, 12832 (1996).

***(E)*-[β-Trimethylsilylacryloyl]-*t*-butyldimethylsilane.**

[4+3]Cycloaddition.[1] Lithium enolates of enones, including o-bromoacetophenone, condense with the acylsilane (**1**) to give 3-*t*-butyldimethylsiloxy-5-trimethylsilyl-3-cycloheptenones. This intriguing transformation likely proceeds by aldol reaction, Brook rearrangement, cyclization to give a conjugate base of 1,2-dialkenyl- 1,2-cyclopropanediol monosilyl ether, and Cope rearrangement.

O
SiMe2Bu^t
Me3Si
(1)
+
OLi
THF
-80° -> -30°
O
TBSO
SiMe3
84%

[1]Takeda, K., Takeda, M., Nakajima, A., Yoshii, E. *JACS* **117**, 6400 (1995).

Trimethylsilyl azide. 13, 24–25; **14**, 25; **15**, 342–343; **16**, 17; **18**, 379–380

Organoazides. Alkyl azides are obtained by a displacement reaction of alkyl halides by using the Me_3SiN_3–Bu_4NF reagent system.[1] 2-Azidoimidazoles are obtained from the corresponding bromo derivatives in a reaction catalyzed by $(Ph_3P)_2PdCl_2$.[2]

A cyclic allylic azide is required for a synthesis of (+)-pancratistatin.[3] This intermediate is obtainable from a Pd-catalyzed reaction of Me_3SiN_3 with a *meso*-2-cyclohexene-1,4-diol dicarbonate in the presence of a chiral ligand.

OCOOMe
O
O
OCOOMe
(Pd-Cl)2
Me3SiN3
ligand*
CH2Cl2, rt
OCOOMe
O
O
N3
82%
OH
HO
OH
H
OH
O
NH
O
OH O
(+)-pancratistatin

α-Azido ketones.[4] Alkenes give these products with a mixture of Me_3SiN_3 and CrO_3 in dichloromethane at room temperature (9 examples, 51–82%).

β-Azido alcohols. The opening of epoxides by Me_3SiN_3 is catalyzed by (salen)Cr(III)[5] or Ph_4SbOH.[6] The products may be hydrolyzed or the silyl ethers isolated.

β-Azido N-tosylamines.[7] Opening of aziridines on reaction with Me_3SiN_3 is catalyzed by imidochromium complexes, and the regioselectivity of this reaction is enhanced by molecular sieves.

Ph-aziridine-NTs → (Me_3SiN_3; t-BuNCrCl$_3$(dme), CH_2Cl_2) → PhCH(N_3)CH$_2$NHTs + PhCH(NHTs)CH$_2N_3$ (40 : 1) 86%

Schmidt rearrangement.[8] With pyridinium tosylate as a catalyst, *gem*-azido silyl ethers are formed from silyl enol ethers. The products undergo photoinduced rearrangement to give amides or lactams.

i-Pr$_3$SiO-cyclohexene → (Me_3SiN_3; pyHOTs / CH_2Cl_2) → 1-(i-Pr$_3$SiO)-1-N_3-cyclohexane 77% → (hv) → caprolactam (NH, O) 89%

α-Iodocycloalkenones.[9] This preparation involves iodination of the adducts obtained after the conjugate addition of Me_3SiN_3 to enones and subsequent elimination. It is suitable for the introduction of iodine into β-substituted enones.

α-Azido hydroximoyl chlorides.[10] These polyfunctional compounds are obtained when conjugated nitroalkenes are treated with Me_3SiN_3 and $TiCl_4$.

PhCH=CHNO$_2$ → (Me_3SiN_3 - $TiCl_4$; CH_2Cl_2) → PhCH(N_3)C(Cl)=NOH 78%

[1] Ito, M., Koyakumaru, K.-i., Ohta, T., Takaya, H. *S* 376 (1995).
[2] Kawasaki, I., Taguchi, N., Yoneda, Y., Yamashita, M., Ohta, S. *H* **43**, 1375 (1996).
[3] Trost, B.M., Pulley, S.R. *JACS* **117**, 10143 (1995).
[4] Reddy, M.V.R., Kumareswaran, R., Vankar, Y.D. *TL* **36**, 6751 (1995).
[5] Martinez, L.E., Leighton, J.L., Carsten, D.H., Jacobsen, E.N. *JACS* **117**, 5897 (1995).
[6] Fujiwara, M., Tanaka, M., Baba, A., Ando, H., Souma, Y. *TL* **36**, 4849 (1995).
[7] Leung, W.-H., Yu, M.-T., Wu, M.-C., Yeung, L.-L. *TL* **37**, 891 (1996).
[8] Evans, P.A., Modi, D.P. *JOC* **60**, 6662 (1995).
[9] Sha, C.-K., Huang, S.-J. *TL* **36**, 6927 (1995).
[10] Kumaran, G., Kulkarni, G.H. *SC* **25**, 3735 (1995).

1-Trimethylsilylbenzotriazole.

β-Alkylation of enones.[1] The adducts with enones undergo alkylation and then elimination of benzotriazole.

LDA / THF ; BnX ; H_3O^+

58%

[1]Katritzky, A.R., Soloducho, J., Musgrave, R.P., Breytenbach, J.C. *TL* **36**, 5491 (1995).

Trimethylsilyl bromide. 15, 51; **16**, 50; **18**, 380

Phosphonic esters from phosphoranes.[1] The facile transformation of phosphoranes by using Me_3SiBr has found application in the synthesis of α-aminophosphonic esters.

$(EtO)_3P$, $PhCH_3$, Δ, 7 h; Me_3SiBr, 10°, 5 min ; H_2O

~ 100%

Bromohydrins.[2] A highly regioselective ring opening of 2,3-epoxy alcohol derivatives to give products having a *vic*-diol pattern is mediated by $Me_3SiBr–SnBr_2$ (6 examples, 68–99%).

Heterocyclization. The generation of iminium ions from alkoxymethylamines is readily achieved on treatment with Me_3SiBr. The intermediates undergo cyclization in the presence of a juxtaposed triple bond.[3]

Me_3SiBr, MeCN

83%

[1]Seki, M., Kondo, K., Iwasaki, T. *JCS(P1)* 3 (1996).
[2]Oriyama, T., Ishiwata, A., Hori, Y., Yatabe, T., Hasumi, N., Koga, G. *SL* 1004 (1995).
[3]Murata, Y., Overman, L.E. *H* **42**, 549 (1996).

Trimethylsilyl chloride. **15**, 89; **16**, 85–86; **18**, 381

Alkyl chlorides. A preparation of alkyl chlorides from the corresponding bromides is accomplished by reaction with Me_3SiCl in the presence of imidazole.[1] Similarly, alcohols are converted to chlorides with Me_3SiCl in DMSO.[2]

Trimethylsilyl chloride is a convenient source of HCl for hydrochlorination of alkenes (15 examples, 65–99%).[3] The generation of HCl in situ is accomplished by the decomposition of Me_3SiCl with water.

N-(1-Sulfonylalkyl)formamides.[4] These compounds are prepared by reacting aldehydes with formamide and sulfonic acids in the presence of Me_3SiCl. They are important precursors of substituted tosylmethyl isocyanides.

CHO, F, + $HCONH_2$, Me_3SiCl - TsH, MeCN - PhMe, $TolSO_2$, NHCHO, F, 93%

Allylation. Carbonyl compounds are converted into homoallylic alcohols with allyl bromides–tin,[5] diallyldibutylstannane,[6] or allyltributylstannane–dibutyltin chloride.[7] All three protocols require the presence of Me_3SiCl (although acid chlorides may replace Me_3SiCl in the last reaction).

Conjugate additions. In the presence of Me_3SiCl and $BF_3{\cdot}OEt_2$, secondary and tertiary alkylzinc bromides add to enones without a copper catalyst.[8] Enhancement of yields and selectivity is also achieved by using Me_3SiCl for the conjugate addition of stabilized organolithiums such as $(PhS)_3CLi$.[9]

O, COOEt, +, ZnBr, Me_3SiCl - $BF_3{\cdot}OEt_2$, pentane , -30°, O, COOEt, 54%

[1]Peyrat, J.-F., Figadere, B., Cave, A. *SC* **26**, 4563 (1996).
[2]Snyder, D.C. *JOC* **60**, 2638 (1995).
[3]Boudjouk, P., Kim, B.-K., Han, B.-H. *SC* **26**, 3479 (1996).

[4]Sisko, J., Mellinger, M., Sheldrake, P.W., Baine, N.H. *TL* **37**, 8113 (1996).
[5]Zhou, J.-Y., Yao, X.-B., Chen, Z.-G., Wu, S.-H. *SC* **25**, 3081 (1995).
[6]Yasuda, M., Fujibayashi, T., Shibata, I., Baba, A., Matsuda, H., Sonoda, N. *CL* 167 (1995).
[7]Whitesell, J.K., Apodaca, R. *TL* **37**, 3955 (1996).
[8]Hanson, M.V., Rieke, R.D. *JACS* **117**, 10775 (1995).
[9]Liu, H., Cohen, T. *TL* **36**, 8925 (1995).

Trimethylsilyl cyanide. **13**, 87–88; **14**, 107; **15**, 102–104; **17**, 89; **18**, 381–382

Cyanohydrins and ethers. The reaction of Me_3SiCN with aldehydes,[1] is non-catalyzed, so that it can be used to differentiate aldehydes and ketones.

Dibutyltin dichloride can be used as a catalyst in the addition reaction without solvent.[2] Trimethylsilyl bis(fluorosulfonyl)imide is also an efficient catalyst promoting the derivatization at –78° (11 examples, 84–98%).[3]

Ritter reaction.[4] A combination of Me_3SiCN and H_2SO_4 can be used to convert a hydroxy group to a formamido function.

Cyanation. 1-Aza-1,3-dienes are cyanated at C-2 by treatment with Me_3SiCN and chromium(VI) oxide.[5]

1-Cyano-3(1*H*)-1,2-benziodoxols are obtained from 2-iodosobenzoic acid by treatment with Me_3SiCN. The reagent is useful for cyanation at the α-carbon of an *N*-alkyl group.[6]

[1]Manju, K., Trehan, S. *JCS(P1)* 2383 (1995).
[2]Whitesell, J.K., Apodaca, R. *TL* **37**, 2525 (1996).
[3]Kaur, H., Kaur, G., Trehan, S.. *SC* **26**, 1925 (1996).
[4]Chen, H.G., Goel, O.P., Kesten, S., Knobelsdorf, J. *TL* **37**, 8129 (1996).
[5]Prajapati, D., Karmakar, D., Sandhu, J.S. *JCR(S)* 538 (1996).
[6]Zhdankin, V.V., Kuehl, C.J., Krasutsky, A.P., Boltz, J.T., Mismash, B., Woodward, J.K., Simonsen, A.J. *TL* **36**, 7975 (1995).

Trimethylsilyl(diethyl)amine. 18, 382

Deesterification.[1] Esters are cleaved to give acids by heating with this silylamine and MeI in toluene, followed by aqueous quenching.

Michael additions.[2] 1,5-Ketoaldehydes are formed on the admixture of vinyl ketones, aldehydes, and Me_3SiNEt_2 at room temperature and without solvent. It is unusual that the aldehydes behave as Michael donors.

[1]Yamamoto, Y., Shimizu, H., Hamada, Y. *JOMC* **509**, 119 (1996).
[2]Hagiwara, H., Kato, M. *TL* **37**, 5139 (1996).

1-[2-(Trimethylsilyl)ethoxy]ethyl vinyl ether.

Protection of alcohols.[1] The EVE–ether analogues are readily prepared from alcohols and vinyl ether by using pyridinium tosylate as a catalyst at room temperature. Deprotection is achieved on treatment with a fluoride ion source.

[1]Wu, J., Shull, B.K., Koreeda, M. *TL* **37**, 3647 (1996).

Trimethylsilyl iodide. 16, 188–189; **18**, 383

Benzylic deoxygenation.[1,2] Benzylic alcohols are reduced by using Me_3SiI in MeCN at room temperature.

γ-Alkoxyallylstannanes.[3] Mixed acetals of 3-tributylstannylpropanal eliminate a small alkoxy group (e.g., MeO) on exposure to Me_3SiI, leading to the enol ethers.

OTBS; MeO; $SnBu_3$; Me_3SiI - HMDS / CH_2Cl_2; $NaHCO_3$ / H_2O; OTBS; $SnBu_3$; 71%

Reactions of α,β-epoxy silanes.[4] Either silyl enol ethers or iodoalkenes can be prepared by the reaction. Short reaction times favor formation of the enol ethers.

Ph; Ph; $SiMe_3$; Me_3SiI / $CHCl_3$; 2 min; Ph; Ph; $OSiMe_3$; 98%

$SiMe_3$; Me_3SiI / $CHCl_3$; 12 h; I; 78%

***cis*-Carbosilylation of alkynes.**[5] A dialkylzinc provides the carbon residue for this reaction, which is catalyzed by $(Ph_3P)_4Pd$. The trimethylsilyl group becomes attached to the unsubstituted terminus of the alkyne.

[1]Stoner, E.J., Cothron, D.A., Balmer, M.K., Roden, B.A. *T* **51**, 11043 (1995).
[2]Perry, P.J., Pavlidis, V.H., Coutts, I.G.C. *SC* **26**, 101 (1996).
[3]Kadota, I., Sakaihara, T., Yamamoto, Y. *TL* **37**, 3195 (1996).
[4]Bassindale, A.R., Soobramanien, M.-C., Taylor, P.G. *BSCF* **132**, 604 (1995).
[5]Chatani, N., Amishiro, N., Morii, T., Yamashita, T., Murai, S. *JOC* **60**, 1834 (1995).

Trimethylsilyl ketene.

***α,β*-Unsaturated acids.**[1] Homologation of carbonyl compounds (aldehydes, ketones, enals, and enones) by [2+2]-cycloaddition with the ketene and subsequent ring opening of the intermediates offers an alternative method to the Wittig reaction.

$$\mathrm{RCHO} \quad + \quad \mathrm{Me_3SiCH{=}C{=}O} \xrightarrow[\mathrm{Et_2O}]{\mathrm{BF_3 \cdot OEt_2}} \mathrm{RCH{=}CHCOOH} \quad (41-99\%)$$

[1]Black, T.H., Zhang, Y., Huang, J., Smith, D.C., Yates, B.E. *SC* **25**, 15 (1995).

1-[Trimethylsilyl(methoxy)methyl]benzotriazole.

Homologous acids.[1] One-carbon homologation of aromatic aldehydes and ketones is achieved by reaction with the anion of reagent (**1**). The enol ether products are easily hydrolyzed.

(**1**) —BuLi / THF, -78°; RCOAr→ —$ZnBr_2$ / dioxane ; HCl / H_2O→ HOOC–CH(R)(Ar)

[1]Katritzky, A.R., Toader, D., Xie, L. *S* 1425 (1996).

(Trimethylsilyl)methylmagnesium chloride. 15, 343; **18**, 384

Acetal cleavage.[1] Chemoselective deprotection has been demonstrated with this reagent.

Me_3SiCH_2MgCl / PhH; Δ 48 h; 68%

[1]Chen, Y.-H., Tseng, Y.-T., Luh, T.-Y. *CC* 327 (1996).

Trimethylsilylmethyl phenyl sulfide.

1-Trimethylsilylalk-l-en-3-ols.[1] The homologation process of α,β-epoxy silanes with Me_3SiCH_2SPh to give the allylic alcohols can be repeated to construct 1,2-diols stereoselectively.

Me_3SiCH_2SPh - BuLi; -40° -> 0°; 65%; Me_3SiCH_2SPh - BuLi; $HClO_4$; 60%

[1]Raubo, P., Wicha, J. *TA* **7**, 763 (1996).

Trimethylsilyl nitrate–chromium(VI) oxide.

α-Nitro ketones.[1] Functionalization of alkenes with this reagent system is most convenient. Me_3SiNO_3 is formed by adding Me_3SiCl to a solution of $AgNO_3$ in MeCN at 0°.

[1]Reddy, M.V.R., Kumaeswaran, R., Vankar, Y.D. *TL* **36**, 7149 (1995).

B-[γ-(Trimethylsilyl)propargyl]diisopinocampheylborane.

α-Silylallenylation.[1] The reagent reacts enantioselectively with aldehydes at low temperatures by a S_E2' pathway. The products have high ee values.

MeCHO / -100°; H_2O_2 - NaOH; 72% (87% ee)

[1]Brown, H.C., Khire, U.R., Narla, G. *JOC* **60**, 8130 (1995).

Trimethylsilyl(tributyl)stannane–tris(diethylamino)sulfur trimethylsilyldifluoride.

Elimination of dihalides.[1] Elimination of 1,1-, 1,2-, and 1,4-dihalides is observed with this combination of reagents. For 1,1-dibromoalkenes, rearrangement leads to alkynes. 1,2-Dibromobenzene gives benzyne, and α,α′-dibromo-*o*-xylene gives quinodimethane.

Ph, Br / $PivO(CH_2)_5$, Br — $Me_3SiSnBu_3$ / TASF / THF → $PivO(CH_2)_5$—≡—Ph 74%

[1]Sato, H., Isono, N., Miyoshi, I., Mori, M. *T* **52**, 8143 (1996).

Trimethylsilyl trifluoromethanesulfonate.

Protecting-group manipulations. Tetrahydropyranyl ethers are switched to *t*-butyldimethylsilyl ethers[1] on reaction of R_3SiOTf and Et_3N in dichloromethane at room temperature. *t*-Butyl esters also undergo group exchange in the presence of *t*-butyl ethers.[2]

N-t-Butoxycarbonyl derivatives of primary amines are further silylated with Me_3SiOTf to give the base stable products.[3] Rapid acetylation of alcohols is promoted by the same catalyst.[4]

Schmidt reaction. Acetals or enol ethers of azidoalkyl ketones undergo cyclization, which is typified by a synthesis of *N*-acetylpyrrolidine.[5]

OEt, OEt, N_3 — Me_3SiOTf ; / NaI / Me_2CO → N, O 81%

Oxonium-ene reaction.[6] Lactol allylic ethers undergo this reaction under the influence of Me_3SiOTf. The transformation defines a new route to substituted oxacycles.

O, O — Me_3SiOTf / CH_2Cl_2 , -78° → O, H, H, O 58%

Alkylation of acetals. When applied to mixed acetals bearing an α-stannylalkyl branch, the catalyzed reaction with alkynylstannanes gives propargyl ethers, which are precursors of allenyl carbinols.[7]

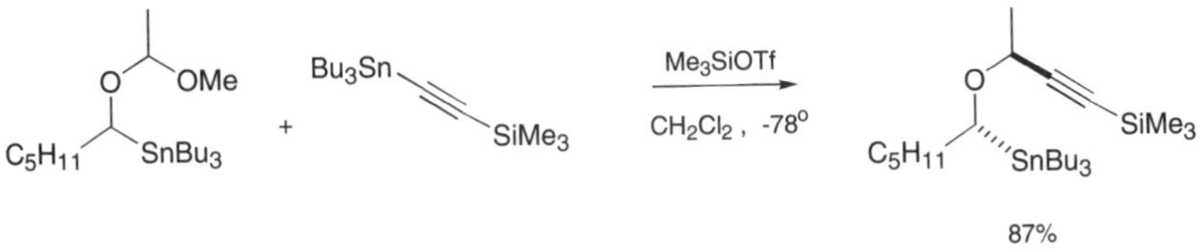

Me_3SiOTf, CH_2Cl_2, -78°

87%

A stereoselective construction of 1,3-diol systems is based on the reaction of lactol acetates with allylsilanes or silyl enol ethers.[8] Formation of the product is subject to 1,3-asymmetric induction by one or more substituents in the ring. Note that $BF_3{\cdot}OEt_2$ is not a suitable catalyst.

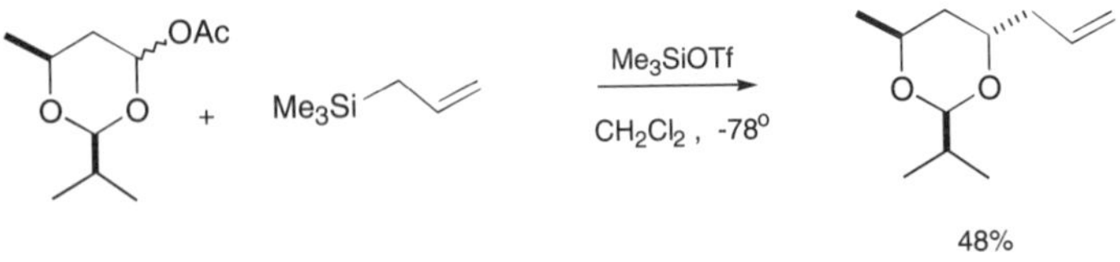

Me_3SiOTf, CH_2Cl_2, -78°

48%

Intramolecular substitutions. A ring expansion at the expense of cyclopropane ring scission has been shown.[9] A glycosyl carbonate extrudes CO_2 on exposure to Me_3SiOTf, as a result of ionization, decarboxylation of the carbonic acid monoester, and the return of the alcohol residue to the anomeric site.[10]

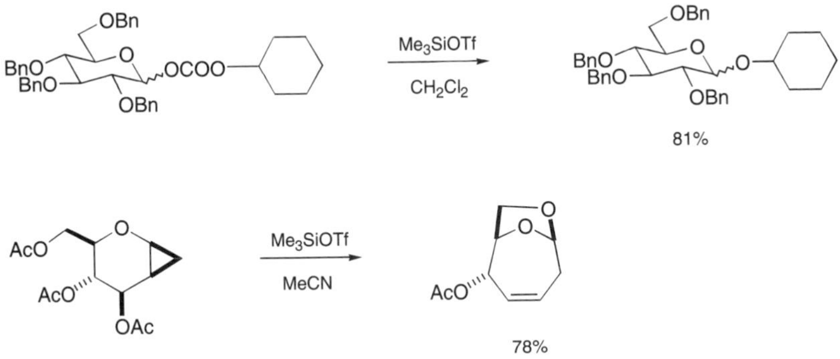

Me_3SiOTf, CH_2Cl_2

81%

Me_3SiOTf, MeCN

78%

Nazarov cyclization.[11] The cyclization of 1,1-difluoro-1,4-pentadien-3-ones is regioselective due to direction by the fluorine atoms.

Me$_3$SiOTf
CH$_2$Cl$_2$ - (CF$_3$)$_2$CHOH
73%

Annulation.[12] In situ prepared *N*-acetoacetylated alkenyl amides undergo cyclization to give tricyclic lactams. The catalyzed reaction is considered to proceed by 1,4-dipolar cycloaddition involving cross-conjugated heteroaromatic betaines.

Me$_3$SiOTf
PhCH$_3$
63%

[1]Oriyama, T., Yatabe, K., Sugawara, S., Machiguchi, Y., Koga, G. *SL* 523 (1996).
[2]Trzeciak, A., Bannwarth, W. *S* 1433 (1996).
[3]Roby, J., Voyer, N. *TL* **38**, 191 (1997).
[4]Procopiou, P.A., Baugh, S.P.D., Flack, S.S., Inglis, G.G.A. *CC* 2625 (1996).
[5]Mossman, C.J., Aube, J. *T* **52**, 3403 (1996).
[6]Mikami, K., Kishino, H. *TL* **37**, 3705 (1996).
[7]Linderman, R.J., Chen, S. *TL* **37**, 3819 (1996).
[8]Boons, G.-J., Eveson, R., Smith, S., Stauch, T. *SL* 536 (1996).
[9]Hoberg, J.O., Bozell, J.J. *TL* **36**, 6831 (1995).
[10]Iimori, T., Shibazaki, T., Ikegami, S. *TL* **37**, 2267 (1996).
[11]Ichikawa, J., Miyazaki, S., Fujiwara, M., Minami, T. *JOC* **60**, 2320 (1995).
[12]Padwa, A., Harring, S.R., Semones, M.A. *JOC* **60**, 2952 (1995).

(Triphenylarsonio)acetaldehyde bromide.

Homologation of aldehydes.[1] Extension of an aldehyde by a butadiene unit can be brought about in one operation by the reagent $[Ph_3AsCH_2CHO]^+Br^-$.

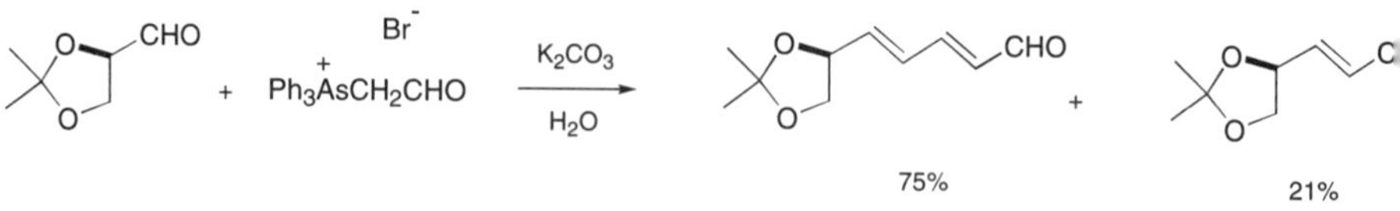

[1]Peng, Z.-H., Li, Y.-L., Wu, W.-L., Liu, C.-X., Wu, Y.-L. *JCS(P1)* 1057 (1996).

Triphenylbismuth difluoride.

Phenylation of 1-alkynes.[1] The formation of 1-phenylalkynes with Ph_3BiF_2 is a reaction catalyzed by copper(I) chloride.

[1]Lermontov, S.A., Rakov, I.M., Zefirov, N.S., Stang, P.J. *TL* **37**, 4051 (1996).

Triphenylbismuthonium ylides.

Acyl carbene equivalents.[1] The moderately stabilized RCOCH=BPh$_3$ ylides are generated from the tetrafluoroborate salts. These ylides react with aldehydes to give α,β-epoxy ketones, in contrast to the stibonium, arsonium, and phosphonium ylides, which afford enones. The weaker Bi–O bond disfavors decomposition of the adducts into enones and the bismuthine oxides.

[1]Matano, Y. *JCS(P1)* 2703 (1994).

Triphenylphosphine. 18, 385–386

1,5-Additions.[1] The umpolung mode of addition to 2,3-butadienoic esters is made possible in the presence of Ph_3P. Zwitterionic intermediates formed by conjugate addition of the phosphine are poised to perform proton abstraction from the true nucleophiles, and the subsequent combination of the ensuing ion pairs follows a course favored by the formation of ylides.

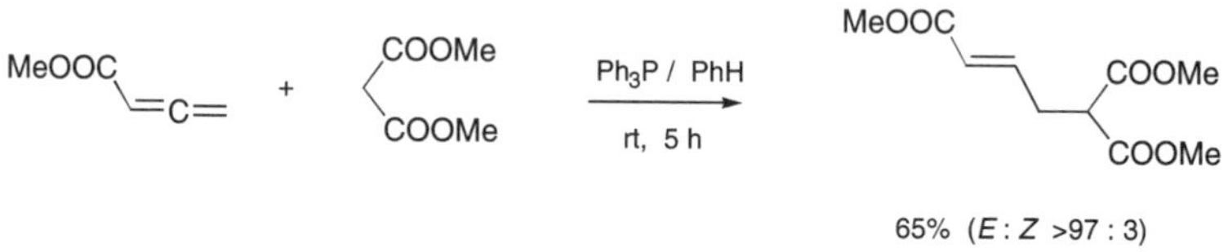

2-Amino-3-acyl-1,4-naphthoquinones. The redox transformation of 2-azido-3-alkenyl-1,4-naphthoquinones by Ph_3P is quite unexpected (4 examples, 45–60%).[2]

Ph$_3$P / CH$_2$Cl$_2$, -30° -> rt; 45%

Chiral aziridines.[3] Chiral epoxides that are readily available can be converted to configurationally inverted aziridines in two steps: ring opening with NaN_3 and treatment with Ph_3P. The latter step involves a unique decomposition process of the Staudinger reaction products.

Ph$_3$P / MeCN; 83%

[1]Zhang, C., Lu, X. *SL* 645 (1995).
[2]Molina, P., Tarraga, A., Jose del Bano, M., Espinosa, A. *LA* 223 (1994).
[3]Shao, H., Zhu, Q., Goodman, M. *JOC* **60**, 790 (1995).

Triphenylphosphine–carbon tetrahalide. 13, 331–332; **15**, 352; **16**, 366–368; **18**, 386–387

4-Pentenenitriles.[1] A mild dehydration of *N*-allyl amides mediated by Ph_3P-CX_4 in the presence of Et_3N yields 3-aza-1,2,5-trienes (*N*-allyl ketenimines). A [3,3]-sigmatropic rearrangement delivers 4-pentenenitriles.

Ph$_3$P - CCl$_4$, Et$_3$N / CH$_2$Cl$_2$; 94%

A similar rearrangement apparently involving allyl cyanate derivatives generates transposed *N*-allyl ureas when the reaction is quenched with amines.[2]

Dichloromethylenation.[3] Olefination of lactones and acetates is conveniently achieved in refluxing THF.

Deacetalization.[4] Essentially neutral conditions are maintained with Ph_3P–CBr_4.

[1]Walters, M.A., Hoem, A.B., McDonough, C.S. *JOC* **61**, 55 (1996).
[2]Ichikawa, Y., Kobayashi, C., Isobe, M. *SL* 919 (1994).

[3]Lakhrissi, M., Chapleur, Y. *JOC* **59**, 5752 (1994).
[4]Johnstone, C., Kerr, W.J., Scott, J.S. *CC* 341 (1996).

Triphenylphosphine–diethyl azodicarboxylate. 13, 322; **14**, 336–337; **17**, 389–390; **18**, 387

Perfluoro-t-butyl ethers.[1] Under the Mitsunobu reaction conditions and with the use of $(CF_3)_3COH$ as nucleophile, the mixed ethers are obtained.

2-Benzoyloxymethyl-2-alkenoic esters.[2] These esters are formed when the Baylis–Hillman reaction products are subjected to the reaction of Ph_3P–DEAD and a carboxylic acid. An S_N2' pathway is followed. The addition of Et_3N is beneficial, because it undergoes quaternization. Thus, the subsequent displacement by the carboxylate ion affords much more rearranged (S_N2') products.

PhCOOH - Et_3N / Ph_3P - DEAD / THF, 0° ; [RCOO⁻ ⁺NEt_3 ... COOMe] → OBz ... COOMe 90%

N-Alkylations. Various amine derivatives are obtained by using *N*-alkyltriflamides,[3] *N,O*-bisbenzyloxycarbonylhydroxylamine,[4] and 5-substituted tetrazoles[5] as nucleophiles in the Mitsunobu reaction. *N*-Tosylserine *t*-butyl ester gives an aziridine in 60% yield.[6]

HO, TsNH ... → Ph_3P - DEAD / THF, 23° → Ts-N ... 60%

Mixed phosphites.[7] Dialkyl phosphites $HP(=O)(OR)_2$ are converted into mixed phosphites of the type $(RO)_2POR'$ by successive treatment with Ph_3P–DEAD and R′OH, including phenol.

C-Alkylations. Triethyl methanetricarboxylate is readily alkylated by alcohols.[8] Propargylic alcohols are homologated to give 3-alkynenitriles with acetone cyanohydrin to supply the cyano group.[9]

Transpositional deoxygenation. *o*-Nitrobenzenesulfonylhydrazine supplies a hydride source to effect the transpositional deoxygenation of allylic alcohols via the *N*-sulfonyl-*N*-allylhydrazines, which are formed by an S_N2 reaction.[10] This reaction has been used to synthesize allenes from propargylic alcohols in one step.[11]

Ph₃P - DEAD

THF

85%

Ph₃P - DEAD

THF

76%

Configuration inversion. During a synthesis of taxusin,[12] stereochemical considerations indicated the importance of installing a bulky β-oriented substituent at the subangular position in ring C prior to closure of the eight-membered ring by a Mukaiyma-type reaction. The configurational adjustment was made by a Misunobu reaction using pivalic acid as the nucleophile. A long reaction time was required.

Ph₃P - DEAD

Me_3CCOOH

taxusin

[1]Sebesta, D.P., O'Rourke, S.S., Pieken, W.A. *JOC* **61**, 361 (1996).
[2]Charette, A.B., Cote, B., Monroc, S., Prescott, S. *JOC* **60**, 6888 (1995).
[3]Edwards, M.L., Stemerick, D.M., McCarthy, J.R. *T* **50**, 5579 (1994); Bell, K.E., Knight, D.W., Gravestock, M.B. *TL* **36**, 8681 (1995).
[4]Hanessian, S., Yang, R.-Y. *SL* 633 (1995).
[5]Purchase, C.F., White, A.D. *SC* **26**, 2687 (1996).
[6]Solomon, M.E., Lynch, C.L., Rich, D.H. *SC* **26**, 2723 (1996).
[7]Grice, I.D., Harvey, P.J., Jenkins, I.D., Gallagher, M.J., Ranasinghe, M.G. *TL* **37**, 1087 (1996).
[8]Cravotto, G., Giovenzana, G.B., Sisti, M., Palmisano, G. *T* **52**, 13007 (1996).
[9]Aesa, M.C., Baan, G., Novak, L., Szantay, C. *SC* **25**, 2575 (1995).
[10]Myers, A.G., Zheng, B. *TL* **37**, 4841 (1996).
[11]Myers, A.G., Zheng, B. *JACS* **118**, 4492 (1996).
[12]Hara, R., Furukawa, T., Horiguchi, Y., Kuwajima, I. *JACS* **118**, 9186 (1996).

Triphenylphosphine–diisopropyl azodicarboxylate. 15, 352–353; **17**, 390; **18**, 387–388

Allylic amines.[1] With phthalimide as nucleophile in the modified Mitsunobu reaction, high yields of the *N*-allylated phthalimides are prepared. The phthaloyl group can be removed with methylamine in methanol.

[1]Sen, S.E., Roach, S.L. *S* 756 (1995).

Triphenylphosphine–*N*-halosuccinimide.

Esters and amides. General methods for the derivatization of carboxylic acids employ the Ph_3P-NXS combinations (NBS, NCS).[1,2]

[1]Froyen, P. *PSS* **91**, 145 (1994).
[2]Froyen, P. *SC* **25**, 959 (1995).

Triphosgene. 18, 388

Oxazolidin-2-ones and thiazolidin-2-ones. 2-Amino alcohols, such as serine, undergo cyclization while incorporating a CO unit on reaction with triphosgene $(Cl_3CO)_2C{=}O$. An improved procedure for preparation of the required triphosgene using an internal cooling system has been developed.[1]

[1]Falb, E., Nudelman, A., Hassner, A. *SC* **23**, 2839 (1993).

Triruthenium dodecacarbonyl.

Reduction.[1] A synthesis of 1-pyrrolines involves the reduction of γ-nitro carbonyl compounds, using a mixture of $Ru_3(CO)_{12}$, CO, and 1,10-phenanthroline.

Carbonylation. Homologation of allylic compounds,[2] stitching of an alkene with an

Ph, O, O_2N; $Ru_3(CO)_{12}$ - CO; phenanthroline; dioxane, 140°; Ph, N; 91%

imidazole ring (at C-4),[3] and cycloaromatization of 1,6-diynes incorporating two CO molecules to give indane-5,6-diol derivatives[4] are some of the examples of $Ru_3(CO)_{12}$-catalyzed reactions. $Ru_3(CO)_{12}$–$Co_2(CO)_8$ is an effective bimetallic catalyst for the production of carboxylic acids[5] from iodoalkanes, CO_2, and hydrogen.

MeN, N; +; $Ru_3(CO)_{12}$ - CO; PhMe, 160°; MeN, N, O; 50%

Ac, Ac; $Ru_3(CO)_{12}$ - CO; t-BuMe$_2$SiH - PCy$_3$; MeCN, 140°; t-BuMe$_2$SiO, t-BuMe$_2$SiO, Ac, Ac; 71%

Enamides.[6] The addition of *N*-aryl amides to alkynes affords enamides. A phosphine is present in the reaction medium.

PhNHCHO + ≡—C_6H_{13} —[$Ru_3(CO)_{12}$ - PCy_3 / PhMe, 180°]→ Ph(CHO)N–CH=CH–C_6H_{13} 67% (*E* : *Z* 93 : 7)

[1]Watanabe, Y., Yamamoto, J., Akazome, M., Kondo, T., Mitsudo, T. *JOC* **60**, 8328 (1995).
[2]Mitsudo, T., Suzuki, N., Kondo, T., Watanabe, Y. *JOC* **59**, 7759 (1994).
[3]Chatani, N., Fukuyama, T., Kakiuchi, F., Murai, S. *JACS* **118**, 493 (1996).
[4]Chatani, N., Fukumoto, Y., Ida, T., Murai, S. *JACS* **115**, 11614 (1993).
[5]Fukuoka, A., Gotoh, N., Kobayashi, N., Hirano, M., Komiya, S. *CL* 567 (1995).
[6]Kondo, T., Tanaka, A., Kotachi, S., Watanabe, Y. *CC* 413 (1995).

Tris(4-bromophenyl)aminium hexafluoroantimonate. 14, 338; **16**, 369–370; **17**, 391; **18**, 389

Mukaiyama aldol and related reactions. In the condensation with this catalyst to prepare β-alkoxy α,α-difluoro ketones,[1] the results are far superior to employing other Lewis acids, such as $SbCl_5$ and $Cu(OTf)_2$, whereas some (e.g., Me_3SiOTf, $TiCl_4$, and $BF_3 \cdot OEt_2$) are not effective at all. The reactions of silylated nucleophiles with acetals afford excellent yields of β-alkoxy ketones.[2]

4-MeOC_6H_4C(OMe)=CF_2 + PhCHO —[Ar_3NSbCl_6 / CH_2Cl_2, 0°]→ 4-MeOC_6H_4CO–CF_2–CH(OMe)Ph 80%

Ar = 4-BrC_6H_5

Diquinane synthesis.[3] 2-(4-Alkenyl)cyclopropyl sulfides are transformed into diquinanes on oxidation with tris(4-bromophenyl)aminium ions in dichloromethane at room temperature.

STol, Bn, Ar cyclopropane substrate —[Ar_3NSbX_6 / CH_2Cl_2]→ diquinane (STol, Bn, Ar) + diquinane (STol, Bn, Ar) (2.6-4.6 : 1)

[1] Kodama, Y., Yamane, H., Okumura, M., Shiro, M., Taguchi, T. *T* **51**, 12217 (1995).
[2] Kamata, M., Yokoyama, Y., Karasawa, N., Kato, M., Hasegawa, E. *TL* **37**, 3483 (1996).
[3] Takemoto, Y., Furuse, S.-i., Koike, H., Ohra, T., Iwata, C., Ohishi, H. *TL* **36**, 4085 (1995).

Tris(dibenzylideneacetone)dipalladium. 14, 339; **15**, 353–355; **16**, 372; **17**, 394; **18**, 389–393

N-Arylations. A general catalytic system for *N*-arylation with aryl bromides[1,2] (including bromopyridines)[3] consists of $Pd_2(dba)_3$, *t*-BuONa, and a phosphine, such as BINAP. Alternatively, diaryliodonium salts can be used as electrophiles.[4]

Nitromethylation.[5] Allylic derivatives, such as carbonates, undergo chain extension to give *(E)*-1-nitroalk-3-enes.

BnO–CH2CH=CHCH2–OCOOEt → ($MeNO_2$; $(dba)_3Pd_2$ - PPh_3) BnO–CH2CH=CHCH2CH2–NO_2 62%

Coupling of allylic compounds. Cross-coupling of alkenylstannanes with allylic carbonates provides 1,4-dienes.[6] Allyltriethylgermane is obtained from $Et_3GeSnBu_3$ and allylic halides.[7]

Dihydropyrans.[8] Allenylpalladium species generated from a propargyl carbonate may react intramolecularly with a remote nucleophile. Thus, a new method for alkylidenedihydropyran synthesis is developed.

Ph, OCOOMe, OH → ($(dba)_3Pd_2$ - dppb; THF, 50°) Ph 40%

Aryl halide–organozinc halide coupling. The Pd-catalyzed coupling is applicable to metal-complexed aryl halides, allowing a facile preparation of phenylalanine derivatives and related compounds.[9]

$(OC)_3Cr$–Ar–Cl + IZn, NHBoc, COOMe → ($(dba)_3Pd_2$ - $(o\text{-Tol})_3P$; 50°, 3 h) $(OC)_3Cr$–Ar, NHB[oc], COOM[e] 40%

Stille coupling. The coupling of arylstannanes with 4-chlorocyclobutenones leads directly to annulated aromatic compounds.[10] Apparently, the coupled products undergo electrocyclic ring opening and reclosure.

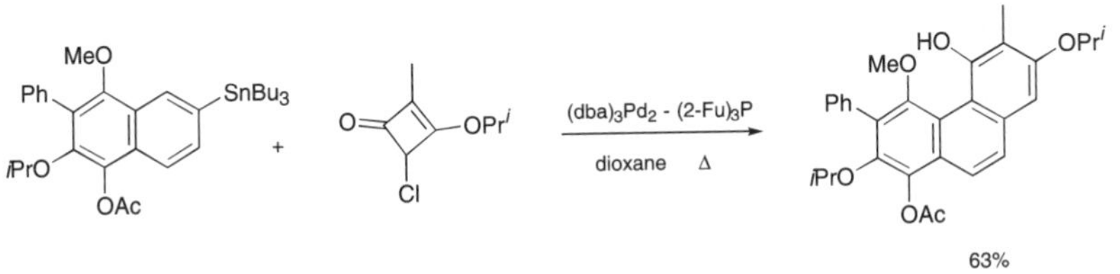

(dba)$_3$Pd$_2$ - (2-Fu)$_3$P
dioxane Δ
63%

Intramolecular Heck reaction. A significant application of this process is reported in the elaboration of the 5β-hydroxy-10β-hydroxymethyl structural motif[11] of the cardiac glycosides such as strophanthidin and ouabagenin.

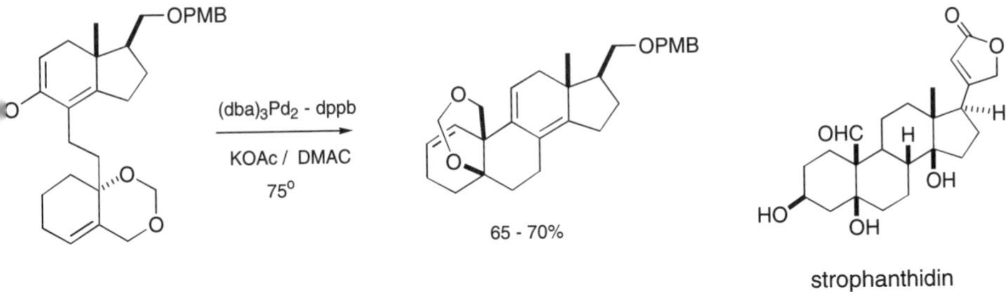

(dba)$_3$Pd$_2$ - dppb
KOAc / DMAC
75°
65 - 70%
strophanthidin

Reactions of methylenecyclopropanes. Intramolecular [3+2]-cycloaddition[12] and a more unusual ring-opening coupling to provide 1,4-dienes[13] are catalyzed by Pd$_2$(dba)$_3$.

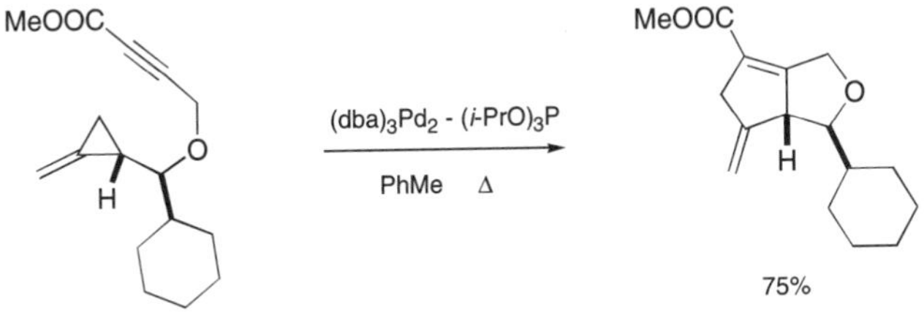

(dba)$_3$Pd$_2$ - (i-PrO)$_3$P
PhMe Δ
75%

[1]Wolfe, J.P., Buchwald, S.L. *JOC* **61**, 1133 (1996).
[2]Wolfe, J.P., Wagaw, S., Buchwald, S.L. *JACS* **118**, 7215 (1996).
[3]Wagaw, S., Buchwald, S.L. *JOC* **61**, 7240 (1996).

[4]Kang, S.-K., Lee, H.-W., Choi, W.-K., Hong, R.-K., Kim, J.-S. *SC* **26**, 4219 (1996).
[5]Deardorff, D.R., Savin, K.A., Justman, C.J., Karanjawala, Z.E., Sheppeck, J.E., Hager, D.C., Aydin, N. *JOC* **61**, 3616 (1996).
[6]Castano, A.M., Echavarren, A.M. *TL* **37**, 6587 (1996).
[7]Nakano, T., Ono, K., Migita, T. *CL* 697 (1996).
[8]Fournier-Nguefack, C., Lhoste, P., Sinou, D. *SL* 553 (1996).
[9]Jackson, R.F.W., Turner, D., Block, M.H. *SL* 862 (1996).
[10]Koo, S., Liebeskind, L.S. *JACS* **117**, 3389 (1995).
[11]Deng, W., Jensen, M.S., Overman, L.E., Rucker, P.V., Vionnet, J.-P. *JOC* **61**, 6760 (1996).
[12]Lautens, M., Ren, Y. *JACS* **118**, 9597 (1996).
[13]Corlay, H., Fouquet, E., Motherwell, W.B. *TL* **37**, 5983 (1996).

Tris(dibenzylideneacetone)dipalladium–chloroform.

Carbonylations. Two different types of dienoic amides are generated in the carbonylation of propargylic amines[1] and 2,3-dienylamines.[2]

N–Bn
$(dba)_3Pd_2 \cdot CHCl_3$
CO - dppp - TsOH
CH_2Cl_2, 75°
N–Bn, O
83%

PhNH
$(dba)_3Pd_2 \cdot CHCl_3$
CO - dppp - TsOH
CH_2Cl_2, 100°
PhNH, O
66%

Coumarins. A new synthesis of coumarins[3] is directly from phenols and propynoic esters (8 examples, 46–79%).

OMe, MeO, OH + COOEt
$(dba)_3Pd_2 \cdot CHCl_3$
NaOAc - HCHO
OMe, MeO, O, O
77%

Ene reaction. A convenient cyclization of 1,7-enynes[4] is useful for the synthesis of cassiol.

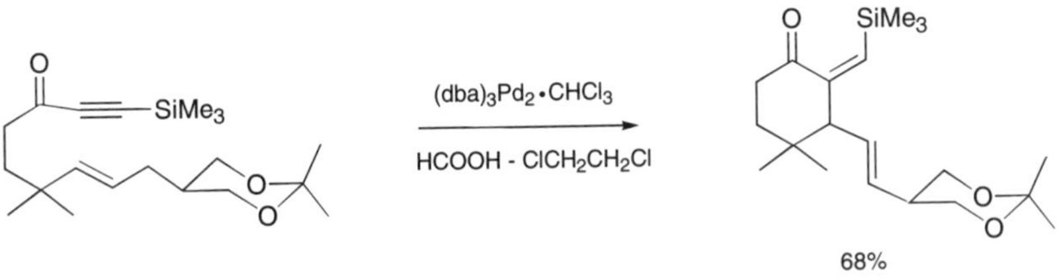

(dba)$_3$Pd$_2$•CHCl$_3$

HCOOH - ClCH$_2$CH$_2$Cl

68%

Allylations. Both the electrophilic allylation of carbonyl compounds with allylic phosphates in the presence of Et_2Zn[5] and the nucleophilic allylation of malononitriles and related compounds with allylstannanes[6] are catalyzed by $Pd_2(dba)_3 \cdot CHCl_3$.

Addition to allenes. The addition of malononitriles and cyanoacetates to substituted allenes shows regioselectivities according to the nature of the substituents. Thus, α-addition[7] and γ-addition[8] are respectively observed in the cases of alkoxy and sulfenyl derivatives. Addition and substitution occur with allenylstannanes.[9]

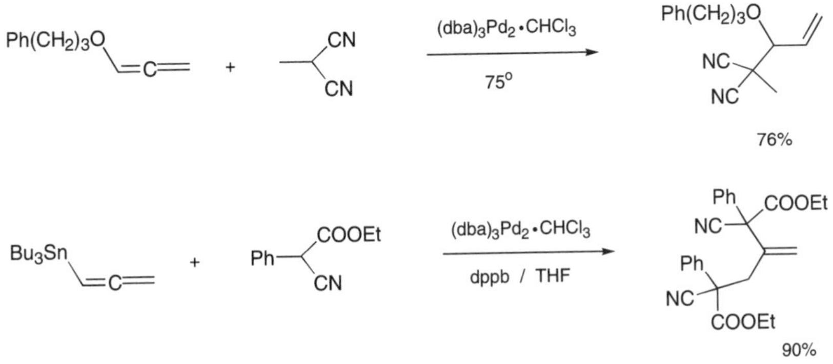

(dba)$_3$Pd$_2$•CHCl$_3$

75°

76%

(dba)$_3$Pd$_2$•CHCl$_3$

dppb / THF

90%

Of some interest is the fact that enynes give allenes[10] in the addition.

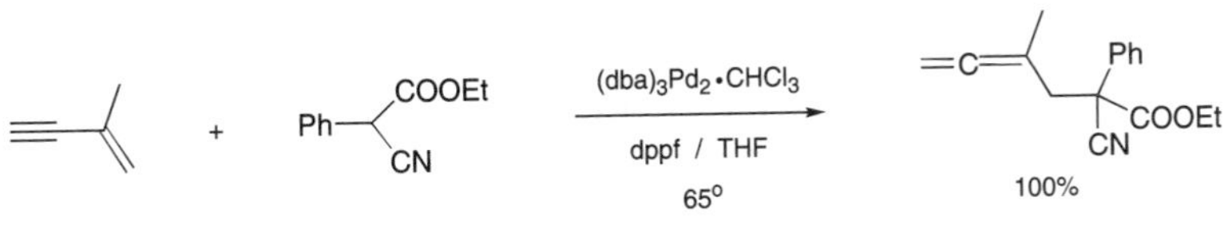

(dba)$_3$Pd$_2$•CHCl$_3$

dppf / THF

65°

100%

Furan formation.[11] In a synthesis of halenaquinone, both the furan ring and the cyclohexanone unit fused to it are formed in the same step by an intramolecular coupling involving an alkyne side chain and an iodoalkene moiety.

[1]Imada, Y., Alper, H. *JOC* **61**, 6766 (1996).
[2]Imada, Y., Vasapollo, G., Alper, H. *JOC* **61**, 7982 (1996).
[3]Trost, B.M., Toste, F.D. *JACS* **118**, 6305 (1996).
[4]Trost, B.M., Li, Y. *JACS* **118**, 6625 (1996).
[5]Kang, S.-K., Kim, D.-Y., Hong, R.-K., Ho, P.-S. *SC* **26**, 1493 (1996).
[6]Yamamoto, Y., Fujiwara, N. *CC* 2013 (1995).
[7]Yamamoto, Y., Al-Masum, M. *SL* 969 (1995).
[8]Yamamoto, Y., Al-Masum, M., Takeda, A. *CC* 831 (1996).
[9]Yamamoto, Y., Al-Masum, M., Fujiwara, N. *CC* 381 (1996).
[10]Salter, M.M., Gevorgyan, V., Saito, S., Yamamoto, Y. *CC* 17 (1996).
[11]Kojima, A., Takemoto, T., Sodeoka, M., Shibasaki, M. *JOC* **61**, 4876 (1996).

Tris(ethylenedioxyboryl)methane.

Homologation of carbonyl compounds. The reagent (**1**) is prepared from dimethoxychloroborane by treatment with lithium and chloroform, followed by ethylene glycol.[1] Although the yield is low (16%), the method is expedient. The homologation simply involves deprotonation and addition of the carbonyl substrates.

[1]Schummer, D., Höfle, G. *T* **51**, 11219 (1995).

Tris(2-methoxyphenyl)bismuthane.

Macrocyclic diesters.[1] Very high template ability of the reagent enables the efficient synthesis of the diesters from cyclic anhydrides and glycols in one step.

[1] Ogawa, T., Yoshikawa, A., Wada, H., Ogawa, C., Ono, N., Suzuki, H. *CC* 1407 (1995).

Tris(pentafluorophenyl)borane. 18, 393–394

Hydrosilylation.[1] The reductive silylation of aromatic carbonyl compounds and esters is catalyzed by $(C_6F_5)_3B$ at room temperature (9 examples, 80–96%). Aldehydes react faster.

[1]Parks, D.J., Piers, W.E. *JACS* **118**, 9440 (1996).

Tris[2-(perfluorohexyl)ethyl]tin hydride.

Reductions.[1] This reagent, $(C_6F_{13}CH_2CH_2)_3SnH$, reduces many organic substances via pathways involving radicals. For example, 1-adamantyl bromide gives adamantane in 90% yield. Nitro compounds, selenides, and dithiocarbonates are also reduced. The advantage associated with this reagent is that while it has reactivities comparable to those of Bu_3SnH, its derivatives can be separated from the organic products by liquid–liquid extraction. More importantly, the demonstrated possibility of employing catalytic quantities of this tin hydride in combination with $NaBH_3CN$ makes its applications very attractive.

[1]Curran, D.P., Hadida, S. *JACS* **118**, 2531 (1996).

Tris(trimethylsilyl)germane.

Reductions.[1] $(Me_3Si)_3GeH$ is a novel reducing agent based on a mechanism involving radicals. It replaces carbon-bound halogen atoms, xanthate functions, isocyanides, phenylseleno, and tertiary nitro groups (10 examples, 96–100%).

[1]Chatgilialoglu, C., Ballestri, M. *OM* **14**, 5017 (1995).

Tris(trimethylsilyl)silane.

Radical cyclization. Acyl radicals are generated from selenoesters on treatment with $(Me_3Si)_3SiH$ in the presence of Et_3B at or below room temperature. Such a radical can add to a juxtaposed conjugated ester to form a cyclic ketone with an acetic ester side chain. Substituents at the γ and δ positions of the ester unit have great influences on the cyclization mode.[1,2] Triorganotin hydrides can also be used, but a higher temperature is required.

SePh
COOMe
$(Me_3Si)_3SiH$ - Et_3B
PhH, air, rt
COOMe

[1]Evans, P.A., Roseman, J.D. *JOC* **61**, 2252 (1996).
[2]Evans, P.A., Roseman, J.D., Garber, L.T. *JOC* **61**, 4880 (1996).

Tris(trimethylsilyl)titanacyclobutene.

Alkenylsilanes.[1] A very efficient conversion of carbonyl compounds to alkenylsilanes is based on reaction with the reagent (10 examples, 71–95%).

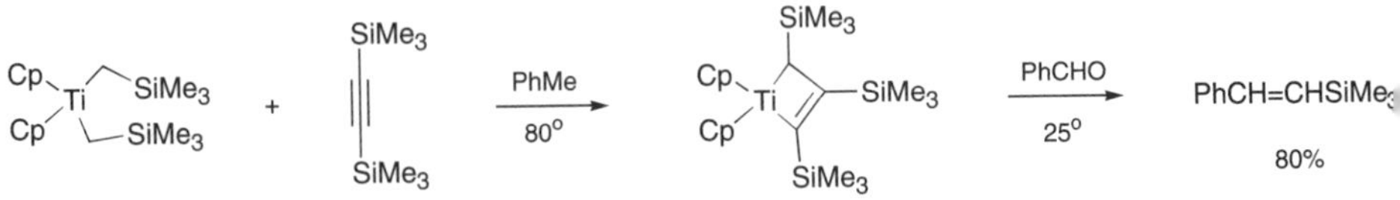

[1]Petasis, N.A., Staszewski, J.P., Fu, D.-K. *TL* **36**, 3619 (1995).

Tungsten carbene complexes.

Olefination.[1] Complexes of type (**1**) effect the olefination of ketones. Direction by a hydroxy group separated from the ketone by 2- or 3-carbons is observed. The reaction overwhelmingly favors the *(E)*-isomers.

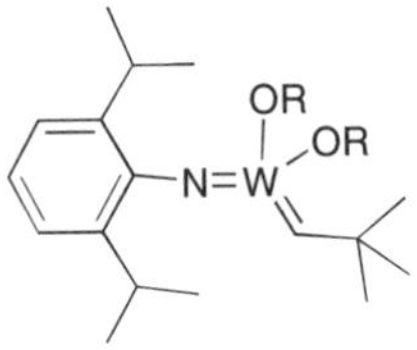

(**1**)

From 1,4-diyn-3-ols.[2] Highly unsaturated complexes are readily formed as shown below.

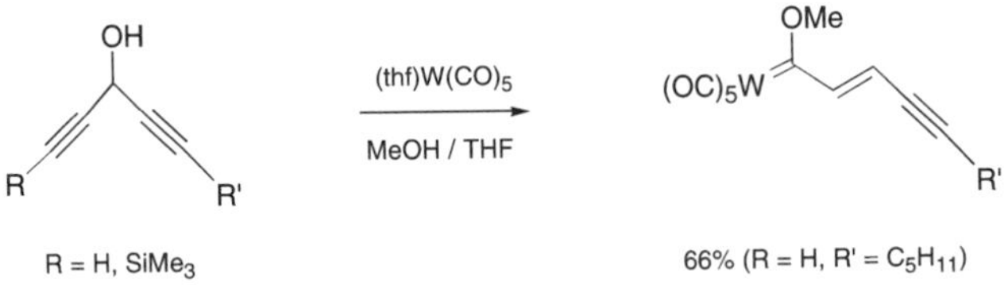

[1]Fujimura, O., Fu, G.C., Rothemund, P.W.K., Grubbs, R.H. *JACS* **117**, 2355 (1995).
[2]Cosset, C., Del Rio, I., Peron, V., Windmüller, B., Le Bozec, H. *SL* 435 (1996).

Tungsten(VI) chloride.

Chlorides.[1] With WCl_6 in refluxing dichloromethane, oxygen compounds are chlorinated. Thus, alcohols are converted into chlorides, benzaldehydes into *gem*-dichlorides, and epoxides into *vic*-dichlorides.

[1]Firouzabadi, H., Shiriny, F. *T* **52**, 14929 (1996).

Tungsten(VI) oxychloride–2,6-dibromophenol.

Ring-closing metathesis.[1] $WOCl_2(OAr)_2$ formed in situ with Et_4Pb form a practical catalyst system for the synthesis of cycloalkenes (11 examples, 58–91%).

[1]Nugent, W.A., Feldman, J., Calabrese, J.C. *JACS* **117**, 8992 (1995).

U

Ultrasound. 15, 363; **16**, 377–379; **18**, 395

Miscellaneous. *N*-Acylation of unprotected amino acids,[1] desilylation of TBS-ethers,[2] alkylation of ketone enolates with (iso)/quinolinium salts,[3] Wolff rearrangement,[4] bromination of active arenes with NBS,[5] and dichloroketene formation from Zn and Cl_3CCOCl (for cycloaddition to alkynes)[6] have all been promoted with ultrasound.

Tertiary alkyl acetates. Usually considered very difficult is the conversion of tertiary alkyl halides to acetates without the complication of elimination. The use of both a phase-transfer catalyst and ultrasound apparently favors the displacement by $Zn(OAc)_2$.[7]

2-Furanyl ketones. A synthesis of 2-furanyl ketones using the Barbier technique takes advantage of the facile lithiation of furan by organolithiums. Thus, sonication of a mixture of lithium, *t*-butyl chloride, a lithium carboxylate, and excess furan in THF accomplishes the tandem reaction.[8]

COOH + O → (Li - t-BuCl; THF))))) → O O

76%

[1]Anuradha, M.V., Ravindranath, B. *T* **53**, 1123 (1997).
[2]Lee, A.S.-Y., Yeh, H.-C., Tsai, M.-H. *TL* **36**, 6891 (1995).
[3]Diaba, F., Lewis, I., Grignon-Dubois, M., Navarre, S. *JOC* **61**, 4830 (1996).
[4]Winum, J.-Y., Kamal, M., Leydet, A., Roque, J.-P., Montero, J.-L. *TL* **37**, 1781 (1996).
[5]Paul, V., Sudalai, A., Daniel, T., Srinivasan, K.V. *SC* **25**, 2401 (1995).
[6]Parker, M.S.A., Rizzo, C.J. *SC* **25**, 2781 (1995).
[7]Jayasree, J., Rao, J.M. *SC* **26**, 1103 (1996).
[8]Aurell, M.J., Einhorn, C., Einhorn, J., Luche, J.L. *JOC* **60**, 8 (1995).

V

Vanadium(II/III) chloride.

Reductive coupling.[1] Treating an intermediate prepared by a Grignard reaction of a ketone with $VCl_2(tmeda)_2$ in the presence of a trace of oxygen leads to a hydrocarbon dimer.

Homoallylic alcohols.[2] $[V_2Cl_3(thf)_6]_2[Zn_2Cl_6]$ is effective in inducing a reaction between allyl bromides and carbonyl compounds in THF–HMPA.

[1]Kataoka, Y., Akiyama, H., Makihira, I., Tani, K. *JOC* **61**, 6094 (1996).
[2]Kataoka, Y., Makihira, I., Tani, K. *TL* **37**, 7083 (1996).

Vanadyl bis(acetylacetonate).

Self-arylation of alkyl aryl sulfides.[1] Oxidative dimerization gives sulfonium salts of alkylthioaryl aryl sulfides in excellent yields (5 examples, 94–100%). The products undergo dealkylation on heating with pyridine.

Ar-S-R —[$VO(acac)_2$ - MsOH; $(CF_3CO)_2O$, Bu_4NClO_4]→ Ar(R)S⁺–C₆H₃(R')–SR ClO_4^- —[py, Δ]→ ArS–C₆H₃(R')–SR

94 - 100% (two steps)

[1] Yamamoto, K., Kobayashi, S., Shouji, E., Tsuchida, E. *JOC* **61**, 1912 (1996).

Vinyldimethylsilyl chloride.

1,3-Dipolar cycloaddition.[1] This reagent acts as a 1,2-ethanediol equivalent when its reaction with 4-hydroxy-2-isoxaline 2-oxides is followed by oxidative desilylation of the tricyclic adducts.

(isoxazoline N-oxide with R^2, R^1, OH, COOEt) + Cl–Si(Me)₂–CH=CH₂ —[imidazole, MeCN]→ tricyclic adduct

R = COOEt

79 - >99%

[1]Righi, P., Marotta, E., Landuzzi, A., Rosini, G. *JACS* **118**, 9446 (1996).

Vinyltriphenylphosphonium bromide.

Carbon chain stitching.[1] The consecutive CuBr-catalyzed addition of a Grignard reagent[1] or phosphonate anion[2] and Wittig olefination performs stitching of two building blocks with a vinyl group as the linchpin.

PhMgBr + $CH_2=CH-PPh_3^+\ Br^-$ $\xrightarrow{\text{CuBr - }Ag_2CO_3}$ (OHC–C_6H_4–NMe_2, THF) Ph–CH_2–CH=CH–C_6H_4–NMe_2 73%

[1]Shen, Y., Yao, J. *JOC* **61**, 8659 (1996).
[2]Shen, Y., Yao, J. *JCR(S)* 428 (1996).

X

Xenon(II) fluoride.

α-Fluoro ethers. The sulfur moiety of *O,S*-acetals is replaced by fluorine on treatment with XeF_2 (7 examples, 72–85%).

Alkenyl fluorides.[2] These compounds can be prepared from alkenylstannanes by reaction with XeF_2 in the presence of AgOTf.

Decarboxylative rearrangement.[3] Trimethylsilyl benzoates undergo this radical reaction with XeF_2 in CH_2Cl_2. Interestingly, the reaction in MeCN does not proceed with rearrangement, and the main products are arenes derived from aryl radicals.

Me_3SiOOC–C_6H_4–$C(CH_3)_3$ $\xrightarrow[CH_2Cl_2]{XeF_2}$ FCOO–C_6H_4–$C(CH_3)_3$ 26%

[1]Lu, Q., Benneche, T. *ACS* **50**, 850 (1996).
[2]Tius, M.A., Kawakami, J.K. *T* **51**, 3997 (1995).
[3]Nongkunsarn, P., Ramsden, C.A. *JCS(P1)* 121 (1996).

Y

Ytterbium. 14, 348; **15**, 366; **16**, 384; **18**, 401

vic-Dialkylation of thioketones.[1] Diaryl thioketones are reduced by Yb in HMPA–THF, and the subsequent addition of RX introduces alkyl groups at sulfur and the sp^2-carbon.

Reductive coupling of carbonyl compounds. A divalent ytterbium thiolate, $Yb(SPh)_2$, is prepared from Yb and $PhSSiMe_3$. This species promotes the pinacolization of aromatic aldehydes in acetonitrile or propionitrile (*dl:meso*, 3–4:1).[2]

Amides are converted to 1,2-diaminoalkenes by Yb–$YbI_2$3 and acylsilanes to alkynes.[4]

Yb - YbI_2, THF Δ — 43%

Yb / HMPA - THF, -10°, 20 min — Ph—≡—Ph 67%

α-Hydroxy imines.[5] The C–C bond formation between diaryl ketones and isonitriles is promoted by Yb.

Yb / HMPA - THF, -35°; H^+ — 69%

[1]Makioka, Y., Uebori, S.-Y., Tsuno, M., Taniguchi, Y., Takaki, K., Fujiwara, Y. *JOC* **61**, 372 (1996).
[2]Taniguchi, Y., Nagata, K., Kitamura, T., Fujiwara, Y., Deguchi, D., Maruo, M., Makioka, Y., Takaki, K. *TL* **37**, 3465 (1996).
[3]Ogawa, A., Nanke, T., Takami, N., Sekiguchi, N., Kambe, N., Sonoda, N. *AOMC* **9**, 461 (1995).
[4]Taniguchi, Y., Fujii, N., Takaki, K., Fujiwara, Y. *AOMC* **9**, 377 (1995).
[5]Makioka, Y., Tsuno, M., Takaki, K., Taniguchi, Y., Fujiwara, Y. *CL* 821 (1995).

Ytterbium(III) triflate. 18, 402–403

Cleavage of methoxyacetates.[1] These esters undergo methanolysis in preference to others (e.g., acetates) in the presence of $Yb(OTf)_3$. Accordingly, the synthetic use of methoxyacetate esters as protected alcohols is broadened.

Allylation of carbonyl compounds.[2] Ytterbium triflate promotes the allylation reaction, which is mediated by indium in aqueous media.

Enamination.[3] The condensation of amines with ketones is greatly facilitated by $Yb(OTf)_3$. The catalyst and high pressure have a synergistic effect.

Aldol and imino-aldol reactions. A Yb complex prepared from $Yb(OTf)_3$ and a C_2-symmetric α,α'-bistriflamidobibenzyl has been used in the Mukaiyama aldol reaction,[4] resulting in moderate asymmetric induction. Imines are activated toward enol derivatives, such as ketene silyl ethers.[5] *N*-(α-aminoalkyl)benzotriazoles are suitable surrogates of imines.[6] One-pot syntheses of β-amino esters[7] and ketones[8] can also be achieved.

Apparent exceptions to simple condensations are the formation of quinoline,[9] β-lactam,[10] and the imino Diels–Alder reactions,[11] due to a change in reactivity of one reaction partner or the other.

N / Ph + MeO — $Yb(OTf)_3$, MeCN, rt, 10 min → N / Ph

74%

NPh / Ph + OMe / $OSiMe_3$ — $Yb(OTf)_3$, MeCN, rt → PhN / Ph / O

93%

Conjugate additions. Indoles react with electron-deficient alkenes at the β-carbon.[12] Michael addition in water using β-ketoesters as donors gives quantitative yields.[13] The addition of amines to unsaturated esters is favored by both the catalyst and pressure.[14] Interestingly, $Yb(OTf)_3$ also promotes radical addition to *N*-enoyloxazolidinones.[15]

O / O / N / O / Ph / Ph — $Yb(OTf)_3$, *i*-PrI, Bu_3SnH, Et_3B, O_2, CH_2Cl_2 - THF - hexane → O / O / N / O / Ph / Ph

93%

Glycosylation.[16] Using glucose 1,2-cyclic sulfite as the donor the catalyzed glycosylation of alcohols is S_N2-like and β-selective.

Opening of epoxides and aziridines. A very favorable effect of $Yb(OTf)_3$–$LiOPr^i$ [the active catalyst is $Yb(OPr^i)_3$; the triflate itself has too high Lewis acidity] on the ring openings is noted. Thus, 2-azido alcohols and 1,2-diamines are readily prepared from epoxides[17] and aziridines,[18] respectively.

[1]Hanamoto, T., Sugimoto, Y., Yokoyama, Y., Inanaga, J. *JOC* **61**, 4491 (1996).
[2]Wang, R., Lim, C.-M., Tan, C.-H., Lim, B.-K., Sim, K.-Y., Loh, T.-P. *TA* **6**, 1825 (1995).
[3]Jenner, G. *TL* **37**, 3691 (1996).
[4]Uotsu, K., Sasai, H., Shibasaki, M. *TA* **6**, 71 (1995).
[5]Kobayashi, S., Araki, M., Ishitani, H., Nagayama, S., Hachiya, I. *SL* 233 (1995).
[6]Kobayashi, S., Ishitani, H., Komiyama, S., Oniciu, D.C., Katritzky, A.R. *TL* **37**, 3731 (1996).
[7]Kobayashi, S., Araki, M., Yasuda, M. *TL* **36**, 5773 (1995).
[8]Kobayashi, S., Ishitani, H. *CC* 1379 (1995).
[9]Y. Makioka, Shindo, T., Taniguchi, Y., Takaki, K., Fujiwara, Y. *S* 801 (1995).
[10]Annunziata, R., Cinquini, M., Cozzi, F., Molteni, V., Schupp, O. *JOC* **61**, 8293 (1996).
[11]Kobayashi, S., Ishitani, H., Nagayama, S. *S* 1195 (1995).
[12]Harrington, P.E., Kerr, M.A. *SL* 1047 (1996).
[13]Keller, E., Feringa, B.L. *TL* **37**, 1879 (1996).
[14]Jenner, G. *T* **52**, 13557 (1996).
[15]Sibi, M.P., Jasperse, C.P., Ji, J. *JACS* **117**, 10779 (1995).
[16]Sanders, W.J., Kiessling, L.L. *TL* **35**, 7335 (1994).
[17]Meguro, M., Asao, N., Yamamoto, Y. *CC* 1021 (1995).
[18]Meguro, M., Yamamoto, Y. *H* **43**, 2473 (1996).

Yttrium(III) isopropoxide.

Cyanohydrin silyl ethers.[1] Yttrium isopropoxide–more likely, the $(i\text{-PrO})_{13}Y_5O$ species—complexes with 1,3-bis(2-methylferrocenyl)propane-1,3-dione (**1**) to afford a highly efficient catalyst for the asymmetric silylcyanation of electron-rich aromatic aldehydes with Me_3SiCN.

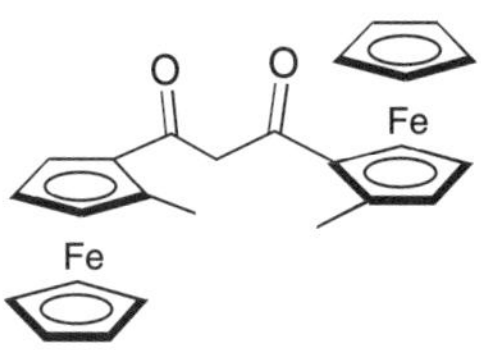

(**1**)

[1]Abiko, A., Wang, G. *JOC* **61**, 2264 (1996).

Z

Zeolites. 15, 367; **18**, 405–406

Dehydration-hydration.[1] Alkynols provide either conjugated enynes or enones over zeolite HSZ-360 at lower and higher temperatures, respectively.

HSZ-360, PhCl, 65°

HSZ-360, PhCl, 130°

gem-Diacetates.[2,3] Aldehydes are converted to $RCH(OAc)_2$ in the presence of zeolites.

Cyclization.[4] Epoxy polyenes undergo cyclization mediated by zeolites.

4A mol sieves, $CHCl_3$ Δ

70%

Electrophilic alkenylation.[5] Arenes react with phenylacetylene over HZS-360 zeolite in 1,2-dichlorobenzene at 110° to give 1,1-diarylethenes (10 examples, 40–94%).

β-Ketoesters. Both transesterification[6] and synthesis from aldehydes and ethyl diazoacetate[7] are catalyzed by zeolites.

Benzoxazoles.[8] Beckmann rearrangement of *o*-hydroxyacetophenone oximes proceeds on heating with HY zeolite at 160° (8 examples, 84–95%).

Oxidation. Epoxidation of chiral allylic alcohols[9] with urea–H_2O_2 catalyzed by ZSM-5 zeolite (titanium silicate-1) and ene-type hydroperoxidation of alkenes[10] photosensitized by thiazine dye cation-exchanged zeolites have been reported. In the latter reaction, only one product is formed.

[1]Sartori, G., Pastorio, A., Maggi, R., Bigi, F. *T* **52**, 8287 (1996).
[2]Pereira, C., Gigante, B., Marcelo-Curto, M.J., Carreyre, H., Perot, G., Guisnet, M. *S* 1077 (1995).
[3]Kumar, P., Hegde, V.R., Kumar, T.P. *TL* **36**, 601 (1995).
[4]Sen, S.E., Zhang, Y.Z., Roach, S.L. *JOC* **61**, 9534 (1996).
[5]Sartori, G., Bigi, F., Pastorio, A., Porta, C., Anenti, A., Maggi, R., Moretti, N., Gnappi, G. *TL* **36**, 9177 (1995).
[6]Balaji, B.S., Sasidharan, M., Kumar, R., Chanda, B. *CC* 707 (1996).
[7]Sudrik, S.G., Balaji, B.S., Singh, A.P., Mitra, R.B., Sonawane, H.R. *SL* 369 (1996).
[8]Bhawal, B.M., Mayabhate, S.P., Likhite, A.P., Deshmukh, A.R.A.S. *SC* **25**, 3315 (1995).
[9]Adam, W., Kumar, R., Reddy, T.I., Renz, M. *ACIEE* 35, 880 (1996).
[10]Li, X., Ramamurthy, V. *JACS* **118**, 10666 (1996).

Zinc. **13**, 346–347; **14**, 349–350; **16**, 386–387; **17**, 406–407; **18**, 406–408

Elimination. The preparation of the bis[methyl *(S)*-lactyl] acetylenedicarboxylate[1] involves alcoholysis of dibromofumaryl chloride and debromination with Zn in refluxing THF. Asymmetric induction in the Diels–Alder reactions of the chiral diester has been probed.

Alkynylsilanes. Silylation of alkynes[2] and copper(I) alkynides[3] is promoted by Zn in MeCN (sealed-tube reactions).

Coupling reactions. The hexacarbonyldicobalt complexes of cycloocta-1,5-diynes and cyclooct-3-ene-1,5-diynes have been acquired by reductive cyclization of the dication precursors.[4]

$Co_2(CO)_6$ Ph + + Ph $Co_2(CO)_6$ — Zn, CH_2Cl_2, rt, 2 h → $Co_2(CO)_6$ Ph Ph $Co_2(CO)_6$

48%

The coupling of benzyl bromides with organotin derivatives with Zn powder leads to benzyltin compounds.[5]

Allylation and propargylation. Aqueous media are suitable for the allylation of β-ketoesters,[6] and when a modified β-cyclodextrin is present, the allylation of cyclohexenone shows asymmetric induction (30% ee).[7] The analogous Barbier-type propargylation of aldehydes also proceeds at room temperature (10 examples, 41–85%).[8]

Allylation of imines[9] using Zn is comparable to that using Mg. The Barbier-type allylation of aldehydes can be carried out in liquid ammonia.[10]

Allyl ketones are rapidly formed by the reaction of allyl bromide with acid chlorides in ether.[11]

γ-Nitro esters.[12] The conjugate addition to nitroalkenes with Reformatsky reagents represents a new way for the preparation of the nitro esters.

Ether cleavage.[13] In the presence of acid chlorides, ethers are transformed into esters and alkyl chlorides.

Reduction of sulfoxides.[14] A combination of Zn, Ac_2O, and DMAP in MeCN reduces sulfoxides to sulfides at room temperature (8 examples, 50–97%).

[1]Charlton, J.L., Chee, G., McColeman, H. *CJC* **73**, 1454 (1995).
[2]Sugita, H., Hatanaka, Y., Hiyama, T. *TL* **36**, 2769 (1995).
[3]Sugita, H., Hatanaka, Y., Hiyama, T. *CL* 379 (1996).
[4]Melikyan, G.G., Khan, M.A., Nicholas, K.M. *OM* **14**, 2170 (1995).
[5]Marton, D., Russo, U., Stivanello, D., Tagliavini, G. *OM* **15**, 1645 (1996).
[6]Ahonen, M., Sjoholm, R. *CL* 341 (1995).
[7]Fornasier, R., Marcuzzi, F., Piva, M., Tonellato, U. *G* **126**, 633 (1996).
[8]Yavari, I., Riazi-Kermani, F. *SC* **25**, 2923 (1995).
[9]Wang, D.-K., Dai, L.-X., Hou, X.-L., Zhang, Y. *TL* **37**, 4187 (1996).
[10]Makosza, M., Grela, K. *SC* **26**, 2935 (1996).
[11]Ranu, B.C., Majee, A., Das, A.R. *TL* **37**, 1109 (1996).
[12]Menicagli, R., Samaritani, S. *T* **52**, 1425 (1996).
[13]Bhar, S., Ranu, B.C. *JOC* **60**, 745 (1995).
[14]Wang, Y., Koreeda, M. *SL* 885 (1996).

Zinc, activated.

A useful technique for the preparation of active submicronic Zn powder (as well as Co and Cu powders) is by electrolysis with pulsed ultrasonic irradiation.[1] Greatly enhanced efficiency in allylation achieved by using such Zn powders is noted. Another method involves reduction of $ZnCl_2$ with Na in liquid ammonia.[2]

Ketones. Rieke zinc produced by the reduction of $Zn(CN)_2$ with Li naphthalenide reacts with alkyl halides, and the organozinc halides can be used to form ketones on treatment with CuCN and acid chlorides.[3] Secondary and tertiary alkylzinc bromides are readily prepared in this direct manner.[4] α-Chloromethyl ketones have been prepared by this method using chloroacetyl chloride in the coupling reaction.[5]

3-Thienylzinc halides.[6] These reagents are obtained from 3-thienyl halides and Rieke zinc. They undergo Ni-catalyzed coupling with aryl iodides.

[1]Durant, A., Delplancke, J.-L., Winand, R., Reisse, J. *TL* **36**, 4257 (1995).
[2]Makosza, M., Grela, K., Fabianowski, W. *T* **52**, 9575 (1996).
[3]Hanson, M., Rieke, R.D. *SC* **25**, 101 (1995).
[4]Rieke, R.D., Hanson, M.V., Brown, J.D., Niu, Q.J. *JOC* **61**, 2726 (1996).
[5]Rieke, R.D., Brown, J.D., Wu, A.X. *SC* **25**, 3923 (1995).
[6]Wu, X., Rieke, R.D. *JOC* **60**, 6658 (1995).

Zinc, amalgamated.

Styrenes.[1] A modified Clemmensen reduction of aryl ketones using Zn(Hg) and some ethanol in refluxing formic acid gives styrenes (*E*-form major; 5 examples, 69–83%).

Fragmentation. Some 3-acetylcycloalkanones undergo fragmentation[2] under the Clemmensen reduction conditions.

Zn-Hg, HCl / H_2O, Δ; R = H: 44%, 10%, 14%

[1]Hiegel, G.A., Carney, J.R. *SC* **26**, 2625 (1996).
[2]Bailey, K.E., Davis, B.R. *AJC* **48**, 1827 (1995).

Zinc–acetic acid.

Reduction of α-oxoketene dithioacetals. The double bond is saturated to give dithioacetals of β-keto aldehydes.[1] Note that different sets of products are obtained on reduction by the Zn–$ZnCl_2$/TMEDA system.[2]

Zn / HOAc-H_2O, 0°: 94%
Zn - $ZnCl_2$, EtOH Δ: TMEDA (1.5 equiv) 85%; TMEDA (3 equiv) 65%; TMEDA (5 equiv) 55%

Desulfurization. Reductive desulfurization of α-(2-pyridinethio)sulfones[3] occurs on treatment with Zn-HOAc, removing the thiopyridyl group.

Benzyl acetates.[4] Aryl carbonyl compounds (but not others) undergo reductive acetylation at room temperature.

Reduction of nitroalkenes. γ-Acetoxy nitroalkenes are converted to unsaturated oximes, and the method is applicable to the synthesis of 2,3-unsaturated sugars.[5]

AcO–CH(R)–CH=CHNO$_2$ → (Zn - HOAc (H_2O), Et_2O) → R–CH=CH–CH=NOH

Ordinary nitroalkenes give saturated carbonyl compounds on treatment with Zn–CF_3COOH in organic solvents.[6]

Amino acids from N-trichloroethoxycarbonyloxazolidin-5-ones.[7] Dechlorinative fragmentation of the *N*-protecting group by Zn–HOAc triggers ring cleavage to afford iminium species that are hydrolyzed in situ.

Zn / HOAc -H_2O

80%

[1]Rao, C.S., Patro, B., Ila, H., Junjappa, H. *IJC(B)* **35B**, 57 (1996).
[2]Yadav, K.M., Suresh, J.R., Patro, B., Ila, H., Junjappa, H. *T* **52**, 4679 (1996).
[3]Boivin, J., Lallemand, J., Schmitt, A., Zard, S.Z. *TL* **36**, 7243 (1995).
[4]Rani, B.R., Ubukata, M., Osada, H. *BCSJ* **68**, 282 (1995).
[5]Koos, M. *TL* **37**, 415 (1996).
[6]Saikia, A.K., Barua, N.C., Sharma, R.P., Ghosh, A.C *JCR(S)* 124 (1996).
[7]Chollet, J.-F., Miginiac, L., Rudelle, J., Bonnemain, J.-L. *SC* **23**, 2101 (1993).

Zinc–nickel chloride.

Reduction of conjugated double bonds.[1] This selective reduction method is valuable in situations in which other double bonds must be retained. Such a reaction, further facilitated with ultrasound, has been employed during a synthesis of penitrem-D.

Zn, $NiCl_2 \cdot 6H_2O$

$MeO(CH_2)_2OH$

))))

90%

penitrem-D

[1]Smith, A.B., III, Nolen, E.G., Shirai, R., Blasé, F.R., Ohta, M., Chida, N., Hartz, R.A., Fitch, D.M., Clark, W.M., Sprengeler, P.A. *JOC* **60**, 7837 (1995).

Zinc–copper couple. 13, 348; **15**, 367–368; **16**, 387–388; **17**, 407; **18**, 408–409

RCOCl → RCHO.[1] The reduction system consists of Zn–Cu, Bu_3P, and MsOH.

1-Alkoxy-1-siloxycyclopropanes.[2] Trapping of the organozinc species derived from β-haloalkanoic esters with a silyl chloride provides the cyclopropanone acetals.

Zn-Cu / THF Δ ; t-$BuMe_2SiCl$

$OSiMe_2Bu^t$

56%

vic-Diamines.[3] Imines undergo reductive coupling. In the presence of (+)–10-camphorsulfonic acid, the *(R,R)*-1,2–diamines are formed.

Dehalogenation. The use of Zn-CuI in DMF to eliminate [ClF] from α-allyloxy-α-chloroperfluoroalkanoic esters initiates the synthesis of δ,ε-unsaturated β,β-difluoro-α-keto esters by a Claisen rearrangement.[4]

COOEt F_3C Cl

Zn - CuI / DMF, 25° ; PhH Δ

F F COOEt O

86%

[1]Maeda, H., Maki, T., Ohmori, H. *TL* **36**, 2247 (1995).
[2]Yasui, K., Tanaka, S., Tamaru, Y. *T* **51**, 6881 (1995).
[3]Shimizu, M., Iida, T., Fujisawa, T. *CL* 609 (1995).
[4]Shi, G.-Q., Cai, W.-L. *JOC* **60**, 6289 (1995).

Zinc borohydride. 14, 351; **16**, 388–389; **18**, 409

$RCOOH \rightarrow RCH_2OH$.[1] The reduction in refluxing THF proceeds in good yields (16 examples, 70–95%), including amino acids, which give 1,2-amino alcohols.[2]

Alkylferrocenes.[3] Ionic deoxygenation of ferrocenyl ketones with $Zn(BH_4)_2$ in THF at room temperature is catalyzed by $ZnCl_2$ (11 examples, 88–92%).

[1]Narasimhan, S., Madhavan, S., Prasad, K.G. *JOC* **60**, 5314 (1995).

[2]Narasimhan, S., Madhavan, S., Prasad, K.G. *SC* **26**, 703 (1996).
[3]Bhattacharyya, S. *OM* **15**, 1065 (1996).

Zinc borohydride–aluminum phosphate.

Epoxide cleavage.[1] The supported borohydride reagent cleaves epoxides at room temperature, giving alcohols.

Hydration of styrenes.[2] Mixture of alcohols are obtained. Arylalkynes also give secondary alcohols.

[1]Campelo, J.M., Chakraborty, R., Marinas, J.M. *SC* **26**, 415 (1996).
[2]Campelo, J.M., Chakraborty, R., Marinas, J.M. *SC* **26**, 1639 (1996).

Zinc bromide. 13, 349; **15**, 368; **16**, 389–391; **18**, 409

Dehydrogenation.[1] The aromatization of imidazolines has been observed.

Hydrolysis of gem-bis(benzotriazolyl)methanes.[2] A method for *p*-formylation of nitroarenes consists of vicarious nucleophilic substitution with tris(benzotriazol-1-yl)methane and successive treatment with $ZnBr_2$ and aqueous acid.

O_2N–C_6H_4–CH(Bt)$_2$ → ($ZnBr_2$ / dioxane ; HCl - H_2O) → O_2N–C_6H_4–CHO 86%

1,3-Diols.[3] The stereoselective addition of organometallic reagents to β-hydroxy ketones to give predominantly *syn*-1,3-diols is mediated by $ZnBr_2$.

[1]Katritzky, A.R., Zhu, L., Lang, H., Denisko, O., Wang, Z. *T* **51**, 13271 (1995).
[2]Katritzky, A.R., Xie, L. *TL* **37**, 347 (1996).
[3]Ruano, J.L.G., Tito, A., Culebras, R. *T* **52**, 2177 (1996).

Zinc chloride. 13, 349–350; **15**, 368–371; **16**, 391–392; **18**, 410–411

Allylation of arenes.[1] $ZnCl_2$ supported on SiO_2–K_2CO_3/Al_2O_3 is an effective catalyst system.

Aziridines from epoxides.[2] A direct access to aziridines is provided by the reaction of epoxides with iminotriphenylphosphoranes under the influence of $ZnCl_2$. However, the yields are variable (< 5–84%).

Ph-epoxide + PhN=PPh$_3$ → ($ZnCl_2$; $ClCH_2CH_2Cl$, 80°, 1 h) → Ph-aziridine-NPh 72%

Coupling of allyl trimethylsilyl ethers and allylsilanes.[3] Zinc chloride and several other Lewis acids (but not $BF_3 \cdot OEt_2$) promote the formation of 1,5-dienes from the two components. The reaction generally occurs at the less hindered site of the allyl cation.

α-(Trimethylsilyl)allenyl ketones.[4] The reaction products of acylsilanes with 3-chloropropynyllithium expel chloride to form the allenyl ketones in 40–68% yield (5 examples).

MeLi ; AcSiMe3 → [Cl–CH2–C≡C–C(OLi)(SiMe3)CH3] → ZnCl2, Et2O → allenyl ketone (SiMe3) 68%

Claisen rearrangement.[5] The ester enolate Claisen rearrangement of allyl esters of protected amino acids proceeds after replacing the lithium ion with Zn.

4,4-Dimethyloxazolines.[6] An expedient synthesis of these compounds from nitriles and 2-amino-2-methylpropanol involves microwave heating of the mixture (without solvent) with $ZnCl_2 \cdot 2H_2O$ for 15–60 min.

RCN + H2N–C(CH3)2–CH2OH → ZnCl2, microwave → oxazoline (R) up to 94%

[1]Kodomari, M., Nawa, S., Miyoshi, T. *CC* 1895 (1995).
[2]Kühnau, D., Thomsen, I., Jorgensen, K.A. *JCS(P1)* 1167 (1996).
[3]Yokozawa, T., Furuhashi, K., Natsume, H. *TL* **36**, 5243 (1995).
[4]Cunico, R.F., Nair, S.K. *SC* **26**, 803 (1996).
[5]Kazmaier, U., Schneider, C. *SL* 975 (1996).
[6]Clarke, D.S., Wood, R. *SC* **26**, 1335 (1996).

Zinc chloride–hydrogen bromide.

Cleavage of MEM ethers.[1] For substrates that chelate to zinc salts, but do not undergo deprotection, the use of $H_2ZnCl_2Br_2$ often solves the problem.

[1]Herbert, J.M., Knight, J.G., Sexton, B. *T* **52**, 15257 (1996).

Zirconacyclopentadienes.

Fused aromatics.[1] Symmetrical and unsymmetrical zirconacyclopentadienes are readily prepared from Cp_2ZrBu_2 and alkynes or diynes. A CuCl-mediated reaction of these metallocycles with 1,2-dihaloarenes and heteroarenes gives fused aromatic products. The presence of DMPU or HMPA is important, and for best results, 2.5–3 equivalents of these additives are required. Anthracenes are obtained from 1,2,4,5-tetrahalobenzenes.

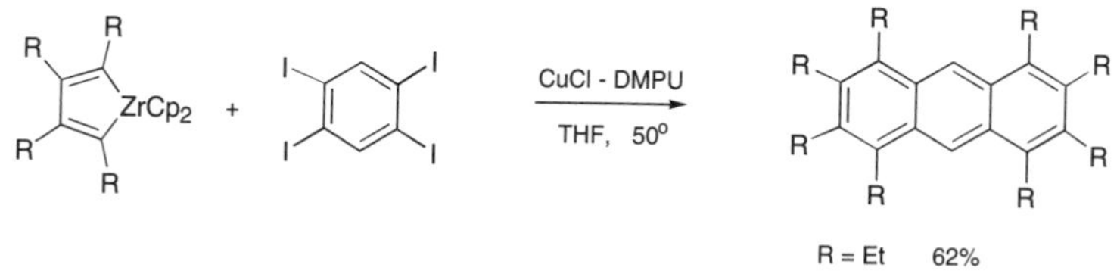

R = Et 62%

[1]Takahashi, T., Hara, R., Nishihara, Y., Kotora, M. *JACS* **118**, 5154 (1996).

Zirconium(IV) *t*-butoxide.

Oppenauer oxidation.[1] Using chloral as the hydride acceptor and $Zr(OBu^t)_4$ as a mediator, the oxidation is accomplished at room temperature (17 examples, 67–99%).

[1]Krohn, K., Knauer, B., Kupke, J., Seebach, D., Beck, A.K., Hayakawa, M. *S* 1341 (1996).

Zirconium(IV) chloride. 18, 413

Dithioacetalization.[1] Practically quantitative yields of dithioacetals are obtained with $ZrCl_4$ on silica as a catalyst.

Hydrostannylation and allylstannylation of alkynes. The addition of tin hydrides to alkynes furnishes predominantly *(Z)*-alkenylstannanes.[2] Allylstannanes also add in a *trans* manner that is γ-selective with respect to the allyl moiety.[3]

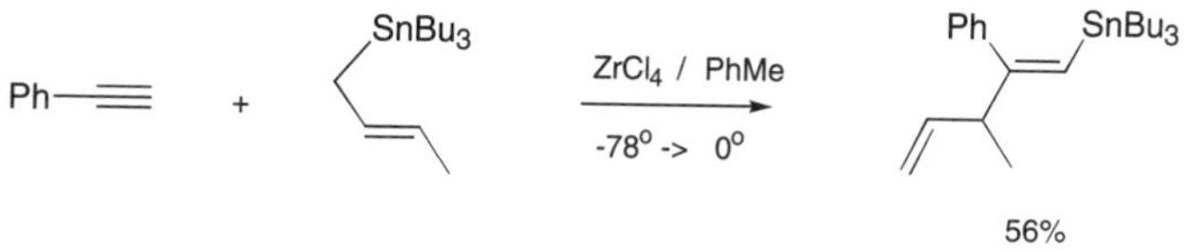

56%

Oxetanes.[4] Allylsilanes form [2+2]-adducts with aldehydes. Unlike reactions catalyzed by many other Lewis acids, this reaction retains the silyl group.

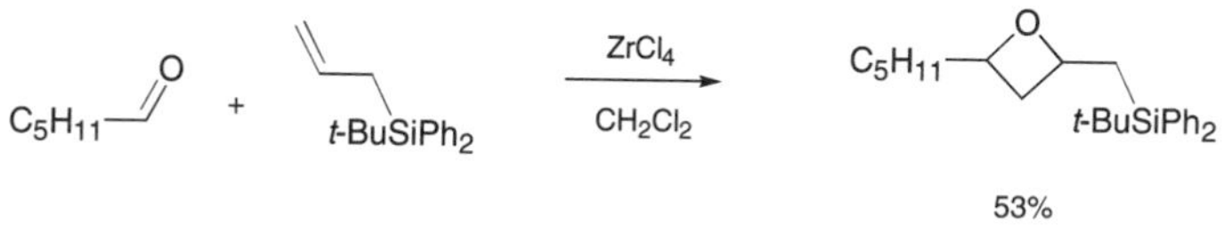

53%

(Z)-3-Iodo-2-(hydroxyalkyl)acrylates.[5] $ZrCl_4$ mediates the reaction of methyl propynoate and a carbonyl compound in the presence of Bu_4NI. This process is analogous to the Baylis–Hillman reaction, with the difference being that iodine is incorporated.

≡—COOMe + O=CH—C5H11 → ($ZrCl_4$ - Bu_4NI, CH_2Cl_2, 0°, 5 h) → 84% (Z:E 73 : 27)

o-Fries rearrangement.[6] The rearrangement takes place at ambient temperature in the presence of $ZrCl_4$ (8 examples, 31–97%).

[1]Patney, H.K., Margan, S. *TL* **37**, 4621 (1996).
[2]Asao, N., Liu, J.-X., Sudoh, T., Yamamoto, Y. *JOC* **61**, 4568 (1996).
[3]Asao, N., Matsukawa, Y., Yamamoto, Y. *CC* 1513 (1996).
[4]Akiyama, T., Yamanaka, M. *SL* 1095 (1996).
[5]Zhang, C., Lu, X. *S* 586 (1996).
[6]Harrowven, D.C., Dainty, R.F. *TL* **37**, 7659 (1996).

Zirconium(IV) triflate.

Azidolysis of epoxides.[1] The opening of epoxide to give 2-azido alcohols may employ 1,1,3,3-tetramethylguanidinium azide in the presence of $Zr(OTf)_4$.

[1]Crotti, P., Di Bussolo, V., Favero, L., Macchia, F., Pineschi, M. *TL* **37**, 1675 (1996).

Zirconocene, Zr-alkylated. 15, 81; **18**, 414

Alkylation–oxidation sequence.[1] Organozirconocene chlorides add to aliphatic and aromatic aldehydes ($ZnBr_2$ catalyzed). Oppenauer oxidation occurs when the reaction time is prolonged.

$C_8H_{17}Zr(Cp)_2Cl$ + PhCHO → ($ZnBr_2$, THF, 25°, 3 h) → [$C_8H_{17}CH(OZr(Cp)_2Cl)Ph$] → ($H_2O$) → $C_8H_{17}CH(OH)Ph$ 88%

[$C_8H_{17}CH(OZr(Cp)_2Cl)Ph$] → (PhCHO / 6 h) → $C_8H_{17}COPh$ 83%

1,3-Dienes.[2] Treatment of an alkyne with Cp_2ZrR_2 and then ethyl vinyl ether furnishes a diene. If iodine is used in the finial quenching, an iododiene results.

Cyclofunctionalization. 1,n-Dienes form zirconabicycles, which allow insertion by isonitriles. Most significantly, *trans*-diquinanes are obtained from 1,6-dienes.[3] Cyclization with *N*-allyl-2-bromoanilines[4] as a building block is also a useful transformation.

Cp_2ZrBu_2, -78° -> rt; $TsCH_2NC$, R—≡—R; 46%

Cyclization of a 1,7-diene derived from α-terpineol, followed by oxidative cleavage of the zirconacycle, leads to a diol, which can be converted to (+)-elemol.[5] Interestingly, the kinetic product is the isomeric *cis*-fused zirconabicycle.

Cp_2ZrCl_2 - BuLi, PhMe Δ; O_2; H_3O^+ — OBn, OH, OH → (+)-elemol

Cp_2ZrCl_2 - BuLi, hexane Δ; O_2; H_3O^+ — OBn, OH, OH

(Alkenyl)chlorozirconocenes.[6] The reaction of 2-halo-1-alkenes with Cp_2ZrBu_2 gives good yields of alkenylzirconocenes. These are regioisomers of those products obtained by hydrozirconation methods. Such zirconocenes are useful for coupling with ArI, acyl chlorides, and allyl halides in the presence of a Pd(0) catalyst and $ZnCl_2$. It is possible to achieve a direct coupling in a one-pot protocol.

Cp_2ZrBu_2; 92%

Allyl ethers.[7] Alkenyl(chloro)zirconocenes that are readily obtained from alkynes and $Cp_2Zr(H)Cl$ react with α-chloro ethers, when $ZnCl_2$ is used as a transmetallating agent to increase the reactivity. Allyl ethers are produced.

[1]Zheng, B., Srebnik, M. *JOC* **60**, 3278 (1995).
[2]Takahashi, T., Kondakov, D.Y., Xi, Z., Suzuki, N. *JACS* **117**, 5871 (1995).
[3]Davis, J.M., Whitby, R.J., Jaxa-Chamiec, A. *TL* **35**, 1445 (1994).
[4]Tietze, L.F., Grote, T. *JOC* **59**, 192 (1994).
[5]Taber, D.F., Wang, Y. *TL* **36**, 6639 (1995).
[6]Takahashi, T., Kotora, M., Fischer, R., Nishihara, Y., Nakajima, K. *JACS* **117**, 11039 (1995).
[7]Pereira, S., Zheng, B., Srebnik, M. *JOC* **60**, 6260 (1995).

Zirconocene dichloride. **14**, 122; **15**, 120–121; **18**, 415

Alkylalumination. With Cp_2ZrCl_2 to promote the *cis*-alkylalumination of alkynes with R_3Al, alkenylaluminums are obtained. Reaction of the latter with α-acetoxy-α-benzylideneaminoacetic esters constitutes a key step in the synthesis of *(Z)*-β,γ-unsaturated amino acid derivatives.[1]

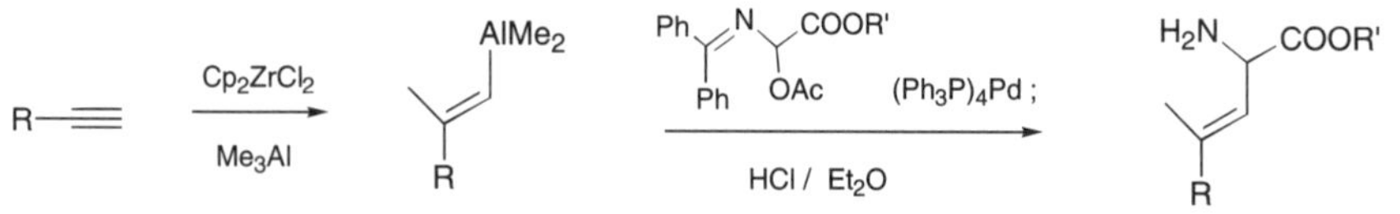

When a modified zirconocene dichloride (**1**) in which each Cp is replaced by the chiral 1-neomenthylindenyl group is engaged in the methylalumination, the alcohols isolated after oxidation are optically active.[2]

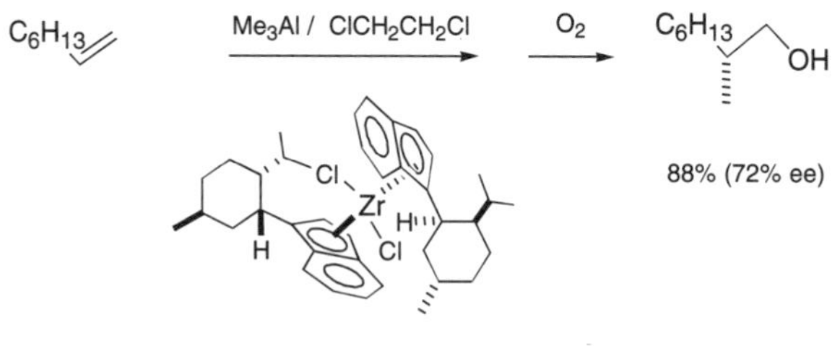

88% (72% ee)

(**1**)

3-Alkylidenetetrahydrofurans.[3] The low-valent zirconium species generated by the reduction of Cp_2ZrCl_2 with Mg reacts with allyl propargyl ethers to form the zirconabicycles. On protonolysis, the alkylidenetetrahydrofurans are obtained.

[1]O'Donnell, M.J., Li, M., Bennett, W.D., Grote, T. *TL* **35**, 9383 (1994).
[2]Kondakov, D.Y., Negishi, E.-I. *JACS* **117**, 10771 (1995).
[3]Miura, K., Funatsu, M., Saito, H., Ito, H., Hosomi, A. *TL* **37**, 9059 (1996).

Zirconocene hydrochloride. 14, 81; **15**, 80–81; **18**, 416–417

Reduction of lactones and anhydrides.[1] In this reaction bisalkoxides are produced for subsequent conversion into bisalkoxyphosphines.

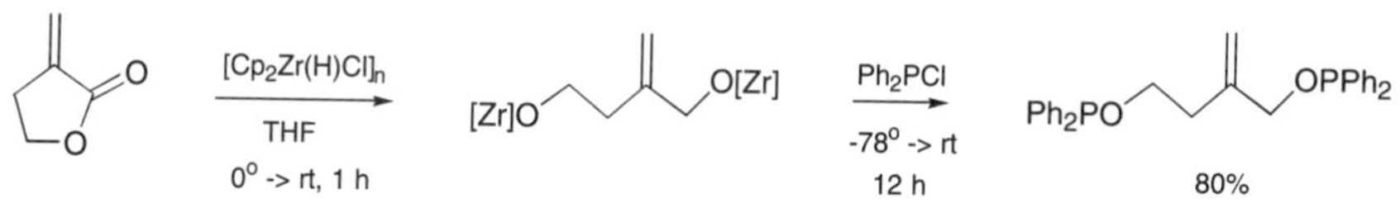

[Cp$_2$Zr(H)Cl]$_n$, THF, 0° -> rt, 1 h; Ph$_2$PCl, -78° -> rt, 12 h; 80%

Allyl ethers.[2] Hydrozirconation of alkynes followed by reaction with α-chloroethers in the presence of $ZnCl_2$ gives *(E)*-allyl ethers (17 examples, 45–92%).

Thioallylation.[3] Allylzirconium species derived from allenyl sulfides react with carbonyl compounds in an *anti*-selective fashion to give 4-hydroxy-3-sulfenyl-1-alkenes.

(E)-2-Alkenyl-1,3,2-dioxaborolidines. Prepared by $Cp_2Zr(H)Cl$-catalyzed hydroboration of 1-alkynes with pinacolborane,[4] these compounds are useful for the synthesis of α-bromo ketones.[5]

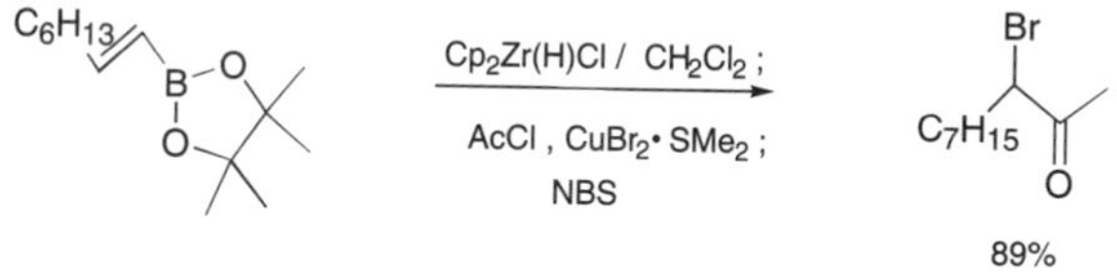

Cp$_2$Zr(H)Cl / CH$_2$Cl$_2$; AcCl , CuBr$_2$• SMe$_2$; NBS; 89%

Hydroboration of alkenes is similarly accomplished. When the solvent is changed from CH_2Cl_2 to CCl_4, the products from 1-alkenes are the homologous 1,1,1,3-tetrachloroalkanes.[6] These two reactions are more efficiently catalyzed by $(Ph_3P)_3RhCl$.

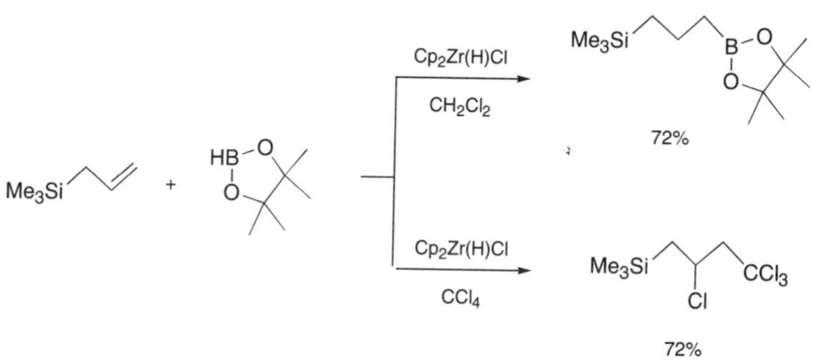

Cp$_2$Zr(H)Cl, CH$_2$Cl$_2$: 72%

Cp$_2$Zr(H)Cl, CCl$_4$: 72%

The alkenylborolidines undergo facile boron migration during hydrozirconation.[7] Thus, the *B*-isopropenyl derivative gives the 1-propylborolidine on aqueous quenching and the 3-borolidinylpentyl methyl ketone on reaction with MVK.

Cp$_2$Zr(H)Cl / 25°

87%

α-Allenic boronic esters.[8] Hydrozirconation of alkenylboronic esters in CH_2Cl_2 gives the *gem*-borazircona alkanes, which can be used to couple with propargyl bromide. The α-allenic boronic esters thus obtained react with aldehydes to afford 2-(1-hydroxyalkyl)-1,3-dienes.

Cp$_2$Zr(H)Cl / 25°

CuCN

R'CHO

Primary amines.[9] Hydrozirconation of 1-alkenes followed by reaction with *O*-(mesitylenesulfonyl)hydroxylamine provides primary amines (10 examples, 62–88%).

(E)-Alkenyl chalcogenides. The alkenylzirconocene chlorides are converted into selenides[10] and tellurides[11] by reaction with organochalcogen halides.

[1]Cenac, N., Zablocka, M., Igau, A., Majoral, J.-P., Skowronska, A. *JOC* **61**, 796 (1996).
[2]Pereira, S., Zheng, B., Srebnik, M. *JOC* **60**, 6260 (1995).
[3]Chino, M., Liang, G.H., Matsumoto, T., Suzuki, K. *CL* 231 (1996).
[4]Pereira, S., Srebnik, M. *OM* **14**, 3127 (1995).
[5]Zheng, B., Srebnik, M. *TL* **36**, 5665 (1995).
[6]Pereira, S., Srebnik, M. *JACS* **118**, 909 (1996).
[7]Pereira, S., Srebnik, M. *JOC* **60**, 4316 (1995).
[8]Zheng, B., Srebnik, M. *JOC* **60**, 486 (1995).
[9]Zheng, B., Srebnik, M. *JOC* **60**, 1912 (1995).
[10]Huang, X., Zhu, L.-S. *JCS(P1)* 767 (1996).
[11]Sung, J.W., Lee, C.-W., Oh, D.Y. *TL* **36**, 1503 (1995).

AUTHOR INDEX

SUBJECT INDEX